APPLIED TRIGONOMETRY

THOMAS J. McHALE

PAUL T. WITZKE

Addison-Wesley Publishing Company
Reading, Massachusetts • Menlo Park, California
London • Amsterdam • Don Mills, Ontario • Sydney

Library of Congress Cataloging in Publication Data

McHale, Thomas J., 1931-
Applied Trigonometry.

1. Trigonometry--Programmed instruction.
I. Witzke, Paul T. II. Title.
QA531.M46 1984 516.2'4'077 84-2825
ISBN 0-201-04723-3

Reproduced by Addison-Wesley from camera-ready copy supplied by the authors.

Reprinted with corrections, May 1985

ISBN 0-201-04723-3
21 22-CRS-02 01

PREFACE

APPLIED TRIGONOMETRY is part of an Applied Mathematics Series that also includes CALCULATION AND CALCULATORS, APPLIED ALGEBRA I, and APPLIED ALGEBRA II. Though individual texts can be used in other courses, the full Series was developed for a one-year Technical Mathematics course.

APPLIED TRIGONOMETRY is designed to teach the trigonometric concepts and skills needed in basic science, technology, pre-engineering, and mathematics itself. The word "applied" in the title does not mean that major emphasis is given to applied problems, though many basic types of applied problems are included. Rather, the word "applied" means that emphasis is given to topics that are broadly useful, including the solution of triangles, vectors, geometric concepts, complex numbers, and sine waves. A scientific calculator is used in place of trigonometric tables and in all computations. Instruction in the use of a calculator is given when needed.

The content of APPLIED TRIGONOMETRY is presented in a programmed format with assignment self-tests with answers within each chapter and supplementary problems for each assignment at the end of each chapter. Answers for all supplementary problems are given in the back of the text.

The text is accompanied by the book TESTS FOR APPLIED TRIGONOMETRY which contains a diagnostic test, twenty-nine assignment tests, eight chapter tests four multi-chapter tests (every two chapters), and a comprehensive test. Three parallel forms are provided for chapter tests, multi-chapter tests, and the comprehensive test. A full set of answer keys for the tests is included. Because of the large number of tests provided, various options are possible in using them. For example, an instructor can use the chapter tests alone, the multi-chapter tests alone, or some combination of the two types.

> Note: The test book is provided only to teachers. Copies of the tests for student use must be made by some copying process.

The following features make the instruction effective and efficient for students:

1. The instruction, which is based on a task analysis, contains examples of all types of problems that appear in the tests.

2. The full and flexible set of tests can be used as a teaching tool to identify learning difficulties which can be remedied by tutoring or class discussion.

3. Because of the programmed format and the full and flexible set of tests provided, the text is ideally suited for individualized instruction.

The authors wish to thank Patricia R. Mallion of the Addison-Wesley Publishing Company for many general suggestions that improved this text. They wish to thank the following for their suggestions: Paul J. Eldersveld, of the College of Du Page; Donald M. Hindle, of Community College of Rhode Island; Peter Uluave, of Utah Technical College; and Robert A. Yawin, of Springfield Technical Community College. They also wish to thank Arleen D'Amore who typed and proofread the camera-ready copy, Peggy McHale who prepared the drawings and made the corrections, Gail W. Davis who did the final proofreading, and Allan A. Christenson who prepared the Index.

HOW TO USE THE TEXT AND TESTS

This text and the tests available in TESTS FOR APPLIED TRIGONOMETRY can be used in various instructional strategies ranging from paced instruction with all students taking the same content to totally individualized instruction. The general procedure for using the materials is outlined below.

1. The diagnostic test can be administered either to simply get a measure of the entry skills of the students or as a basis for prescribing an individualized program.

2. Each chapter is covered in a number of assignments (see below). After the students have completed each assignment and the assignment self-test (in the text), the assignment test (from the test book) can be administered, corrected, and used as a basis for tutoring. The assignment tests are simply a teaching tool and need not be graded. The supplementary problems at the end of the chapter can be used at the instructor's discretion for students who need further practice.

   Note: Instead of using the assignment self-tests (in the text) as an integral part of the assignments, they can be used at the completion of a chapter as a chapter review exercise.

3. After the appropriate assignments are completed, either a chapter test or a multi-chapter test can be administered. Ordinarily, these tests should be graded. Parallel forms are provided to facilitate the test administration, including the retesting of students who do not achieve a satisfactory score.

4. After all desired chapters are completed in the manner above, the comprehensive test can be administered. Since the comprehensive test is a parallel form of the diagnostic test, the difference score can be used as a measure of each student's improvement.

ASSIGNMENTS FOR APPLIED TRIGONOMETRY

Chapter	Assignment	Pages
Ch. 1:	#1	(pp. 1-12)
	#2	(pp. 13-27)
	#3	(pp. 28-37)
Ch. 2:	#4	(pp. 41-57)
	#5	(pp. 57-68)
	#6	(pp. 68-80)
Ch. 3:	#7	(pp. 84-100)
	#8	(pp. 100-116)
	#9	(pp. 117-131)
	#10	(pp. 132-144)
Ch. 4:	#11	(pp. 149-166)
	#12	(pp. 167-182)
	#13	(pp. 183-198)
	#14	(pp. 198-209)
Ch. 5:	#15	(pp. 213-224)
	#16	(pp. 225-235)
	#17	(pp. 236-249)
	#18	(pp. 250-261)
Ch. 6:	#19	(pp. 264-280)
	#20	(pp. 280-289)
	#21	(pp. 290-301)
Ch. 7:	#22	(pp. 304-314)
	#23	(pp. 315-327)
	#24	(pp. 328-337)
	#25	(pp. 338-344)
	#26	(pp. 345-351)
Ch. 8:	#27	(pp. 356-370)
	#28	(pp. 371-385)
	#29	(pp. 386-395)

CONTENTS

CHAPTER 1: RIGHT TRIANGLES (Pages 1-40)

CHAPTER 2: OBLIQUE TRIANGLES (Pages 41-83)

CHAPTER 3: TRIGONOMETRIC FUNCTIONS (Pages 84-148)

CHAPTER 8: SINE WAVES (Pages 356-398)

Chapter 1 RIGHT TRIANGLES

In this chapter, we will show how the angle-sum principle, the Pythagorean Theorem, and the three basic trigonometric ratios (sine, cosine, and tangent) can be used to find unknown angles and sides in right triangles.

1-1 THE ANGLE-SUM PRINCIPLE FOR TRIANGLES

In any triangle, the sum of the three angles is 180°. We will discuss that principle in this section.

---

1. There are three angles in any triangle. The angles are usually labeled with capital letters.

   Any angle between 0° and 90° is called an acute angle.
   Any angle with exactly 90° is called a right angle.
   Any angle between 90° and 180° is called an obtuse angle.

   In triangle ABC:

   Angle A is an acute angle.

   a) Angle B is an ____________ angle.

   b) Angle C is an ____________ angle.

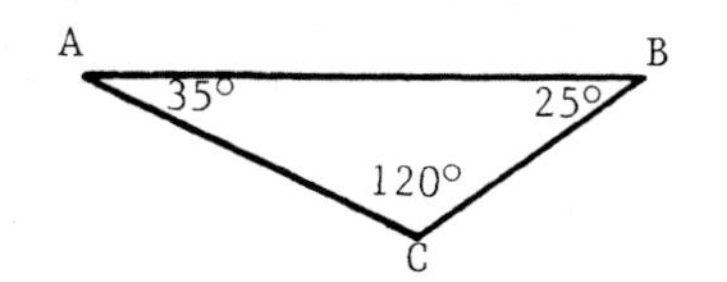

---

2. In triangle MPR:

   a) There are two acute angles, angle ______ and angle ______.

   b) The obtuse angle is angle ______.

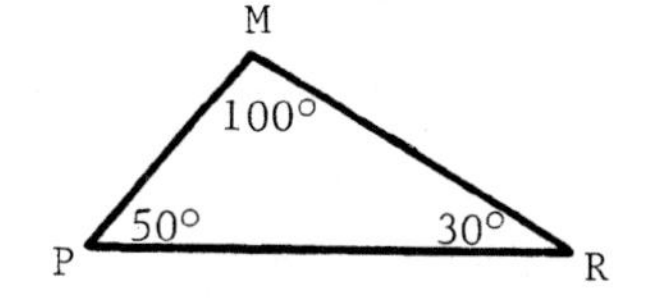

Answers to 1: a) acute b) obtuse

---

3. The angle-sum principle for triangles is this:

THE SUM OF THE THREE ANGLES OF ANY TRIANGLE IS 180°

   By subtracting the total number of degrees of the two known angles from 180°, find the number of degrees of the unknown angle in each triangle below.

   a)

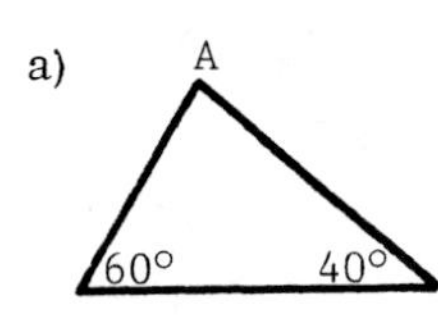

   A = ______

   b)

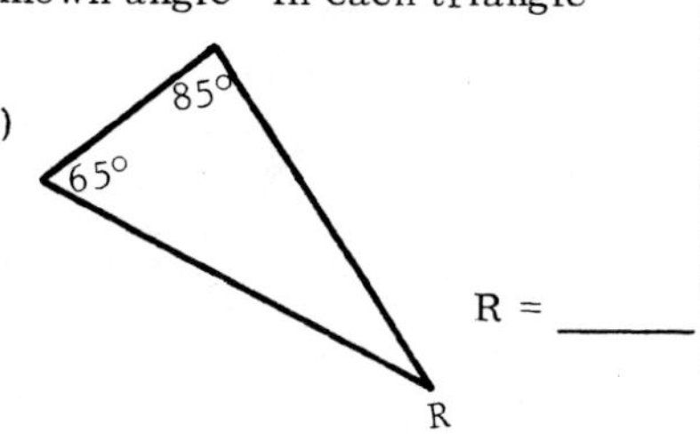

   R = ______

Answers to 2: a) Angles P and R b) Angle M

4. The angle-sum principle also applies to right triangles (those that contain a right angle). Therefore, in right triangle CDE: a) The sum of the angles is ______. b) Angle D = ______. c) Angle E = ______. (Triangle CDE: C, 30°, D, E)	a) A = 80° b) R = 30°
5. In a right triangle, the right angle contains 90°. Since the sum of the other two angles must be 90°, the other two angles must be acute angles. In right triangle TBC: a) Angles T and B are both ________________ angles. b) The sum of angles T and B must be ______°. (Triangle TBC: B, T, C)	a) 180° b) 90° c) 60°
6. Since the sum of the two acute angles in a right triangle is 90°, it is easy to find the size of one acute angle if we know the size of the other. To do so, we simply subtract the known angle from 90°. In right triangle ABC: a) If angle B contains 55°, angle C contains 90° - 55° = ______°. b) If angle C contains 37°, angle B contains 90° - 37° = ______°. (Triangle ABC: A, B, C)	a) acute b) 90°
7. By simply subtracting the known acute angle from 90°, find the unknown acute angle in each right triangle. a) (Triangle: 27°, T) T = ______ b) (Triangle: R, 45°) R = ______	a) 35° b) 53°
	a) T = 63° b) R = 45°

1-2 LABELING ANGLES AND SIDES IN TRIANGLES

In this section, we will discuss the conventional way to label the angles and sides in any triangle.

8. We labeled the angles and sides of the triangle at the right in the conventional way. Notice these points:

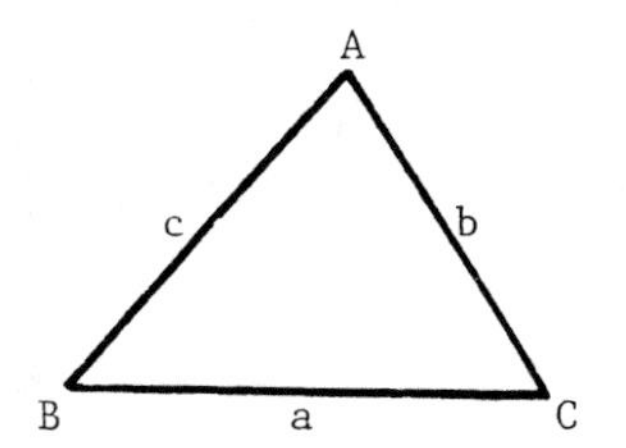

1) Each angle is labeled with a capital letter.

2) Each side is labeled with the small letter corresponding to the capital letter of the angle opposite it. That is:

The side opposite angle A is labeled "a".
The side opposite angle B is labeled "b".
The side opposite angle C is labeled "c".

Sometimes we use the two capital letters at each end of the side to represent the side. For example:

Instead of "a", we use BC (or CB).

a) Instead of "b", we use ______________.

b) Instead of "c", we use ______________.

---

a) AC (or CA)

b) AB (or BA)

9. When the sides of a triangle are not labeled with small letters, we must use two capital letters to represent the sides. For example, in the triangle below.

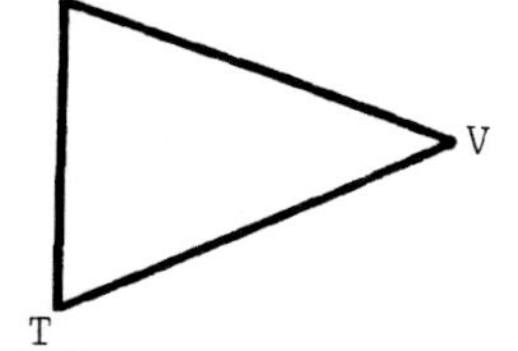

The side opposite angle M is TV (or VT).

a) The side opposite angle T is ________.

b) The side opposite angle V is ________.

---

a) MV (or VM)

b) MT (or TM)

10. In any triangle:

THE LONGEST SIDE IS OPPOSITE THE LARGEST ANGLE.

THE SHORTEST SIDE IS OPPOSITE THE SMALLEST ANGLE.

In triangle DFH:

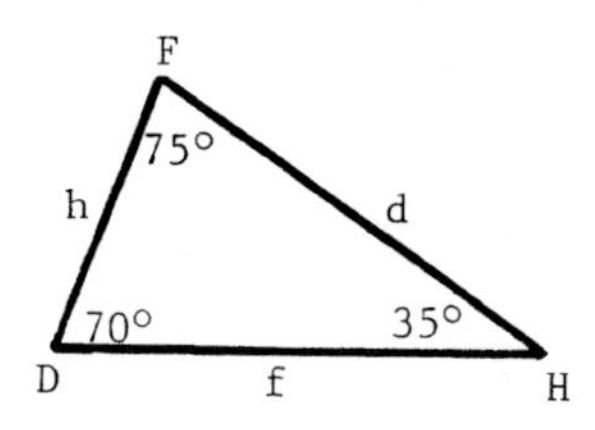

a) Angle F is the largest angle. Therefore, ______ is the longest side.

b) Angle H is the smallest angle. Therefore, ______ is the shortest side.

---

a) f b) h

11. In the triangle at the right:

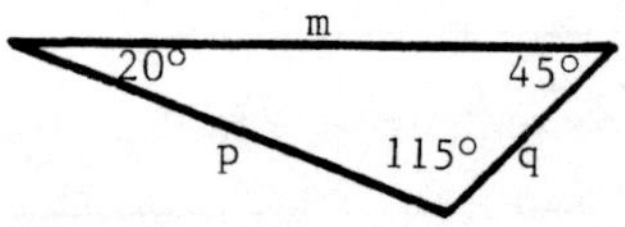

a) The longest side is _____.

b) The shortest side _____.

m, p, q, 20°, 45°, 115°

---

12. In a right triangle, the side opposite the right angle is called the "hypotenuse". The sides opposite the two acute angles are simply called "legs". For example:

a) m	b) q

In right triangle DFH:

"f" is the hypotenuse, and
"d" and "h" are the legs.

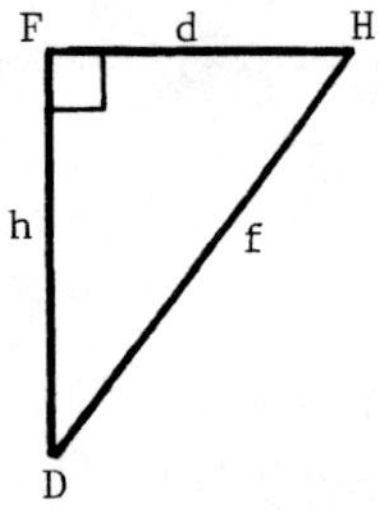

a) The hypotenuse "f" can be labeled in two other ways: _____ or _____.

b) Leg "d" can be labeled in two other ways: _____ or _____

F, d, H, h, f, D

---

13. Ordinarily we use small letters to represent the hypotenuse and legs of a right triangle. Use small letters to complete the questions below.

a) DH or HD

b) FH or HF

In right triangle FET:

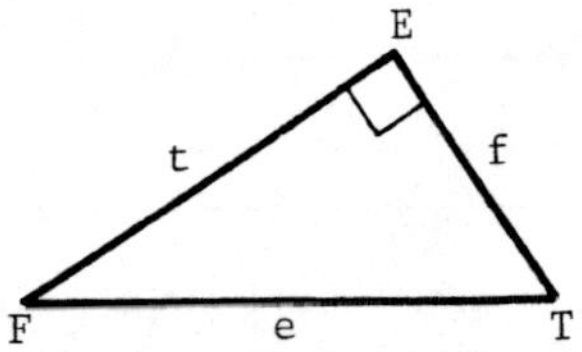

a) The hypotenuse is _____.

b) The legs are _____ and _____.

E, t, f, F, e, T

---

14. Since the right angle is the largest angle in any right triangle, the hypotenuse is the longest side of any right triangle.

a) e

b) t and f

In right triangle DMF:

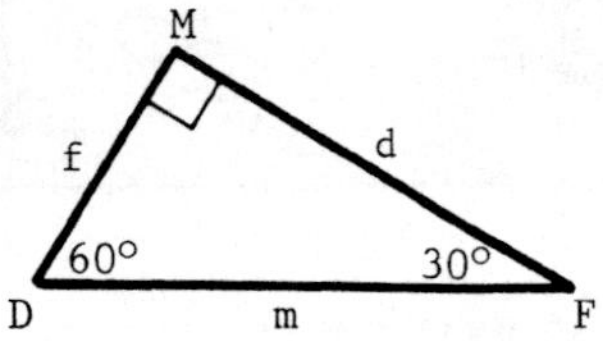

a) The longest side is _____.

b) The shortest side is _____.

M, f, d, 60°, 30°, D, m, F

---

a) m (the hypotenuse)	b) f

1-3 THE PYTHAGOREAN THEOREM

In this section, we will discuss the Pythagorean Theorem and show how it can be used to find the length of an unknown side in a right triangle.

15. The Pythagorean Theorem states the following relationship among the three sides of a right triangle:

> IN ANY RIGHT TRIANGLE, THE SQUARE OF THE LENGTH OF THE HYPOTENUSE IS EQUAL TO THE SUM OF THE SQUARES OF THE LENGTHS OF THE TWO LEGS.

For right triangle ABC, the Pythagorean Theorem says:

$c^2 = a^2 + b^2$

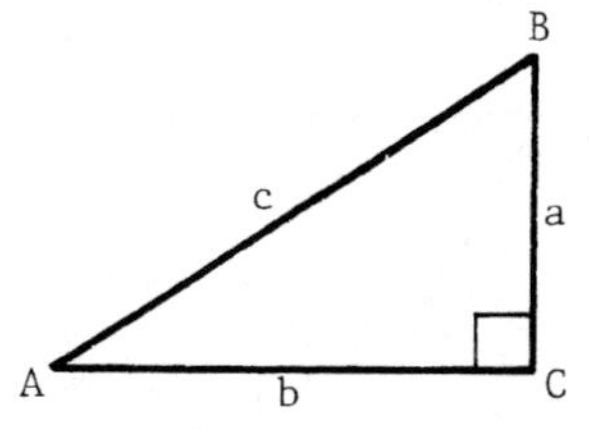

Using small letters, state the Pythagorean Theorem for each right triangle below.

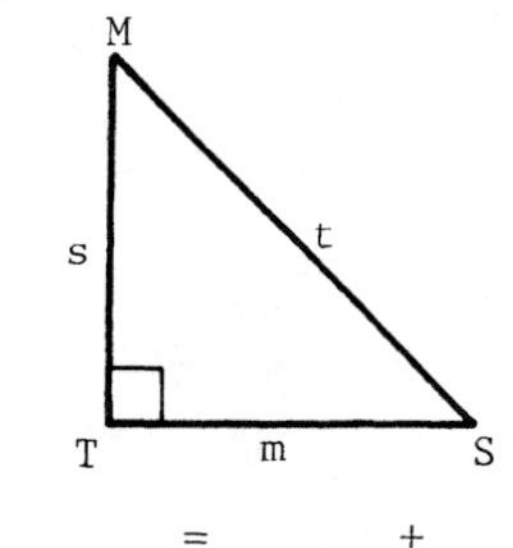

b)

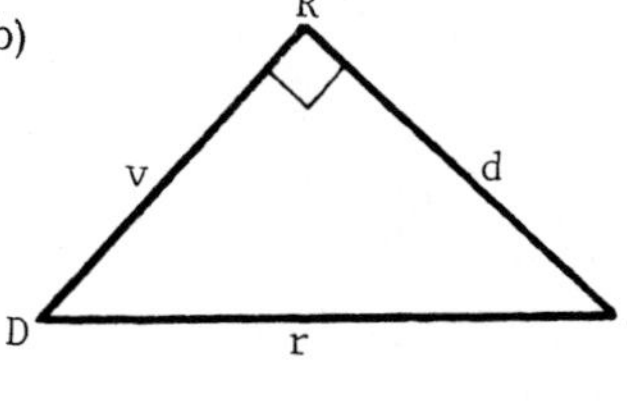

a) ______ = ______ + ______ b) ______ = ______ + ______

---

16. The lengths of the two legs of the right triangle below are given. We can use the Pythagorean Theorem to find the length of the hypotenuse. The steps are:

Answers to 15: a) $t^2 = s^2 + m^2$ b) $r^2 = v^2 + d^2$

1) Write the Pythagorean Theorem.

$m^2 = d^2 + p^2$

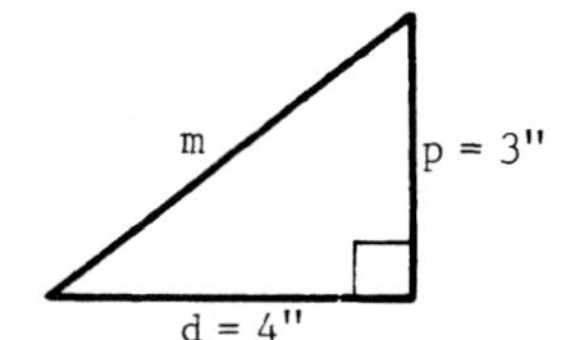

2) Substitute the known values and simplify.

$$m^2 = (4'')^2 + (3'')^2$$
$$= 16 \text{ in}^2 + 9 \text{ in}^2$$
$$= 25 \text{ in}^2$$

3) Find "m" by taking the square root of the right side.

$m = \sqrt{25 \text{ in}^2} =$ ______

---

17. We can use the same steps to find the length of the hypotenuse below.

Answer to 16: m = 5 in

$$h^2 = t^2 + b^2$$
$$= (7\text{m})^2 + (4 \text{ m})^2$$
$$= 49 \text{ m}^2 + 16 \text{ m}^2$$
$$= 65 \text{ m}^2$$

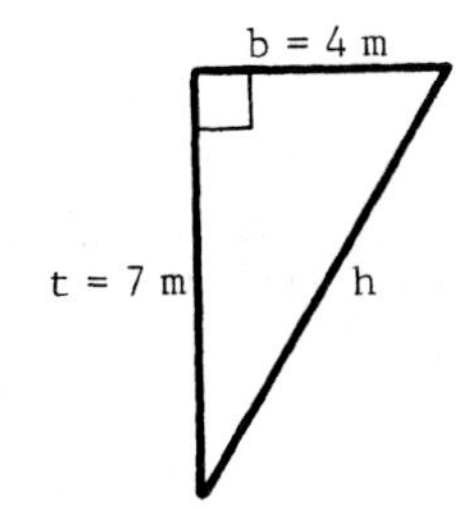

Use a calculator to find "h" by taking the square root of the right side. Round to tenths.

$h = \sqrt{65 \text{ m}^2} =$ ______

18. When using the Pythagorean Theorem to find the hypotenuse of a right triangle, all of the calculations can be done in one process on a calculator. To prepare to do so, we solve for the hypotenuse by taking the square root of the other side before substituting. That is:

$$\text{If } c^2 = a^2 + b^2, \quad c = \sqrt{a^2 + b^2}$$

Using the method above, we set up the solution below so that "t" can be found in one calculator process.

$$t = k^2 + d^2$$

$$t = \sqrt{k^2 + d^2}$$

$$t = \sqrt{(5.68')^2 + (9.75')^2}$$

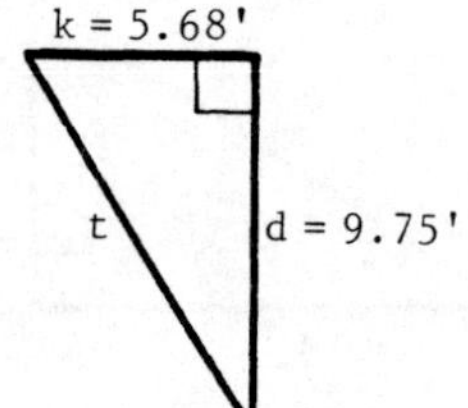

The calculator steps for the solution are shown. Notice that we added the squares of 5.68 and 9.75, pressed [=] to complete that addition, and then pressed [$\sqrt{x}$].

Enter	Press	Display
5.68	[x^2] [+]	32.2624
9.75	[x^2] [=] [$\sqrt{x}$]	11.283834

Rounding to tenths, we get: t = ________

h = 8.1 m

---

19. In right triangle TMB, the length of the two legs and the size of two angles are given.

a) Using the angle-sum principle, find the size of angle B. ______

b) Using the Pythagorean Theorem, find the length of the hypotenuse "m". Round to tenths.

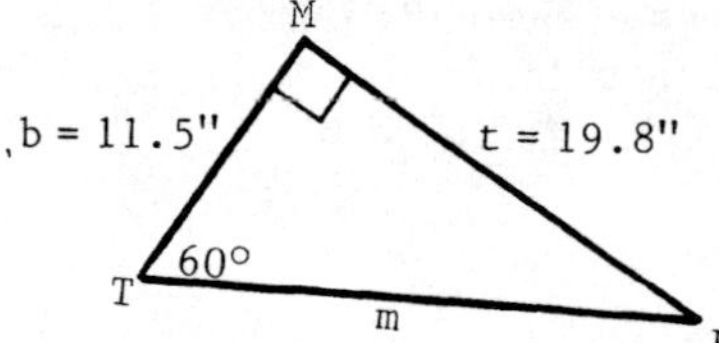

m = ________

t = 11.3 ft

---

20. The distance between the opposite corners of a rectangle or square is called the "diagonal" of the rectangle or square. The diagonal of each figure is the hypotenuse of a right triangle. Use the Pythagorean Theorem to find the diagonal of the rectangle and square below. Round to tenths.

a)

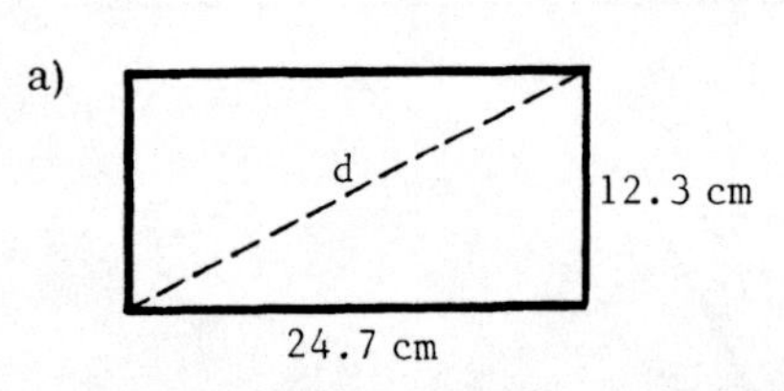

b)

d
2.5 m
2.5 m

d = ______ d = ______

a) Angle B = 30°

b) m = 22.9 in

a) $d = 27.6$ cm

b) $d = 3.5$ m

21. The lengths of the hypotenuse and one leg are given for the triangle below. We can use the Pythagorean Theorem to find the length of the unknown leg "t". The steps are:

1) Write the Pythagorean Theorem.

$$d^2 = t^2 + k^2$$

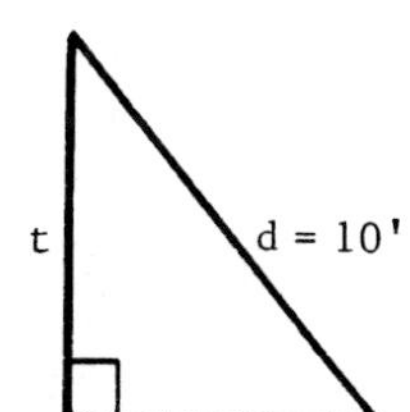

2) Solve for "t^2".

$$t^2 = d^2 - k^2$$

3) Substitute and simplify.

$$t^2 = (10')^2 - (6')^2$$

$$= 100 \text{ ft}^2 - 36 \text{ ft}^2$$

$$= 64 \text{ ft}^2$$

4) Find "t" by taking the square root of the right side.

$$t = \sqrt{64 \text{ ft}^2} = \_\_\_\_\_\_\_\_$$

---

8 ft

22. We can use the same steps to find the length of leg "v" below.

$$h^2 = v^2 + b^2$$

$$v^2 = h^2 - b^2$$

$$v^2 = (9 \text{ cm})^2 - (7 \text{ cm})^2$$

$$v^2 = 81 \text{ cm}^2 - 49 \text{ cm}^2$$

$$v^2 = 32 \text{ cm}^2$$

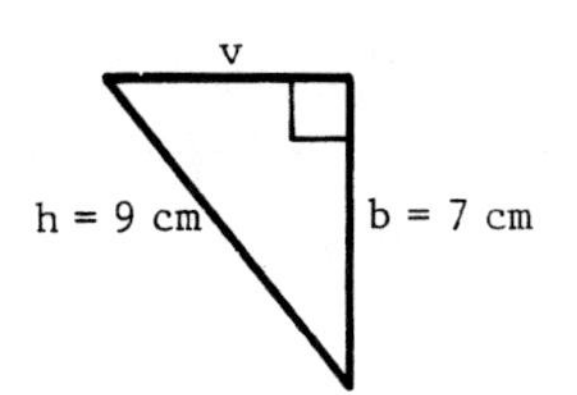

Find "v" by taking the square root of the right side. Round to tenths.

$$v = \sqrt{32 \text{ cm}^2} = \_\_\_\_\_\_\_\_$$

$v = 5.7$ cm

23. When using the Pythagorean Theorem to find an unknown leg in a right triangle, we can also do all of the calculations in one process on a calculator. To prepare to do so, we solve for the leg before substituting. As an example, let's solve for leg "y" below.

$$z^2 = y^2 + x^2$$

$$y^2 = z^2 - x^2$$

$$y = \sqrt{z^2 - x^2}$$

$$y = \sqrt{(18 \text{ m})^2 - (12 \text{ m})^2}$$

z = 18 m

y

x = 12 m

The calculator steps for the solution are shown. Notice again that we pressed [=] before pressing [$\sqrt{x}$].

Enter	Press	Display
18	[x^2] [-]	324.
12	[x^2] [=] [$\sqrt{x}$]	13.416408

Rounding to tenths, we get: $y = \_\_\_\_\_\_\_\_$

24. We set up the calculator solution for leg "a" below. Complete the solution. Round to tenths.

$r^2 = m^2 + a^2$

$a^2 = r^2 - m^2$

$a = \sqrt{r^2 - m^2}$

$a = \sqrt{(27.5')^2 - (21.9')^2}$

$a =$ ________

R
m = 21.9'
a
A
r = 27.5'
M

y = 13.4 m

---

25. In right triangle RQS, the length of the hypotenuse and one leg and the size of two angles are given.

a) Using the angle-sum principle, find the size of angle S. ______

b) Using the Pythagorean Theorem, find the length of leg "s". Round to tenths.

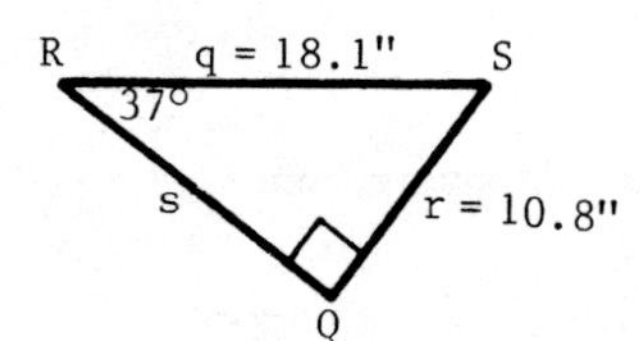

s = ______

a = 16.6 ft

---

26. Use the Pythagorean Theorem to find the width of the rectangle at the right. Round to tenths.

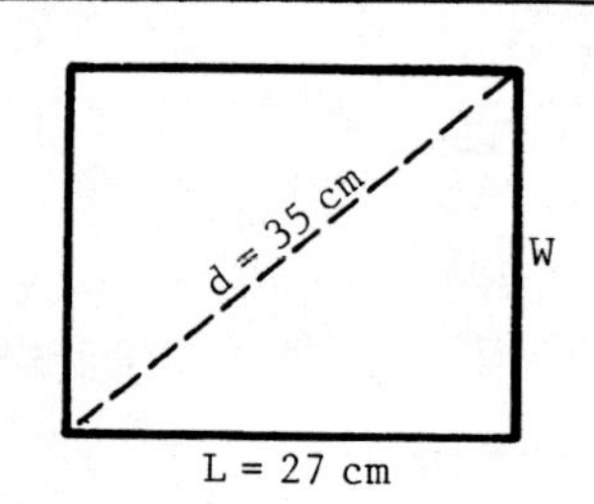

W = ________

a) Angle S = 53°

b) s = 14.5 in

---

27. There are two holes in the metal plate at the right. Use the Pythagorean Theorem to find the distance between the holes, measured center-to-center. Round to tenths.

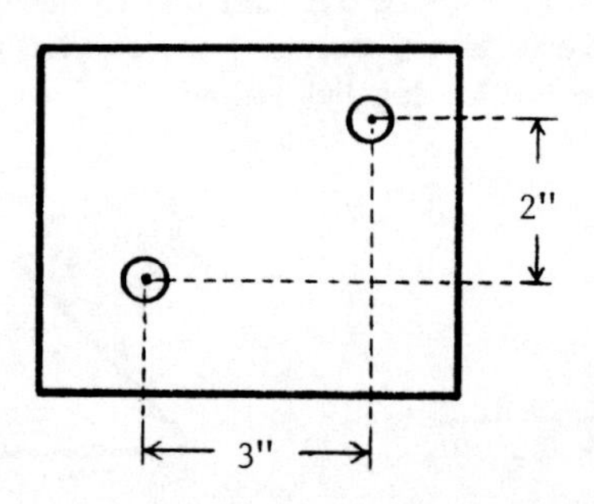

W = 22.3 cm

3.6 in

28. In the rectangular prism below, BD is the diagonal of the prism and BC is the diagonal of the base of the prism. To find BD, we must find BC first. The Pythagorean Theorem can be used to find both diagonals.

a) ABC is a right triangle. Find BC, the diagonal of the base.

BC = _______

b) BCD is a right triangle. Find BD, the diagonal of the prism. Round to tenths.

BD = 

a) BC = 50 m

b) BD = 53.9 m

---

29. We can find the <u>area</u> of this right triangle in two steps.

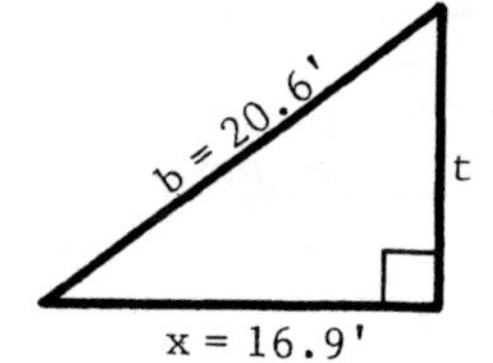

a) First find the length of leg "t". Round to tenths.

t = _______

b) Then use the area formula below, rounding to tenths.

$$A = \frac{(\text{leg 1})(\text{leg 2})}{2}$$

a) t = 11.8 ft b) A = 99.7 ft^2

1-4 ISOSCELES AND EQUILATERAL TRIANGLES

In this section, we will discuss <u>isosceles</u> and <u>equilateral</u> triangles and solve problems involving triangles of those types.

---

30. Triangles with two equal sides are called <u>isosceles</u> triangles. In an isosceles triangle, <u>the angles opposite the equal sides are equal.</u>

The triangle at the right is an isosceles triangle because sides "c" and "d" are equal.

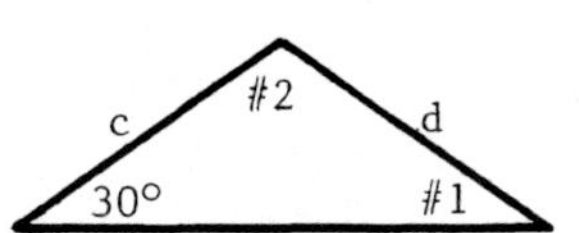

a) Since the angles opposite "c" and "d" are equal, angle #1 = ______.

b) Therefore, angle #2 = ______.

31. In the isosceles triangle on the right, sides "t" and "p" are equal.

Angles #1 and #2 must be equal because they are opposite the equal sides. How many degrees are there in each of these two equal angles? _______

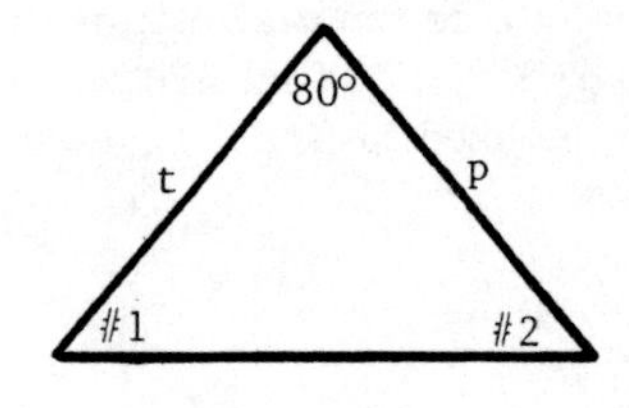

a) 30° b) 120°

---

32. The triangle at the right is an isosceles right triangle since it contains a right angle, and sides "c" and "k" are equal.

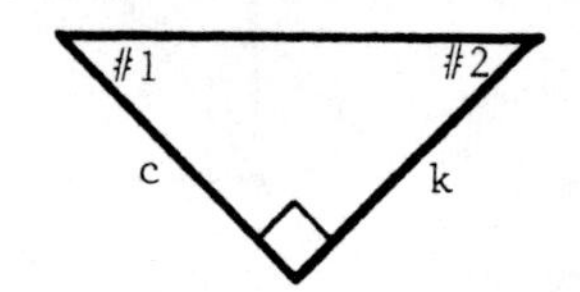

a) Angle #1 = _______°

b) Angle #2 = _______°

50°

---

33. Triangles with three equal sides are called equilateral triangles. In an equilateral triangle, all three angles are equal.

The triangle at the right is an equilateral triangle because all three sides are equal. How many degrees are there in each of the three equal angles? _______

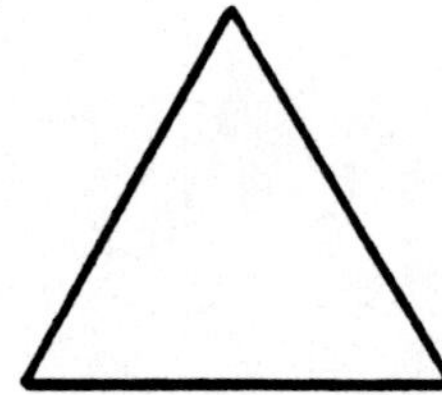

a) 45°

b) 45°

---

34. When a height is drawn to the unequal side in an isosceles triangle, it bisects that side. That is, it cuts that side into two equal parts. We can use that fact to find the height and the area of the triangle when all three sides are known.

Triangle ABC is an isosceles triangle. AD is a height drawn to the unequal side BC.

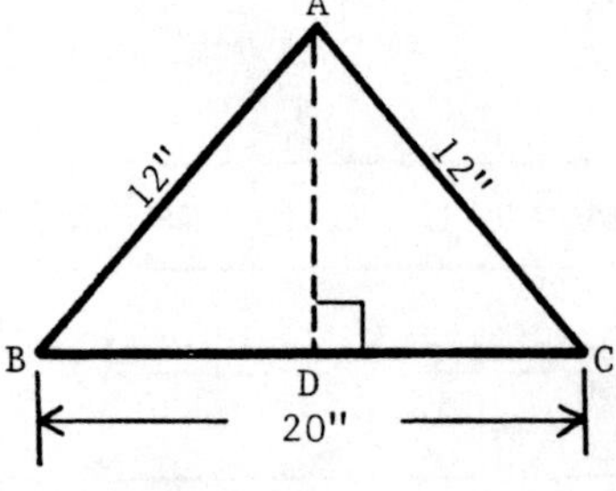

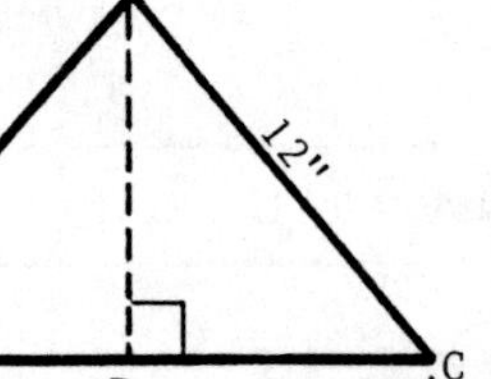

a) Since AD bisects BC, how long are BD and DC? _______

b) Using the Pythagorean Theorem, find the length of AD to the nearest tenth.

AD = _______

c) The area of the triangle is  $\frac{(BC)(AD)}{2}$ = _______

60°

---

a) 10 in

b) 6.6 in

c) 66 in^2

35. When a height is drawn to any side in an equilateral triangle, it bisects that side. We can use that fact to find the height and the area of the triangle when the length of the sides is known.

Triangle DEF is an equilateral triangle. FG is a height drawn to DE.

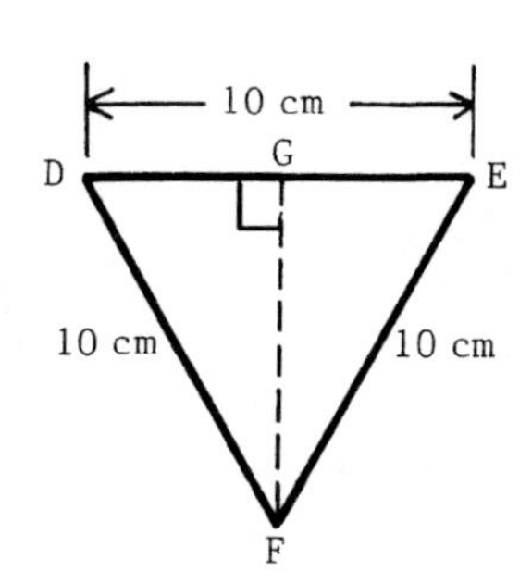

a) Since FG bisects DE, how long are DG and GE? _______

b) Using the Pythagorean Theorem, find the length of FG to the nearest tenth.

FG = _______

c) The area of the triangle is $\frac{(DE)(FG)}{2}$ = _______

---

a) 5 cm

b) 8.7 cm

c) 43.5 cm^2

36. In the rectangular figure at the right, the shaded part is an isosceles triangle. Find the area of the triangle.

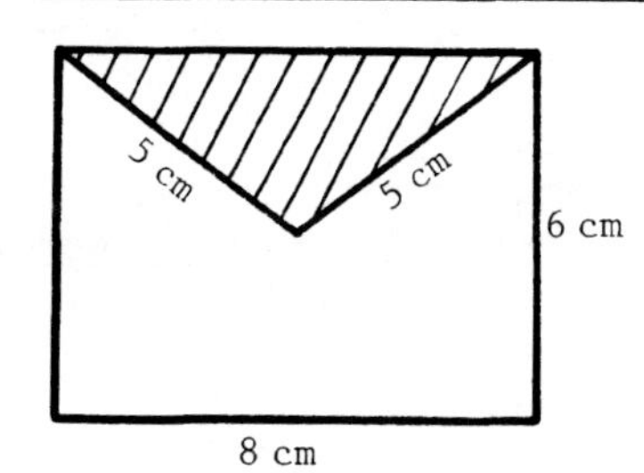

A = _______

A = 12 cm^2

37. In the figure at the right, a V-shaped cut has been made in the top of a rectangle. The cut is an isosceles triangle whose base is 3" and whose height is 2".

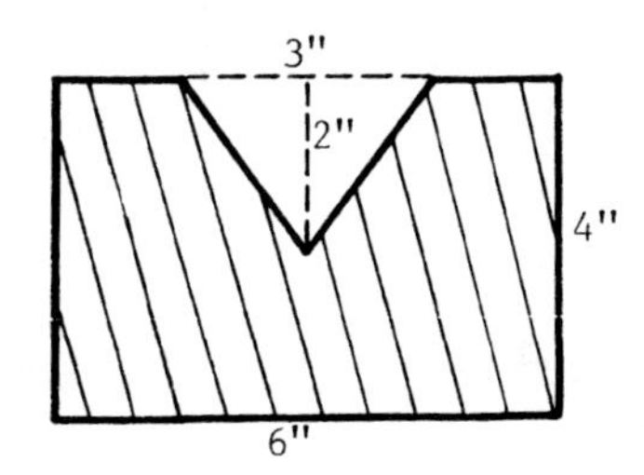

To find the area of the shaded figure, we must subtract the area of the triangle from the area of the rectangle.

a) The area of the rectangle is _______.

b) The area of the triangle is _______.

c) The area of the shaded figure is _______.

---

a) 24 in^2

b) 3 in^2

c) 21 in^2

SELF-TEST 1 (pp. 1-12)

1. Angle P = ______
2. Which side is shortest? ______

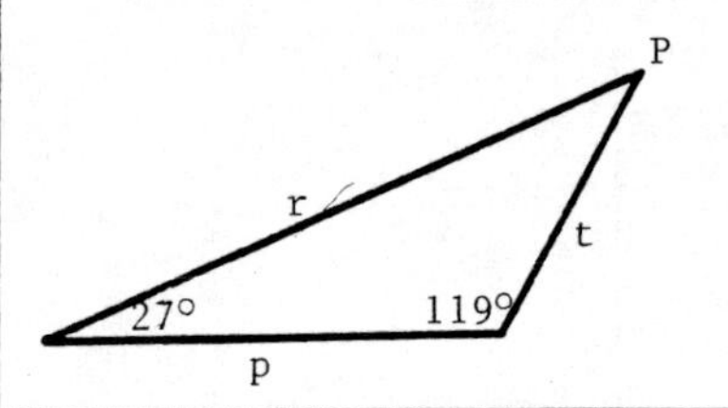

3. Angle F = ______
4. Which side is longest? ______

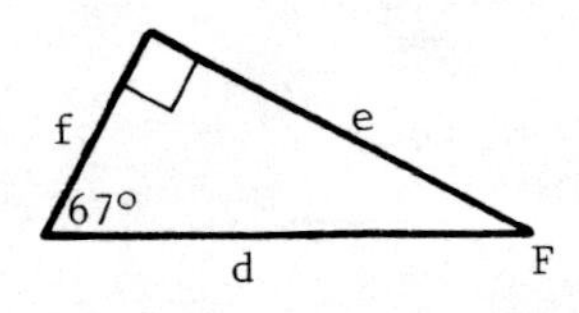

5. Find side "c". Round to tenths.

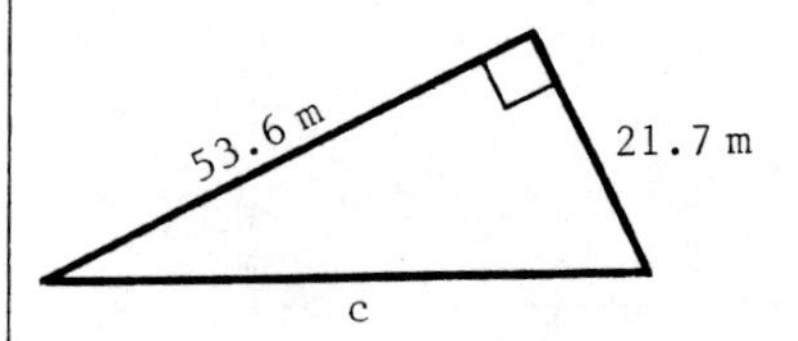

c = ________

6. Find side "p". Round to hundredths.
7. Find the area. Round to hundredths.

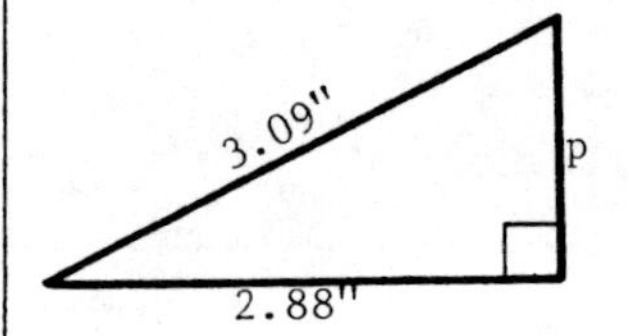

p = ________

A = ________

8. FHG is an isosceles triangle. Find angle G and angle H.

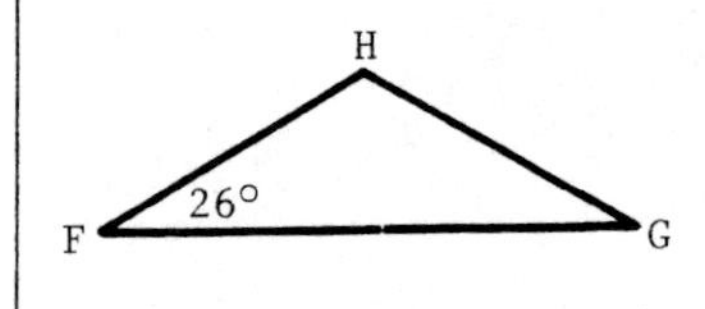

G = ______

H = ______

9. ABC is an isosceles right angle. Find angle A and angle B.

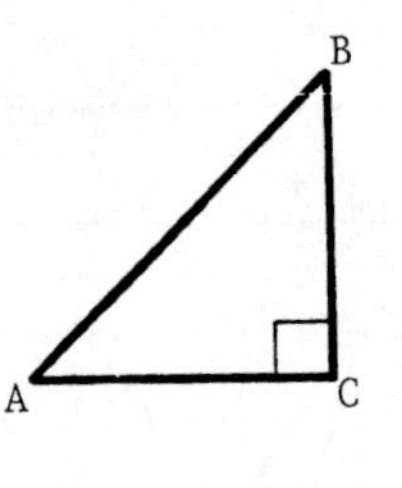

A = ______

B = ______

10. PQR is an equilateral triangle. Find angle P, angle Q, and angle R.

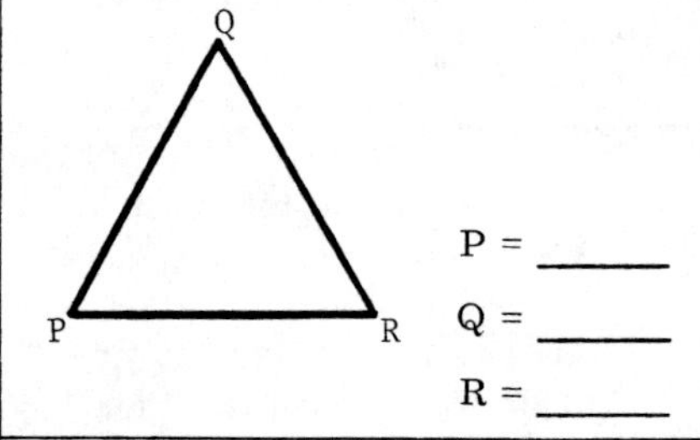

P = ______

Q = ______

R = ______

11. Find "d", the diagonal of the rectangle. Round to the nearest whole number.

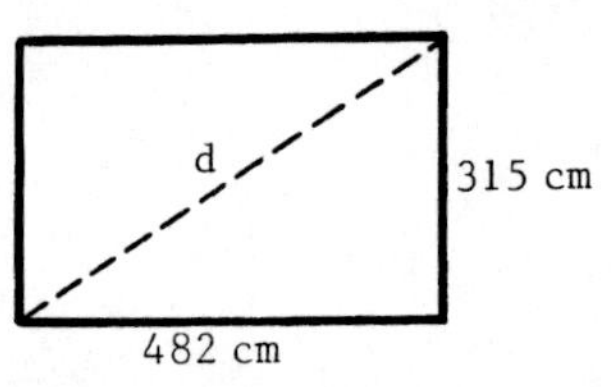

d = ________

12. Find the area of the shaded figure. Round to the nearest whole number.

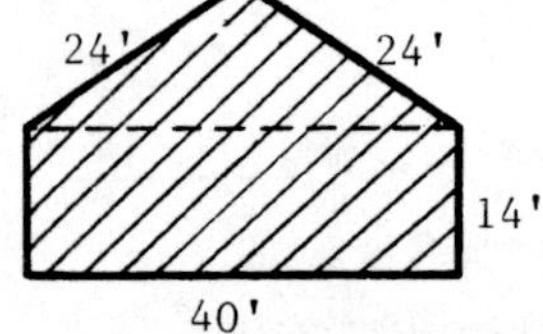

A = ________

ANSWERS:

1. 34°
2. t
3. 23°
4. d
5. c = 57.8 m
6. p = 1.12 in
7. A = 1.61 in^2
8. G = 26°
   H = 128°
9. A = 45°
   B = 45°
10. P = 60°
    Q = 60°
    R = 60°
11. d = 576 cm
12. A = 825 ft^2

1-5 THE TANGENT RATIO

When the Pythagorean Theorem or angle-sum principle cannot be used to find an unknown side or unknown angle in a right triangle, we can usually use one of the three basic trigonometric ratios to do so. In this section, we will define the tangent ratio and show how it can be used to find an unknown side or angle in a right triangle.

---

38. In right triangle ABC, "c" is the hypotenuse, and "a" and "b" are the two legs.

"a" is the side opposite angle A.

"b" is the side adjacent to angle A.

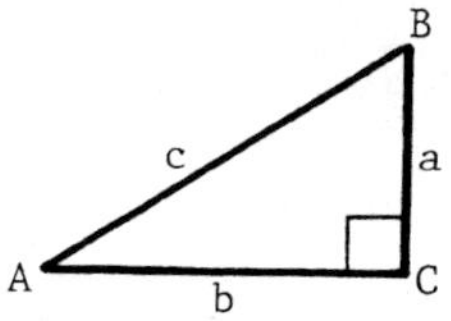

Note: The word "adjacent" means "next to". Though both "c" and "b" are "next to" angle A, only "b" is the "side adjacent" because "c" is the hypotenuse.

In the same triangle: a) the side opposite angle B is ______.

b) the side adjacent to angle B is ______.

---

39. In righ

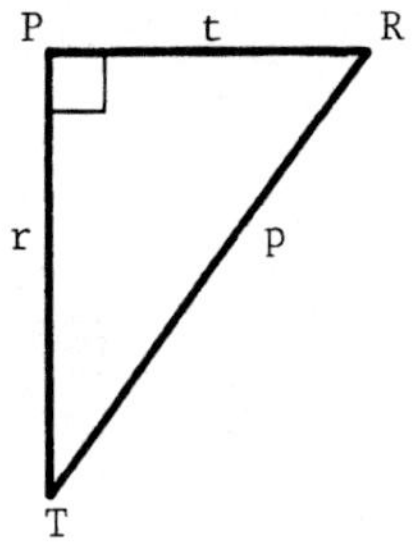

a) The hypotenuse is ______.

b) The side opposite angle T is ______.

c) The side adjacent to angle T is ______.

d) The side opposite angle R is ______.

e) The side adjacent to angle R is ______.

Answers: a) b b) a

---

40. In right triangle HSV:

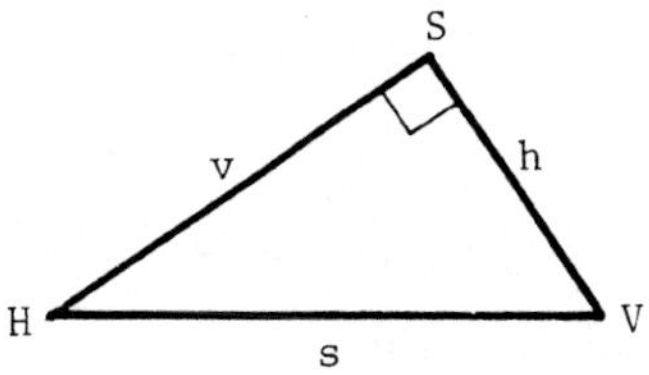

a) The hypotenuse is ______.

b) The side opposite angle V is ______.

c) The side adjacent to angle V is ______.

d) The side opposite angle H is ______.

e) The side adjacent to angle H is ______.

Answers: a) p d) r b) t e) t c) r

---

41. The tangent ratio is one of the three basic trigonometric ratios. In a right triangle, the "tangent of an acute angle" is a comparison of the "length of the side opposite" to the "length of the side adjacent" to the angle. That is:

$$\text{THE TANGENT OF AN ANGLE} = \frac{\text{SIDE OPPOSITE}}{\text{SIDE ADJACENT}}$$

Answers: a) s d) h b) v e) v c) h

Continued on following page.

41. Continued

In right triangle ABC:

The side opposite angle A is "a".

The side adjacent to angle A is "b".

Therefore, the <u>tangent of angle A</u> is $\frac{a}{b}$.

The side opposite angle B is "b".

The side adjacent to angle B is "a".

Therefore, the tangent of angle B is ______.

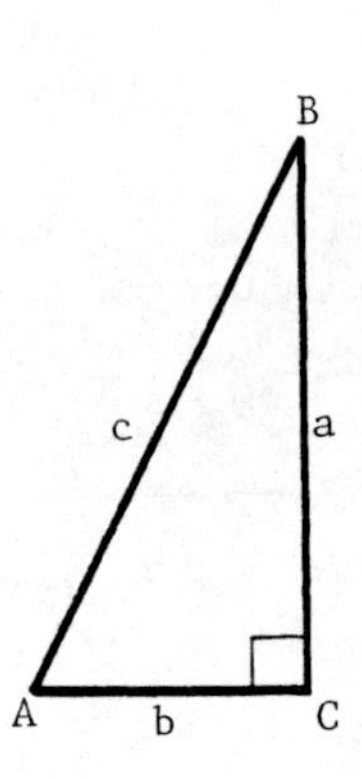

---

42. In right triangle MPT:

a) The tangent of angle T is ______

b) The tangent of angle M is ______.

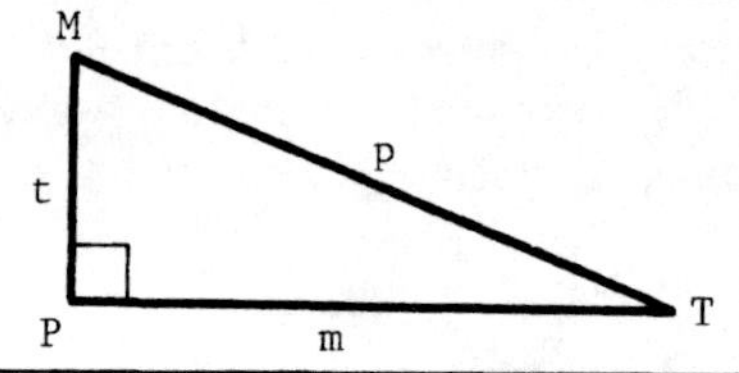

Answer to 41: $\frac{b}{a}$

---

43. Using capital letters to label the sides, complete these for right triangle CDF.

a) The tangent of angle D is ______.

b) The tangent of angle F is ______.

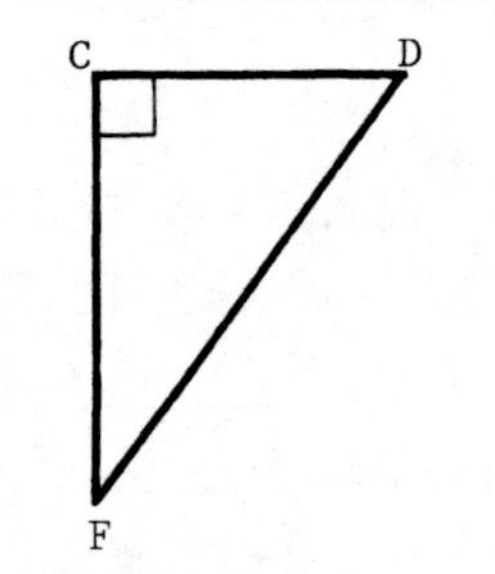

Answer to 42: a) $\frac{t}{m}$ b) $\frac{m}{t}$

---

44. In right triangle ADV:

a) The tangent of angle A is ______.

b) The tangent of angle V is ______.

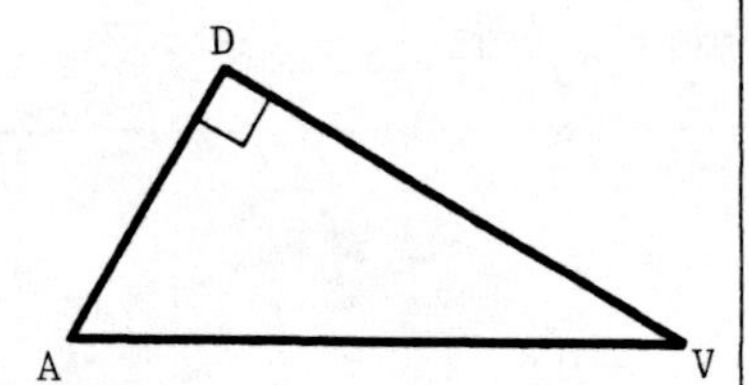

Answer to 43: a) $\frac{CF}{CD}$ b) $\frac{CD}{CF}$

---

45. The tangent of any specific angle has the same numerical value in right triangles of any size. For example, in the diagram below, there are three right triangles: ABG, ACF, and ADE. Each right triangle contains the common angle A which is a 37° angle.

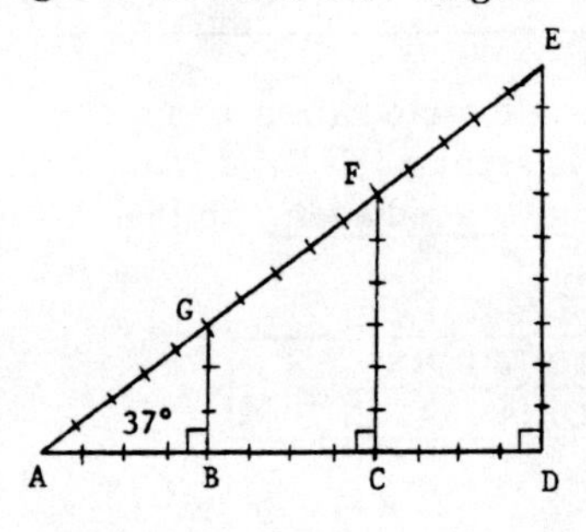

Answer to 44: a) $\frac{DV}{AD}$ b) $\frac{AD}{DV}$

Continued on following page.

45. Continued

To show that the tangent of a 37° angle is the same in right triangles of any size, we computed the tangent of angle A in each of the three right triangles by counting units to approximate the lengths of the sides. As you can see, the tangent of 37° is approximately $\frac{3}{4}$ in each triangle.

In triangle ABG, the tangent of A is $\frac{BG}{AB} = \frac{3}{4}$

In triangle ACF, the tangent of A is $\frac{CF}{AC} = \frac{6}{8}$ or $\frac{3}{4}$

In triangle ADE, the tangent of A is $\frac{DE}{AD} = \frac{9}{12}$ or $\frac{3}{4}$

Tangents are usually expressed as decimal numbers. Therefore, instead of saying that the tangent of 37° is approximately $\frac{3}{4}$, we would say that the tangent of 37° is approximately ________.

---

46. Instead of "the tangent of a 37° angle", the abbreviation "tan 37°" is used. Therefore:

tan 15° means: the tangent of a 15° angle

tan 74° means: ____________________

0.75

---

47. A partial table of tangents is shown at the right. Each tangent is rounded to four decimal places. Notice this fact:

As the size of the angle increases from 0° to 89°, the size of the tangent increases from 0.0000 to 57.2900 .

Angle A	tan A
0°	0.0000
10°	0.1763
20°	0.3640
30°	0.5774
40°	0.8391
50°	1.1918
60°	1.7321
70°	2.7475
80°	5.6713
89°	57.2900

Using the table, complete these:

a) The tangent of any 70° angle is ________.

b) 0.1763 is the tangent of any ______ angle.

the tangent of a 74° angle

---

48. A calculator can be used to find the tangent of any angle. To do so, we simply enter the angle and press [tan].

Note: Some calculators are designed to give tangents of angles measured in degrees, radians, and even grads. Be sure that your calculator is set to give the tangents of angles measured in degrees.

Continued on following page.

a) 2.7475

b) 10°

48. Continued

Following the steps below, find tan 15°, tan 47°, and tan 86°.

Enter	Press	Display
15	tan	.26794919
47	tan	1.0723687
86	tan	14.300666

Tangents are usually rounded to four decimal places. Therefore:

a) tan 15° = ______ b) tan 47° = ______ c) tan 86° = ______

---

49. Use a calculator for these. Round each tangent to four decimal places.

a) tan 34° = ____________ b) tan 78° = ____________

Answers to 48: a) 0.2679 b) 1.0724 c) 14.3007

---

50. A calculator can also be used to find the size of an angle whose tangent is known. Either INV tan , arc tan , or $\tan^{-1}$ is used. Following the steps below, find the angle whose tangent is 0.3173.

Enter	Press	Display
0.3173	INV tan	17.604233
	or arc tan	17.604233
	or $\tan^{-1}$	17.604233

The angle is usually rounded to the nearest whole number. Therefore:

The angle whose tangent is 0.3173 is ______.

Answers to 49: a) 0.6745 b) 4.7046

---

51. Complete these. Round each angle to the nearest whole number.

a) If tan F = 0.7541, F = ______ b) If tan A = 2.3555, A = ______

Answer to 50: 18°

---

52. We cannot use the Pythagorean Theorem to find side MR in the right triangle below because the length of only one side is known. However, we can use tan P to do so. The steps are:

$$\tan P = \frac{MR}{PR}$$

$$\tan 25° = \frac{MR}{52.5}$$

$$MR = (52.5)(\tan 25°)$$

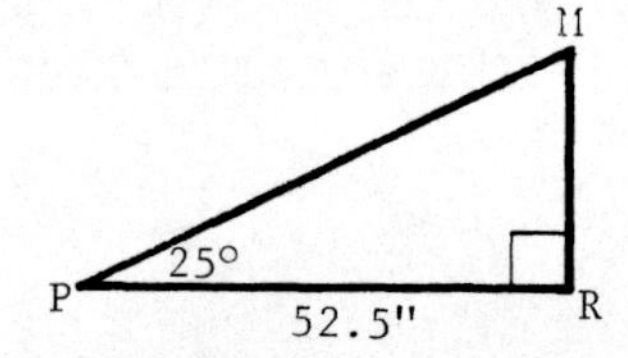

Answers to 51: a) 37° b) 67°

Continued on following page.

52. Continued

To evaluate (52.5)(tan 25°) on a calculator, follow the steps below. Wait until tan 25° appears on the display before pressing [=].

Enter	Press	Display
52.5	[x]	52.5
25	[tan] [=]	24.481152

Rounding to the nearest tenth of an inch, MR = ________

24.5 in

53. We cannot use the Pythagorean Theorem to find side CE in the right triangle below because only one side is known. However, we can use tan C to do so. The steps are:

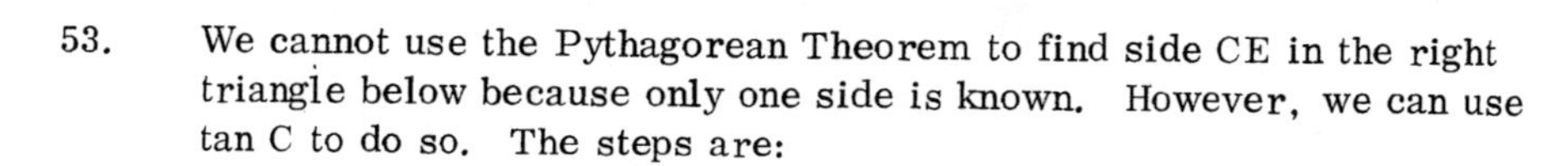

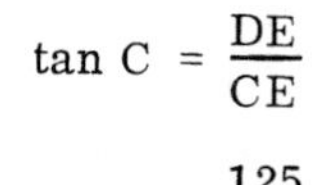

$$\tan C = \frac{DE}{CE}$$

$$\tan 55° = \frac{125}{CE}$$

$$(\tan 55°)(CE) = 125$$

$$CE = \frac{125}{\tan 55°}$$

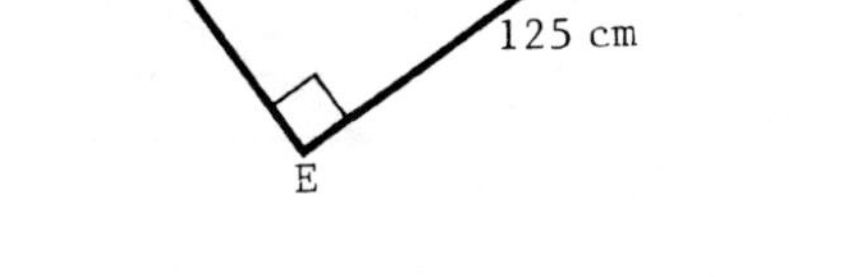

To complete the evaluation on a calculator, follow the steps below. Wait until tan 55° appears on the display before pressing [=].

Enter	Press	Display
125	[÷]	125.
55	[tan] [=]	87.525942

Rounding to the nearest tenth of a centimeter, CE = ________

87.5 cm

54. We cannot use the angle-sum principle to find angle R in the right triangle at the right because angle T is not known. However, we can use tan R to do so. The steps are:

$$\tan R = \frac{ST}{RS}$$

$$\tan R = \frac{19.8}{34.3}$$

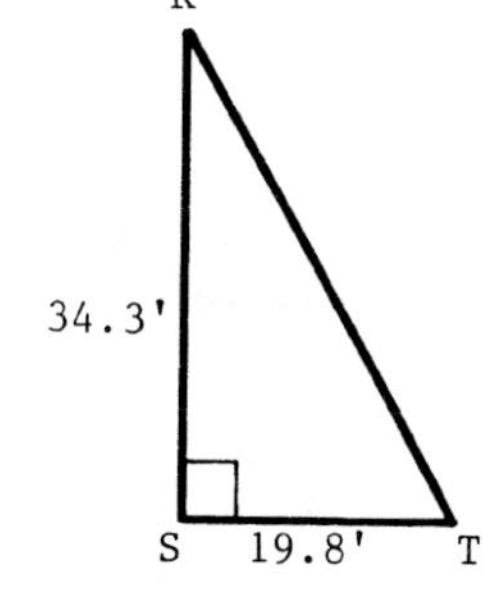

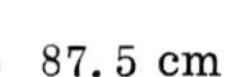

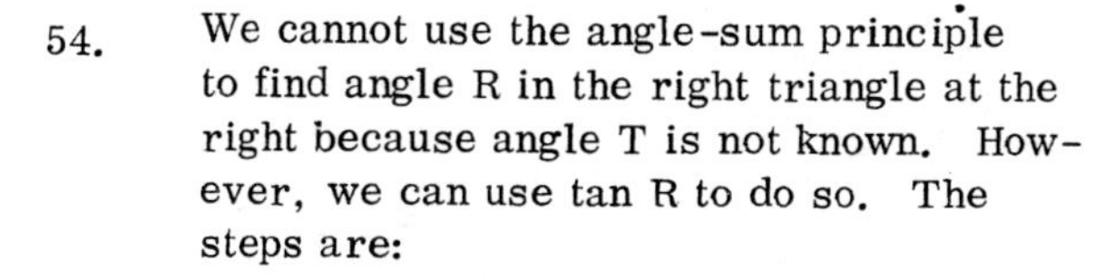

To find R, follow the steps below. Notice that we pressed [=] to complete the division before pressing [INV] [tan], [arc] [tan], or [tan⁻¹].

Enter	Press	Display
19.8	[÷]	19.8
34.3	[=] [INV] [tan]	29.996098
	or [arc] [tan]	
	or [$\tan^{-1}$]	

Rounding to the nearest whole number degree, R = ________

30°

1-6 THE SINE RATIO

The sine ratio is also one of the three basic trigonometric ratios. In this section, we will define the sine ratio and show how it can be used to find an unknown side or angle in a right triangle.

---

55. In a right triangle, the "sine of an acute angle" is a comparison of the "length of the side opposite" the angle to the "length of the hypotenuse". That is:

$$\text{THE SINE OF AN ANGLE} = \frac{\text{SIDE OPPOSITE}}{\text{HYPOTENUSE}}$$

Note: The word "sine" is pronounced "sign". It is not pronounced "sin".

In right triangle ABC:

The side opposite angle A is "a".

The hypotenuse is "c".

Therefore, the sine of angle A is $\frac{a}{c}$.

What is the sine of angle B? ______

---

56. In right triangle XYT:

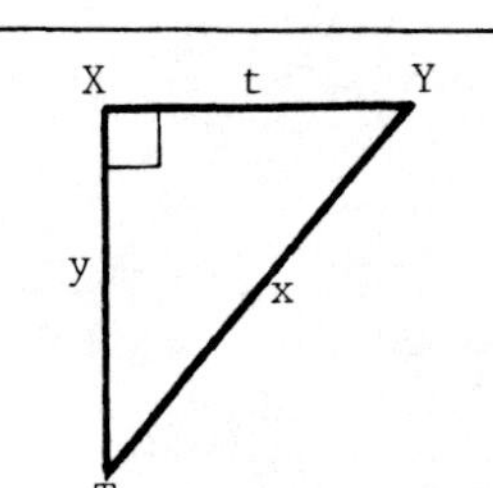

a) The sine of angle T is ______.

b) The sine of angle Y is ______.

Answer to 55: $\frac{b}{c}$

---

57. Using capital letters to label the sides, complete these for right triangle PQS.

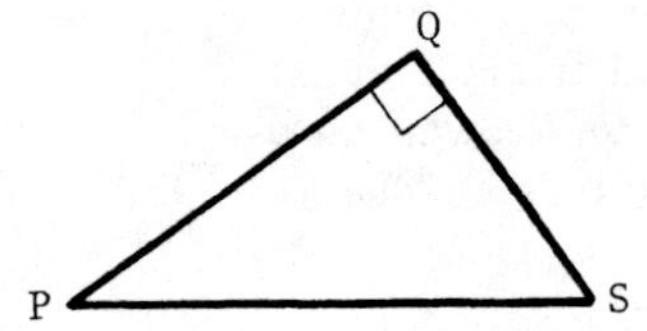

a) The sine of angle P is ______.

b) The sine of angle S is ______.

Answers to 56: a) $\frac{t}{x}$ b) $\frac{y}{x}$

---

58. To show that the sine of a specific angle has the same numerical value in right triangles of any size, we will again use a 37° angle as an example. Let's count units to find the sine of angle A in each of the three triangles at the right.

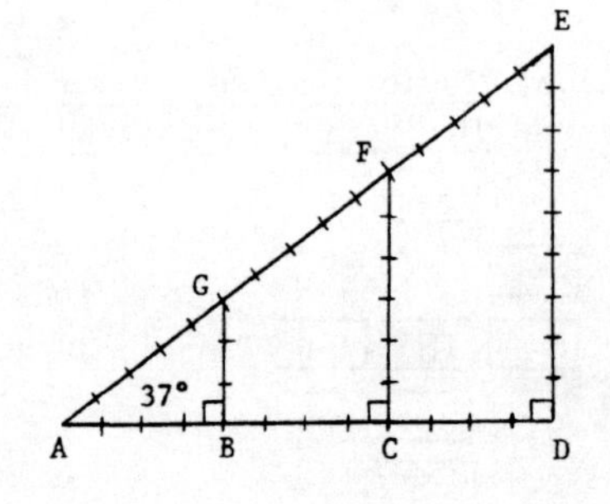

Answers to 57: a) $\frac{QS}{PS}$ b) $\frac{PQ}{PS}$

Continued on following page.

58. Continued

In triangle ABG, the sine of A is $\frac{BG}{AG} = \frac{3}{5}$

In triangle ACF, the sine of A is $\frac{CF}{AF} = \frac{6}{10}$ or $\frac{3}{5}$

In triangle ADE, the sine of A is $\frac{DE}{AE} = \frac{9}{15}$ or $\frac{3}{5}$

Sines are usually expressed as decimal numbers. Therefore, instead of saying that the sine of 37° is approximately $\frac{3}{5}$, we would say that the sine of 37° is approximately ______.

Answer: 0.6

59. Instead of "the sine of a 37° angle", the abbreviation "sin 37°" is used. Though the abbreviation is "sin", it is still pronounced "sign".

sin 15° means: the sine of a 15° angle

sin 48° means: ______________________

Answer: the sine of a 48° angle

60. A partial table of sines is shown at the right. Each sine is rounded to four decimal places. Notice this fact:

As the size of the angle increases from 0° to 90°, the size of the sine increases from 0.0000 to 1.0000.

Angle A	sin A
0°	0.0000
10°	0.1736
20°	0.3420
30°	0.5000
40°	0.6428
50°	0.7660
60°	0.8660
70°	0.9397
80°	0.9848
90°	1.0000

Using the table, complete these:

a) The sine of any 50° angle is ______.

b) 0.9848 is the sine of any ______ angle.

Answers: a) 0.7660 b) 80°

61. To find the sine of an angle on a calculator, we enter the angle and press [sin]. Use a calculator for these. Round to four decimal places.

a) sin 7° = ______ b) sin 41° = ______ c) sin 83° = ______

Answers: a) 0.1219 b) 0.6561 c) 0.9925

62. To find an angle whose sine is known on a calculator, we enter the sine and then press either [INV] [sin] or [arc] [sin] or [sin⁻¹]. Use a calculator for these. Round to the nearest whole-number degree.

a) If sin D = 0.2666, D = ______

b) If sin G = 0.8511, G = ______

Answers: a) 15° b) 58°

63. We cannot use the Pythagorean Theorem to find side ST in the right triangle below because only one side is known. However, we can use sin Q to do so. The steps are:

$$\sin Q = \frac{ST}{SQ}$$

$$\sin 35° = \frac{ST}{96}$$

$$ST = (96)(\sin 35°)$$

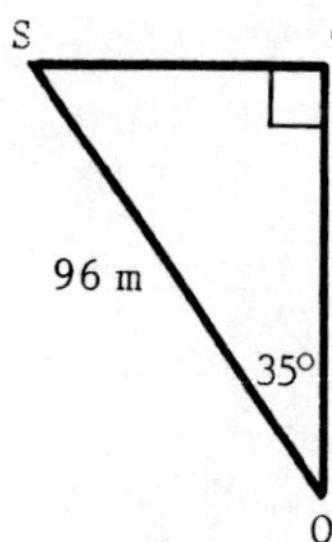

To evaluate (96)(sin 35°) on a calculator, follow the steps below. Wait until sin 35° appears on the display before pressing [=].

Enter	Press	Display
96	[x]	96.
35	[sin] [=]	55.063338

Rounding to the nearest tenth of a meter, ST = ________.

55.1 m

64. We cannot use the Pythagorean Theorem to find hypotenuse CF in the right triangle below because only one side is known. However, we can use sin F to do so. The steps are:

$$\sin F = \frac{CR}{CF}$$

$$\sin 50° = \frac{57.8}{CF}$$

$$(\sin 50°)(CF) = 57.8$$

$$CF = \frac{57.8}{\sin 50°}$$

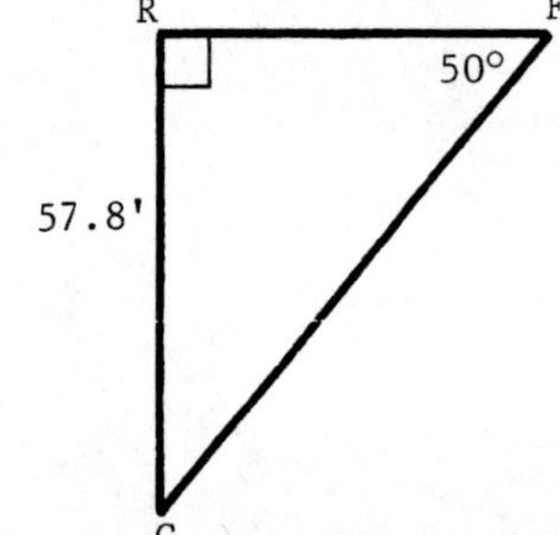

To complete the evaluation on a calculator, follow the steps below. Wait until sin 50° appears on the display before pressing [=].

Enter	Press	Display
57.8	[÷]	57.8
50	[sin] [=]	75.452541

Rounding to the nearest tenth of a foot, CF = ________

75.5 ft

65. We cannot use the angle-sum principle to find angle D below because angle R is not known. However, we can use sin D to do so. The steps are:

$$\sin D = \frac{PR}{DR}$$

$$\sin D = \frac{60.1}{85.8}$$

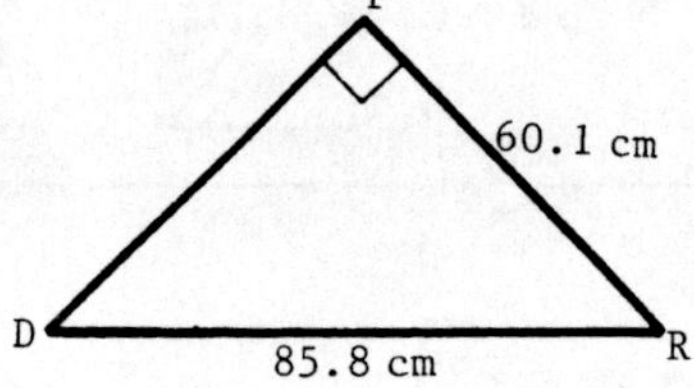

Continued on following page.

65. Continued

To find D on a calculator, following the steps below. Notice that we pressed [=] to complete the division before pressing [INV] [sin], [arc] [sin], or [sin^{-1}]

Enter	Press	Display
60.1	[÷]	60.1
85.8	[=] [INV] [sin]	44.464419
	or [arc] [sin]	
	or [sin^{-1}]	

Rounding to the nearest whole-number degree, D = ______

44°

1-7 THE COSINE RATIO

The cosine ratio is the last of the three basic trigonometric ratios. In this section, we will define the cosine ratio and show how it can be used to find an unknown side or angle in a right triangle.

66. In a right triangle, the "cosine of an acute angle" is a comparison of the "length of the side adjacent" to the angle to the "length of the hypotenuse". That is:

$$\text{THE COSINE OF AN ANGLE} = \frac{\text{SIDE ADJACENT}}{\text{HYPOTENUSE}}$$

Note: The word "cosine" is pronounced "co-sign".

In right triangle CDE:

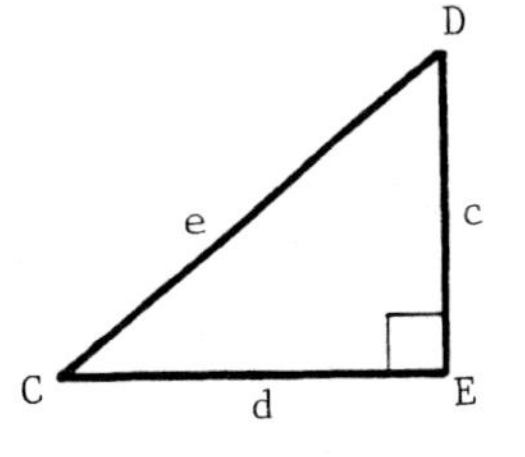

The side adjacent to angle C is "d".

The hypotenuse is "e".

Therefore, the cosine of angle C is $\frac{d}{e}$

What is the cosine of angle D? ______

$\frac{c}{e}$

67. The abbreviation for "cosine" is "cos". Therefore, in right triangle MFT:

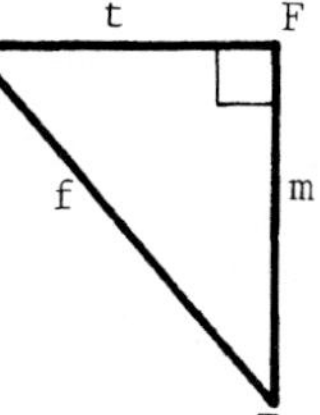

a) cos M = ______

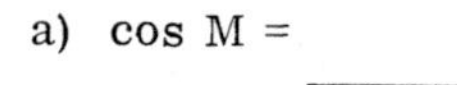

b) cos T = ______

a) $\frac{t}{f}$ b) $\frac{m}{f}$

68. Using capital letters to label the sides, complete these.

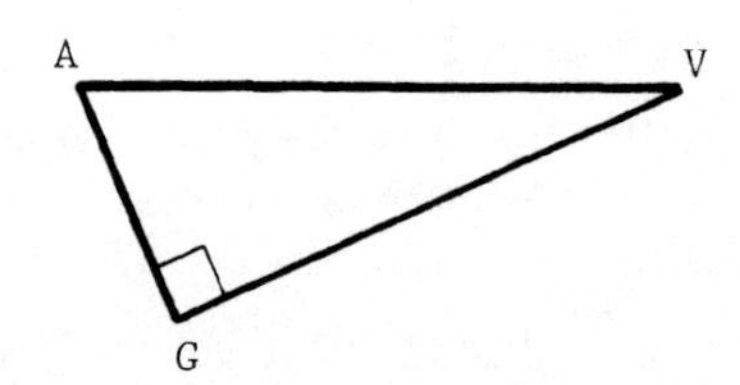

a) cos A = ______

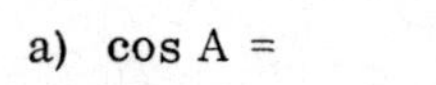

b) cos V = ______

---

a) $\frac{AG}{AV}$ b) $\frac{GV}{AV}$

69. To show that the cosine of a specific angle has the same numerical value in right triangles of any size, we will again use a 37° angle as the example. Let's count units to find the cosine of angle A in each of the three triangles at the right.

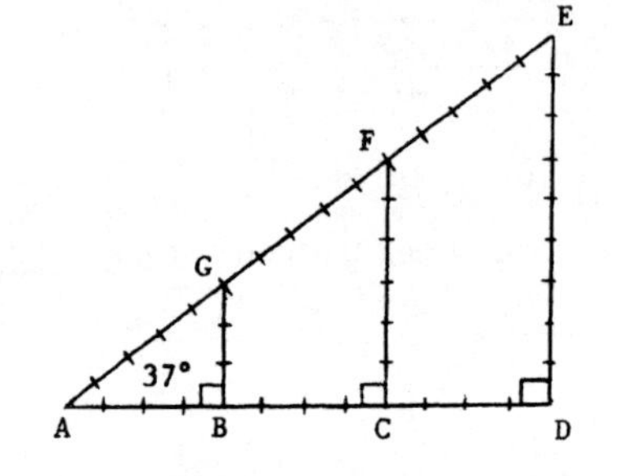

In triangle ABG, $\cos A = \frac{AB}{AG} = \frac{4}{5}$

In triangle ACF, $\cos A = \frac{AC}{AF} = \frac{8}{10}$ or $\frac{4}{5}$

In triangle ADE, $\cos A = \frac{AD}{AE} = \frac{12}{15}$ or $\frac{4}{5}$

Expressed as a decimal number, cos 37° in each of the three triangles is approximately ________.

0.8

70. A partial table of cosines is shown at the right. Each cosine is rounded to four decimal places. Notice this fact:

As the size of the angle increases from 0° to 90°, the size of the cosine decreases from 1.0000 to 0.0000.

Angle A	cos A
0°	1.0000
10°	0.9848
20°	0.9397
30°	0.8660
40°	0.7660
50°	0.6428
60°	0.5000
70°	0.3420
80°	0.1736
90°	0.0000

Using the table, complete these:

a) cos 50° = ________

b) If cos D = 0.1736, D = ________

---

a) 0.6428

b) 80°

71. To find the cosine of an angle on a calculator, we enter the angle and press cos . Use a calculator for these. Round to four decimal places.

a) cos 9° = ________ b) cos 27° = ________ c) cos 81° = ________

---

a) 0.9877

b) 0.8910

c) 0.1564

72. To find an angle whose cosine is known on a calculator, we enter the cosine and then press either [INV] [cos] or [arc] [cos] or [$\cos^{-1}$]. Use a calculator for these. Round to the nearest whole-number degree.

a) If cos H = 0.7561, H = ______

b) If cos V = 0.2099, V = ______

Answers: a) 41° b) 78°

73. We cannot use the Pythagorean Theorem to find side AC in the right triangle below because only one side is known. However, we can use cos A to do so. The steps are:

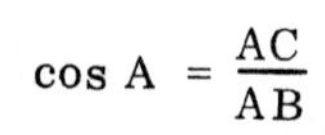

$$\cos A = \frac{AC}{AB}$$

$$\cos 53° = \frac{AC}{60.7}$$

$$AC = (60.7)(\cos 53°)$$

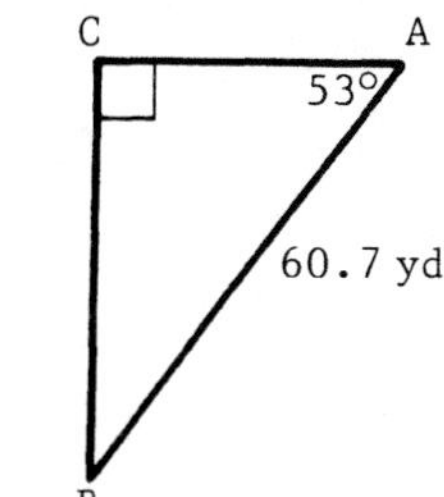

To evaluate (60.7)(cos 53°) on a calculator, follow the steps below. Wait until cos 53° appears before pressing [=].

Enter	Press	Display
60.7	[x]	60.7
53	[cos] [=]	36.530172

Rounding to the nearest tenth of a yard, AC = ______

Answer: 36.5 yd

74. We cannot use the Pythagorean Theorem to find hypotenuse FM in the right triangle below because only one side is known. However, we can use cos F to do so. The steps are:

$$\cos F = \frac{FG}{FM}$$

$$\cos 33° = \frac{256}{FM}$$

$$(\cos 33°)(FM) = 256$$

$$FM = \frac{256}{\cos 33°}$$

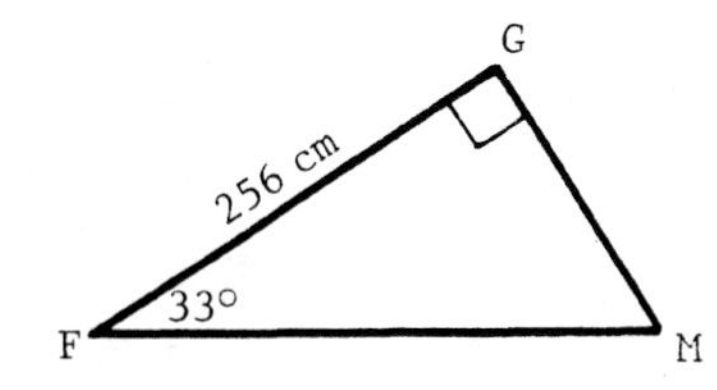

To complete the evaluation on a calculator, follow the steps below. Wait until cos 33° appears before pressing [=].

Enter	Press	Display
256	[÷]	256.
33	[cos] [=]	305.245

Rounding to the nearest centimeter, FM = ______

Answer: 305 cm

75. We cannot use the angle-sum principle to find angle D below because angle S is not known. However, we can use cos D to do so. The steps are:

$$\cos D = \frac{DH}{DS}$$

$$\cos D = \frac{14.7}{18.7}$$

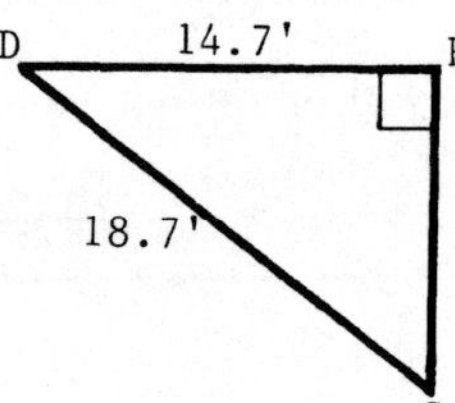

Find D on a calculator. Be sure to press [=] before pressing [INV] [cos] or [arc] [cos] or [$\cos^{-1}$].

To the nearest whole-number degree, D = _______

38°

1-8 CONTRASTING THE BASIC TRIGONOMETRIC RATIOS

In this section, we will give some exercises contrasting the definitions of the three basic trigonometric ratios.

76. The definitions of the three basic trigonometric ratios for angle A in the right triangle are:

$$\tan A = \frac{\text{side opposite angle A}}{\text{side adjacent to angle A}}$$

$$\sin A = \frac{\text{side opposite angle A}}{\text{hypotenuse}}$$

$$\cos A = \frac{\text{side adjacent to angle A}}{\text{hypotenuse}}$$

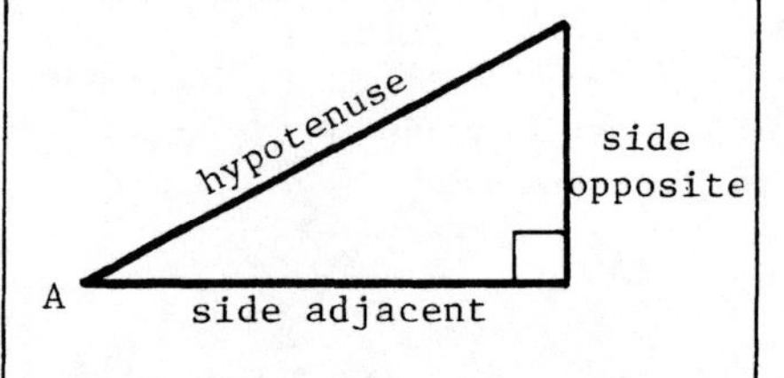

When writing the trig ratios for an angle in a right triangle, it is helpful to locate the hypotenuse first. For example, in triangle BFT:

a) The hypotenuse is _____.

b) sin B = _____

c) cos B = _____

d) tan B = _____

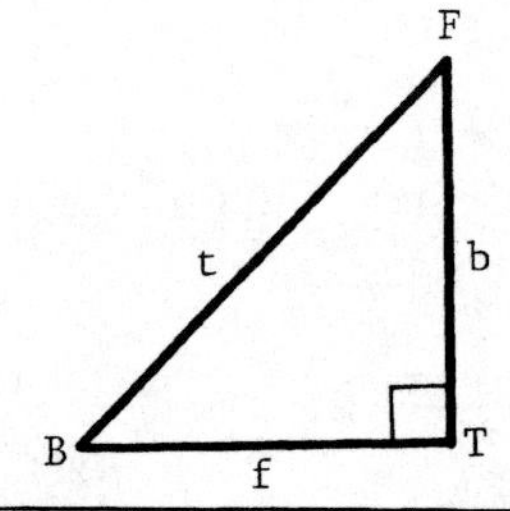

a) t

b) $\frac{b}{t}$

c) $\frac{f}{t}$

d) $\frac{b}{f}$

77. Using capital letters for the sides, complete these:

a) cos P = ______

b) tan S = ______

c) sin P = ______

d) cos S = ______

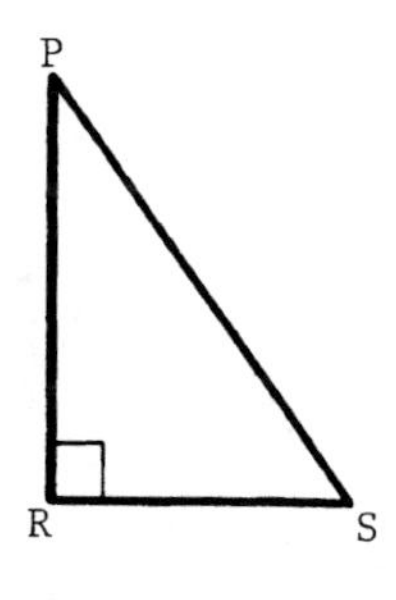

---

78. Using capital letters for the sides, complete these:

a) tan H = ______

b) tan V = ______

c) sin H = ______

d) sin V = ______

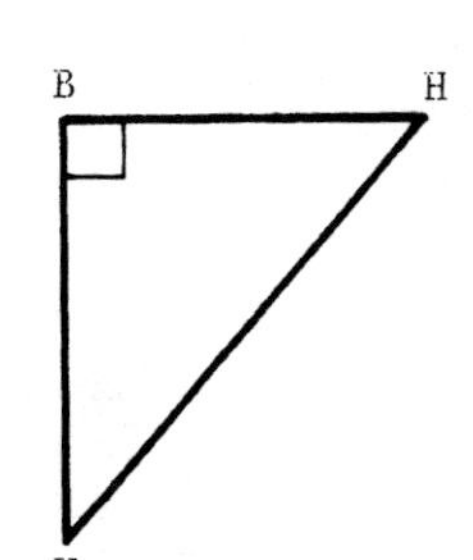

a) $\frac{PR}{PS}$ c) $\frac{RS}{PS}$

b) $\frac{PR}{RS}$ d) $\frac{RS}{PS}$

---

79. Complete:

a) tan T = ______

b) cos F = ______

c) sin T = ______

d) tan F = ______

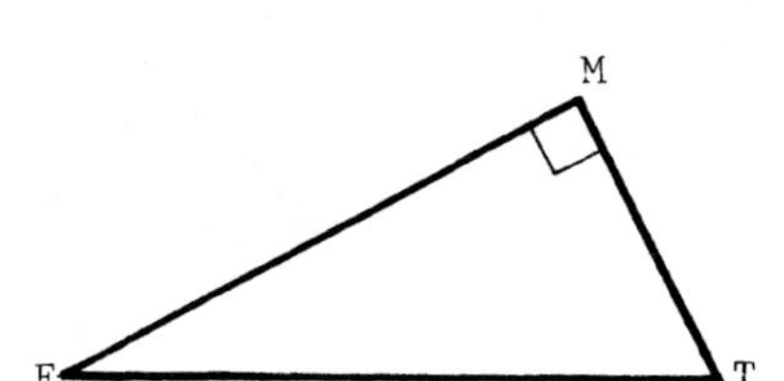

a) $\frac{BV}{BH}$ c) $\frac{BV}{HV}$

b) $\frac{BH}{BV}$ d) $\frac{BH}{HV}$

---

80. Using small letters for the sides, complete these:

a) sin D = ______

b) tan C = ______

c) sin C = ______

d) cos D = ______

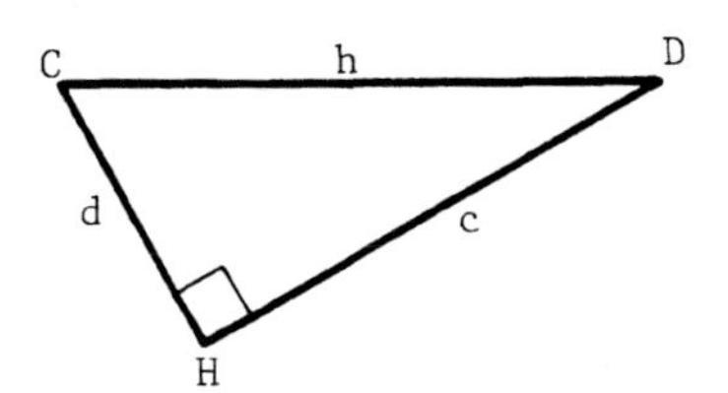

a) $\frac{FM}{MT}$ c) $\frac{FM}{FT}$

b) $\frac{FM}{FT}$ d) $\frac{MT}{FM}$

---

a) $\frac{d}{h}$ c) $\frac{c}{h}$

b) $\frac{c}{d}$ d) $\frac{c}{h}$

81. If a ratio does not involve the hypotenuse, it is a "tangent" ratio. For example, in right triangle FGH:

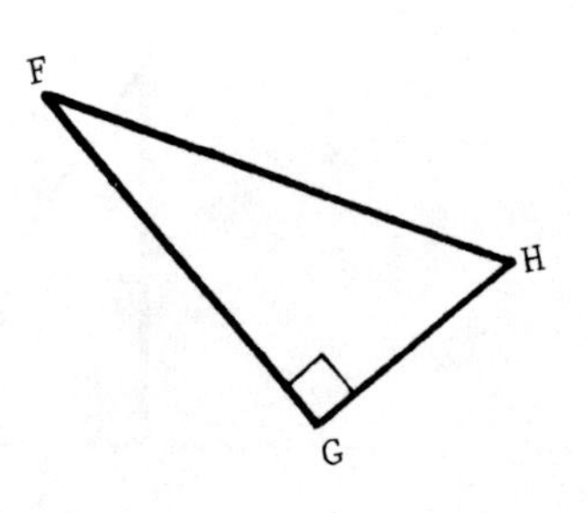

a) The ratio $\frac{FG}{GH}$ is the tangent of angle ______.

b) The ratio $\frac{GH}{FG}$ is the tangent of angle ______.

---

a) H   b) F

82. If a ratio involves the hypotenuse, it is the sine of one angle and the cosine of the other. For example, in the triangle on the right:

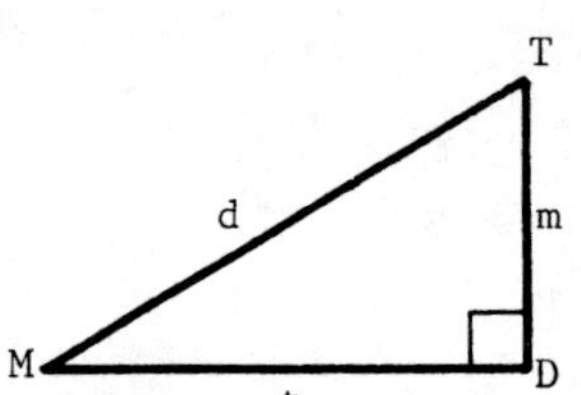

$\frac{m}{d}$ is the sine of angle M and the cosine of angle T.

$\frac{t}{d}$ is the sine of angle ____ and the cosine of angle ____.

---

sine of angle T, cosine of angle M

83. In this right triangle:

$\frac{MN}{NP}$ is tan P

a) $\frac{NP}{MN}$ is ______________

b) $\frac{MN}{MP}$ is either ______________ or ______________

---

a) tan M

b) sin P or cos M

84. In this right triangle:

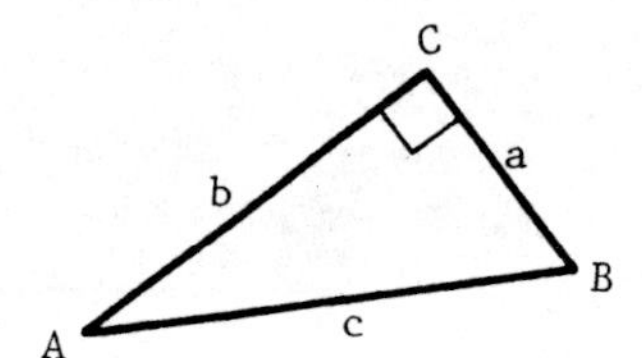

a) $\frac{a}{b}$ = ______________

b) $\frac{b}{a}$ = ______________

c) $\frac{a}{c}$ = ______________ or ______________

---

a) tan A

b) tan B

c) sin A or cos B

85. In this right triangle:

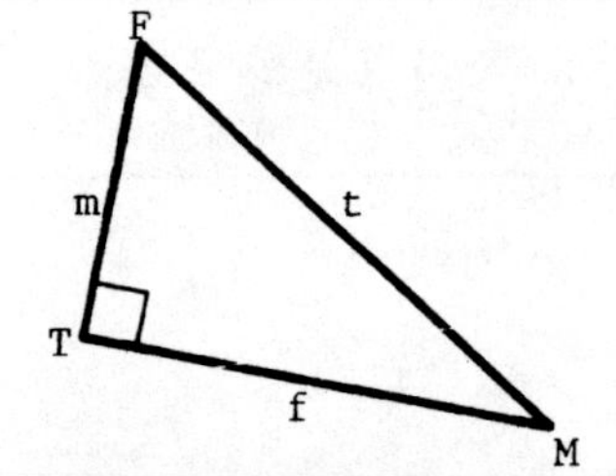

a) $\frac{m}{t}$ = __________ or __________

b) $\frac{f}{t}$ = __________ or __________

---

a) sin M or cos F   b) sin F or cos M

SELF-TEST 2 (pp. 13-27)

Find the numerical value of each trigonometric ratio. Round to four decimal places.

1. cos 35° = ______

2. tan 23° = ______

3. sin 66° = ______

Find each angle. Round to the nearest whole-number degree.

4. If sin B = 0.1614, B = ______

5. If cos F = 0.2597, F = ______

6. If tan R = 1.3318, R = ______

Evaluate each of the following. Round as directed.

7. Round to tenths.

$\frac{27.8}{\cos 19°}$ = ______

8. Round to hundreds.

(42,900)(tan 61°) = ______

9. Round to hundredths.

$\frac{1.53}{\sin 40°}$ = ______

Find each angle. Round to the nearest whole-number degree.

10. $\tan A = \frac{5,730}{1,980}$

A = ______

11. $\sin P = \frac{2.19}{7.55}$

P = ______

12. $\cos H = \frac{66.7}{78.4}$

H = ______

13. In right triangle PRT, define the following trigonometric ratios. Use capital letters for labeling the sides.

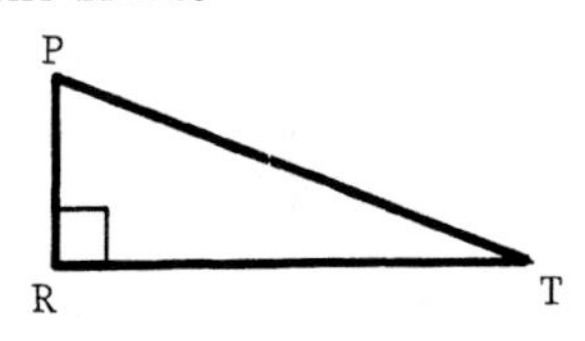

a) sin T = ______

b) tan T = ______

c) tan P = ______

d) cos P = ______

14. In right triangle DEH, define the following trigonometric ratios. Use small letters for labeling the sides.

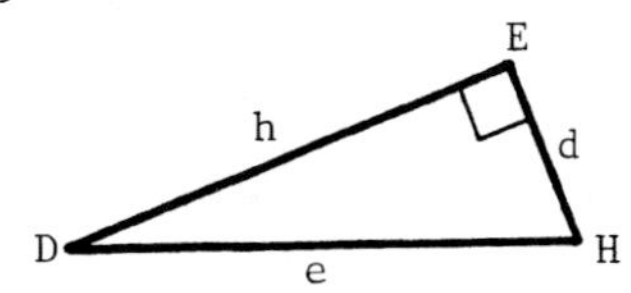

a) cos H = ______

b) sin H = ______

c) sin D = ______

d) tan H = ______

ANSWERS:

1. 0.8192
2. 0.4245
3. 0.9135
4. B = 9°
5. F = 75°
6. R = 53°
7. 29.4
8. 77,400
9. 2.38
10. A = 71°
11. P = 17°
12. H = 32°

13. a) $\frac{PR}{PT}$ b) $\frac{PR}{RT}$ c) $\frac{RT}{PR}$ d) $\frac{PR}{PT}$

14. a) $\frac{d}{e}$ b) $\frac{h}{e}$ c) $\frac{d}{e}$ d) $\frac{h}{d}$

1-9 FINDING UNKNOWN SIDES IN RIGHT TRIANGLES

Either the Pythagorean Theorem or the trig ratios are used to find unknown sides in right triangles. We will discuss the use of both methods in this section.

86. When only two sides of a right triangle are known, we use the Pythagorean Theorem to find the third side.

In right triangle ABC, AB and BC are known. Let's use the Pythagorean Theorem to find AC.

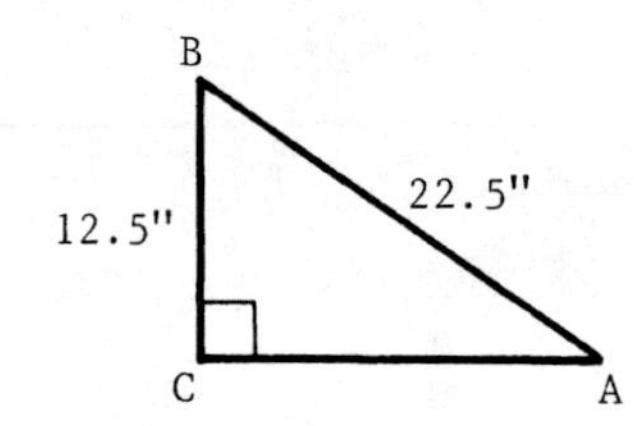

$$(AB)^2 = (BC)^2 + (AC)^2$$

$$(22.5)^2 = (12.5)^2 + (AC)^2$$

$$(AC)^2 = (22.5)^2 - (12.5)^2$$

$$AC = \sqrt{(22.5)^2 - (12.5)^2}$$

Use a calculator to complete the solution. Round to the nearest tenth of an inch. AC = ________

87. When only one side and one acute angle of a right triangle are known, we use a trig ratio to find an unknown side.

Answer to 86: 18.7 in

In right triangle CDF, side CD and angle C are known. We can use the <u>sine ratio</u> to find DF.

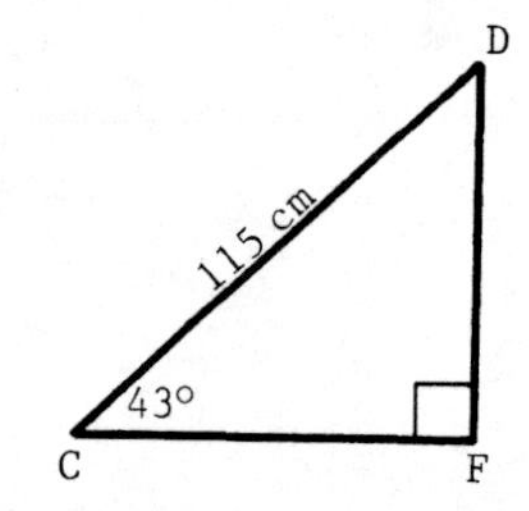

$$\sin C = \frac{DF}{CD}$$

$$\sin 43° = \frac{DF}{115}$$

$$DF = (115)(\sin 43°)$$

Use a calculator to complete the solution. Round to the nearest tenth of a centimeter. DF = ________

88. To find PS in this right triangle, we can use the <u>cosine ratio</u>.

Answer to 87: 78.4 cm

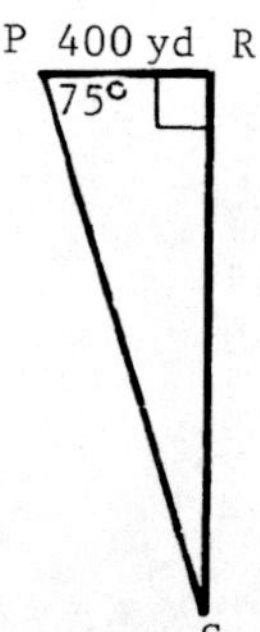

$$\cos P = \frac{PR}{PS}$$

$$\cos 75° = \frac{400}{PS}$$

$$(\cos 75°)(PS) = 400$$

$$PS = \frac{400}{\cos 75°}$$

Use a calculator to complete the solution. Round to the nearest yard.
PS = ________

1,545 yd

89. Before using a trig ratio to find an unknown side in a right triangle, we must decide whether to use the sine, cosine, or tangent of the given angle. The following strategy can be used.

Find "c" in right triangle CHM.

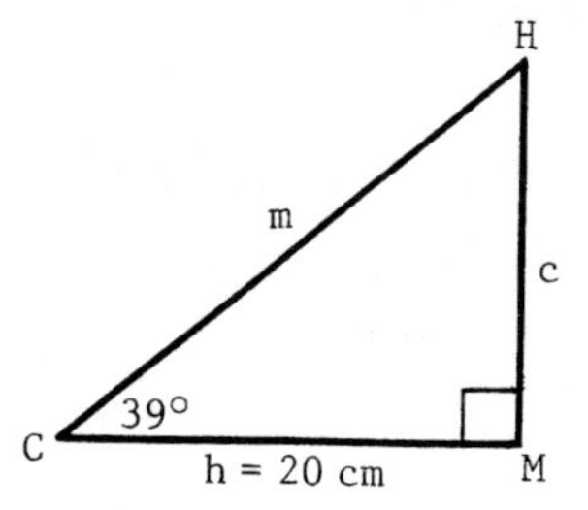

1) Identify the side you want to find ("c") and the known side ("h") in terms of the known angle (C).

"c" is the side opposite angle C.

"h" is the side adjacent to angle C.

2) Identify the ratio that includes both "side opposite" and "side adjacent".

The ratio is the tangent of angle C.

3) Use that ratio to set up an equation.

$$\tan C = \frac{c}{h} \quad \text{or} \quad \tan 39° = \frac{c}{20}$$

Use a calculator to complete the solution. Round to the nearest tenth of a centimeter. c = ________

16.2 cm, from:
$c = (20)(\tan 39°)$

90. Let's use the same steps to solve for DM in this right triangle.

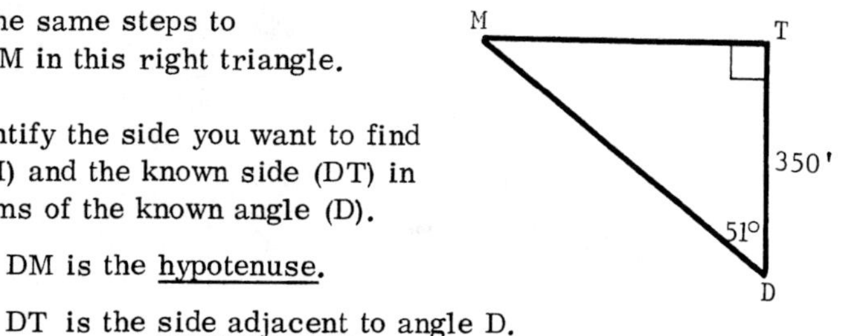

1) Identify the side you want to find (DM) and the known side (DT) in terms of the known angle (D).

DM is the hypotenuse.

DT is the side adjacent to angle D.

2) Identify the ratio that includes both "side adjacent" and "hypotenuse".

The ratio is the cosine of angle D.

3) Use that ratio to set up an equation.

$$\cos D = \frac{DT}{DM} \quad \text{or} \quad \cos 51° = \frac{350}{DM}$$

Use a calculator to complete the solution. Round to the nearest foot. DM = ________

556 ft, from:
$$DM = \frac{350}{\cos 51°}$$

91. We want to find GF in the triangle at the right. The known angle is G.

GF is the side adjacent to angle G.

FH is the side opposite angle G.

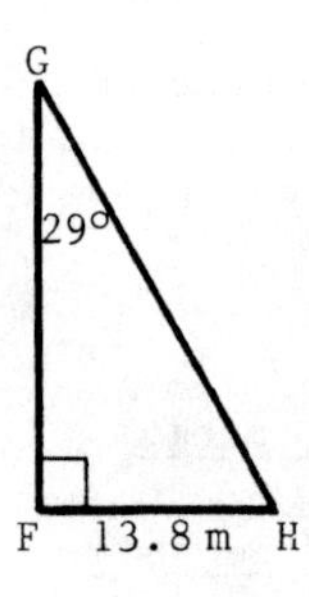

a) Should we use sin 29°, cos 29°, or tan 29° to solve for GF? ________

b) Complete the solution. Round to the nearest tenth of a meter.

GF = ________

---

92. We want to find "d" in the triangle at the right. The known angle is B.

"d" is the hypotenuse.

"b" is the side opposite angle B.

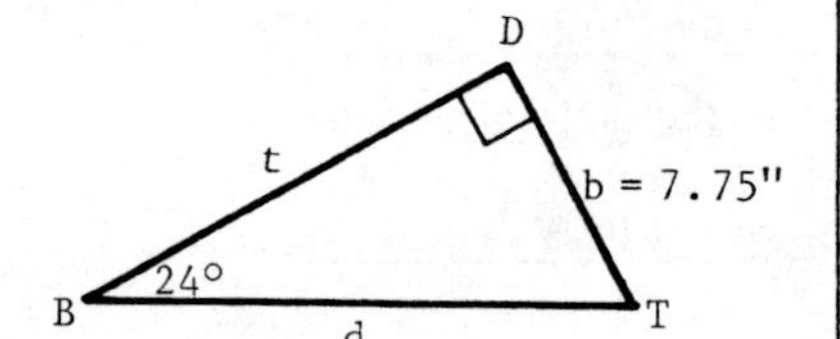

a) Should we use sin 24°, cos 24°, or tan 24° to solve for "d"? ________

b) Complete the solution. Round to the nearest tenth of an inch.

d = ________

Answers to 91:

a) tan 29°

b) 24.9 m, from:

$$\tan 29° = \frac{13.8}{GF}$$

---

93. We want to find "f" in the triangle at the right. The known angle is S.

"f" is the side adjacent to angle S.

"v" is the hypotenuse.

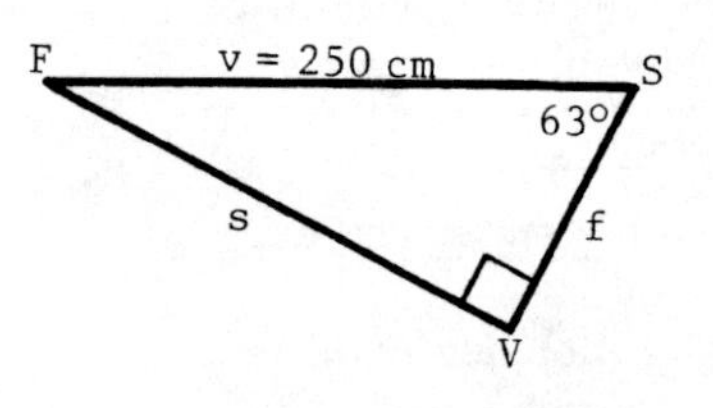

a) Should we use sin 63°, cos 63° or tan 63° to solve for "f"? ________

b) Complete the solution. Round to the nearest centimeter.

f = ________

Answers to 92:

a) sin 24°

b) 19.1 in, from:

$$\sin 24° = \frac{7.75}{d}$$

---

94. Let's solve for "p" in this triangle.

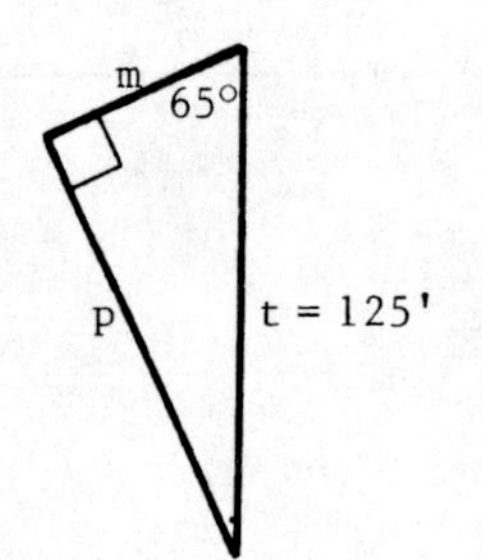

a) Should we use sin 65°, cos 65°, or tan 65°? ________

b) Complete the solution. Round to the nearest foot.

p = ________

Answers to 93:

a) cos 63°

b) 113 cm, from:

$$\cos 63° = \frac{f}{250}$$

95. Let's solve for TV in this triangle.

a) Should we use sin 70°, cos 70°, or tan 70°? ________

b) Complete the solution. Round to the nearest tenth of a meter.

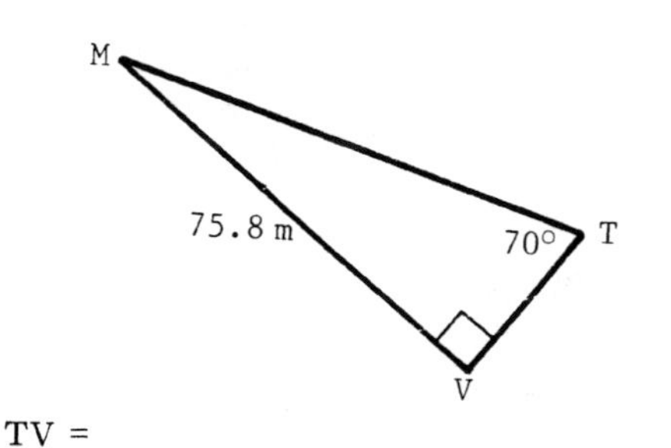

TV = ________

a) sin 65°

b) 113 ft, from:

$$\sin 65° = \frac{p}{125}$$

---

96. Let's solve for AD in this triangle.

a) Should we use sin 25°, cos 25°, or tan 25°? ________

b) Complete the solution. Round to the nearest mile.

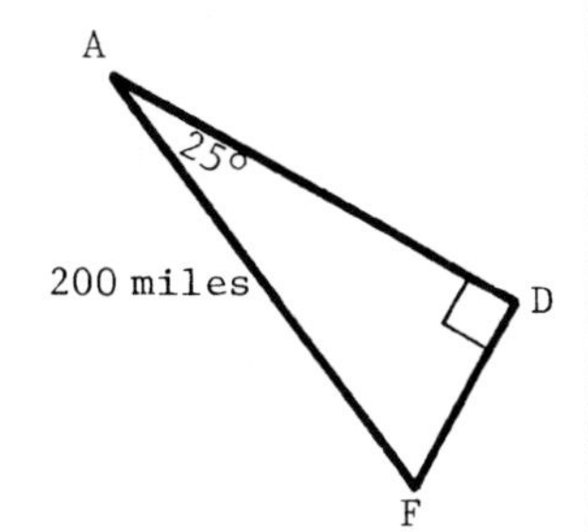

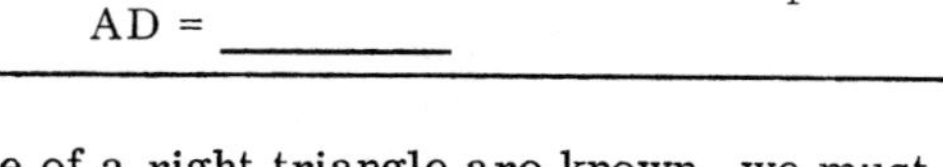

AD = ________

a) tan 70°

b) 27.6 m, from:

$$\tan 70° = \frac{75.8}{TV}$$

---

97. <u>If only one side and an acute angle of a right triangle are known</u>, we must use one of the trig ratios to find an unknown side.

<u>If only two sides of a right triangle are known</u>, we should use the Pythagorean Theorem to find the third side.

In which triangles below would we use the Pythagorean Theorem in order to find MP? ____________

a) 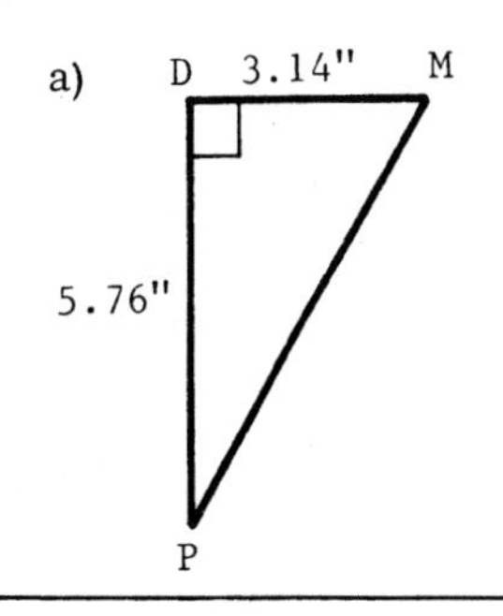

b) 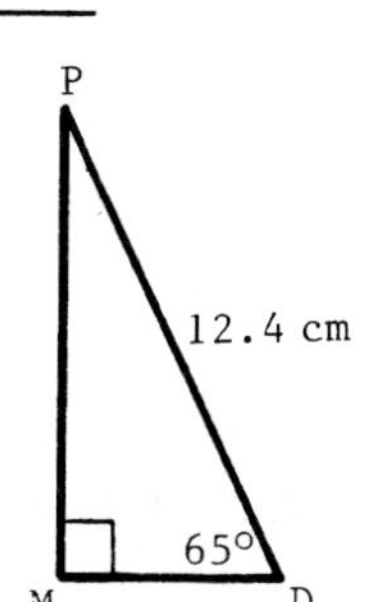

c) 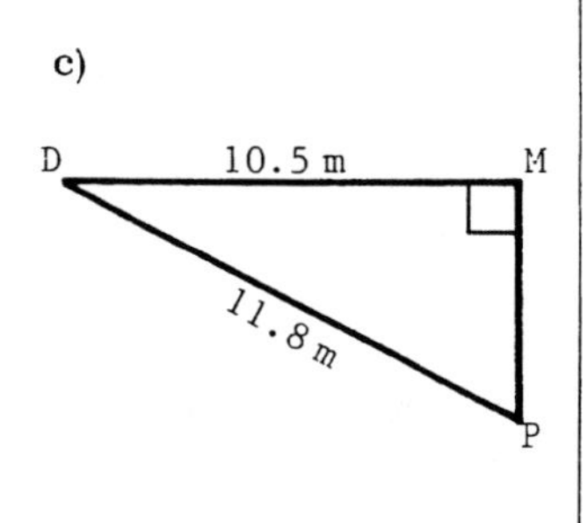

a) cos 25°

b) 181 miles, from:

$$\cos 25° = \frac{AD}{200}$$

---

98. In which triangles below would we have to use a trigonometric ratio to find CD? ____________

a) 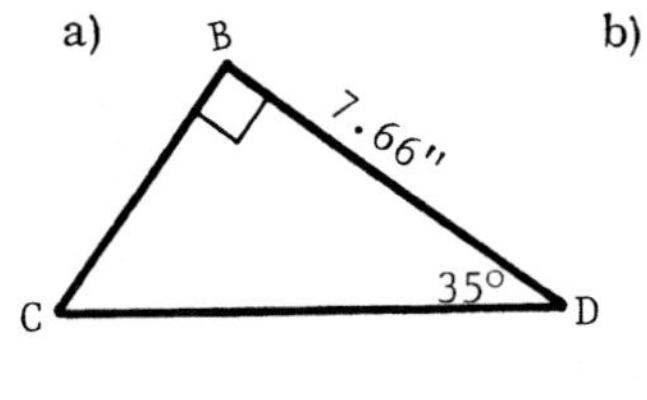

b) 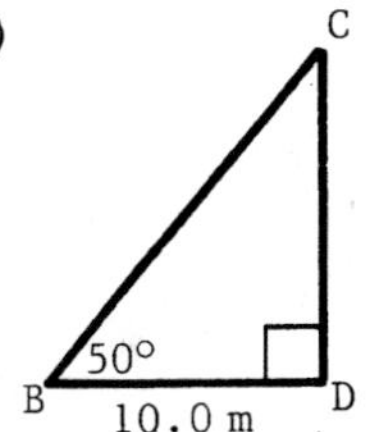

c) 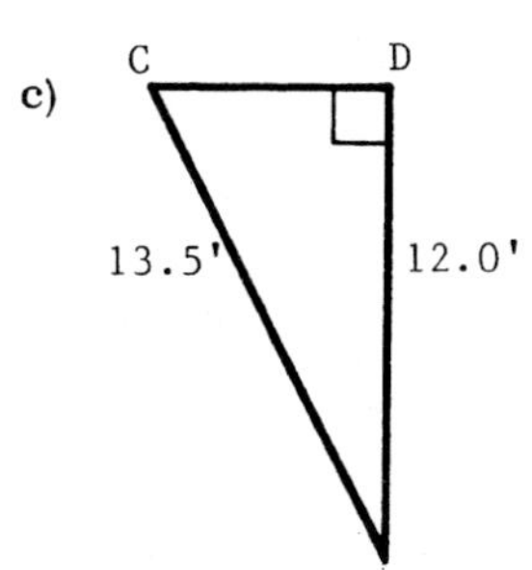

In (a) and (c)

In (a) and (b)

99. If we know two sides and one acute angle of a right triangle, we can use either of two trig ratios or the Pythagorean Theorem to find the third side. Here is an example. Round each answer to the nearest tenth of an inch.

a) Use cos 33° to find "t".

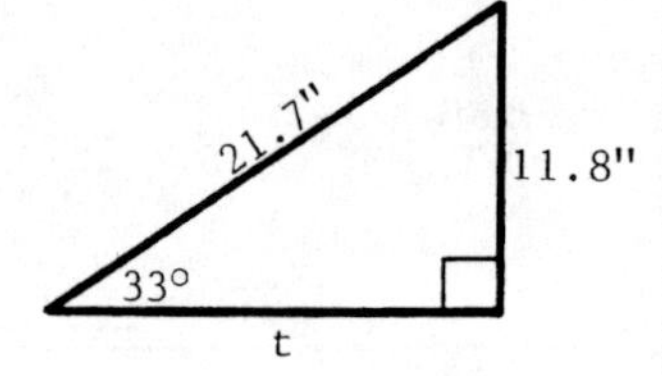

t = ________

b) Use tan 33° to find "t".

t = ________

c) Use the Pythagorean Theorem to find "t".

t = ________

a) t = 18.2 in, from: $\cos 33° = \frac{t}{21.7}$

b) t = 18.2 in, from: $\tan 33° = \frac{11.8}{t}$

c) t = 18.2 in, from: $t = \sqrt{(21.7)^2 - (11.8)^2}$

1-10 FINDING UNKNOWN ANGLES IN RIGHT TRIANGLES

Either the angle-sum principle or the trig ratios are used to find unknown angles in right triangles. We will discuss the use of both methods in this section.

100. If one acute angle in a right triangle is known, we can subtract it from 90° to find the other acute angle. Find angle A in each right triangle below.

a)

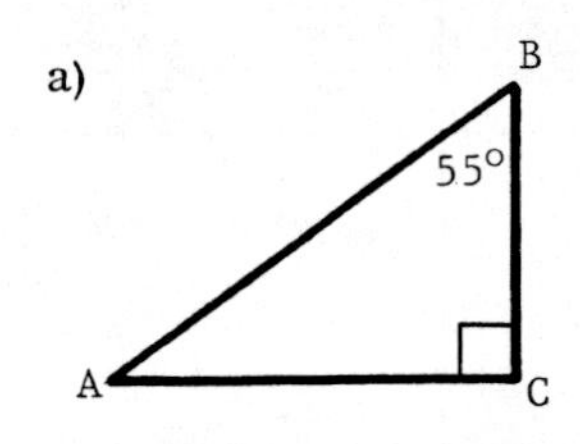

A = ______

b)

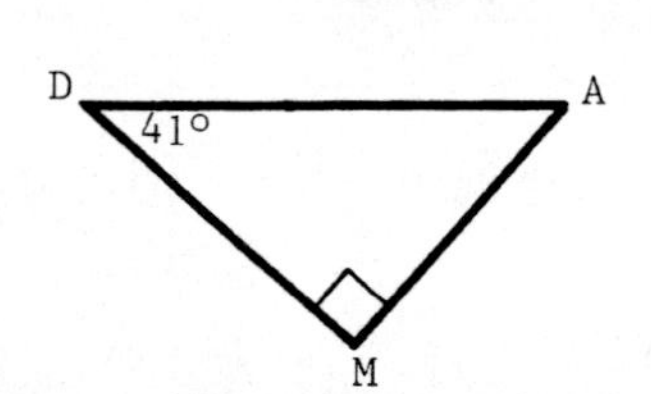

A = ______

a) 35°

b) 49°

101. When neither acute angle in a right triangle is known, we can use a trig ratio to find an acute angle if two sides of the triangle are known. An example is shown.

To find angle A at the right, we can use the sine ratio.

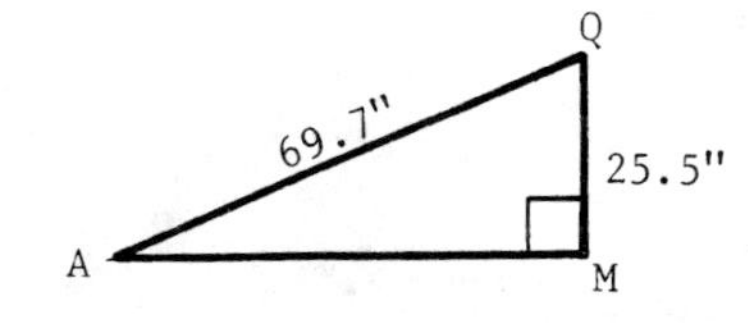

$$\sin A = \frac{MQ}{AQ}$$

$$\sin A = \frac{25.5}{69.7}$$

Use a calculator to complete the solution. Be sure to press [=] to complete the division before pressing [INV] [sin] or [arc] [sin] or [$\sin^{-1}$]. Round to the nearest degree. A = ______

---

102. To find angle P at the right, we can use the cosine ratio.

21°

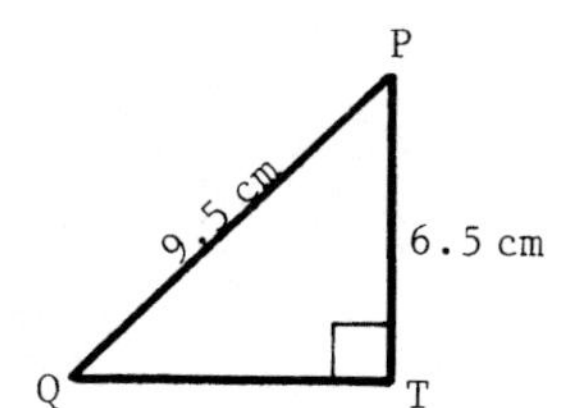

$$\cos P = \frac{PT}{PQ}$$

$$\cos P = \frac{6.5}{9.5}$$

Use a calculator to complete the solution. Press [INV] [cos] or [arc] [cos] or [$\cos^{-1}$] after dividing. Round to the nearest degree. P = ______

---

103. To decide which trig ratio to use to find an unknown angle in a right triangle, we identify the known sides in terms of the desired angle. An example is shown.

47°

To find angle M at the right we identify PV and MP in terms of angle M.

PV is the side opposite angle M.

MP is the side adjacent to angle M.

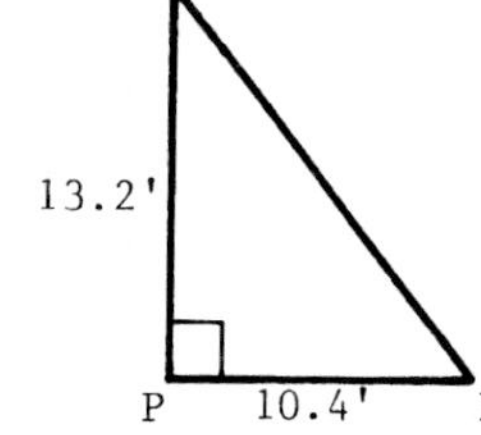

Therefore, we can use tan M to find angle M.

$$\tan M = \frac{PV}{MP}$$

$$\tan M = \frac{13.2}{10.4}$$

Use [INV] [tan] or [arc] [tan] or [$\tan^{-1}$] to complete the solution. Round to the nearest degree. M = ______

52°

104. To find angle T at the right, we identify TV and DT in terms of angle T.

TV is the <u>side adjacent</u> to angle T.

DT is the <u>hypotenuse.</u>

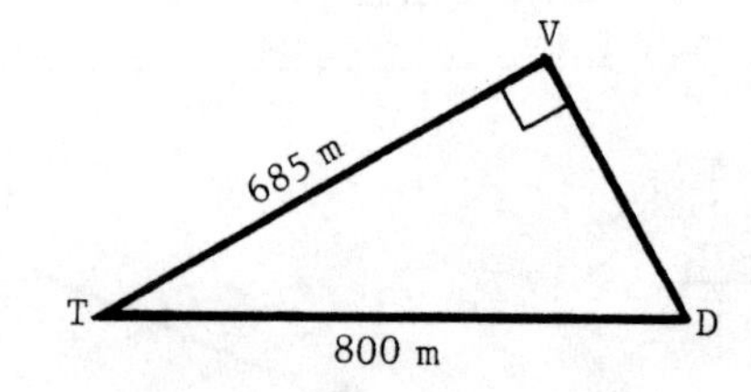

a) Should we use sin T, cos T, or tan T to find angle T? _______

b) To the nearest degree, angle T = _______

---

105. We want to find angle M in this right triangle.

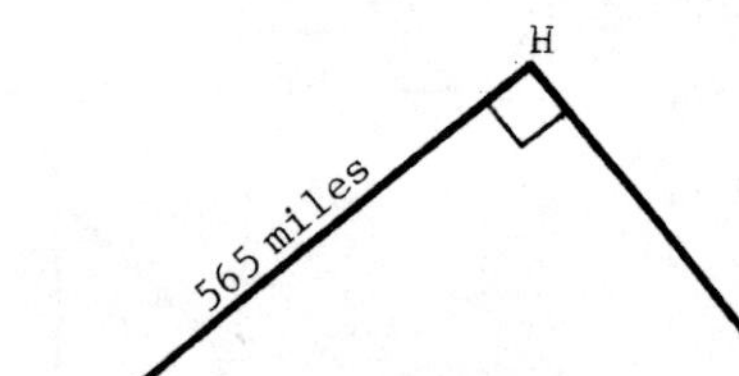

a) Should we use sin M, cos M, or tan M? _______

b) To the nearest degree, angle M = _______

Answers to 104:

a) cos T

b) 31°, from: $\cos T = \frac{685}{800}$

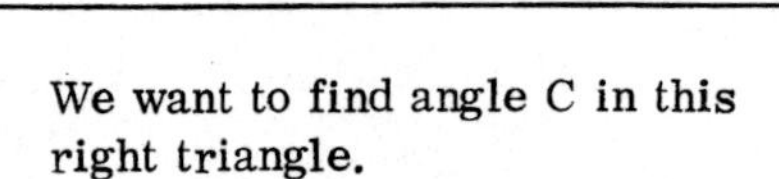

106. We want to find angle C in this right triangle.

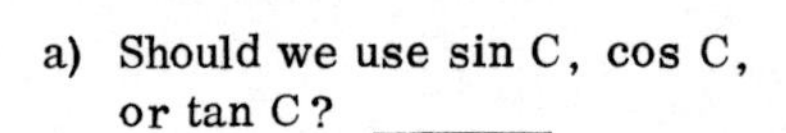

a) Should we use sin C, cos C, or tan C? _______

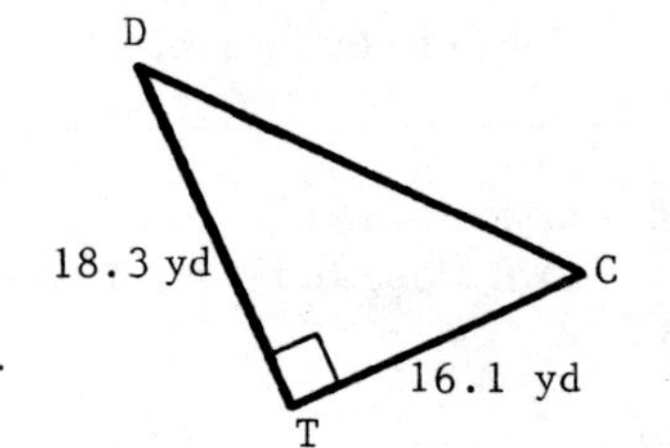

b) To the nearest degree, angle C = _______

Answers to 105:

a) sin M

b) 51°, from: $\sin M = \frac{565}{725}$

---

107. After finding one acute angle in a right triangle, we can find the other acute angle by subtracting from 90°.

In right triangle PDM:

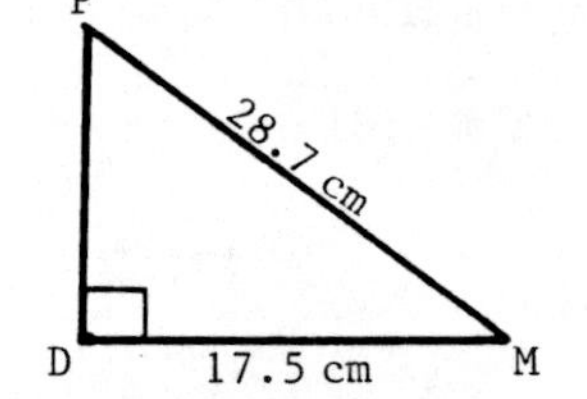

a) To the nearest degree, angle P = _______

b) To the nearest degree, angle M = _______

Answers to 106:

a) tan C

b) 49°, from: $\tan C = \frac{18.3}{16.1}$

Answers to 107:

a) 38°, from: $\sin P = \frac{17.5}{28.7}$

b) 52°, from: 90° - 38°

108. In right triangle FGR:

a) Angle F = _______

b) To the nearest tenth of a foot, FG = _______

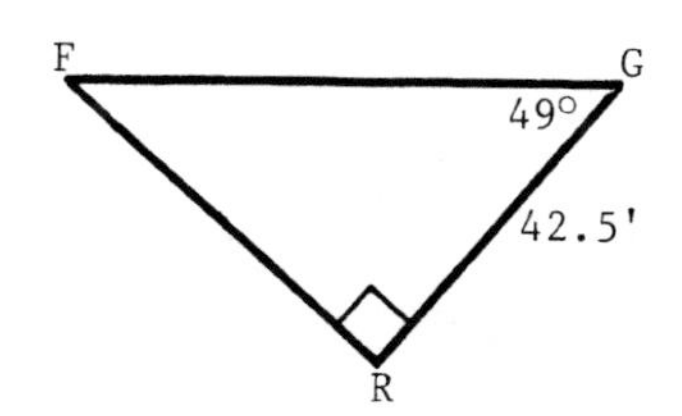

a) 41°

b) 64.8 ft

109. In right triangle ABC:

a) Angle A = _______

b) To the nearest tenth of a centimeter, AB = _______

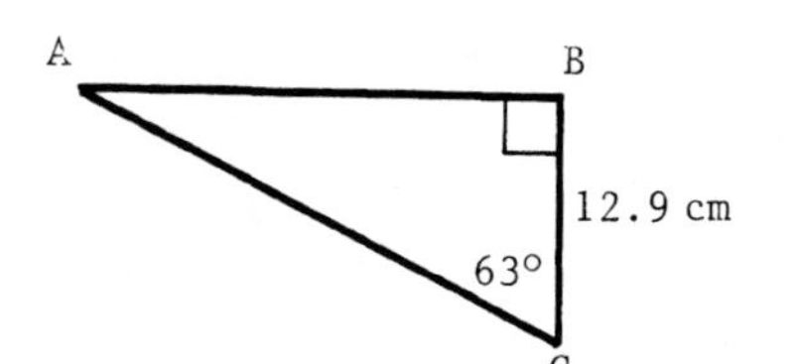

a) 27° b) 25.3 cm

1-11 APPLIED PROBLEMS

In this section, we will discuss some applied problems that involve solving a right triangle.

110. To measure the height of a tall building, a surveyor set his transit (angle-measuring instrument) 100 feet horizontally from the base of the building, measured the angle of elevation to the top of the building, and found it to be 72°.

Find "h", the height of the building. Round to the nearest foot.

h = _______

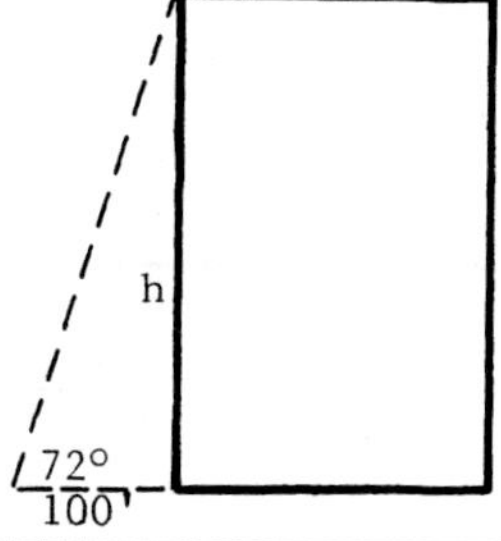

h = 308 ft, from:

$\tan 72° = \frac{h}{100}$

111. A right triangle called an "impedance" triangle is used in analyzing alternating current circuits. Angle A, shown in the diagram, is the "phase angle" of the circuit.

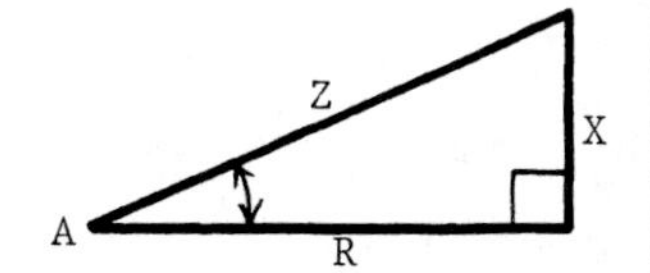

In a particular circuit, X = 2,630 and Z = 6,820. Find angle A to the nearest degree.

A = _______

112. The end-view of the roof of a building is an isosceles triangle, as shown at the right.

Find angle A, the angle of slope of the roof to the nearest degree.

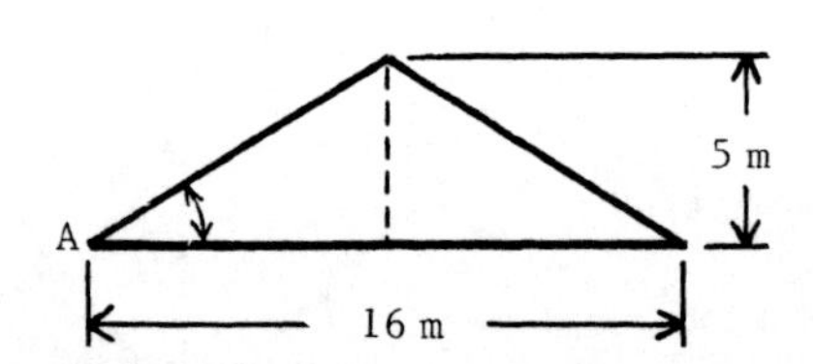

A = ______

A = 23°, from: $\sin A = \frac{2,630}{6,820}$

---

113. In this metal bracket, find D, the distance between the two holes. Round to the nearest hundredth of an inch.

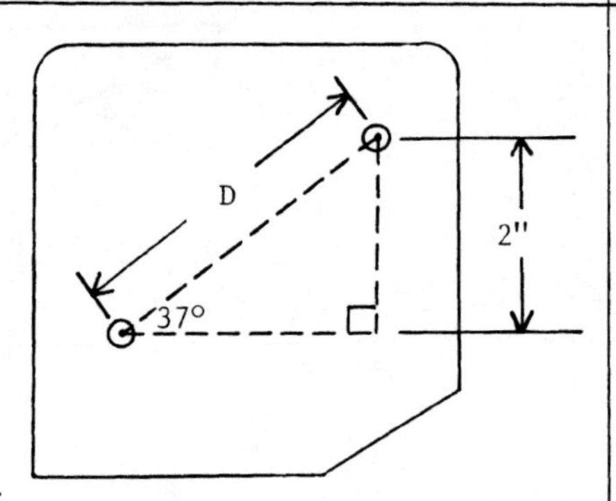

D = ______

A = 32°, from: $\tan A = \frac{5}{8}$

---

114. The metal shape at the right is called a "template". Figure ABCD is a rectangle. We want to find side BE to the nearest tenth of a centimeter. (Note: AE can be found by subtracting BC from DE.)

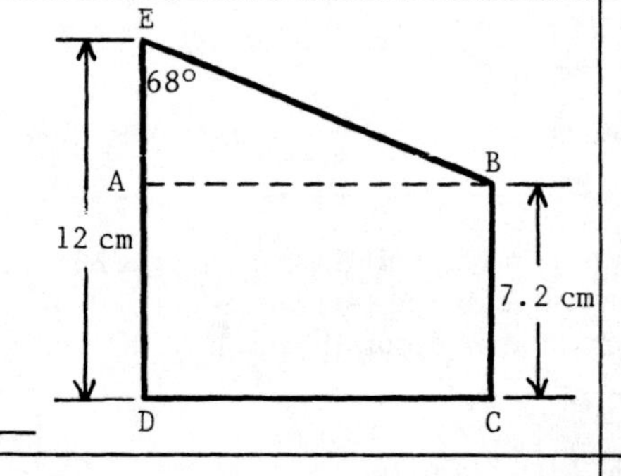

BE = ______

D = 3.32 in, from: $\sin 37° = \frac{2}{D}$

---

115. Here is a cross-sectional view of a metal shaft with a tapered end.

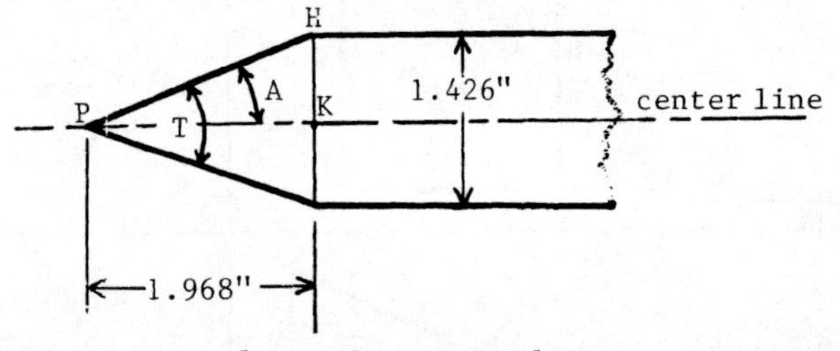

Find T, the taper angle, to the nearest degree. T = ______

Note: Angle A is half of angle T. Angle A is an acute angle in right triangle PHK.

T = ______

BE = 12.8 cm, from: $\cos 68° = \frac{4.8}{BE}$

T = 40°, since A = 20°, from: $\tan A = \frac{0.713}{1.968}$

SELF-TEST 3 (pp. 28-37)

1. Find sides "h" and "v". Round to the nearest tenth of a centimeter.

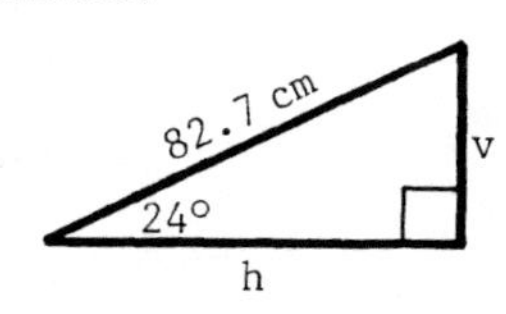

h = ________
v = ________

2. Find angles F and H. Round to the nearest degree.

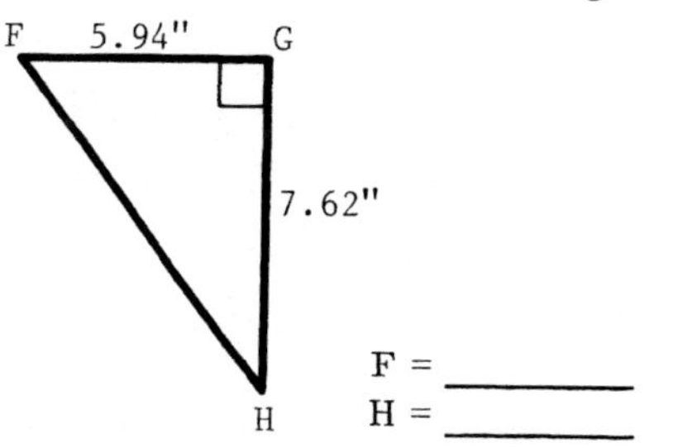

F = ________
H = ________

3. Find angle T. Round to the nearest degree.

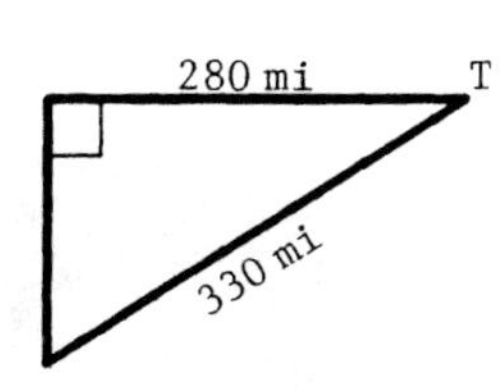

T = ________

4. Find side GP. Round to hundredths.

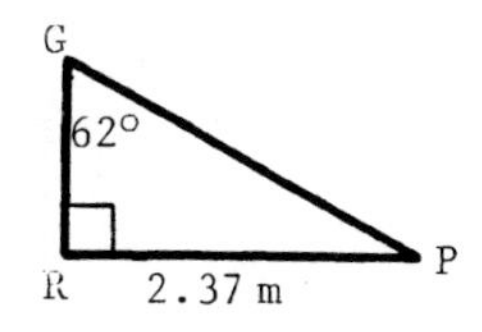

GP = ________

5. Find side "w". Round to tenths.

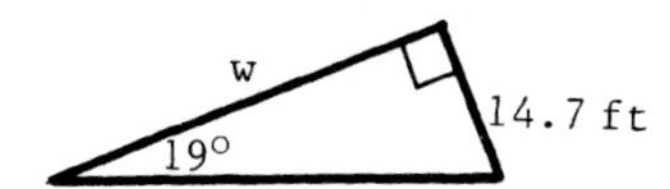

w = ________

6. Find angle A. Round to the nearest degree.

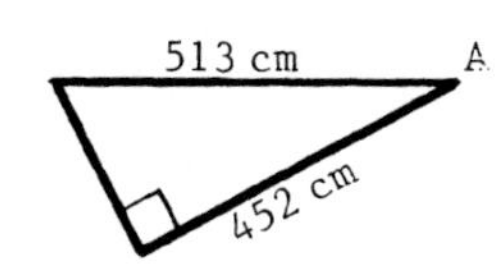

A = ________

7. Find "a" in isosceles triangle DEF. Round to hundredths.

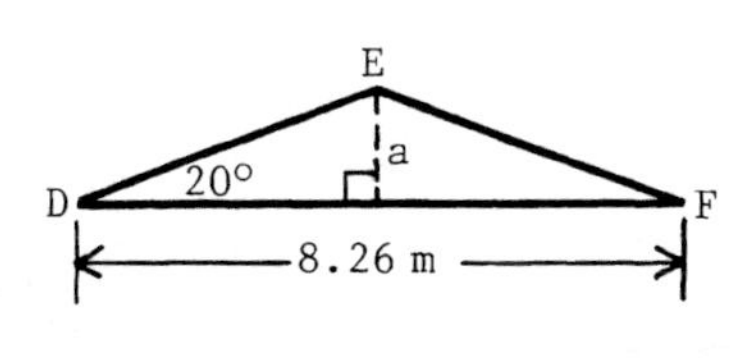

a = ________

8. Find angle P in parallelogram PQRS. Round to the nearest degree.

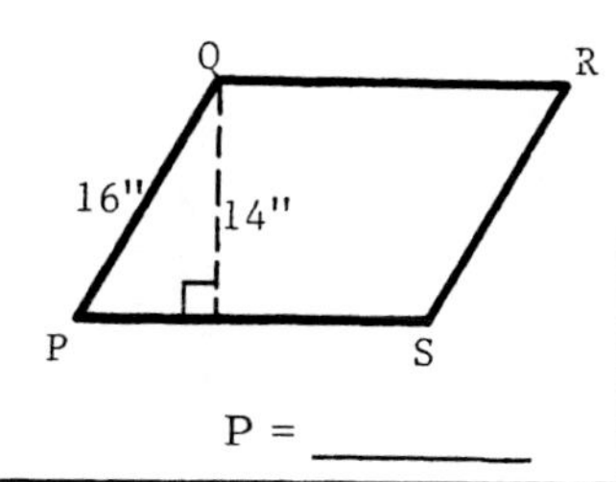

P = ________

ANSWERS:

1. h = 75.6 cm
   v = 33.6 cm
2. F = 52°
   H = 38°
3. T = 32°
4. GP = 2.68 m
5. w = 42.7 ft
6. A = 28°
7. a = 1.50 m
8. P = 61°

SUPPLEMENTARY PROBLEMS - CHAPTER 1

Assignment 1

Use the angle-sum principle to find each unknown angle.

1. 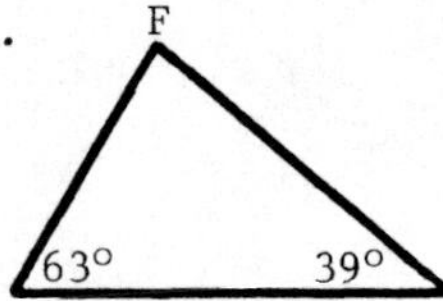

2. 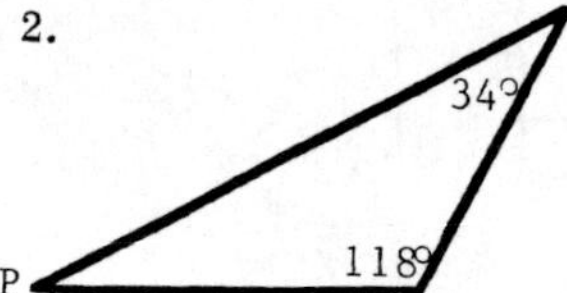

3. 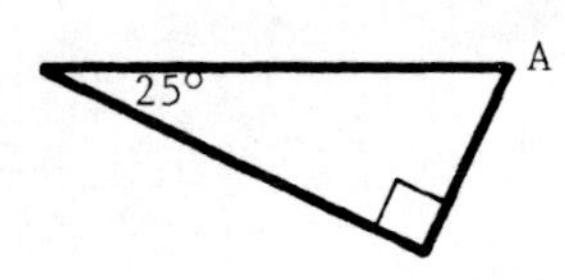

Use the Pythagorean Theorem to find each unknown dimension. Round as directed.

4. Round to hundreds.

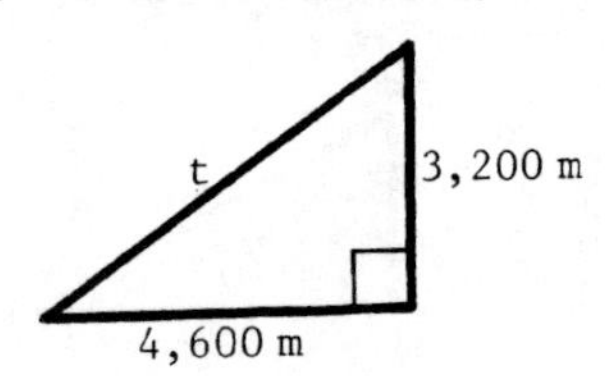

5. Round to tenths.

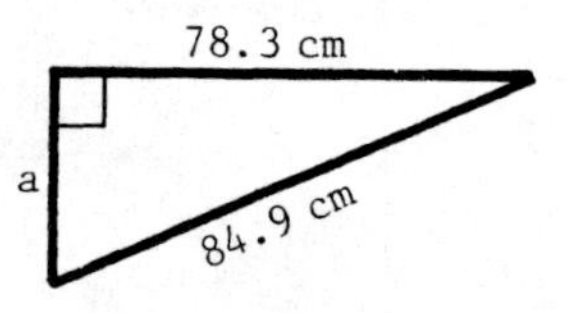

6. Round to hundredths.

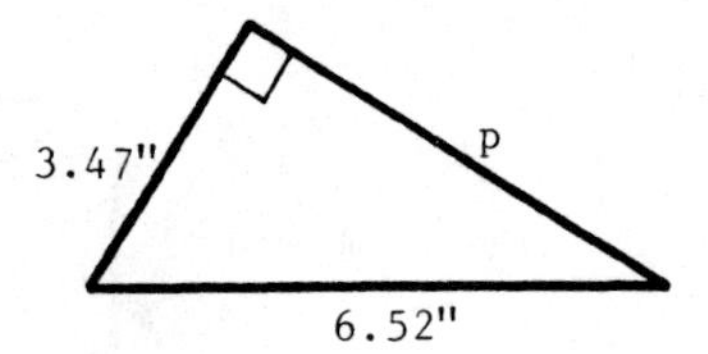

7. Round to tenths.

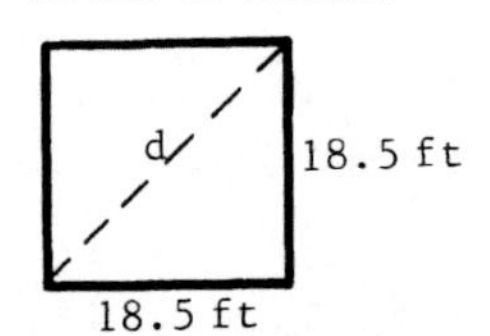

8. Round to the nearest whole number.

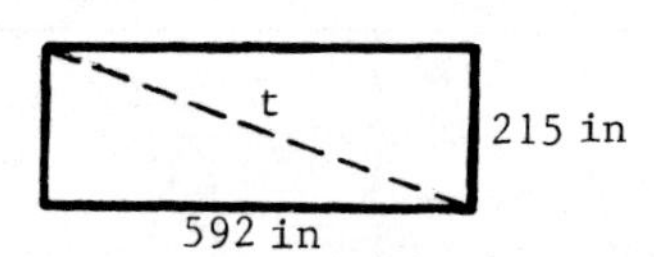

9. Round to thousandths.

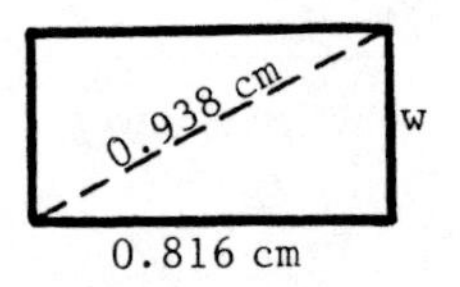

Each triangle is an isosceles triangle. Find each unknown angle.

10. H

37° G

11. 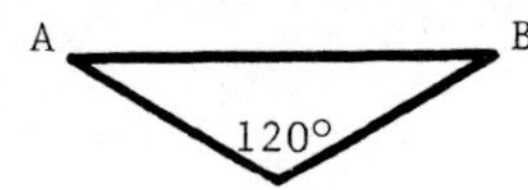

12. 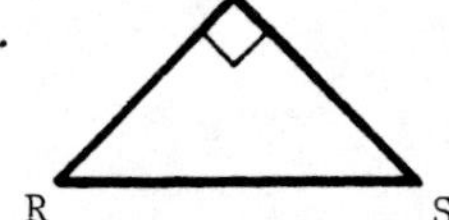

Find the area of each figure. Round as directed.

13. Round to hundreds.

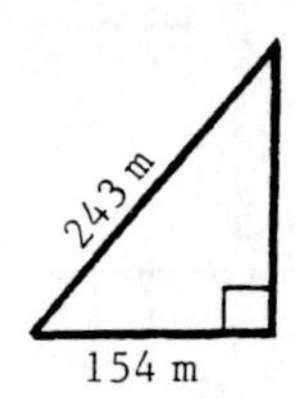

14. Round to the nearest whole number.

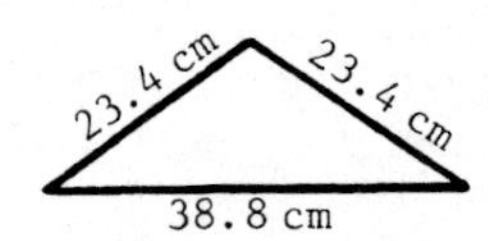

15. Round to tenths.

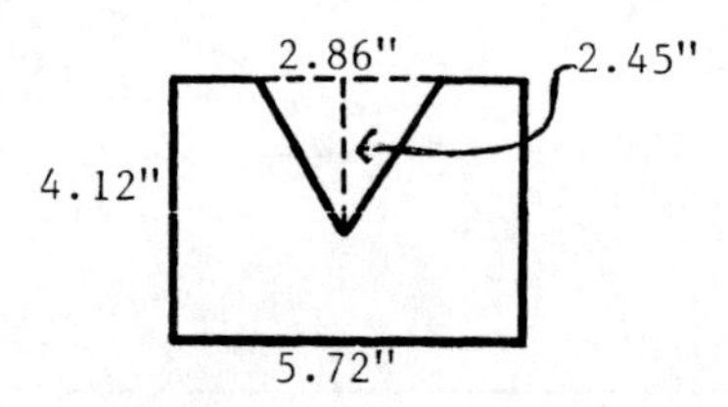

Assignment 2

Find the numerical value of each trig ratio. Round to four decimal places.

1. tan 12°
2. sin 87°
3. cos 71°
4. cos 0°
5. sin 30°
6. cos 30°
7. tan 65°
8. sin 1°

Find each angle. Round to the nearest degree.

9. $\cos A = 0.9715$ 10. $\tan P = 3.2168$ 11. $\sin H = 0.9999$ 12. $\cos Q = 0.3014$

13. $\tan T = 0.1172$ 14. $\cos F = 0.1382$ 15. $\tan A = 1.0000$ 16. $\sin B = 0.2105$

Evaluate each of the following. Round to tenths.

17. $(31.8)(\cos 24°)$ 18. $\dfrac{136}{\tan 85°}$ 19. $\dfrac{51.9}{\sin 42°}$ 20. $(71.3)(\tan 30°)$

Evaluate each of the following. Round to the nearest whole number.

21. $\dfrac{429}{\tan 51°}$ 22. $(1,136)(\sin 14°)$ 23. $(250)(\cos 64°)$ 24. $\dfrac{98.4}{\cos 79°}$

Find each angle. Round to the nearest degree.

25. $\cos A = \dfrac{38.4}{41.7}$ 26. $\sin G = \dfrac{7,850}{9,160}$ 27. $\tan R = \dfrac{5.62}{1.93}$ 28. $\sin E = \dfrac{0.138}{0.897}$

Using sides "r", "s", and "t" in right triangle RST, define the following trig ratios.

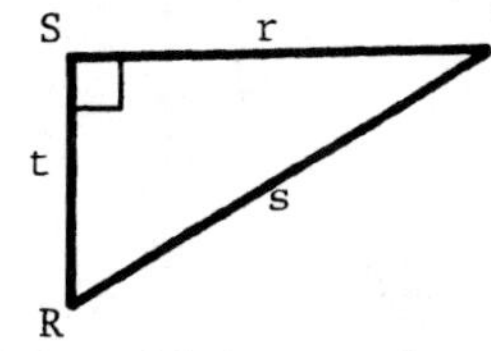

29. sin R 30. cos R 31. tan R

32. sin T 33. cos T 34. tan T

Using sides "a", "b", and "c" in right triangle ABC, define the following trig ratios.

35. cos B 36. sin A 37. tan B

38. tan A 39. cos A 40. sin B

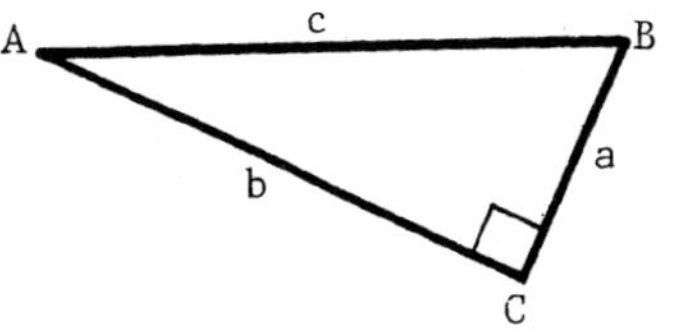

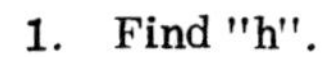

Assignment 3

In problems 1-3, round each answer to hundredths.

1. Find "h". 2. Find "w". 3. Find "p".

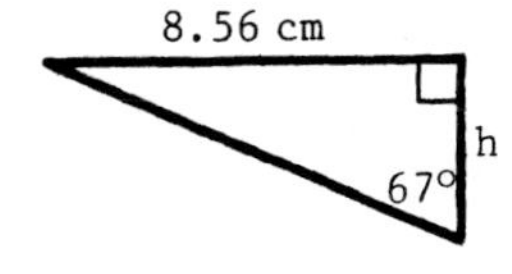

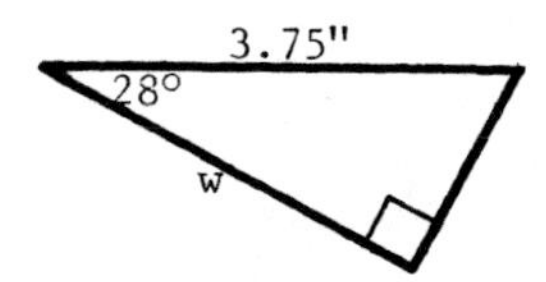

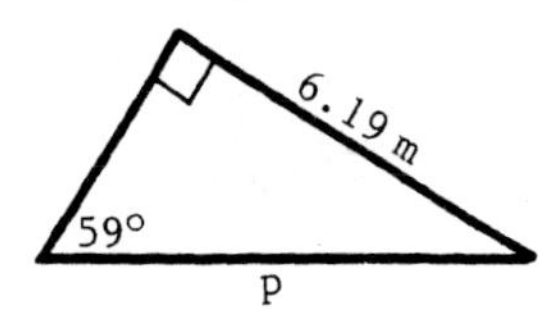

In problems 4-6, round each answer to the nearest whole number.

4. Find "d". 5. Find "t". 6. Find "b".

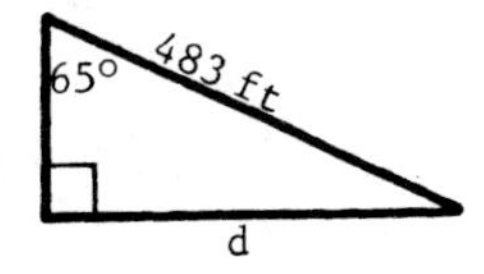

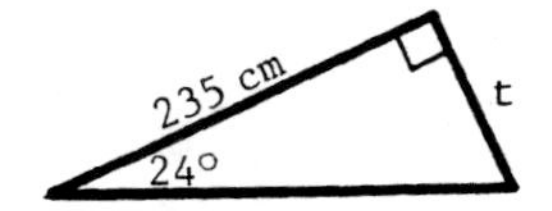

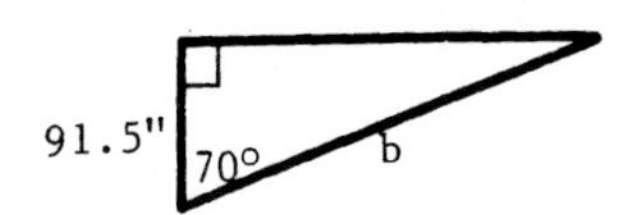

In problems 7-9, round each angle to the nearest degree.

7. Find angles F and G.

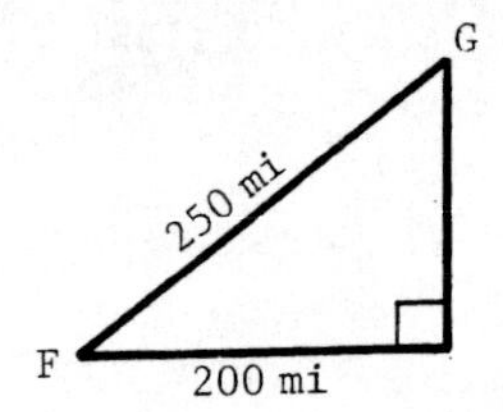

8. Find angles A and B.

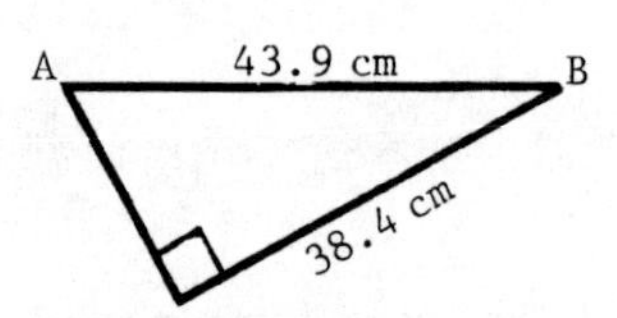

9. Find angles R and S.

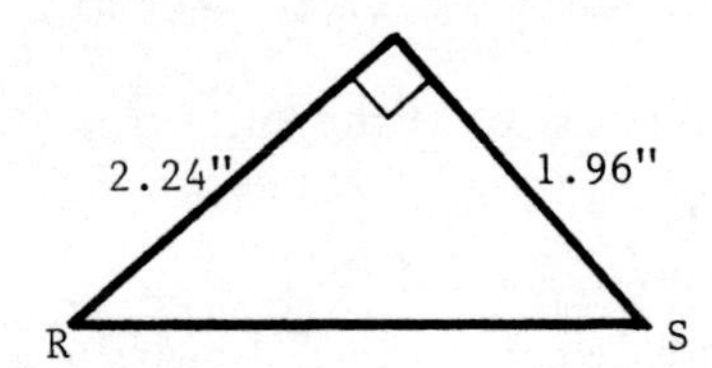

10. Find "h" and "v" in the diagram below. Round to hundredths.

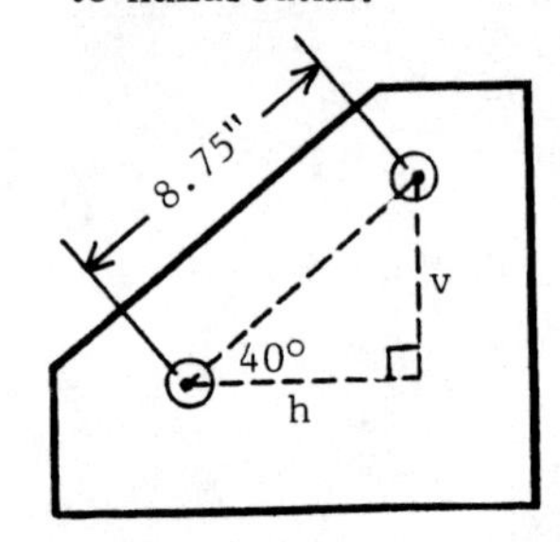

11. The roof diagram below is an isosceles triangle. Find angle A. Round to the nearest degree.

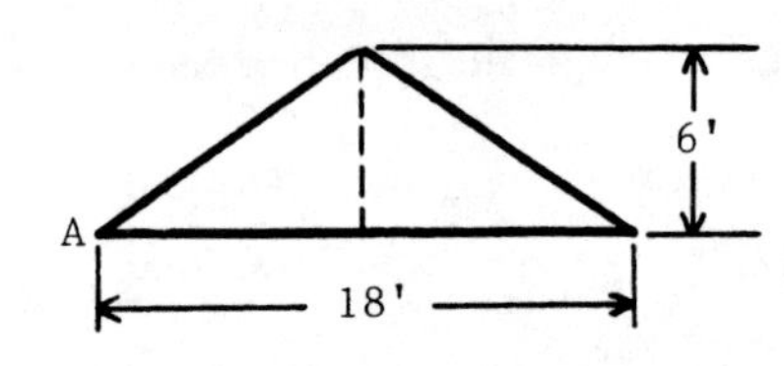

12. In the figure below, the bottom is rectangular. Find "w". Round to tenths.

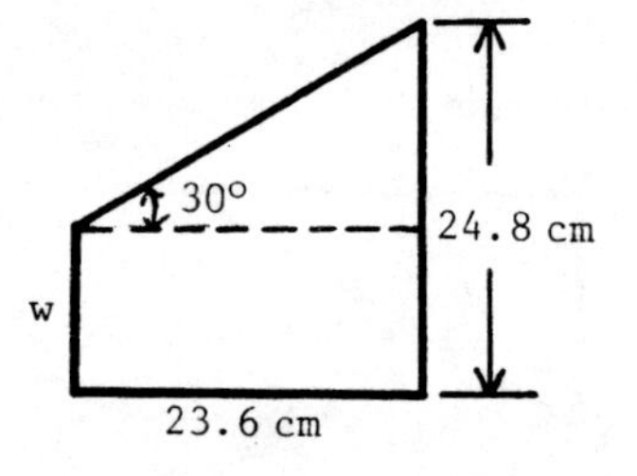

Chapter 2 OBLIQUE TRIANGLES

In this chapter, we will show how the Law of Sines and the Law of Cosines can be used to find unknown sides and angles in oblique triangles. The strategies used to solve oblique triangles are emphasized. Problems involving obtuse angles are delayed until all types of problems involving acute angles are examined.

2-1 OBLIQUE TRIANGLES

In this section, we will define oblique triangles. We will show that the angle-sum principle applies to oblique triangles, but that the Pythagorean Theorem and the three trigonometric ratios do not apply to oblique triangles.

---

1. We have already seen these definitions:

   Any angle between 0° and 90° is called an acute angle.
   Any angle with exactly 90° is called a right angle.
   Any angle between 90° and 180° is called an obtuse angle.

   In triangle CDE:

   Angle C is an acute angle.

   a) Angle D is an ______________ angle.

   b) Angle E is an ______________ angle.

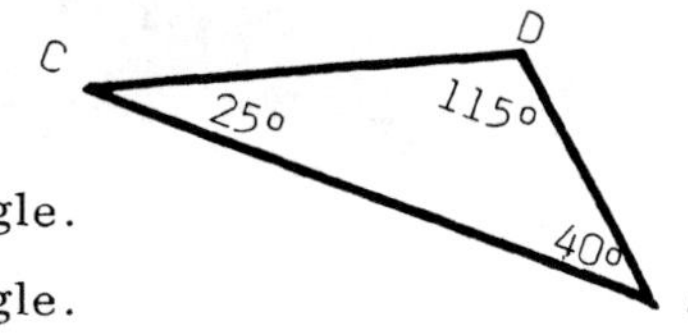

---

2. Any triangle which does not contain a right angle (90°) is called an oblique triangle. (The word "oblique" is pronounced "ō-bleek".) Both triangles below are oblique triangles.

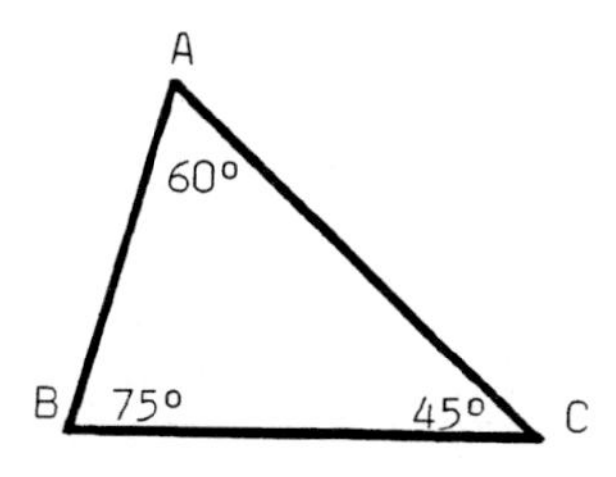

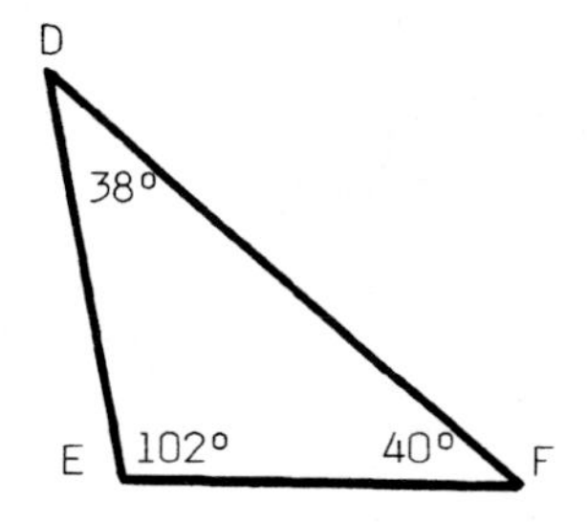

In triangle ABC, all three angles are acute angles. Triangles of that type are called "ACUTE oblique" triangles.

Answers to frame 1: a) obtuse b) acute

Continued on following page.

2. Continued

In triangle DEF, one angle is an obtuse angle. Triangles of that type are called "OBTUSE oblique" triangles.

a) If a triangle contains angles of 20°, 60°, and 100°, it is an _______________ (acute/obtuse) oblique triangle.

b) If a triangle contains angles of 44°, 58°, and 78°, it is an _______________ (acute/obtuse) oblique triangle.

---

3. The angle-sum principle applies to oblique triangles. That is, the sum of the three angles is 180°.

Find the unknown angle in each oblique triangle.

a)

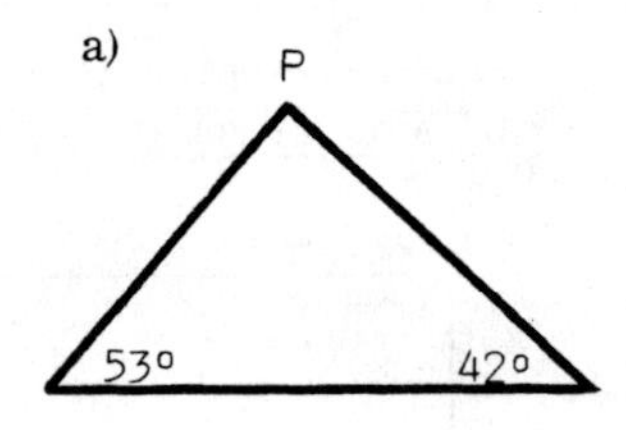

P = ________

b)

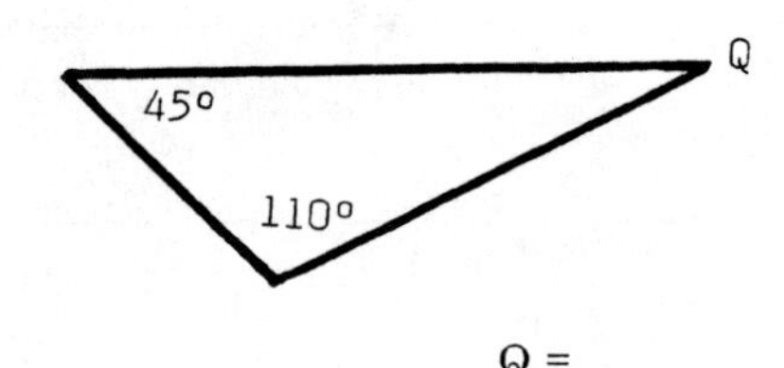

Q = ________

Answers to frame 2: a) obtuse b) acute

---

4. The Pythagorean Theorem applies only to right triangles. It does not apply to oblique triangles.

Triangle DEF is a right triangle. Triangle MPQ is an oblique triangle.

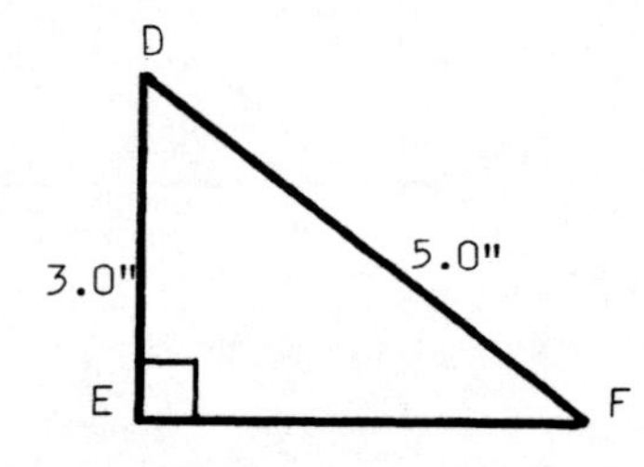

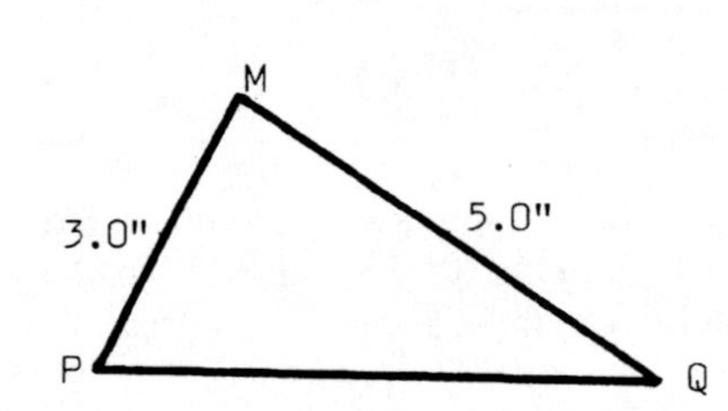

a) Can we use the Pythagorean Theorem to find the length of EF in triangle DEF? ___________

b) Can we use the Pythagorean Theorem to find the length of PQ in triangle MPQ? ___________

Answers to frame 3: a) P = 85° b) Q = 25°

Answers to frame 4: a) Yes b) No

5. The three basic trigonometric ratios (sine, cosine, and tangent) are comparisons of the sides of right triangles. They are not comparisons of the sides of oblique triangles.

Triangle ABC is a right triangle. Triangle FGH is an oblique triangle.

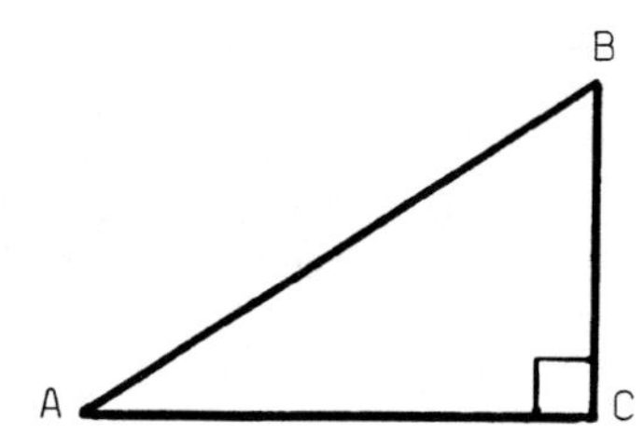

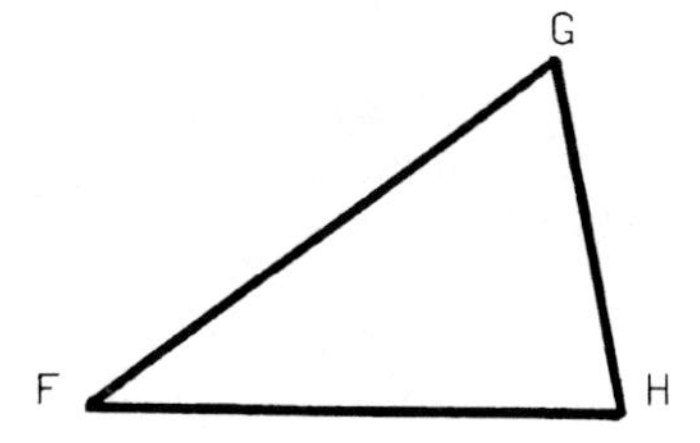

In triangle ABC, sin A is a comparison of two sides of the triangle. That is:

$$\sin A = \frac{BC}{AB}$$

In triangle FGH, is sin F a comparison of two sides of the triangle?

____

No

6. Triangle CDE is a right triangle. Triangle PQR is an oblique triangle.

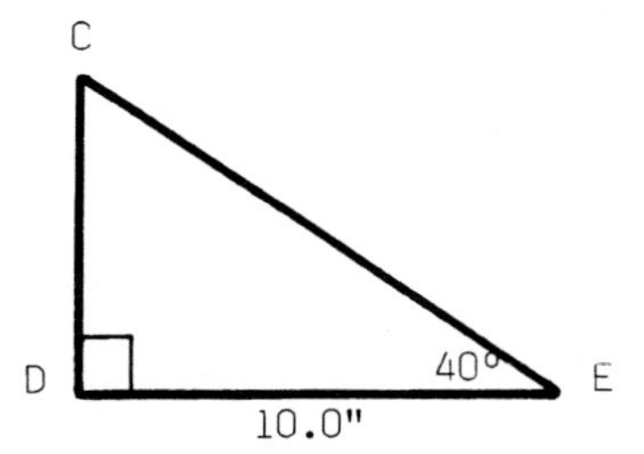

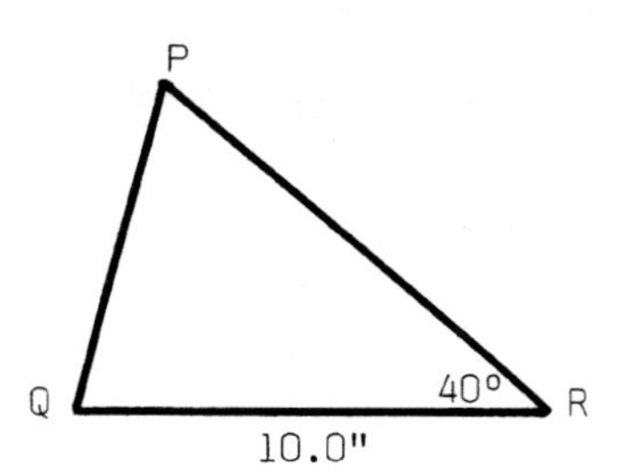

In triangle CDE, we can use the tangent ratio to find the length of CD. We get:

$$\tan E = \frac{CD}{DE}$$

$$\tan 40° = \frac{CD}{10.0''} \quad \text{and} \quad CD = 8.39''$$

Can we use the tangent ratio to find the length of PQ in triangle PQR?

____

No

2-2 RIGHT-TRIANGLE METHODS AND OBLIQUE TRIANGLES

"Solving triangles" means finding the lengths of unknown sides and the sizes of unknown angles. If an oblique triangle is divided into two right triangles, right-triangle methods can be used for the solution. We will discuss that method in this section.

7. If a line drawn from the vertex of an angle of a triangle meets the opposite side at right angles, the line is called an "altitude" of the triangle. In the triangles below, DF and ST are altitudes.

Any oblique triangle can be divided into two right triangles by drawing an altitude from one angle to the opposite side.

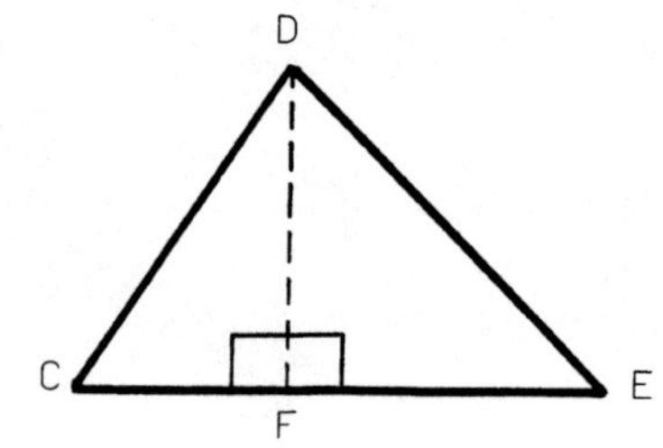

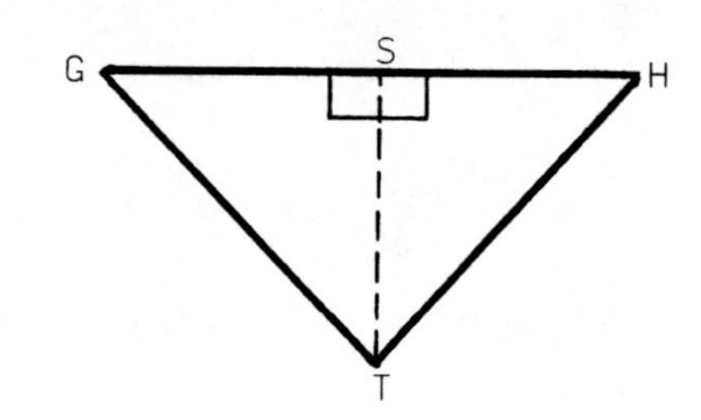

The altitude DF divides oblique triangle CDE into the two right triangles CDF and DEF.

The altitude ST divides oblique triangle GHT into the two right triangles __________ and __________.

8. When an altitude divides an oblique triangle into two right triangles, the Pythagorean Theorem applies to each right triangle.

Answer to 7: GST and HST

Altitude AD divides triangle ABC into the two right triangles ABD and ACD. By using the Pythagorean Theorem twice to find the lengths of BD and CD, we can then add to find the length of BC.

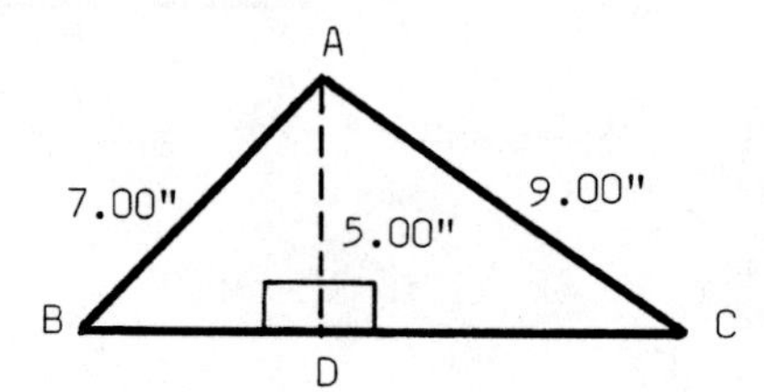

Rounding to hundredths:

a) The length of BD is __________ inches.

b) The length of CD is __________ inches.

c) The length of BC is __________ inches.

Answers to 8:

a) 4.90

b) 7.48

c) 12.38

9. When an altitude divides an oblique triangle into two right triangles, the three basic trigonometric ratios apply to each right triangle.

Altitude RT divides triangle PQR into right triangles PRT and QRT. The length of RT is given.

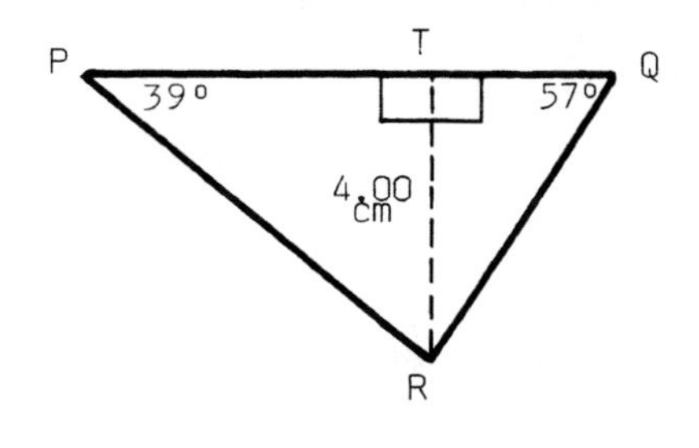

a) In right triangle PRT, we can use sin P to find the length of PR. Round to hundredths.

Since $\sin P = \frac{RT}{PR}$, $\sin 39° = \frac{4.00}{PR}$ and PR = ________.

b) In right triangle QRT, we can use sin Q to find the length of QR. Round to hundredths.

Since $\sin Q = \frac{RT}{QR}$, $\sin 57° = \frac{4.00}{QR}$ and QR = ________.

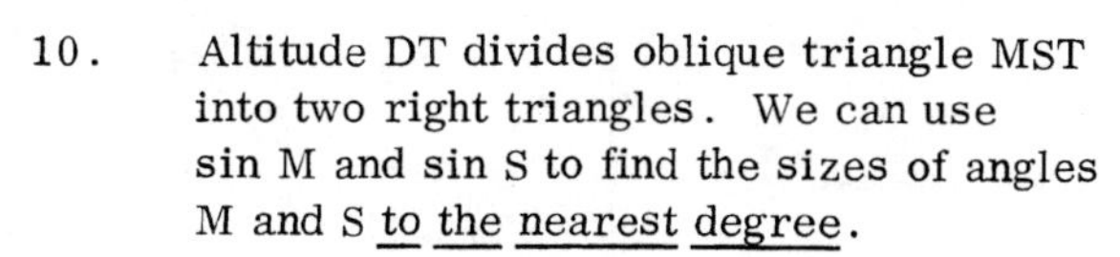

10. Altitude DT divides oblique triangle MST into two right triangles. We can use sin M and sin S to find the sizes of angles M and S to the nearest degree.

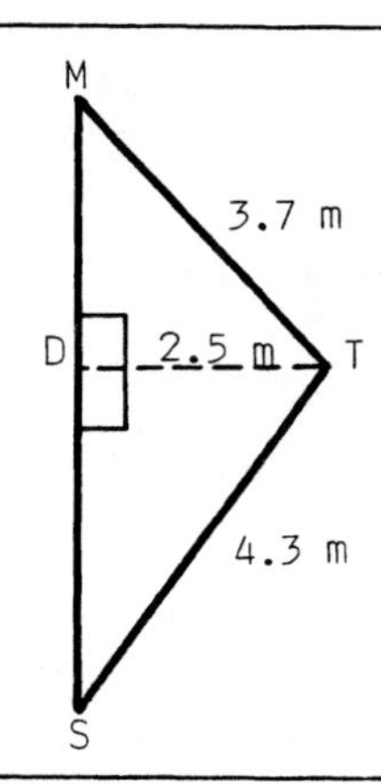

a) Since $\sin M = \frac{DT}{MT} = \frac{2.5}{3.7} = 0.6757$,

angle M = ________

b) Since $\sin S = \frac{DT}{ST} = \frac{2.5}{4.3} = 0.5814$,

angle S = ________

a) 6.36 cm

b) 4.77 cm

---

11. When drawing an altitude to divide an oblique triangle into two right triangles, the length of the altitude is usually not known. Therefore, we usually need a two-step process to find an unknown side or angle. The first step is finding the length of the altitude. An example is given below.

In oblique triangle CDE, the length of altitude EF is not known. Therefore, we need a two-step process to find the length of DE.

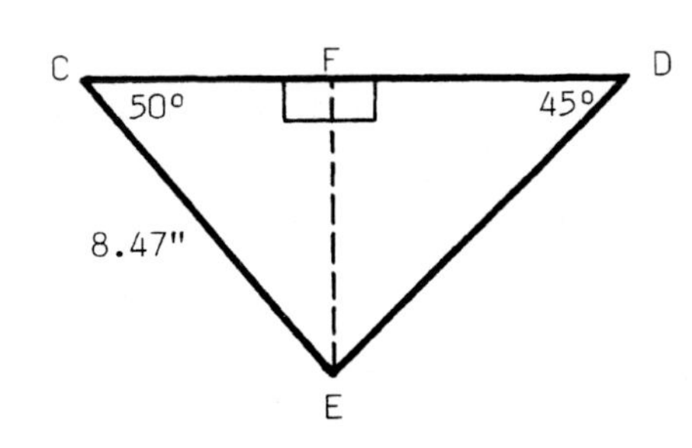

Continued on following page.

a) 43°

b) 36°

11. Continued

a) Find the length of EF. Round to hundredths.

EF = ________

b) Find the length of DE. Round to hundredths.

DE = ________

a) 6.49"

b) 9.18"

---

12. A two-step process is needed to find the size of angle T at the right.

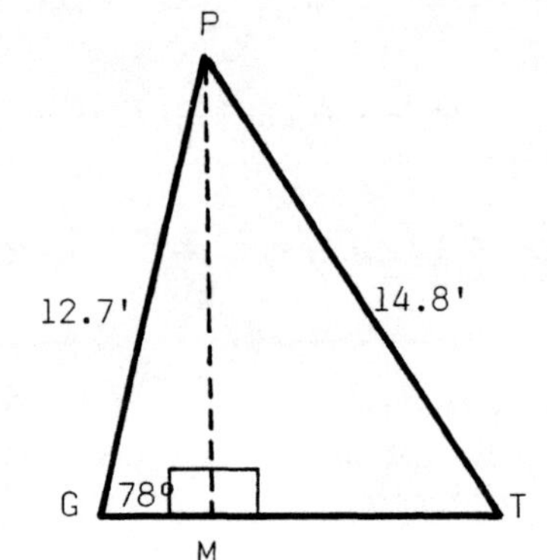

a) Find altitude MP. Round to tenths.

MP = ________

b) Find angle T. Round to the nearest whole-number degree.

Angle T = ________

a) 12.4 feet b) 57°

2-3 THE LAW OF SINES

Instead of solving oblique triangles by right-triangle methods, we use one of two direct methods: the Law of Sines and the Law of Cosines. We will discuss the Law of Sines in this section.

---

13. The Law of Sines involves the ratios of the sides of a triangle to the sines of the angles opposite them. It says that the three ratios of that type for any triangle are equal.

For oblique triangle ABC, the Law of Sines says this:

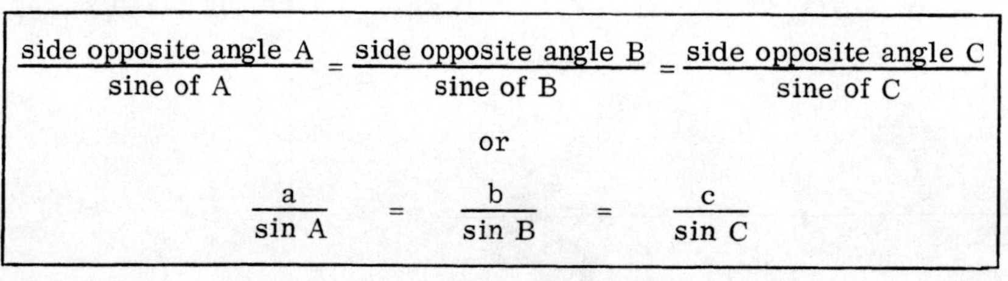

$$\frac{\text{side opposite angle A}}{\text{sine of A}} = \frac{\text{side opposite angle B}}{\text{sine of B}} = \frac{\text{side opposite angle C}}{\text{sine of C}}$$

or

$$\frac{a}{\sin A} = \frac{b}{\sin B} = \frac{c}{\sin C}$$

B, c, a, A, C, b

Continued on following page.

13. Continued

Since the Law of Sines says that all three ratios are equal, we can write the following three proportions:

$$\boxed{\frac{a}{\sin A} = \frac{b}{\sin B}} \qquad \boxed{\frac{a}{\sin A} = \frac{c}{\sin C}} \qquad \boxed{\frac{b}{\sin B} = \frac{c}{\sin C}}$$

---

14. The numerical values of the three angles and three sides of triangle PQR are shown on the figure. The values reported for the sides involve some rounding. We will use this triangle to confirm the Law of Sines which says this:

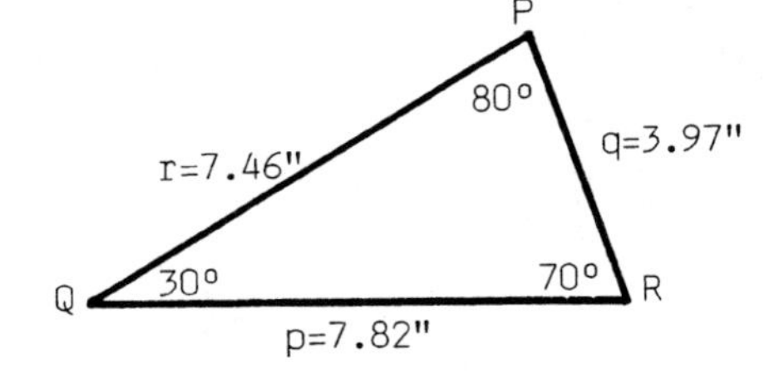

$$\frac{p}{\sin P} = \frac{q}{\sin Q} = \frac{r}{\sin R}$$

Use your calculator to evaluate each ratio. Round to hundredths.

a) $\frac{p}{\sin P} = \frac{7.82}{\sin 80°} =$ ________

b) $\frac{q}{\sin Q} = \frac{3.97}{\sin 30°} =$ ________

c) $\frac{r}{\sin R} = \frac{7.46}{\sin 70°} =$ ________

d) Are the three ratios equal? ________

---

15. According to the Law of Sines for this triangle:

a) $\frac{c}{\sin C} = \frac{d}{\boxed{}}$

b) $\frac{c}{\sin C} = \frac{\boxed{}}{\sin E}$

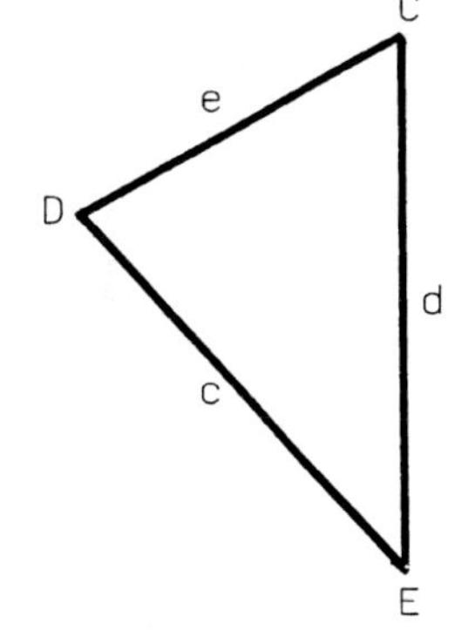

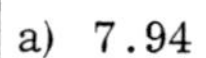

a) 7.94

b) 7.94

c) 7.94

d) Yes

---

a) $\frac{d}{\boxed{\sin D}}$

b) $\frac{\boxed{e}}{\sin E}$

16. According to the Law of Sines for this triangle:

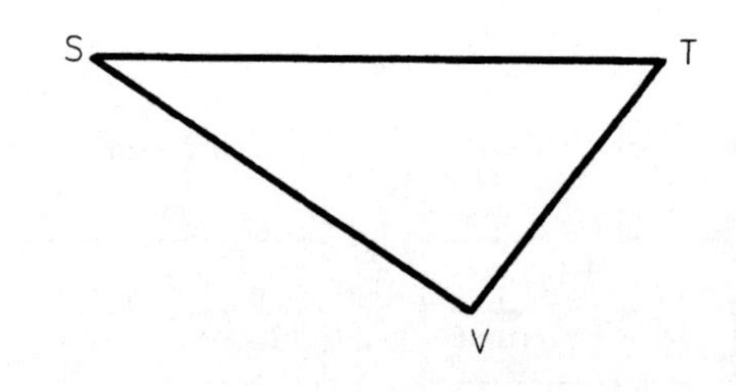

a) $\dfrac{SV}{\sin T} = \dfrac{ST}{\boxed{\quad\quad}}$

b) $\dfrac{SV}{\sin T} = \dfrac{\boxed{\quad\quad}}{\sin S}$

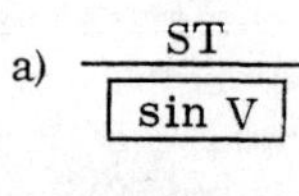

a) $\dfrac{ST}{\boxed{\sin V}}$

b) $\dfrac{\boxed{VT}}{\sin S}$

17. When writing the Law of Sines for the triangle at the right, we have written the ratios with the sides as the numerators. That is:

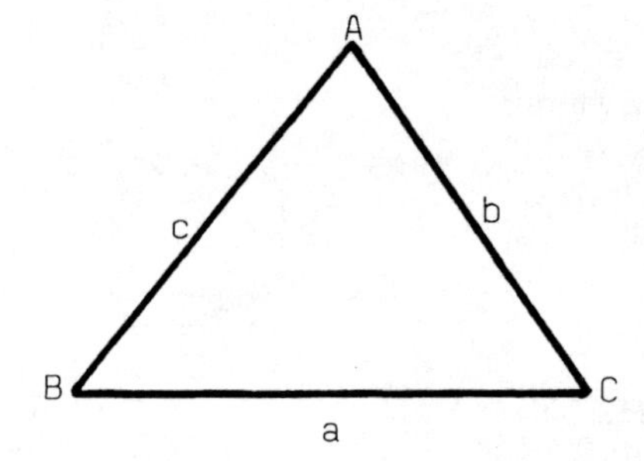

$$\boxed{\dfrac{a}{\sin A} = \dfrac{b}{\sin B} = \dfrac{c}{\sin C}}$$

However, we can also write the Law of Sines with the "sines" as the numerators. We get:

$$\boxed{\dfrac{\sin A}{a} = \dfrac{\sin B}{b} = \dfrac{\sin C}{c}}$$

Write the Law of Sines in two different ways for triangle DFM.

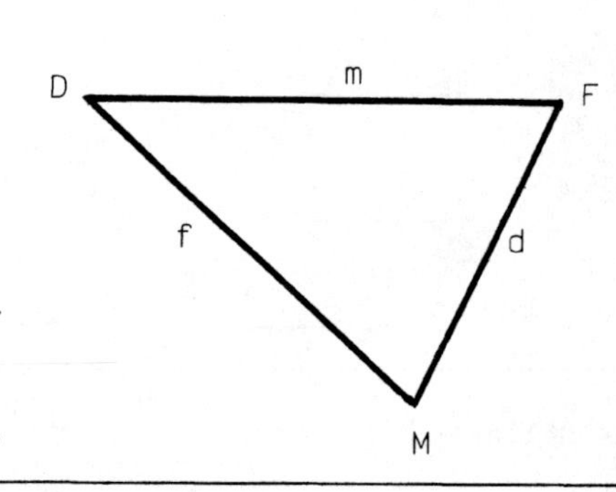

$\_\_\_\_\_ = \_\_\_\_\_ = \_\_\_\_\_$

$\_\_\_\_\_ = \_\_\_\_\_ = \_\_\_\_\_$

$\dfrac{d}{\sin D} = \dfrac{f}{\sin F} = \dfrac{m}{\sin M}$

$\dfrac{\sin D}{d} = \dfrac{\sin F}{f} = \dfrac{\sin M}{m}$

18. According to the Law of Sines for this triangle:

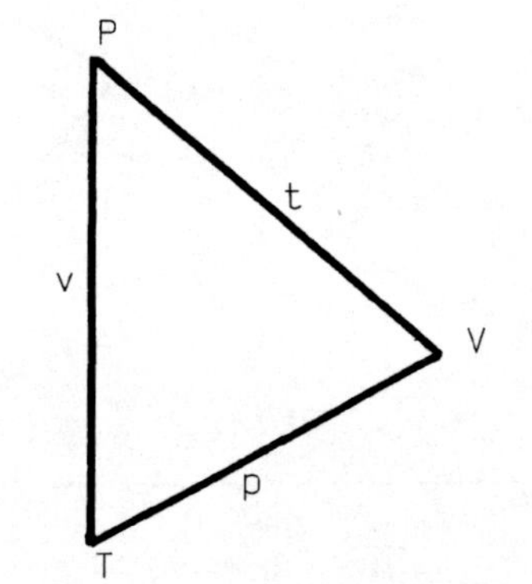

a) $\dfrac{\sin P}{p} = \dfrac{\boxed{\quad\quad}}{t}$

b) $\dfrac{\sin P}{p} = \dfrac{\sin V}{\boxed{\quad\quad}}$

---

a) $\dfrac{\boxed{\sin T}}{t}$ b) $\dfrac{\sin V}{\boxed{v}}$

19. When writing a proportion based on the Law of Sines, we must be consistent. That is, the numerators of both ratios must either be "sides" or "sines".

Using triangle BDF, complete these with one of the two possible ratios. Be consistent.

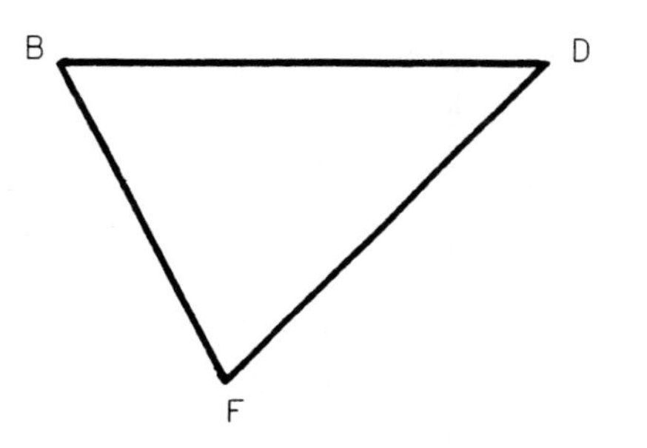

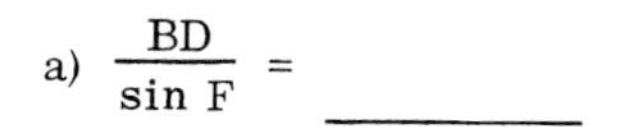

a) $\dfrac{BD}{\sin F} = \_\_\_\_\_\_\_\_$

b) $\dfrac{\sin D}{BF} = \_\_\_\_\_\_\_\_$

Answers to frame 19:

a) $\dfrac{BF}{\sin D}$ or $\dfrac{DF}{\sin B}$

b) $\dfrac{\sin B}{DF}$ or $\dfrac{\sin F}{BD}$

---

20. In this frame we will prove the Law of Sines. We will not expect you to memorize the proof or reproduce it. The proof is given simply to show that a proof is possible.

Using the triangle at the right, we will prove the following proportion.

$$\frac{c}{\sin C} = \frac{b}{\sin B}$$

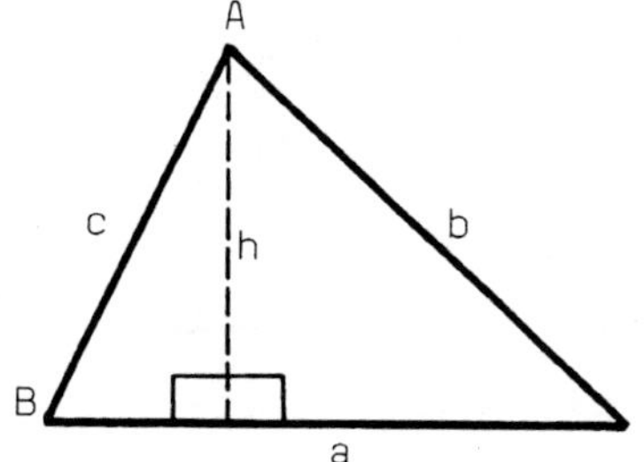

The steps are:

1. Draw an altitude from angle A to side BC.

2. Define sin B and sin C.

$$\sin B = \frac{h}{c} \qquad \sin C = \frac{h}{b}$$

3. Solve for "h" in each equation and equate the solutions.

$$h = c(\sin B) \quad \text{and} \quad h = b(\sin C)$$

$$c(\sin B) = b(\sin C)$$

4. Multiple both sides by $\dfrac{1}{\sin B}$ and $\dfrac{1}{\sin C}$.

$$\left(\frac{1}{\sin B}\right)\left(\frac{1}{\sin C}\right) c(\sin B) = \left(\frac{1}{\sin B}\right)\left(\frac{1}{\sin C}\right) b(\sin C)$$

$$\left(\frac{\sin B}{\sin B}\right)\left(\frac{1}{\sin C}\right)c = \left(\frac{\sin C}{\sin C}\right)\left(\frac{1}{\sin B}\right)b$$

$$\frac{c}{\sin C} = \frac{b}{\sin B}$$

By drawing altitudes from angles B and C, we can prove the following two proportions in a similar manner.

$$\boxed{\frac{a}{\sin A} = \frac{b}{\sin B} \quad \text{and} \quad \frac{a}{\sin A} = \frac{c}{\sin C}}$$

Continued on following page.

20. Continued

We can also write the Law of Sines with the "sines" in the numerator. We get:

$$\frac{\sin A}{a} = \frac{\sin B}{b} = \frac{\sin C}{c}$$

2-4 USING THE LAW OF SINES TO FIND SIDES

In this section, we will use the Law of Sines to find unknown sides in oblique triangles. To simplify the solutions, proportions involving obtuse angles will be delayed until a later section.

---

21. Let's use the Law of Sines to find the length of MP. According to the Law of Sines:

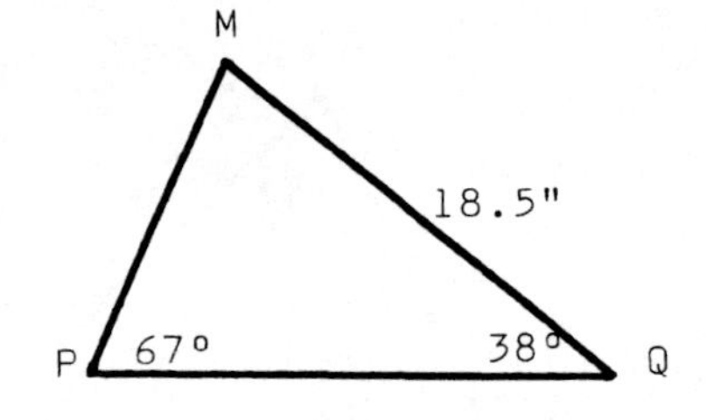

$$\frac{MP}{\sin Q} = \frac{MQ}{\sin P}$$

$$\frac{MP}{\sin 38°} = \frac{18.5''}{\sin 67°}$$

$$MP = \frac{(18.5'')(\sin 38°)}{\sin 67°}$$

To complete the solution on a calculator, follow these steps.

Enter	Press	Display
18.5	[x]	18.5
38	[sin] [÷]	11.389737
67	[sin] [=]	12.373359

Rounding to tenths, MP = ________

---

22. When using the Law of Sines to find an unknown side of a triangle, we set up a proportion like the one below. Notice that we put the unknown side in the numerator on the left side.

Answer (frame 21): 12.4"

$$\frac{RS}{\sin 52°} = \frac{17.9''}{\sin 71°}$$

We can solve for RS by multiplying both sides by sin 52°. We get:

$$\sin 52°\left(\frac{RS}{\sin 52°}\right) = \left(\frac{17.9''}{\sin 71°}\right)\sin 52°$$

$$RS = \frac{(17.9'')(\sin 52°)}{\sin 71°}$$

Having isolated RS, we can find RS in one process on a calculator by following the steps in the last frame. Do so. Round to tenths.

RS = ________

23. Let's use the same process to find MT below.

$$\frac{MT}{\sin 49°} = \frac{6.34 \text{ cm}}{\sin 56°}$$

a) Isolate MT by multiplying both sides by sin 49°.

MT = ______________________

b) Use a calculator to find MT. Round to hundredths.

MT = __________

> 14.9''

---

24. The complete Law of Sines for triangle DFH is:

$$\frac{d}{\sin D} = \frac{f}{\sin F} = \frac{h}{\sin H}$$

or $$\frac{d}{\sin 70°} = \frac{5.00\text{m}}{\sin 50°} = \frac{h}{\sin 60°}$$

If we want to solve for "d" we must use the ratio in which "d" appears and one other ratio. The two possibilities are:

$$\frac{d}{\sin 70°} = \frac{5.00\text{m}}{\sin 50°} \quad \text{and} \quad \frac{d}{\sin 70°} = \frac{h}{\sin 60°}$$

A proportion can be solved only if it contains <u>only one</u> unknown.

The proportion on the left contains <u>one</u> unknown, "d".

The proportion on the right contains <u>two</u> unknowns, "d" and "h".

Which proportion (the one on the left or the one on the right) must we use to solve for "d"? ______________

> a) $MT = \frac{(6.34 \text{ cm})(\sin 49°)}{\sin 56°}$
>
> b) MT = 5.77 cm

---

25. To solve for "s" at the right, we must set up a proportion in which "s" is the only unknown. Let's do so.

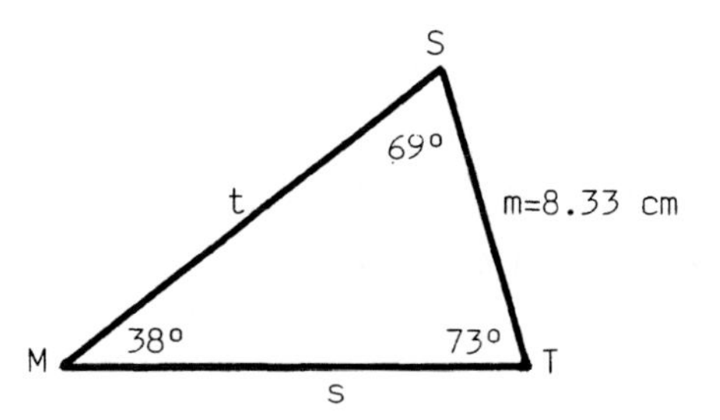

a) Which ratio contains side "s"?

____

b) Which other ratio should we use?

____

c) The proportion is:

__________ = __________

> The one on the left.

26. Sometimes we have to use the angle-sum principle before using the Law of Sines.

a) How large is angle R? ________

Let's set up the proportion needed to find CD at the right.

b) The proportion is:

________ = ________

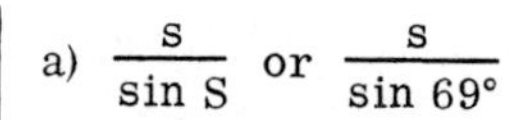
a) $\frac{s}{\sin S}$ or $\frac{s}{\sin 69°}$

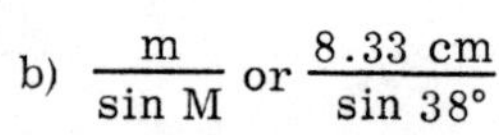
b) $\frac{m}{\sin M}$ or $\frac{8.33 \text{ cm}}{\sin 38°}$

(since we know both "m" and "sin M")

c) $\frac{s}{\sin 69°} = \frac{8.33 \text{ cm}}{\sin 38°}$

---

27. Let's set up the proportion needed to solve for FV at the right.

a) How large is angle F? ______

b) The proportion is:

________ = ________

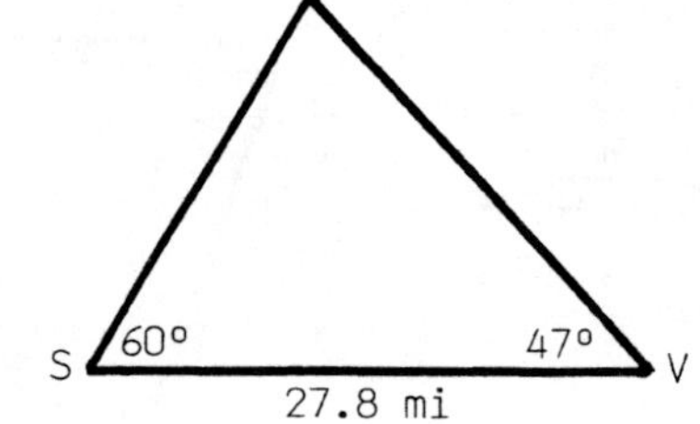

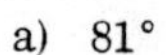
a) 81°

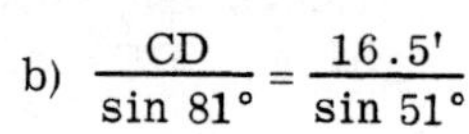
b) $\frac{CD}{\sin 81°} = \frac{16.5'}{\sin 51°}$

---

28. a) Set up the proportion needed to solve for BC.

________ = ________

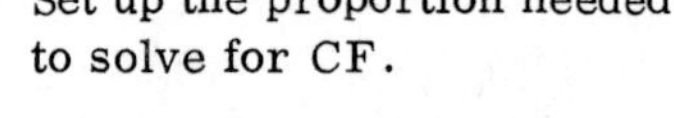
b) Set up the proportion needed to solve for CF.

________ = ________

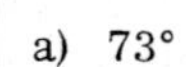
a) 73°

b) $\frac{FV}{\sin 60°} = \frac{27.8 \text{ mi}}{\sin 73°}$

---

29. After using the Law of Sines to calculate an angle or a side, you can check the sensibleness of your answer by applying a relationship involving the sizes of the sides and angles of a triangle. The relationship is:

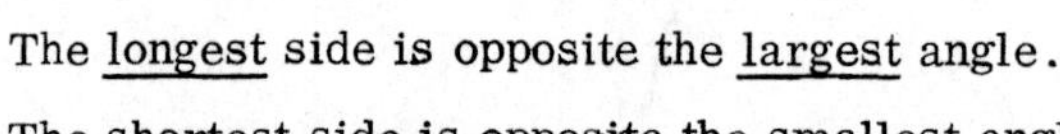
The longest side is opposite the largest angle.

The shortest side is opposite the smallest angle.

We can verify this relationship in triangle FGH.

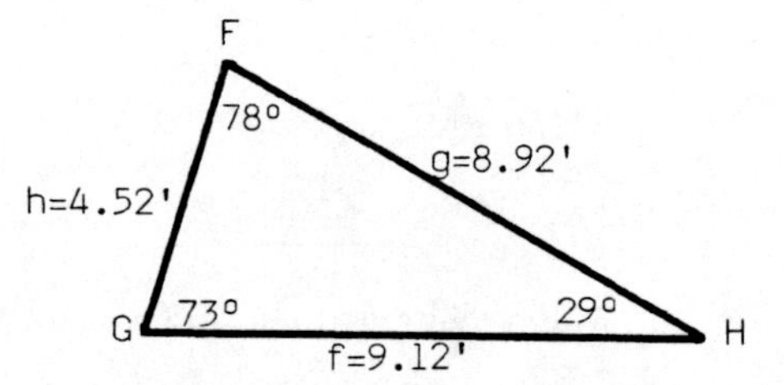

a) The largest angle is ______.

b) The longest side is ______.

c) The smallest angle is ______.

d) The shortest side is ______.

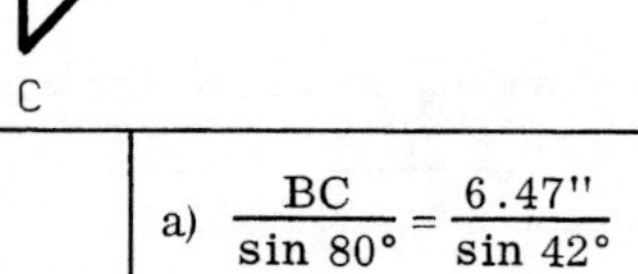
a) $\frac{BC}{\sin 80°} = \frac{6.47''}{\sin 42°}$

b) $\frac{CF}{\sin 58°} = \frac{6.47''}{\sin 42°}$

---

a) F = 78° b) f = 9.12' c) H = 29° d) h = 4.52'

30. In triangle ABC, which side would be larger?

a) AB or AC ______

b) AB or BC ______

c) AC or BC ______

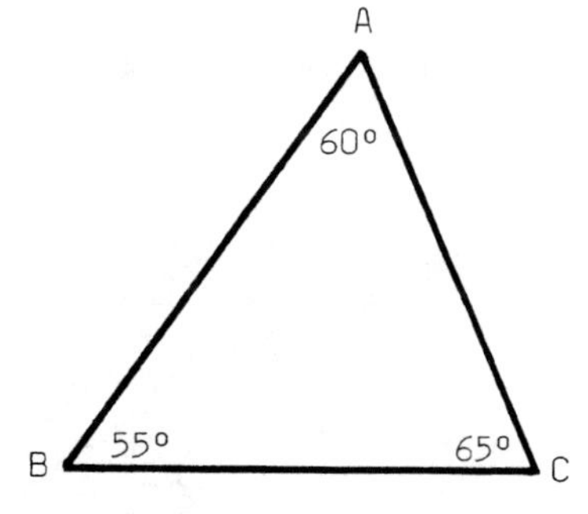

a) AB

b) AB

c) BC

---

31. Let's solve for side FR in triangle FRT.

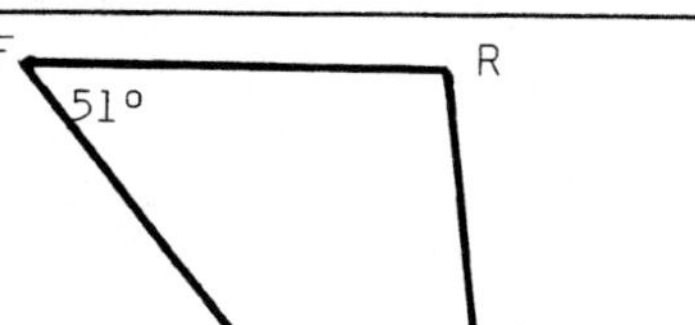

a) Is FR longer or shorter than 49.0? ______

b) Set up the proportion needed to solve for FR.

______ = ______

c) Complete the solution. Round to tenths. FR = ______

a) Shorter, because angle T is smaller than angle F.

b) $\frac{FR}{\sin 32°} = \frac{49.0'}{\sin 51°}$

c) FR = 33.4'

---

32. Let's solve for "d" in triangle CDM.

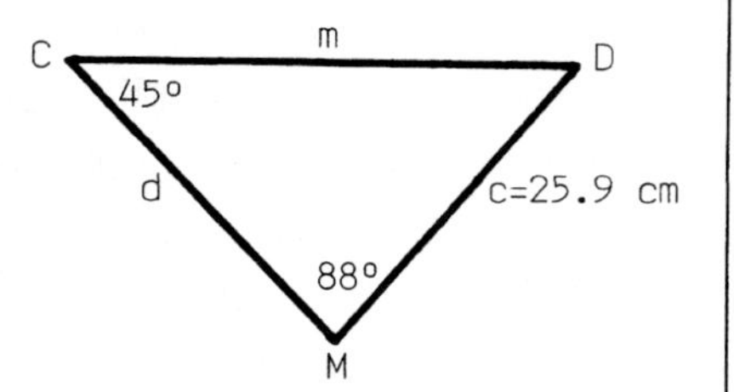

a) How large is angle D? ______

b) Is "d" longer or shorter than 25.9"? ______

c) Set up the proportion needed to solve for "d".

______ = ______

d) Complete the solution. Round to tenths. d = ______

a) D = 47° b) a little longer c) $\frac{d}{\sin 47°} = \frac{25.9 \text{ cm}}{\sin 45°}$ d) d = 26.8 cm

2-5 USING THE LAW OF SINES TO FIND ANGLES

In this section, we will use the Law of Sines to find unknown angles in oblique triangles. To simplify the solutions, proportions involving obtuse angles will be delayed until a later section.

33. Let's use the Law of Sines to find the size of angle T. According to the Law of Sines:

$$\frac{\sin T}{11.2'} = \frac{\sin 64°}{18.5'}$$

$$\sin T = \frac{(11.2')(\sin 64°)}{18.5'}$$

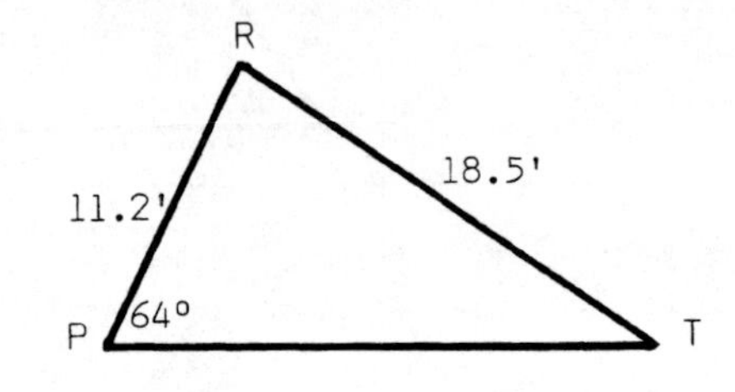

To complete the solution on a calculator, use these steps:

Enter	Press	Display
11.2	[x]	11.2
64°	[sin] [÷]	10.066493
18.5	[=] [INV] [sin]	32.965557
	or [ARC] [sin]	
	or [$\sin^{-1}$]	

Rounded to the nearest whole-number degree, T = ________

Answer: 33°

34. When using the Law of Sines to find an unknown angle of a triangle, we set up a proportion with the "sines" in the numerator and the unknown sine on the left side.

Which proportion below would we use to solve for angle B in this triangle? ______

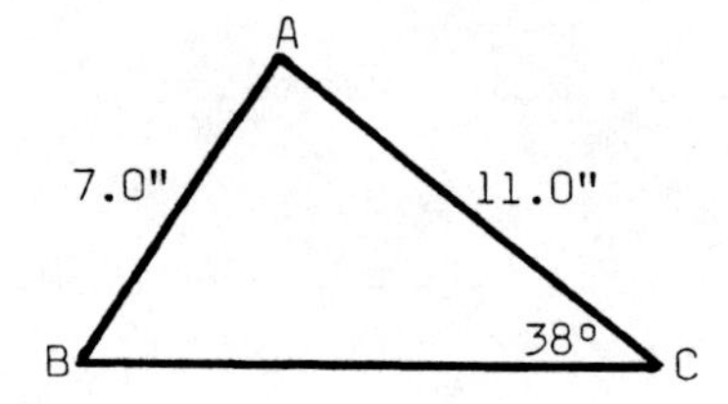

a) $\frac{11.0''}{\sin B} = \frac{7.0''}{\sin 38°}$

b) $\frac{\sin B}{11.0''} = \frac{\sin 38°}{7.0''}$

Answer: b

35. Set up the proportion needed to solve for angle D in each triangle below.

a)

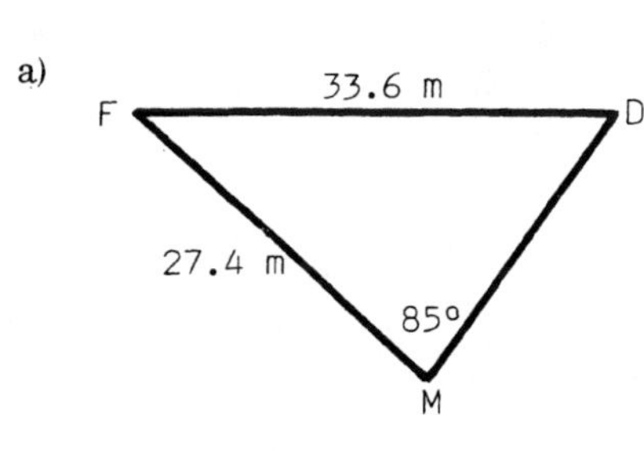

b)

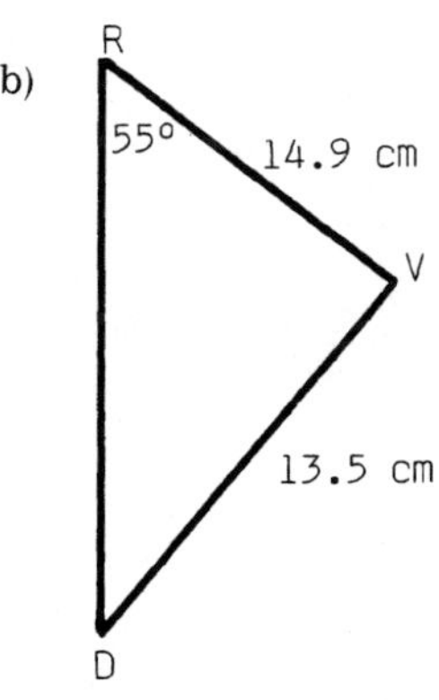

__________ = __________ __________ = __________

---

36. When solving for an angle in a triangle, remember these relations:

1. The <u>largest</u> angle is opposite the <u>longest</u> side.
2. The <u>smallest</u> angle is opposite the <u>shortest</u> side.

In triangle FHT:

a) The largest angle is ______.

b) The smallest angle is ______.

Answers to 35:

a) $\frac{\sin D}{27.4\text{m}} = \frac{\sin 85°}{33.6\text{m}}$

b) $\frac{\sin D}{14.9\text{ cm}} = \frac{\sin 55°}{13.5\text{ cm}}$

---

37. In triangle BKV, angle B contains 52°.

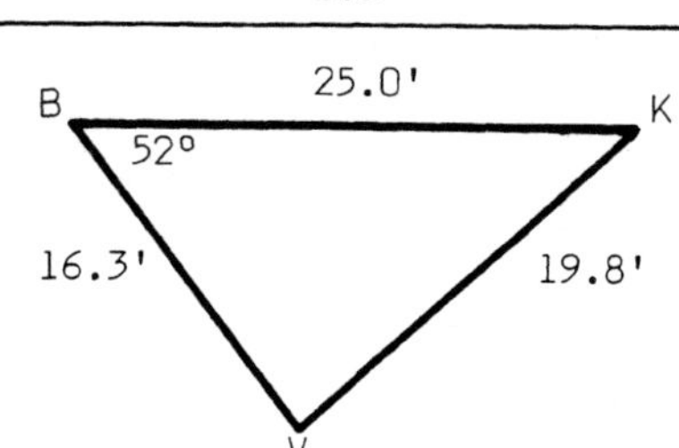

a) Is angle K larger or smaller than 52° ? __________

b) Is angle V larger or smaller than 52°? __________

Answers to 36:

a) T

b) H

---

38. Let's use the Law of Sines to find angle B.

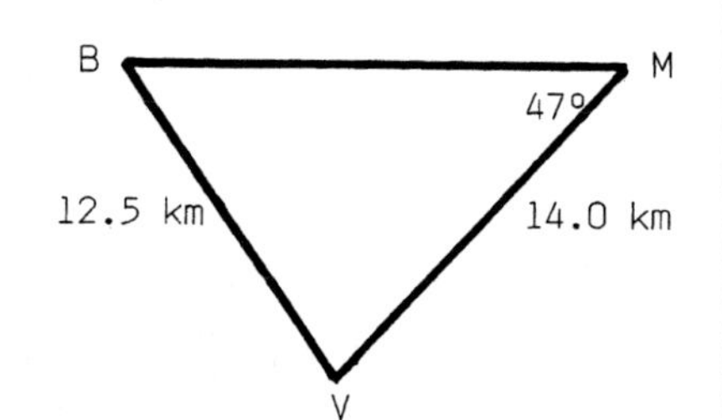

a) Is angle B larger or smaller than 47°? __________

b) Set up the proportion needed to solve for angle B.

__________ = __________

c) Complete the solution. Round to the nearest whole-number degree.

B = ________

Answers to 37:

a) smaller, because BV is shorter than KV.

b) larger, because BK is longer than KV.

39. Let's solve for angle F in this triangle.

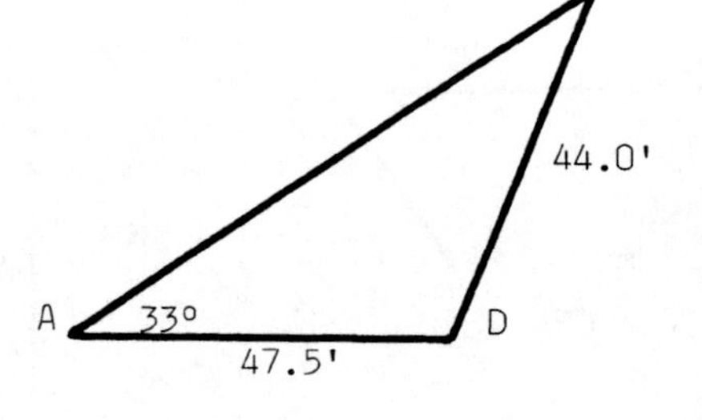

a) Is angle F larger or smaller than 33°? ______________

b) Set up the proportion needed to solve for angle F.

______ = ______

c) Complete the solution. Round to the nearest whole-number degree. F = ______

a) larger, since MV is longer than BV

b) $\frac{\sin B}{14.0 \text{ km}} = \frac{\sin 47°}{12.5 \text{ km}}$

c) 55°

---

a) larger, since AD is longer than DF

b) $\frac{\sin F}{47.5'} = \frac{\sin 33°}{44.0'}$

c) 36°

SELF-TEST 4 (pp. 41-57)

1. Find the unknown angle in each triangle.

a)

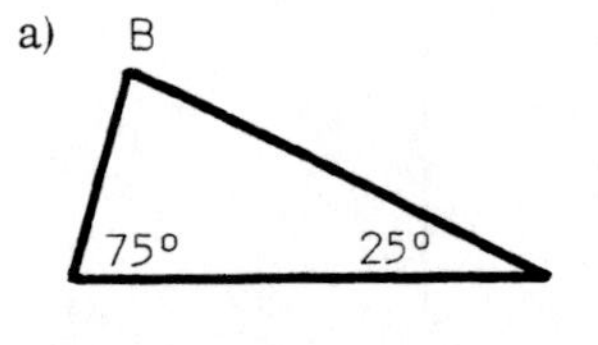

b)

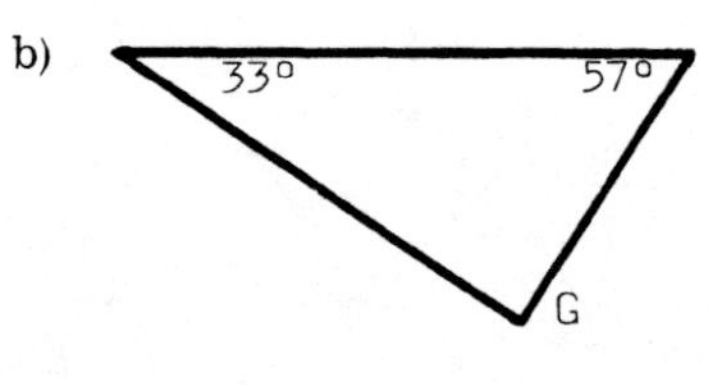

c)

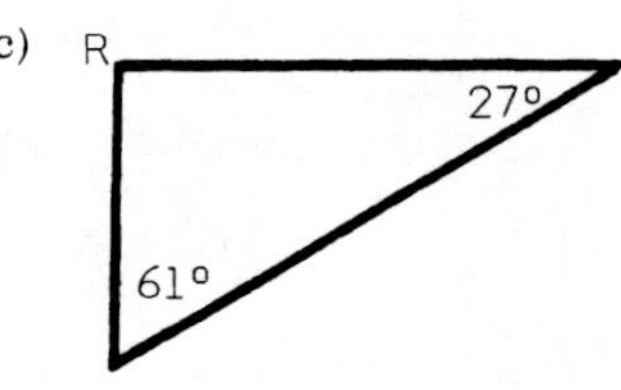

B = ______ G = ______ R = ______

2. Identify each triangle in Problem 1 as "right", "acute oblique", or "obtuse oblique".

a) ______________ b) ______________ c) ______________

3. For triangle PRT at the right, use the Law of Sines to complete these proportions:

a) $\frac{t}{\boxed{}} = \frac{\boxed{}}{\sin P}$ b) $\frac{\sin R}{\boxed{}} = \frac{\boxed{}}{p}$

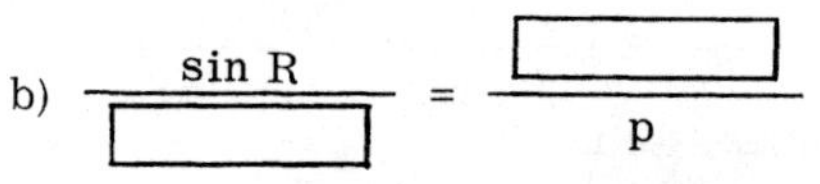

R
t
P
p
r
T

SELF-TEST 4 Continued on following page

SELF-TEST 4 (pp. 41-57) - Continued

4. a) Using the Law of Sines, set up the proportion for finding side DE.

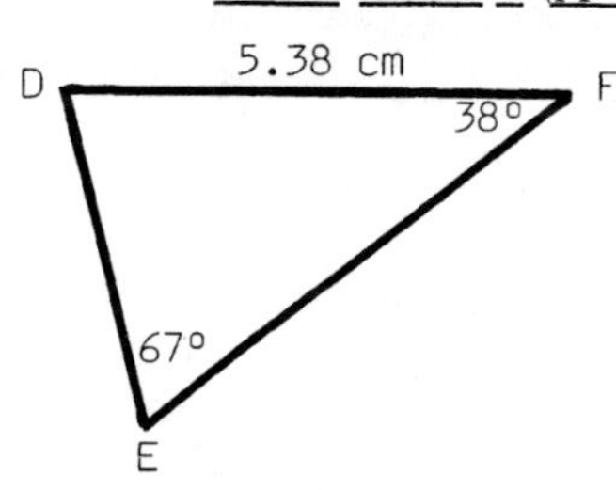

____

b) Find side DE. Round to hundredths.

DE = _______

5. a) Using the Law of Sines, set up the proportion for finding angle N.

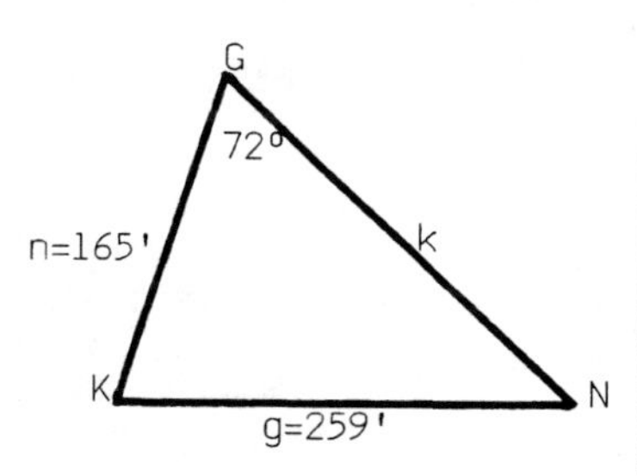

____

b) Find angle N. Round to a whole number.

N = _______

6. Refer to triangle DEF in Problem 4.

a) Which angle (D, E, or F) is largest? _______

b) Which side (DE, DF, or EF) is smallest? _______

c) Which side is larger, DF or EF? _______

7. Refer to triangle GKN in Problem 5.

a) What is the size of angle K? K = _______

b) Which angle is smaller, G or K? _______

c) Which side (g, k, or n is largest? _______

ANSWERS:

1. a) B = 80°
   b) G = 90°
   c) R = 92°

2. a) acute oblique
   b) right
   c) obtuse oblique

3. a) $\frac{t}{\sin T} = \frac{p}{\sin P}$
   b) $\frac{\sin R}{r} = \frac{\sin P}{p}$

4. a) $\frac{DE}{\sin 38°} = \frac{5.38 \text{ cm}}{\sin 67°}$
   b) DE = 3.60 cm

5. a) $\frac{\sin N}{165'} = \frac{\sin 72°}{259'}$
   b) N = 37°

6. a) angle D
   b) side DE
   c) side EF

7. a) K = 71°
   b) angle K
   c) side g

2-6 LIMITATIONS OF THE LAW OF SINES

In this section, we will show some more complex solutions involving the Law of Sines. Then we will discuss some cases in which the Law of Sines cannot be used.

40. To use the Law of Sines to find a side or angle, we must be able to set up a proportion in which that side or the sine of that angle is the only unknown. We have seen two types of solutions.

1. Those in which the proportion can be set up directly.

To solve for angle D in this triangle, we can directly set up a proportion in which sin D is the only unknown. We get:

$$\frac{\sin D}{10.2\text{m}} = \frac{\sin 55°}{8.5\text{m}}$$

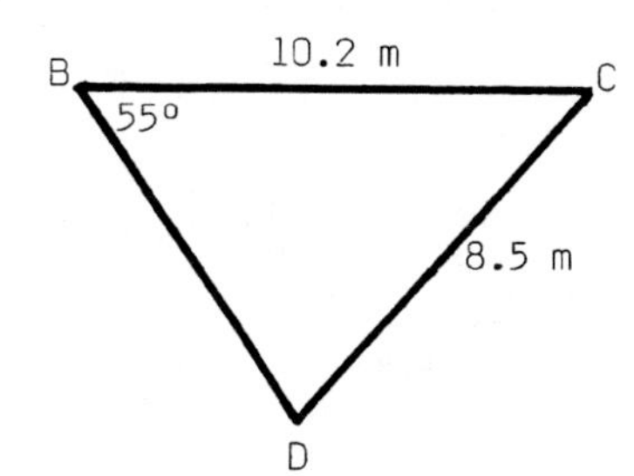

Continued on following page.

40. Continued

2. Those in which the proportion can be set up only after the angle-sum principle is used.

To solve for SV in this triangle, we must use the angle-sum principle to find angle V. Then we can set up a proportion in which SV is the only unknown. We get:

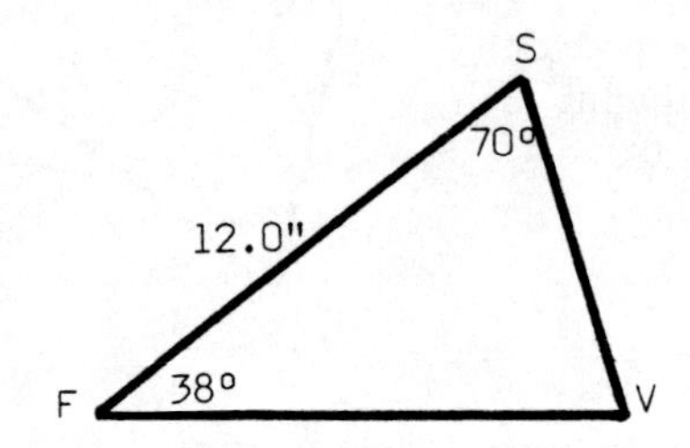

$$\frac{SV}{\sin 38°} = \frac{12.0''}{\sin 72°}$$

---

41. Some solutions involving the Law of Sines are more complex than those in the last frame. Examples are given in this frame and the next frame.

To solve for angle P in this triangle, we cannot use the Law of Sines directly because side "p" is unknown. We would get either of these two proportions:

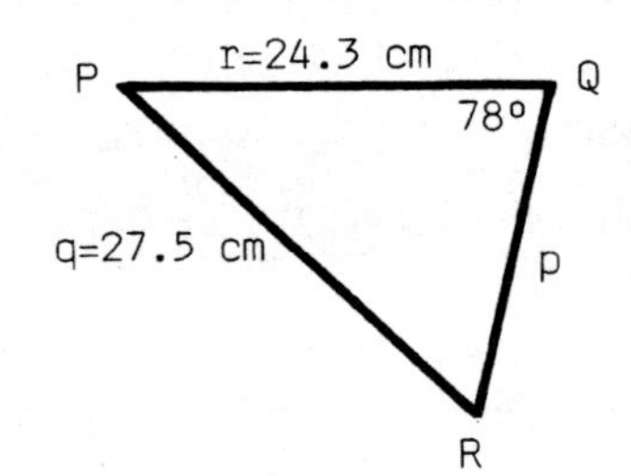

$$\frac{\sin P}{p} = \frac{\sin 78°}{27.5 \text{ cm}}$$

or $$\frac{\sin P}{p} = \frac{\sin R}{24.3 \text{ cm}}$$

However, we can use the Law of Sines to find angle P in two steps. That is:

1. Use the Law of Sines to find angle R.
2. Then use the angle-sum principle to find angle P.

a) Set up the proportion needed to find angle R.

$$\frac{\qquad}{\qquad} = \frac{\qquad}{\qquad}$$

b) Angle R contains ________°. (Round to a whole-number.)

c) Angle P contains ________°.

---

42. To find HS in triangle HST, we cannot use the Law of Sines directly because angle T is unknown. However, we can find HS in three steps. They are:

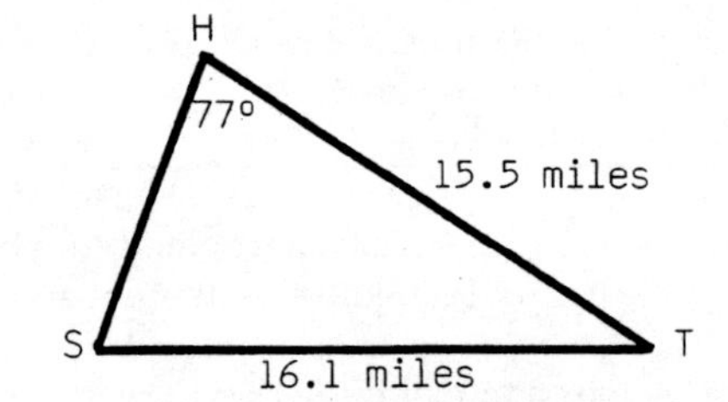

a) Use the Law of Sines to find angle S.

Angle S = ________ (Round to a whole number.)

Continued on following page.

Answers to frame 41:

a) $$\frac{\sin R}{24.3 \text{ cm}} = \frac{\sin 78°}{27.5 \text{ cm}}$$

b) 60°

c) 42°

42. Continued

b) Use the angle-sum principle to find angle T.

Angle T = ________

c) Now use the Law of Sines to find HS. Round to tenths.

HS = ________ miles

---

43. The Law of Sines cannot be used for all solutions of oblique triangles. Examples are given in this frame and the next frame.

In triangle ABD, we are given two sides (AB and AD) and the angle between them called the "included" angle. The three ratios are:

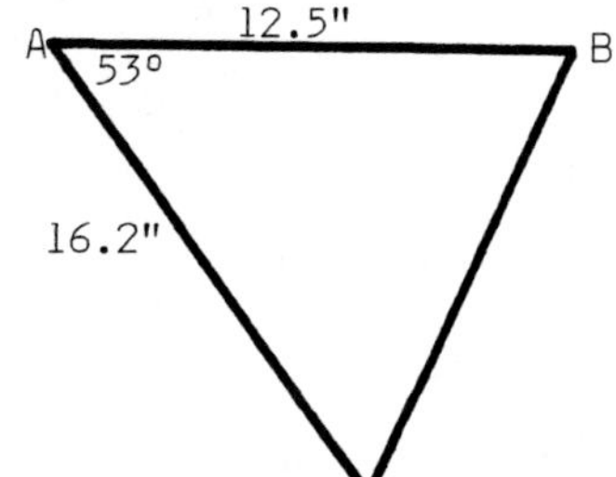

$$\frac{AB}{\sin D} = \frac{AD}{\sin B} = \frac{BD}{\sin A}$$

or

$$\frac{12.5''}{\sin D} = \frac{16.2''}{\sin B} = \frac{BD}{\sin 53°}$$

Can we set up a proportion to solve for side BD or angles B and D? ____________

Answers to 42: a) 70° b) 33° c) 9.0 miles

---

44. In triangle DFK, we are given the lengths of the three sides. Here are the three ratios:

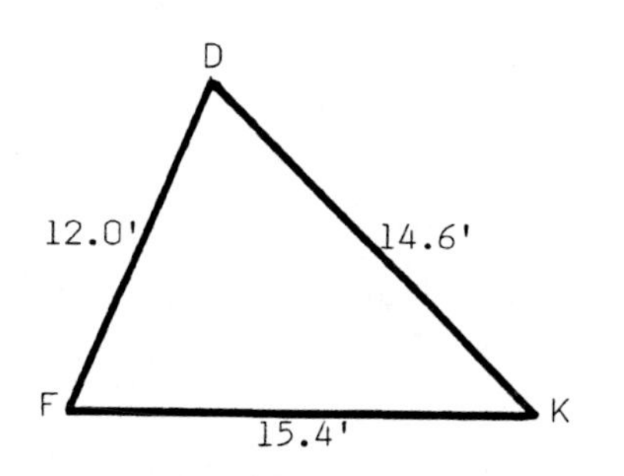

$$\frac{DK}{\sin F} = \frac{DF}{\sin K} = \frac{FK}{\sin D}$$

or

$$\frac{14.6'}{\sin F} = \frac{12.0'}{\sin K} = \frac{15.4'}{\sin D}$$

Can we set up a proportion to solve for any of the three angles? __________

Answer to 43: No, because any proportion has two unknowns.

---

45. There are two cases in which we cannot use the Law of Sines for solutions. They are:

1. When two sides and their included angle are given.
2. When three sides are given but no angles.

In those cases, we must use the Law of Cosines to solve the triangles. We will introduce the Law of Cosines in the next section.

Answer to 44: No, because any proportion has two unknowns.

2-7 THE LAW OF COSINES

When the Law of Sines cannot be used to solve a triangle, we use the Law of Cosines. We will discuss and prove the Law of Cosines in this section.

---

46. In triangle DFP, we are given two sides ("d" and "f") and their included angle (angle P). We want to find side "p". As we saw in the last section, the Law of Sines cannot be used to find "p" in this triangle.

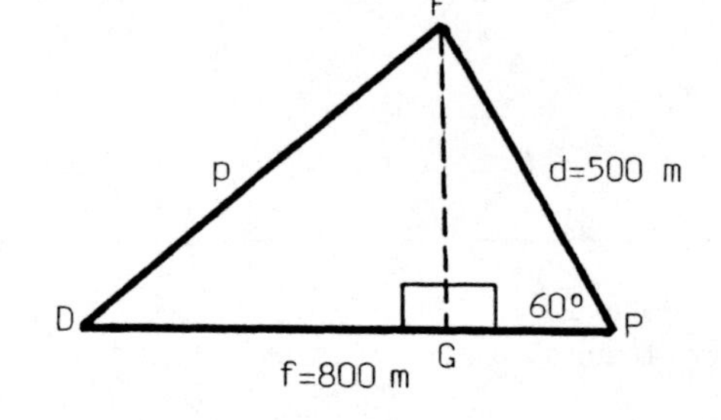

We can find "p" by the right-triangle method. To do so, we draw altitude FG to divide triangle DFP into two right triangles. The steps for finding "p" are:

1. Finding FG. $\sin 60° = \frac{FG}{d}$

   $FG = (500m)(\sin 60°) = 433m$

2. Finding GP. $\cos 60° = \frac{GP}{d}$

   $GP = (500m)(\cos 60°) = 250m$

3. Finding DG. $DG = DP - GP$

   $= 800m - 250m$

   $= 550m$

4. Finding "p".

   In right triangle DFG, we know the lengths of both legs (DG = 550m and FG = 433m). Therefore, we can use the Pythagorean Theorem to find the hypotenuse "p".

   $p^2 = (DG)^2 + (FG)^2$

   $p^2 = (550)^2 + (433)^2$

   $p = \sqrt{(550)^2 + (433)^2} = \sqrt{489,989}$

   $p = 700m$ (This is the solution.)

---

47. In the last frame, we solved for "p" in the triangle DFP by the right-triangle method. We got: $p = 700m$.

There is a relationship called the "Law of Cosines" which can be used to solve for "p" in one step. The Law-of-Cosines formula for "p" is:

$$p^2 = d^2 + f^2 - 2df \cos P$$

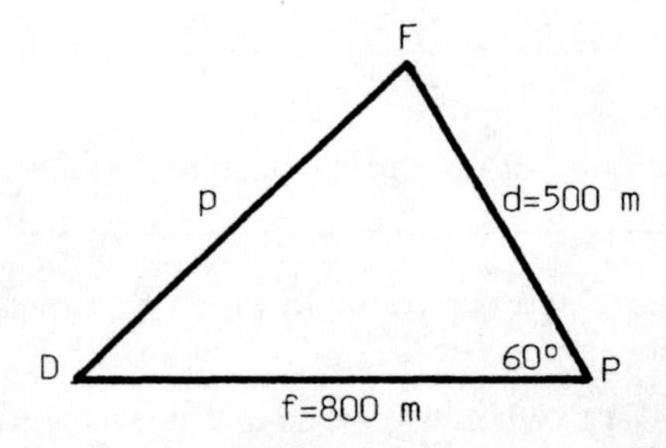

Continued on following page.

47. Continued

Notice these points:

1. "p" is the unknown side. It is opposite the known angle P.
2. "d" and "f" are the two known sides forming angle P.
3. "2df cos P" is a product of four factors. It can be written (2)(d)(f)(cos P). The factors include the two known sides and the known angle.

Using d = 500, f = 800, and cos P = cos 60° = 0.5, we can substitute in the formula and solve for "p". We get:

$p^2 = (500)^2 + (800)^2 - 2(500)(800)(0.5)$

$p^2 = 250,000 + 640,000 - 400,000$

$p^2 = 490,000$

$p = \sqrt{490,000} = 700\text{m}$

Is 700m the same solution we got for "p" in the last frame? ________

Yes

48. For any triangle, we can state three formulas based on the Law of Cosines. The three formulas for triangle BHT are discussed below.

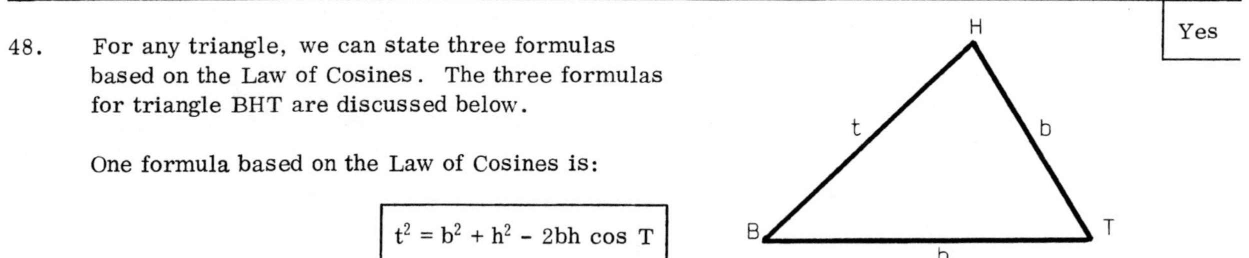

One formula based on the Law of Cosines is:

$$t^2 = b^2 + h^2 - 2bh \cos T$$

Note: 1. "t" is the side opposite angle T, and "cos T" is in the formula.
2. "b" and "h" are the other two sides.

A second formula based on the Law of Cosines is:

$$b^2 = h^2 + t^2 - 2ht \cos B$$

Note: 1. "b" is the side opposite angle B, and "cos B" is in the formula.
2. "h" and "t" are the other two sides.

Complete the third formula based on the Law of Cosines:

$h^2 =$ ________________

$h^2 = b^2 + t^2 - 2bt \cos H$

49. One Law-of-Cosines formula for triangle ACQ is given below. Complete the other two formulas for the triangle.

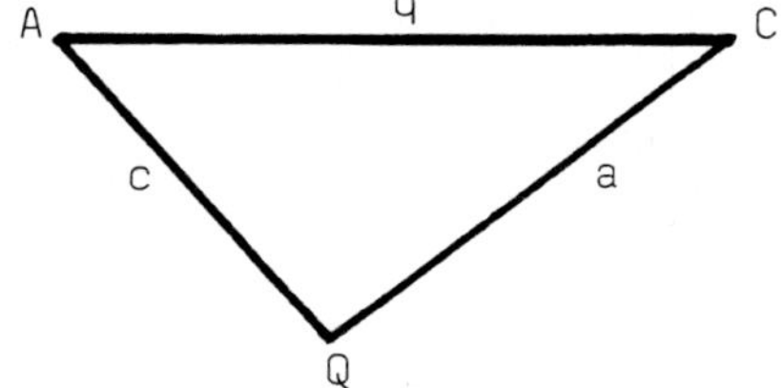

$c^2 = a^2 + q^2 - 2aq \cos C$

$a^2 =$ ________________

$q^2 =$ ________________

---

a) $a^2 = c^2 + q^2 - 2cq \cos A$ b) $q^2 = a^2 + c^2 - 2ac \cos Q$

50. One Law-of-Cosines formula for triangle MRS is given below. Complete the other two formulas for the triangle.

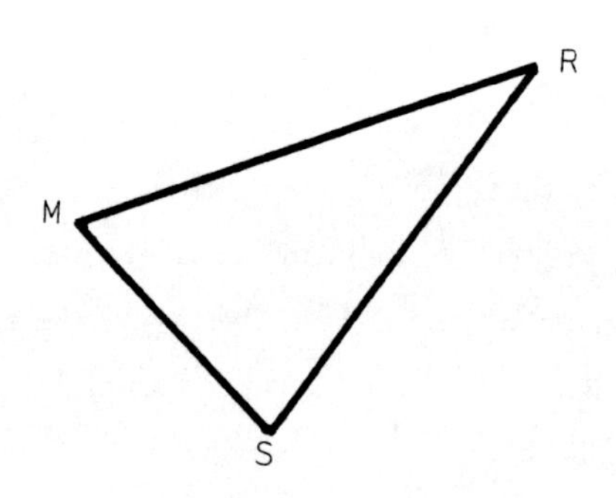

$(MS)^2 = (MR)^2 + (RS)^2 - 2(MR)(RS)(\cos R)$

$(MR)^2 =$ ______________________

$(RS)^2 =$ ______________________

---

a) $(MR)^2 = (MS)^2 + (RS)^2 - 2(MS)(RS)\cos S$

b) $(RS)^2 = (MR)^2 + (MS)^2 - 2(MR)(MS)\cos M$

51. In this frame we will prove the Law of Cosines. We will not expect you to memorize the proof or reproduce it. The proof is given simply to show that a proof is possible.

To understand the proof, you must know the following:

1. The symbol for squaring "sin A" is $\sin^2 A$.
   The symbol for squaring "cos A" is $\cos^2 A$.

   Therefore: $\sin^2 A = (\sin A)(\sin A)$
   $\cos^2 A = (\cos A)(\cos A)$

2. For any angle A, $\underline{\sin^2 A + \cos^2 A = 1}$

   For example, to see that $\sin^2 55° + \cos^2 55° = 1$, follow these steps on your calculator.

Enter	Press	Display
55	[sin] [x^2] [+]	0.6710101
55	[cos] [x^2] [=]	1

3. Just as: $(ab)^2 = (ab)(ab) = a^2b^2$,

   $(a \cos C)^2 = (a \cos C)(a \cos C) = a^2\cos^2 C$

4. Just as: $(a - b)^2 = (a - b)(a - b) = a^2 - 2ab + b^2$,

   $(b - a \cos C)^2 = (b - a \cos C)(b - a \cos C)$
   $= b^2 - 2ab \cos C + a^2\cos^2 C$

We will use triangle ABC to prove the Law of Cosines. We will prove the following formula:

$$c^2 = a^2 + b^2 - 2ab \cos C$$

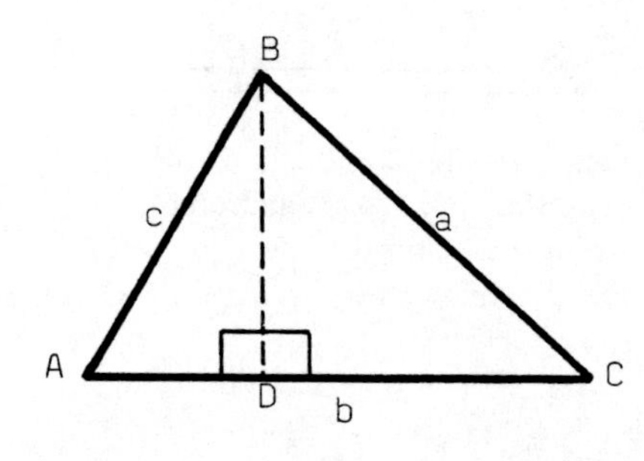

We have drawn altitude BD. Since triangle ABD is a right triangle, the Pythagorean Theorem gives us the following:

$$c^2 = (BD)^2 + (AD)^2$$

Continued on following page.

51. Continued

We will show that the two boxed equations are equivalent by expressing $(BD)^2$ and $(AD)^2$ in terms of "a", "b" and angle C.

Finding $(BD)^2$: Since $\sin C = \dfrac{BD}{a}$,

$$BD = a \sin C$$

$$\text{and } (BD)^2 = \boxed{a^2\sin^2 C}$$

Finding $(AD)^2$: $AD = b - DC$

But $DC = a \cos C$, since $\cos C = \dfrac{DC}{a}$

Therefore: $AD = b - a \cos C$

$$\text{and } (AD)^2 = (b - a\cos C)^2 = \boxed{b^2 - 2ab\cos C + a^2\cos^2 C}$$

Substituting the boxed expressions for $(BD)^2$ and $(AD)^2$ into the Pythagorean Theorem, we get:

If $c^2 = (BD)^2 + (AD)^2$: $\quad c^2 = \boxed{a^2\sin^2 C} + \boxed{b^2 - 2ab\cos C + a^2\cos^2 C}$

Rearranging we get: $\quad c^2 = a^2\sin^2 C + a^2\cos^2 C + b^2 - 2ab\cos C$

Factoring the first two terms we get: $\quad c^2 = a^2(\sin^2 C + \cos^2 C) + b^2 - 2ab\cos C$

Since $\sin^2 C + \cos^2 C = 1$, we get: $\quad c^2 = a^2(1) + b^2 - 2ab\cos C$

or

$$\boxed{c^2 = a^2 + b^2 - 2ab\cos C} \quad \text{(Law of Cosines)}$$

2-8 USING THE LAW OF COSINES TO FIND SIDES AND ANGLES

In this section, we will use the Law of Cosines to find unknown sides and angles in triangles. To simplify the solutions, triangles with obtuse angles will be delayed until a later section.

---

52. In triangle BDV, two sides and their included angle are given. We cannot use the Law of Sines to find side "d". However, we can use the Law of Cosines to do so.

The Law-of-Cosines formula is:

$$d^2 = b^2 + v^2 - 2bv\cos D$$

To solve for "d", we take the square root of each side.

$$d = \sqrt{b^2 + v^2 - 2bv\cos D}$$

Substituting the known values, we get:

$$d = \sqrt{(20.2)^2 + (22.6)^2 - (2)(20.2)(22.6)(\cos 32°)}$$

Continued on following page.

52. Continued

We can then use a calculator to find "d". The steps are:

Enter	Press	Display
20.2	x^2 +	408.04
22.6	x^2 -	918.8
2	x	2
20.2	x	40.4
22.6	x	913.04
32	cos = INV x^2	12.020739

Rounding to tenths, we get: d = ____________

12.0"

53. In triangle CFS, two sides and their included angle are given. To find side CS, we must use the Law of Cosines.

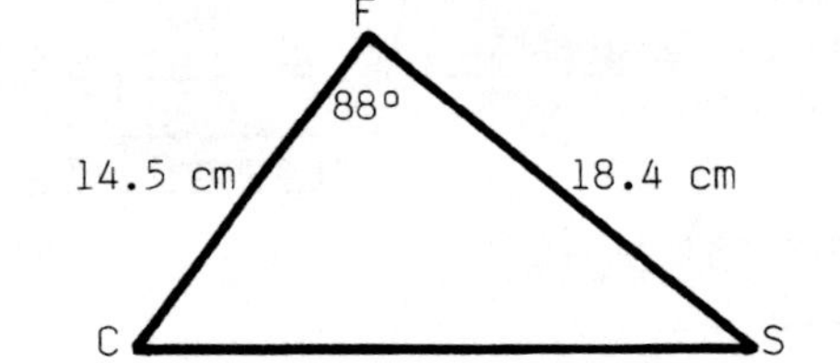

a) Write the Law-of-Cosines formula.

$(CS)^2$ = ______________________

b) Solve for CS by taking the square root of each side.

CS = ______________________

c) Substitute the known values in the formula.

CS = ______________________

d) Use a calculator to complete the solution. Round to tenths.

CS = ____________

a) $(CS)^2 = (CF)^2 + (FS)^2 - 2(CF)(FS)(\cos F)$

b) $CS = \sqrt{(CF)^2 + (FS)^2 - 2(CF)(FS)(\cos F)}$

c) $CS = \sqrt{(14.5)^2 + (18.4)^2 - 2(14.5)(18.4)(\cos 88°)}$

d) CS = 23.0 cm

54. In triangle ATV, three sides and no angles are given. We cannot use the Law of Sines to find angle T. However, we can use the Law of Cosines. The steps are:

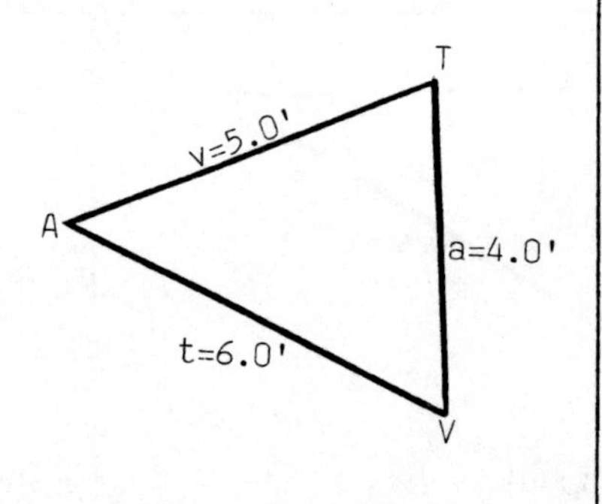

Continued on following page.

54. Continued

1. Write the Law-of-Cosines formula that contains "cos T". It is the one with "t^2" on the left side.

$$t^2 = a^2 + v^2 - 2av \cos T$$

2. Substitute the known values. Notice that "cos T" is the only unknown.

$$(6.0)^2 = (4.0)^2 + (5.0)^2 - 2(4.0)(5.0)(\cos T)$$

3. Simplify to find "cos T".

$$36 = 16 + 25 - 40 \cos T$$

$$36 = 41 - 40 \cos T$$

$$-5 = -40 \cos T$$

$$5 = 40 \cos T$$

$$\cos T = \frac{5}{40}$$

Use a calculator to find angle T. Round to a whole number.

Angle T = ____________

Answer: 83°

55. In the last frame, we substituted into the formula below and then solved for "cos T".

$$t^2 = a^2 + v^2 - 2av \cos T$$

We could rearrange the formula to solve for "cos T" before substituting. The steps are:

1. Add $(-a^2)$ and $(-v^2)$ to both sides.

$$t^2 + (-a^2) + (-v^2) = -2av \cos T$$

2. Take the opposite of each side.

$$(-t^2) + a^2 + v^2 = 2av \cos T$$

or

$$a^2 + v^2 - t^2 = 2av \cos T$$

3. Divide both sides by "2av", the coefficient of "cos T".

$$\cos T = \frac{a^2 + v^2 - t^2}{2av}$$

With "cos T" solved for, we can use a calculator to find angle T. The known values (a = 4.0', v = 5.0', t = 6.0') are substituted in the formula.

$$\cos T = \frac{(4.0)^2 + (5.0)^2 - (6.0)^2}{2(4.0)(5.0)}$$

Continued on following page.

55. Continued

The calculator steps are:

Enter	Press	Display
4	[x^2] [+]	16
5	[x^2] [-]	41
6	[x^2] [=] [÷]	5
2	[÷]	2.5
4	[÷]	0.625
5	[=] [INV] [cos]	82.819244

a) Rounded to a whole number, angle T = ________.

b) Is this the same solution we got in the last frame? ________

---

56. When using the Law of Cosines to find an angle, it is easier to use a calculator if you solve for the cosine of the angle first.

Solve for cos P in this formula.

$$p^2 = b^2 + h^2 - 2bh \cos P$$

Answers to 55:

a) 83°

b) Yes

---

57. Let's use the Law of Cosines to find angle V in this triangle.

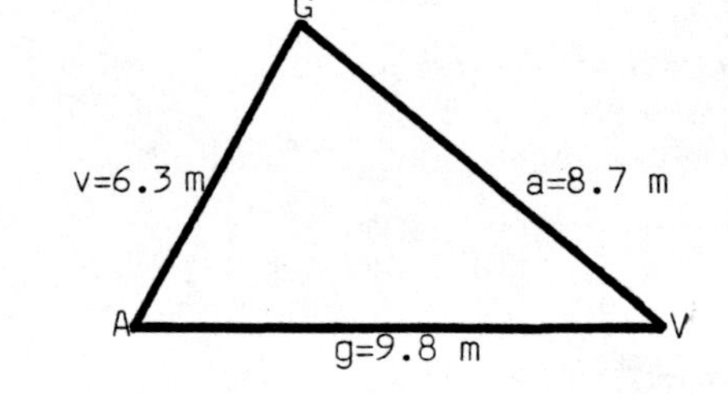

a) Write the formula that contains "cos V".

____

b) Rearrange to solve for cos V.

Answer to 56:

$$\cos P = \frac{b^2 + h^2 - p^2}{2bh}$$

Continued on following page.

57. Continued

c) Substitute the known values in the formula.

cos V = ____________________

d) Use a calculator to complete the solution. Rounded to a whole number, angle V = ______.

---

a) $v^2 = a^2 + g^2 - 2ag\cos V$

b) $\cos V = \dfrac{a^2 + g^2 - v^2}{2ag}$

c) $\cos V = \dfrac{(8.7)^2 + (9.8)^2 - (6.3)^2}{2(8.7)(9.8)}$

d) 39°

---

58. Let's use the Law of Cosines to find angle M in this triangle.

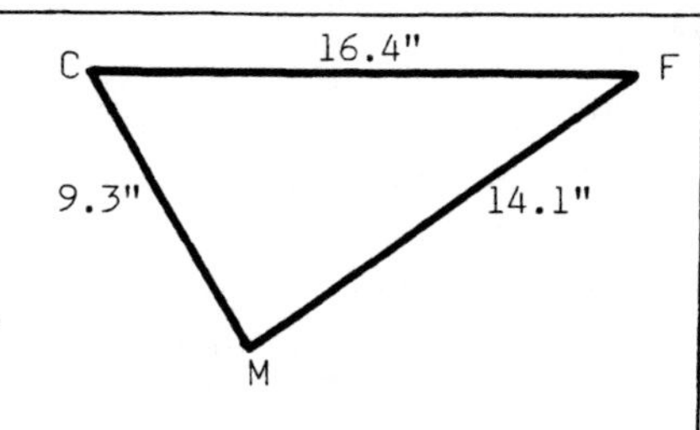

a) Write the formula that contains "cos M".

____

b) Rearrange to solve for cos M.

c) Substitute the known values in the formula.

cos M = ____________________

d) Use a calculator to complete the solution. Rounded to a whole number, angle M = ______.

---

a) $(CF)^2 = (CM)^2 + (MF)^2 - 2(CM)(MF)(\cos M)$

b) $\cos M = \dfrac{(CM)^2 + (MF)^2 - (CF)^2}{2(CM)(MF)}$

c) $\cos M = \dfrac{(9.3)^2 + (14.1)^2 - (16.4)^2}{2(9.3)(14.1)}$

d) M = 86°

SELF-TEST 5 (pp. 57-68)

In triangle DEF, find angle E, angle F, and side DE. Round each angle to a whole number. Round the side to tenths.

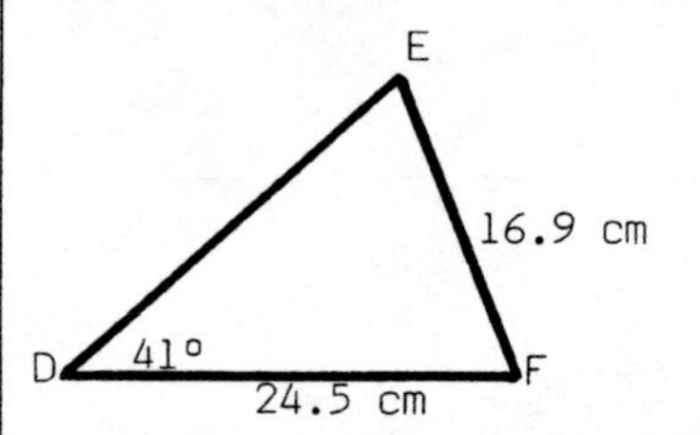

1. Angle E = ______
2. Angle F = ______
3. Side DE = ______

Apply the Law of Cosines to triangle HPT and complete these formulas.

H p T
t h
P

4. t^2 = ______________________________
5. cos H = ______________________________

6. Find side "d". Round to a whole number.

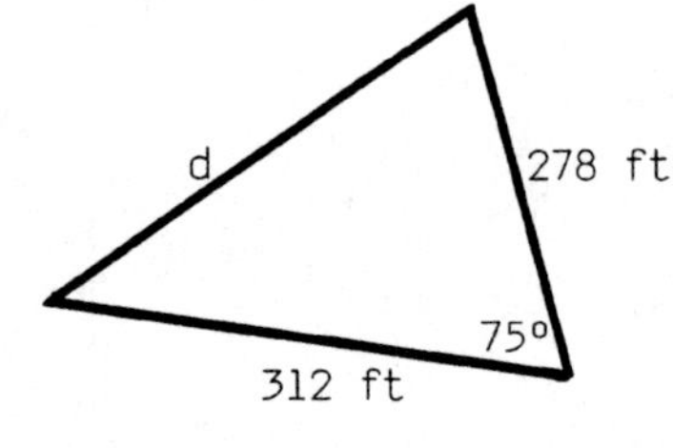

Side "d" = ______

7. Find angle G. Round to a whole number.

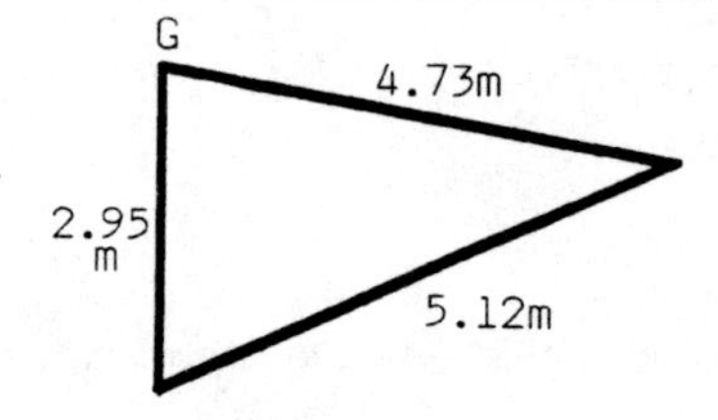

Angle G = ______

ANSWERS:

1. Angle E = 72°
2. Angle F = 67°
3. Side DE = 23.7 cm

4. $t^2 = h^2 + p^2 - 2hp \cos T$

5. $\cos H = \dfrac{p^2 + t^2 - h^2}{2pt}$

(from: $h^2 = p^2 + t^2 - 2pt \cos H$)

6. d = 360 ft
7. G = 80°

2-9 OBTUSE ANGLES AND THE LAW OF SINES

When using the Law of Sines in earlier sections, we avoided triangles with obtuse angles. We will use the Law of Sines with triangles of that type in this section.

59. Angles between 90° and 180° are called "obtuse" angles. Triangle CDF contains an obtuse angle.

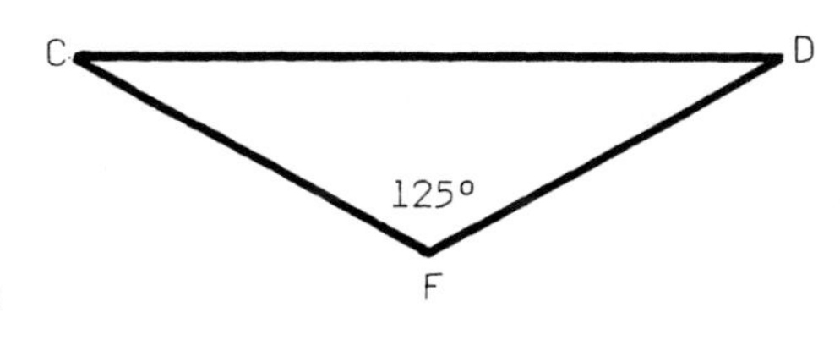

Can a triangle contain more than one obtuse angle? ________

---

60. To find the sine of an obtuse angle on a calculator, we simply enter the angle and press [sin]. For example, we found the sines of four angles (91°, 127°, 154°, 179°) below.

Enter	Press	Display
91	[sin]	0.9998477
127	[sin]	0.7986355
154	[sin]	0.4383711
179	[sin]	0.0174524

The sine of an obtuse angle is a positive number between 0 and 1. As the angle increases from 91° to 179°, does the sine increase or decrease? ____________

Answer to 59: No, because the sum of the angles would then be more than 180°.

---

61. Let's use the Law of Sines to find side "a" in triangle ABC. The proportion is:

$$\frac{a}{\sin 59°} = \frac{19.4''}{\sin 102°}$$

$$a = \frac{(19.4'')(\sin 59°)}{\sin 102°}$$

The calculator steps are:

Enter	Press	Display
19.4	[x]	19.4
59	[sin] [÷]	16.629046
102	[sin] [=]	17.000548

Rounded to the nearest tenth, a = ________

Answer to 60: It decreases.

---

62. Let's use the Law of Sines to find side PR in this triangle. We must find angle R first.

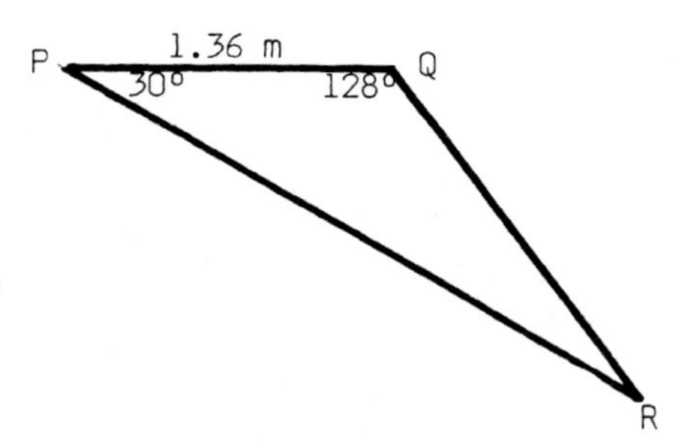

a) Angle R contains ______°.

Answer to 61: 17.0"

Continued on following page.

62. Continued

b) Set up the proportion needed to find PR.

c) Use a calculator to find PR. Round to hundredths.

PR = ________

a) 22°

b) $\frac{PR}{\sin 128°} = \frac{1.36m}{\sin 22°}$

or

$PR = \frac{(1.36m)(\sin 128°)}{\sin 22°}$

c) 2.86m

---

63. Though there is a more extensive treatment of the "sines" of obtuse angles in the next chapter, you should know this fact:

> The sine of an obtuse angle is the same as the sine of the acute angle obtained by subtracting the obtuse angle from 180°.

That is:
sin 150° = sin 30° (from 180°-150°)
sin 135° = sin 45° (from 180°-135°)
sin 93° = sin 87° (from 180°-93°)

The facts above can be confirmed with a calculator. For example:

sin 150° = sin 30° = 0.5
sin 135° = sin 45° = 0.7071068

sin 93° = sin 87° = ________

0.9986295

---

64. When we enter a "sine" between 0 and 1 on a calculator and press [INV] [sin] to find the angle, we always get an acute angle. For example:

Enter	Press	Display
0.9961947	[INV] [sin]	85.000001 (or 85°)
0.8386706	[INV] [sin]	57.000003 (or 57°)
0.3420201	[INV] [sin]	19.999997 (or 20°)

Remember that the above "sines" could also be the "sines" of obtuse angles. To find the obtuse angle, we subtract the acute angle from 180°. That is:

If sin A = 0.9961947, A = 85° or 95° (from 180°-85°)
If sin A = 0.8386706, A = 57° or 123° (from 180°-57°)

If sin A = 0.3420201, A = 20° or ________

160° (from 180°-20°)

65. When using the Law of Sines to find an angle, a calculator always displays an acute angle even when the angle is obtuse. <u>You have to check the diagram to see whether the angle is acute or obtuse</u>.

Let's apply the principle above to this triangle.

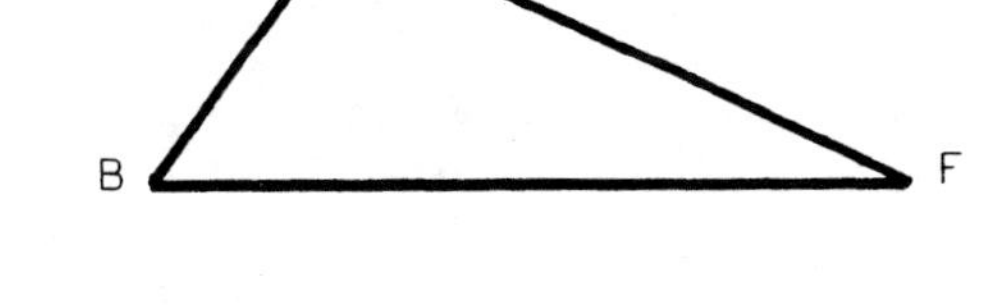

If sin D = 0.9781476 :

a) Angle D could be either ________° or ________°.

b) Since angle D is an <u>obtuse</u> angle, it must be ________°.

If sin B = 0.809017 :

c) Angle B could be either ________° or ________°.

d) Since angle B is an <u>acute</u> angle, it must be ________°.

---

66. Let's use the Law of Sines to find angle T.

a) Set up the proportion needed to solve for angle T.

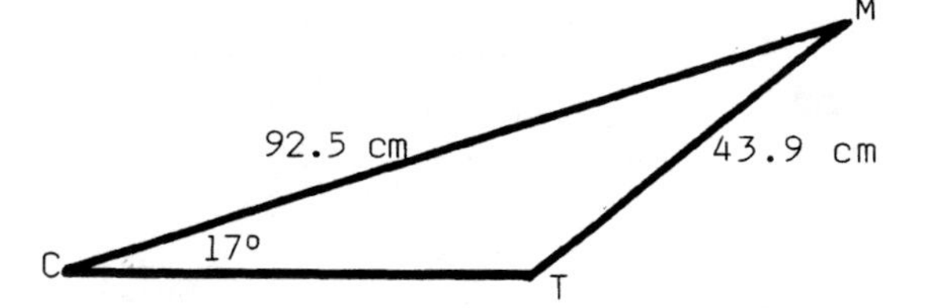

b) Angle T is an __________ (acute/obtuse) angle.

c) Rounded to a whole number, angle T = ________

Answers to 65: a) 78° or 102° b) 102° c) 54° or 126° d) 54°

---

67. We want to find angle P in this triangle. We cannot use the Law of Sines directly because side AK is unknown.

a) What procedure can we use to find P?

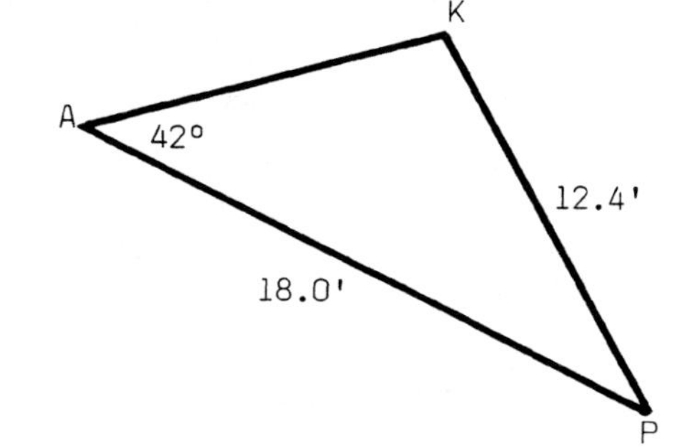

b) Find angle K. Rounded to a whole number, angle K = ________.

c) Find angle P. Angle P = ________

Answers to 66: a) $\frac{\sin T}{92.5\text{ cm}} = \frac{\sin 17°}{43.9\text{ cm}}$ b) obtuse c) 142°

Answers to 67: a) Use the Law of Sines to find angle K° Then use the angle-sum principle to find angle P. b) 104° c) 34°

2-10 OBTUSE ANGLES AND THE LAW OF COSINES

When using the Law of Cosines in an earlier section, we avoided triangles with obtuse angles. We will use the Law of Cosines with triangles of that type in this section.

---

68. To find the cosine of an obtuse angle on a calculator, we simply enter the angle and press [cos]. For example, we found the cosines of four angles (95°, 112°, 139°, and 174°) below.

Enter	Press	Display
95	[cos]	-0.0871557
112	[cos]	-0.3746066
139	[cos]	-0.7547096
174	[cos]	-0.9945219

The cosine of an obtuse angle is a negative number between 0 and -1. We will discuss why the cosines are negative in the next chapter.

---

69. To solve for side FT in this triangle, we must use the Law of Cosines. We get:

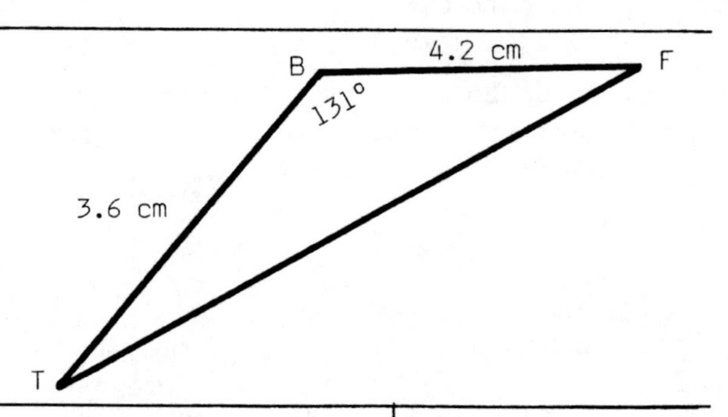

$$(FT)^2 = (BT)^2 + (BF)^2 - 2(BT)(BF)(\cos B)$$

$$FT = \sqrt{(3.6)^2 + (4.2)^2 - 2(3.6)(4.2)(\cos 131°)}$$

Use your calculator to complete the solution. Round to tenths. FT = ________

---

70. There is no confusion between the cosines of obtuse angles and the cosines of acute angles.

Obtuse angles have negative cosines.
Acute angles have positive cosines.

To find the angle corresponding to a negative cosine, we can use a calculator. Some examples are shown. Notice how we use [+/-] to enter a negative number.

Enter	Press	Display
0.1218693	[+/-] [INV] [cos]	96.999997
0.601815	[+/-] [INV] [cos]	127
0.9781476	[+/-] [INV] [cos]	168

Using the above results, complete these. Round to a whole number when necessary.

a) If cos B = -0.1218693, B = ________.

b) If cos D = -0.601815, D = ________.

c) If cos F = -0.9781476, F = ________.

7.1 cm

71. To solve for angle G in this triangle, we must use the Law of Cosines. We get:

$$g^2 = d^2 + h^2 - 2dh \cos G$$

a) Rearrange the formula to solve for cos G.

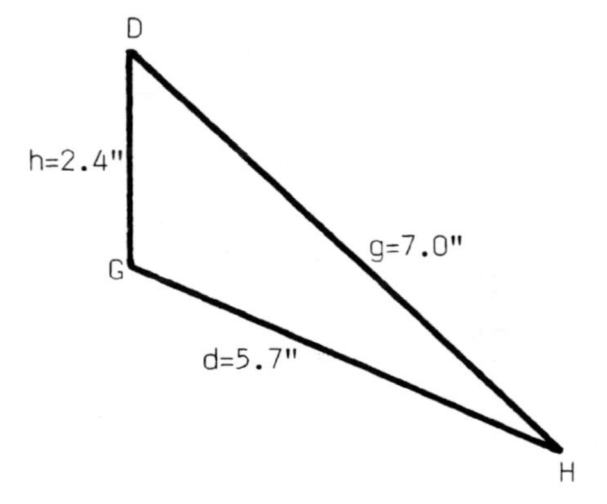

b) Substitute the known values in the formula.

cos G = ______________________

c) Use a calculator to find angle G. Rounded to a whole number, angle G = ________.

a) 97°

b) 127°

c) 168°

---

a) $\cos G = \dfrac{d^2 + h^2 - g^2}{2dh}$

b) $\cos G = \dfrac{(5.7)^2 + (2.4)^2 - (7.0)^2}{(2)(5.7)(2.4)}$

c) 113°

2-11 STRATEGIES FOR SOLVING TRIANGLES

In this section, we will review the strategies (or methods) used to solve right and oblique triangles. Exercises requiring strategy-identification are included.

---

72. Except for the angle-sum principle which can be used with all triangles, different strategies (or methods) are used to solve right and oblique triangles.

For right triangles: use one of the three trig ratios or the Pythagorean Theorem.

For oblique triangles: use either the Law of Sines or the Law of Cosines.

When solving oblique triangles, the Law of Sines is used in all cases except the two below for which the Law of Cosines is used.

1. When only two sides and their included angle are known.
2. When only three sides are known.

If you cannot remember the two special cases above, use this strategy.

WHEN SOLVING OBLIQUE TRIANGLES, TRY THE LAW OF SINES FIRST. IF THE LAW OF SINES DOES NOT WORK, USE THE LAW OF COSINES.

73. For any triangle, check to see whether the angle-sum principle can be used.

In which triangle(s) below can we use the angle-sum principle? ________

a)

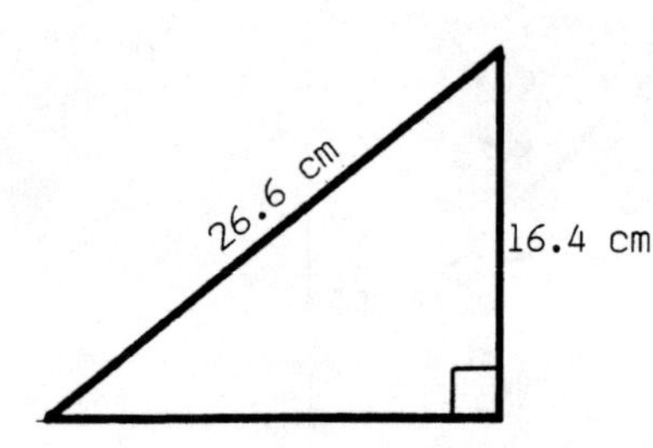

b)

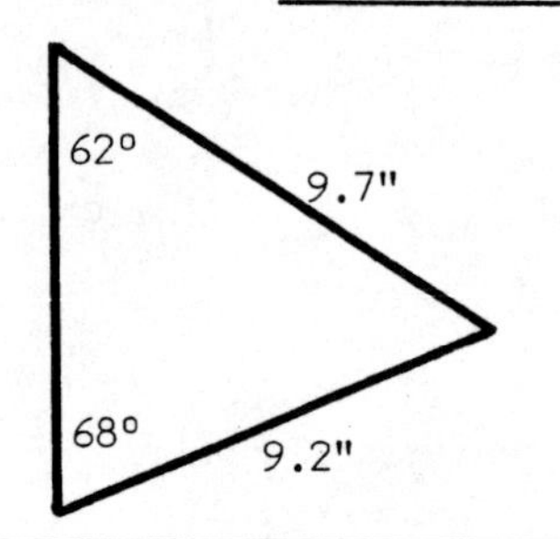

---

74. Check to see whether it is a right or oblique triangle. If it is a <u>right triangle</u>, think: <u>TRIG RATIOS AND PYTHAGOREAN THEOREM.</u>

In right triangle CDE, which trig ratio would you use?

a) To solve for "e"? ________

b) To solve for "d"? ________

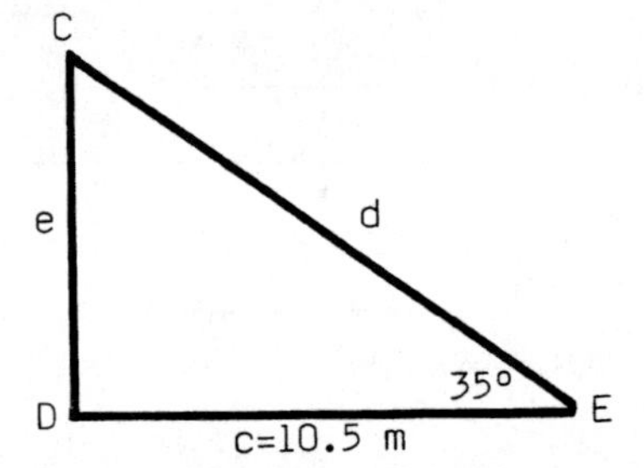

Only (b)

---

75. In triangle BDT, two sides are given.

a) How would you solve for "d"?

____

b) How would you solve for angle B?

____

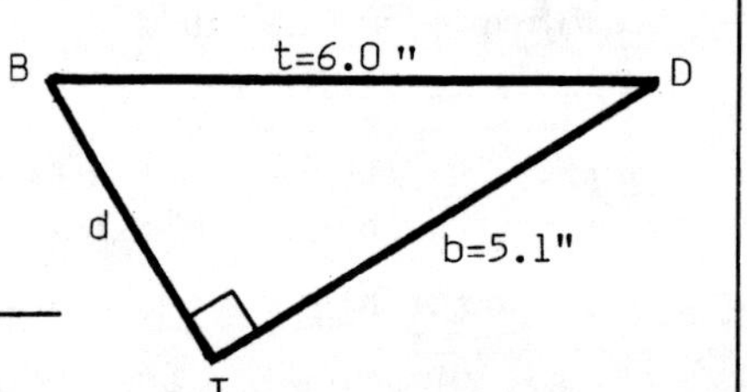

a) tan E or tan 35°

b) cos E or cos 35°

---

76. If it is an oblique triangle, think: <u>LAW OF SINES</u> or <u>LAW OF COSINES.</u>

In this oblique triangle:

a) To solve for side PR, would you use the Law of Sines or the Law of Cosines? ________

b) Write the equation you would use.

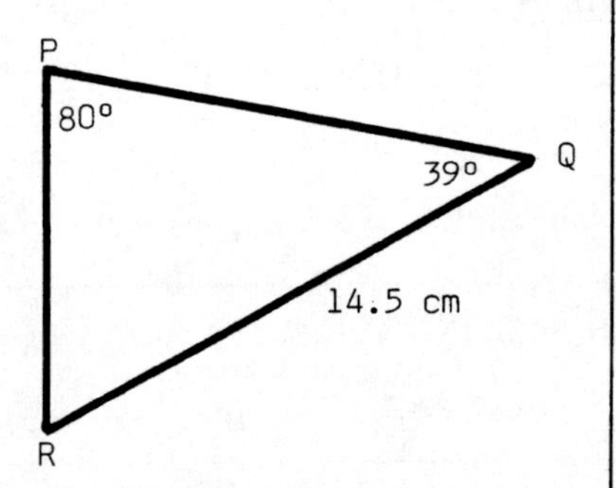

____

a) Use the Pythagorean Theorem.

b) Use sin B.

77. a) To solve for side CP, would you use the Law of Sines or the Law of Cosines? ______________________

b) Write the equation you would use.

____

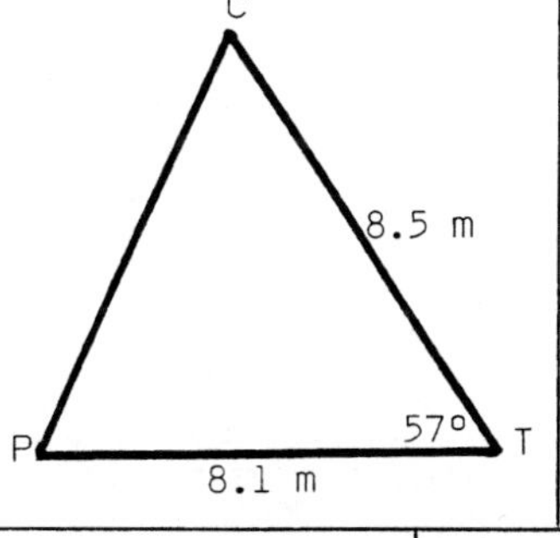

a) Law of Sines

b) $\dfrac{PR}{\sin Q} = \dfrac{QR}{\sin P}$

or

$\dfrac{PR}{\sin 39°} = \dfrac{14.5 \text{ cm}}{\sin 80°}$

---

78. a) To find angle M, which law would you use?

____

b) Write the equation you would use.

____

M t=24.5" P

m=17.8"

113°

T

a) Law of Cosines

b) $(CP)^2 = (CT)^2 + (PT)^2 - 2(CT)(PT)(\cos T)$

or

$(CP)^2 = (8.5)^2 + (8.1)^2 - 2(8.5)(8.1)(\cos 57°)$

---

79. a) To find angle A, which law would you use? ______________________

b) Write the equation you would use.

____

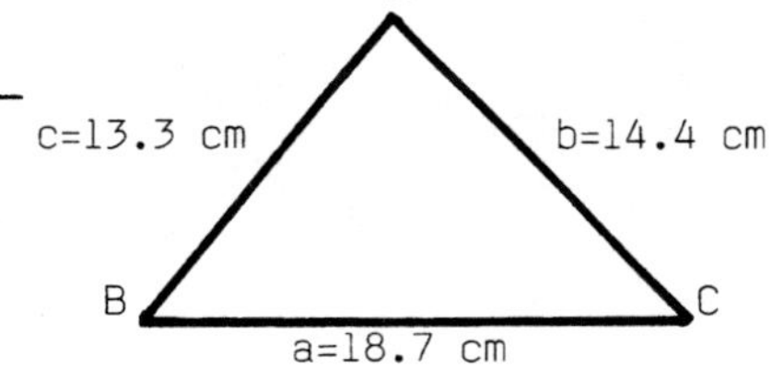

a) Law of Sines

b) $\dfrac{\sin M}{m} = \dfrac{\sin T}{t}$

or

$\dfrac{\sin M}{17.8"} = \dfrac{\sin 113°}{24.5"}$

---

80. To find angle V, two steps are needed. They are:

1. ______________

____

____

2. ______________________

____

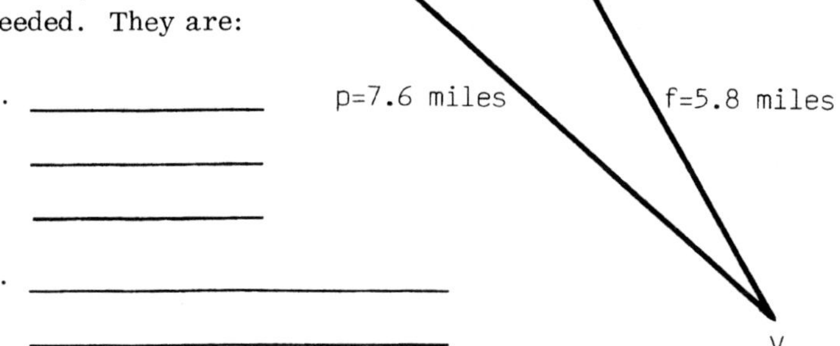

a) Law of Cosines

b) $a^2 = b^2 + c^2 - 2bc \cos A$

or

$\cos A = \dfrac{b^2 + c^2 - a^2}{2bc}$

or

$\cos A = \dfrac{(14.4)^2 + (13.3)^2 - (18.7)^2}{2(14.4)(13.3)}$

---

1. Find angle P using the Law of Sines.

2. Find angle V using the angle-sum principle.

81. To find side "q", three steps are needed. They are:

1. ______________________

____

2. ______________________

3. ______________________

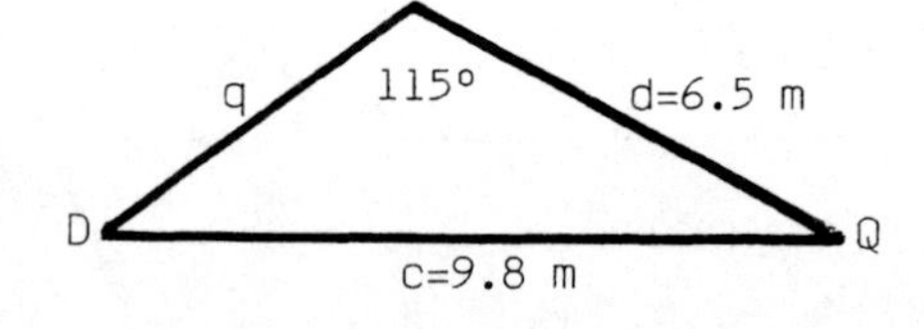

---

82. To find side CK, two steps are needed. They are:

1. ______________________

____

2. ______________________

____

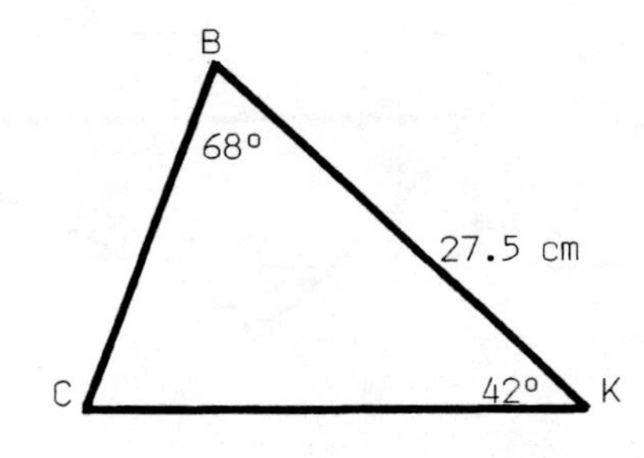

1. Find angle D using the Law of Sines.
2. Find angle Q using the angle-sum principle.
3. Find side "q" using the Law of Sines.

---

83. To find side PM, would we use the Law of Sines or the Law of Cosines?

____

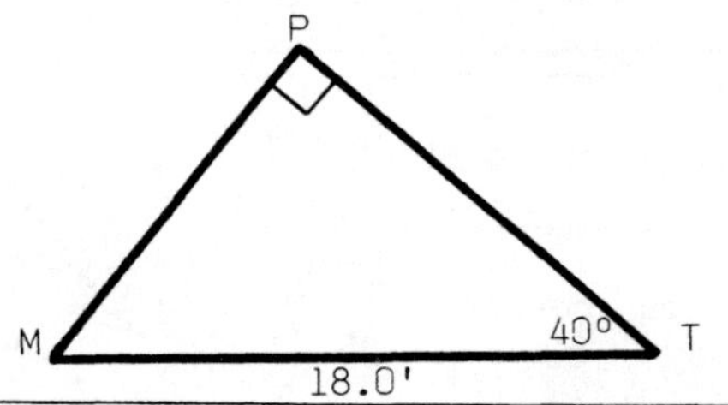

1. Find angle C using the angle-sum principle.
2. Then find CK using the Law of Sines.

---

84. We want to find two angles, A and T, in this triangle.

a) Let's find angle A first. Write the equation we should use.

____

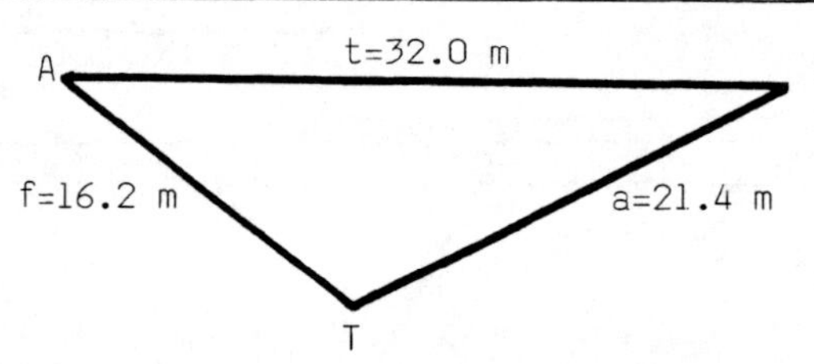

b) Having found angle A, we can find angle T in either of two ways. Write the two equations we could use.

1. ______________________

or 2. ______________________

Neither. It is a <u>right</u> triangle. We would use <u>sin T</u>.

---

a) $a^2 = f^2 + t^2 - 2ft \cos A$

or

$$\cos A = \frac{f^2 + t^2 - a^2}{2ft}$$

b) 1. Law of Sines:

$$\frac{\sin T}{t} = \frac{\sin A}{a}$$

2. Law of Cosines:

$$t^2 = a^2 + f^2 - 2af \cos T$$

or

$$\cos T = \frac{a^2 + f^2 - t^2}{2af}$$

<u>Note</u>: It is preferable to use the Law of Sines because it is simpler.

85. To find angle V, two steps are needed.

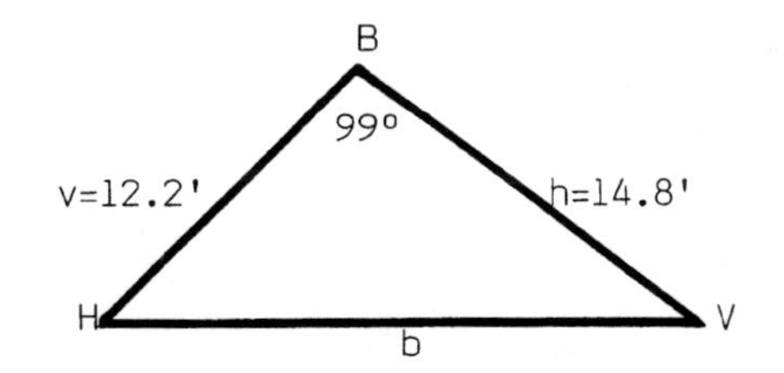

a) What is the first step?

____

b) Write the equation that would be used for the first step?

____

c) Knowing side "b", we can find angle V in either of two ways. Write the two equations we could use.

1. ______________________

or 2. ______________________

---

a) Find side "b".

b) $b^2 = h^2 + v^2 - 2hv \cos B$

c) 1. Law of Sines: $\frac{\sin V}{v} = \frac{\sin B}{b}$

2. Law of Cosines: $\cos V = \frac{b^2 + h^2 - v^2}{2bh}$

Note: It is preferable to use the Law of Sines because it is simpler.

2-12 APPLIED PROBLEMS

In this section we will discuss some applied problems that involve solving an oblique triangle.

---

86. A surveyor has to find the distance EF across a river. First he laid out line DE. Then he used a transit to measure angles D and E.

Find EF. Round to a whole number.

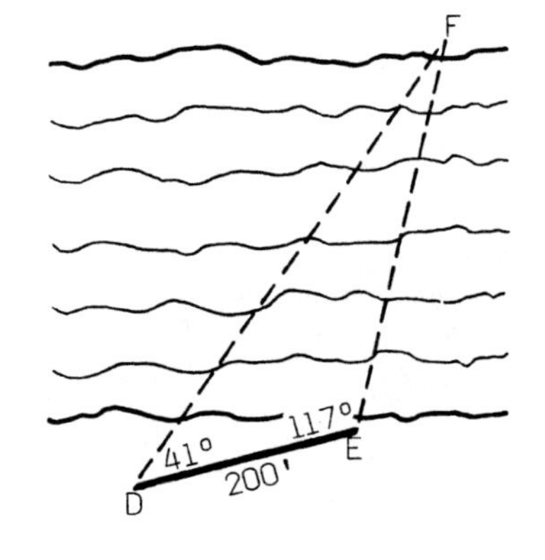

EF = ______________

87. The shape at the right is formed by steel girders.

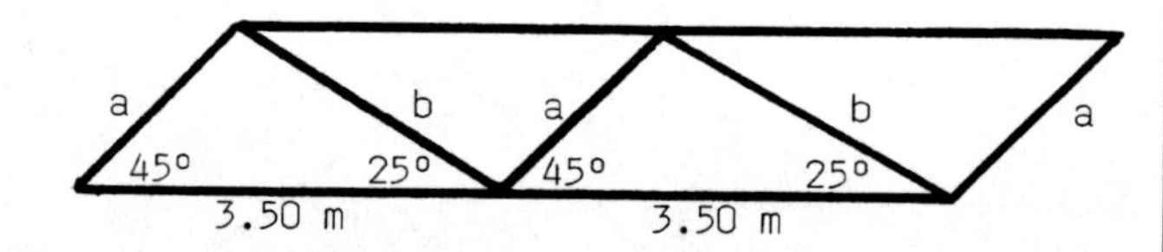

a) Find "a". Round to hundredths.

a = ________

b) Find "b". Round to hundredths.

b = ________

EF = 350 feet

The steps are:

1. Use the angle-sum principle to find angle F.
2. Then use the Law of Sines to find EF.

$$\frac{EF}{\sin D} = \frac{DE}{\sin F}$$

---

88. Two holes are drilled in a circular metal disc whose center is at point C. Find "d", the center-to-center distance between the two holes. Round to hundredths.

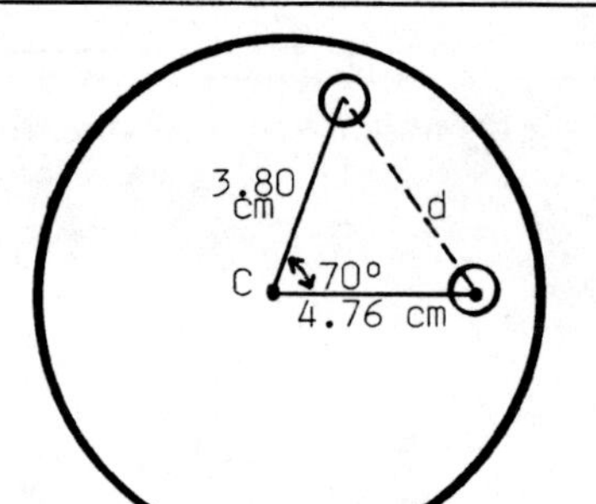

d = ________

a) a = 1.57m, from:

$$\frac{a}{\sin 25°} = \frac{3.50\text{m}}{\sin 110°}$$

b) b = 2.63m, from:

$$\frac{b}{\sin 45°} = \frac{3.50\text{m}}{\sin 110°}$$

---

89. Three steps are needed to find PT, the length of a lake. They are:

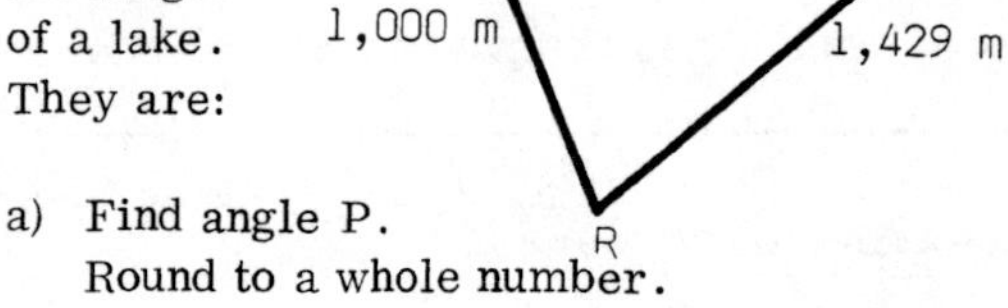

a) Find angle P. Round to a whole number.

P = ________

b) Find angle R. Round to a whole number.

R = ________

c) Find PT. Round to a whole number.

PT = ________

d = 4.97 cm, from:

$$d = \sqrt{(3.80)^2 + (4.76)^2 - 2(3.80)(4.76)(\cos 70°)}$$

90. Three holes are drilled in a rectangular plate. Their center-to-center distances are given. Find angle A. Round to a whole number.

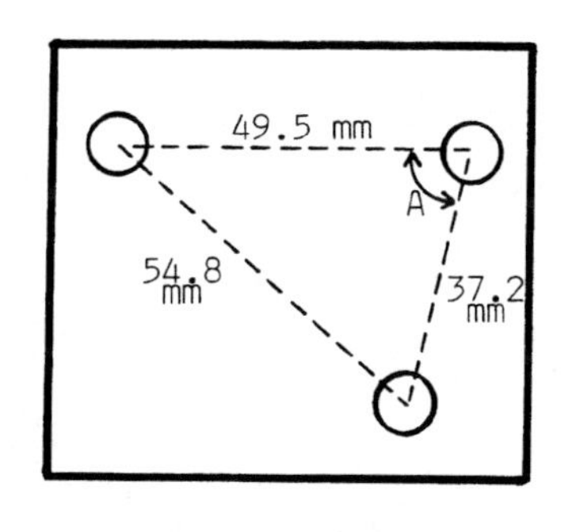

A = ________

a) P = 64°, from:

$$\frac{\sin P}{1429\text{m}} = \frac{\sin 39°}{1000\text{m}}$$

b) R = 77°, from:

180° - (64° + 39°)

c) PT = 1,548 meters, from:

$$\frac{PT}{\sin 77°} = \frac{1000}{\sin 39°}$$

---

91. The top and bottom edges of the metal template at the left below are parallel. We want to find side "s". To do so, we draw a dashed line parallel to "s" in the figure at the right. By doing so, we formed an oblique triangle.

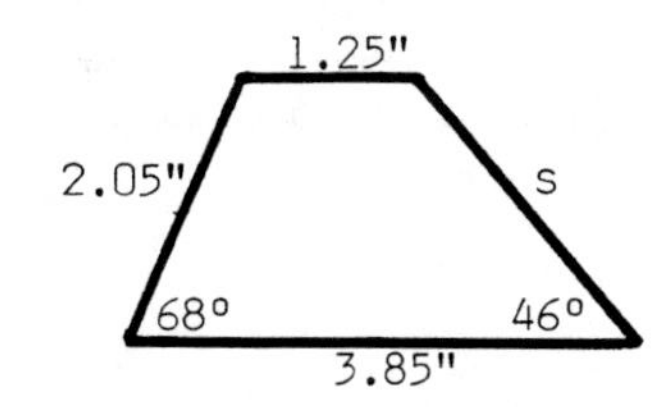

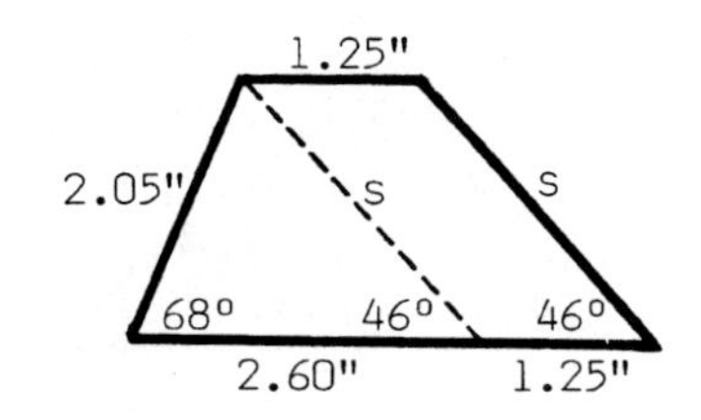

Find side "s". Round to hundredths.

s = ________

A = 77°, from:

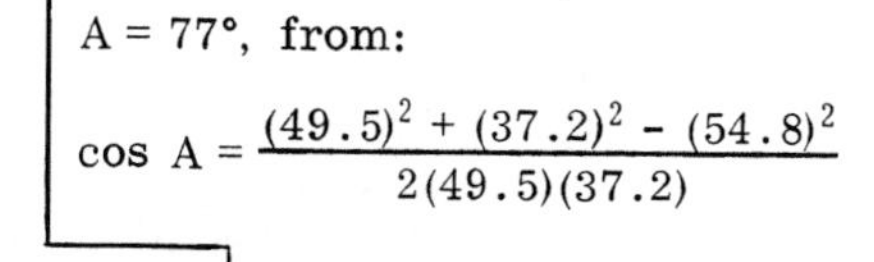

$$\cos A = \frac{(49.5)^2 + (37.2)^2 - (54.8)^2}{2(49.5)(37.2)}$$

---

s = 2.64", from:

$$s^2 = (2.05)^2 + (2.60)^2 - 2(2.05)(2.60)(\cos 68)°$$

or

$$\frac{s}{\sin 68°} = \frac{2.05''}{\sin 46°}$$

Note: Using the Law of Sines is preferable.

SELF-TEST 6 (pp. 68-80)

1. Find angle P. Round to a whole number.

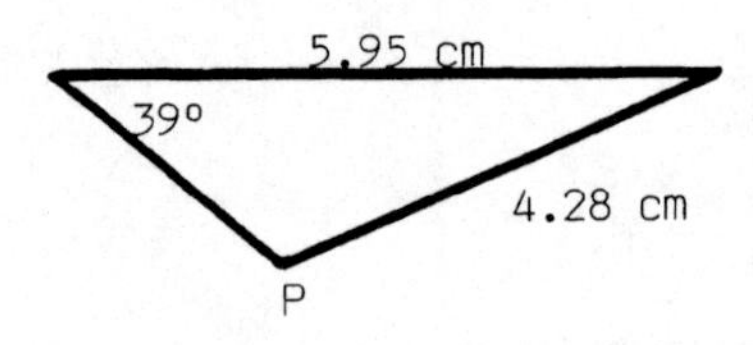

P = ________

2. Find side "d". Round to tenths.

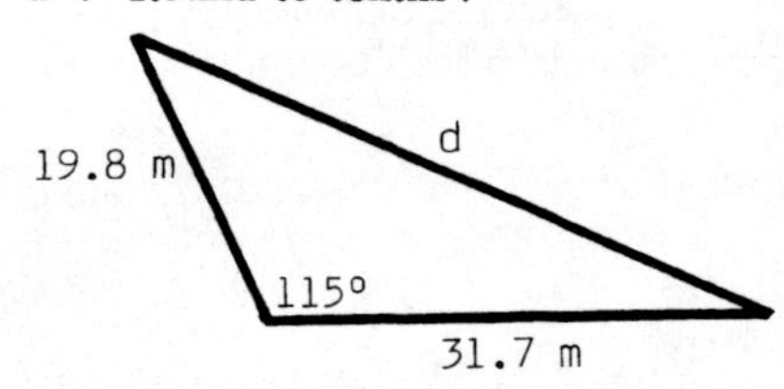

d = ________

Which principle at the right (a, b, or c) would be used to find the unknown angle or side represented by a single letter in each triangle below?

a) Law of Sines
b) Law of Cosines
c) Angle-Sum Principle

3.

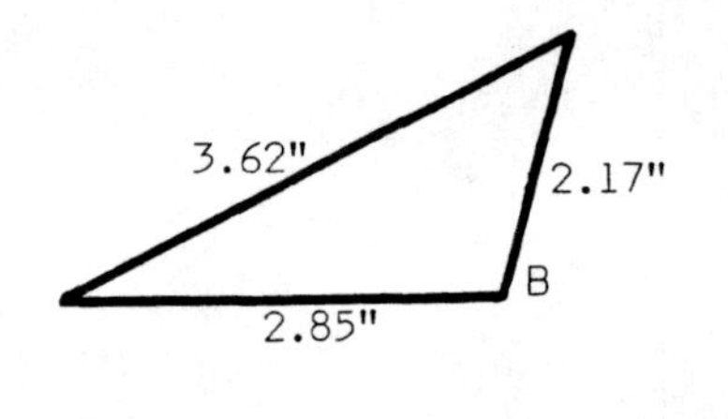

______

4.

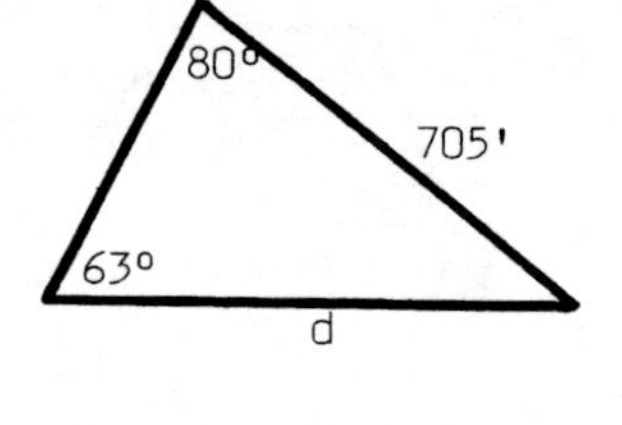

______

5.

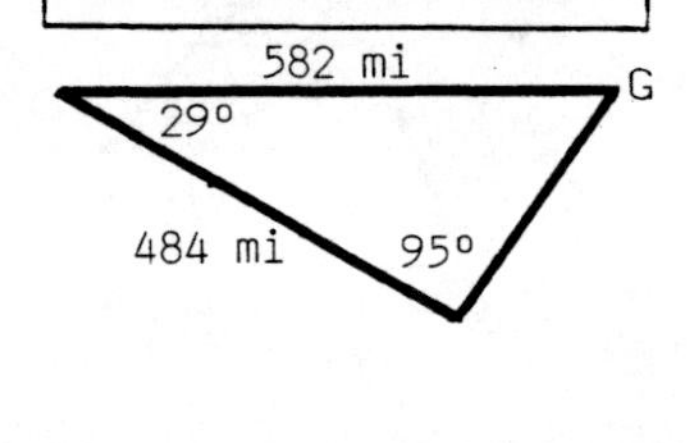

______

6. Find "h", the height of the cliff shown in the diagram. Round to a whole number.

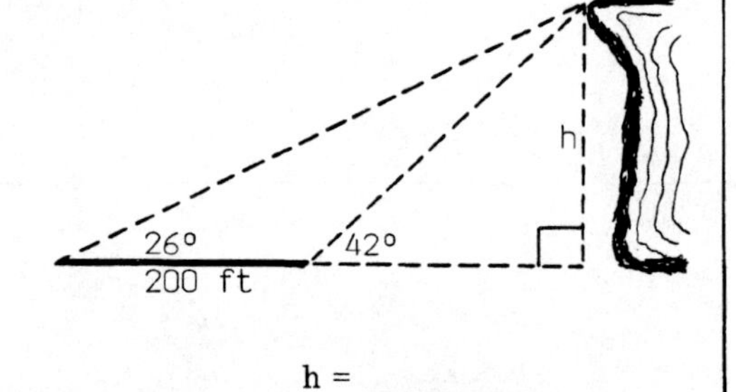

h = ________

ANSWERS:

1. P = 119°
2. d = 43.9m
3. b
4. a
5. c
6. h = 213 ft

SUPPLEMENTARY PROBLEMS - CHAPTER 2

Assignment 4

1. Which of the following are obtuse angles? ________

   a) 100° b) 96° c) 87° d) 90° e) 24° f) 172°

Find the third angle in each triangle.

2. Triangle #1: 58°, 25°, ________
3. Triangle #2: 19°, 71°, ________
4. Triangle #3: 88°, 44°, ________
5. Triangle #4: 24°, 21°, ________

State whether each triangle above is: a) acute oblique, b) obtuse oblique, or c) right.

6. #1: ______ 7. #2: ______ 8. #3: ______ 9. #4: ______

In the oblique triangle below:

10. The longest side is side ______.
11. The shortest side is side ______.

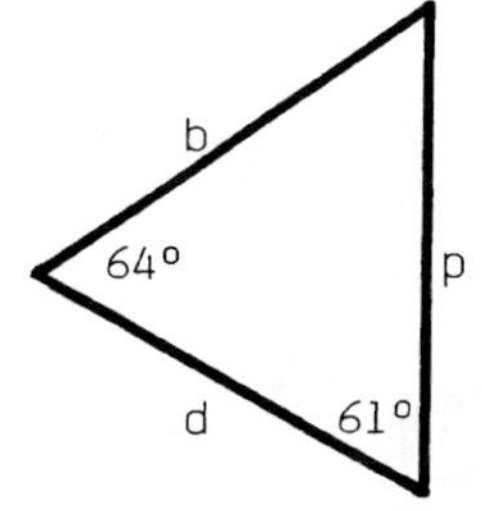

In the oblique triangle below:

12. The largest angle is angle ______.
13. The smallest angle is angle ______.

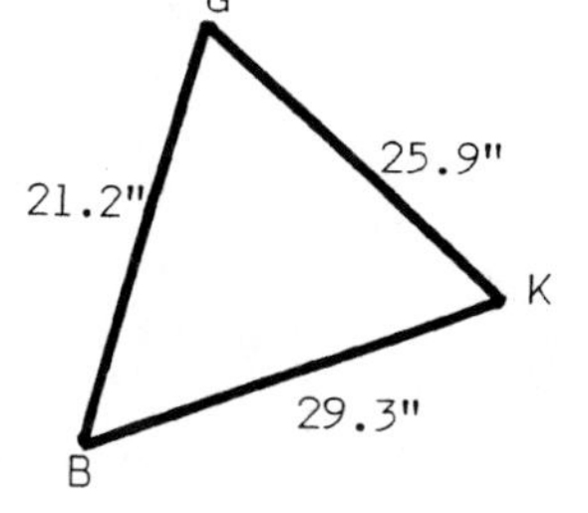

Referring to triangle RST, use the Law of Sines to complete these proportions.

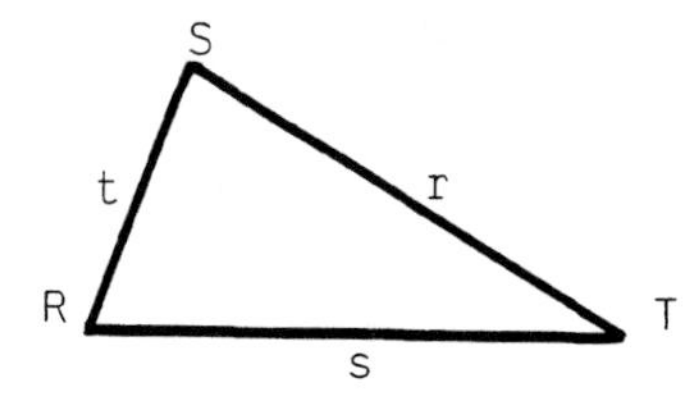

14. 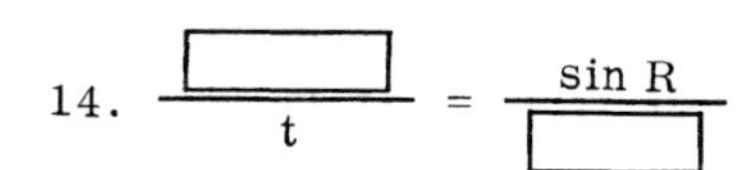

15. 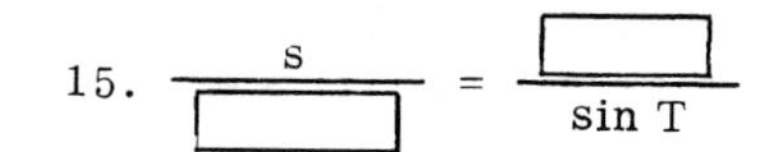

In triangle ABC, find side "a", angle B, and side "b". Round each to a whole number.

16. a = ______
17. B = ______
18. b = ______

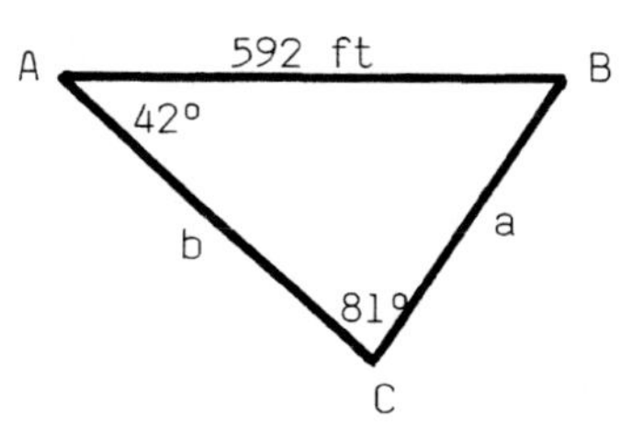

In triangle HKP, find angle H, angle K, and side "k". Round each angle to a whole number. Round the side to tenths.

19. H = ______
20. K = ______
21. k = ______

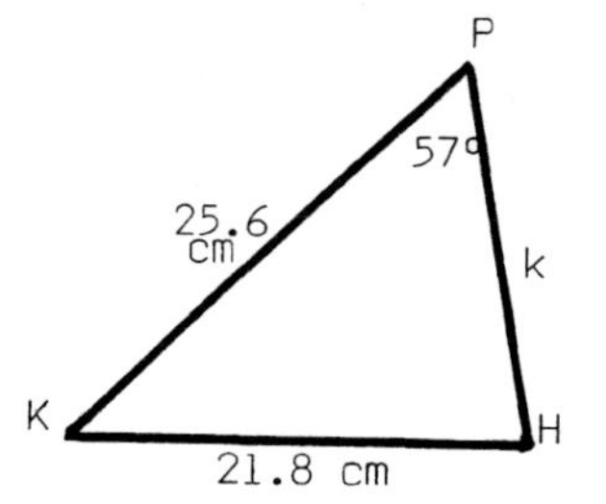

Assignment 5

State whether the Law of Sines or the Law of Cosines would be used to find the unknown angle or side represented by a single letter in each triangle below.

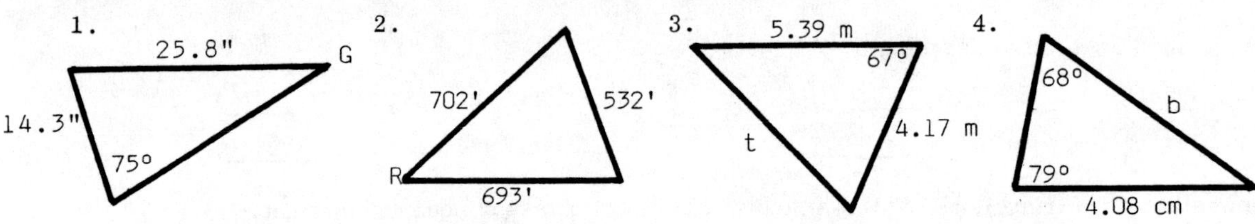

Referring to triangle PRV, complete these formulas using the Law of Cosines.

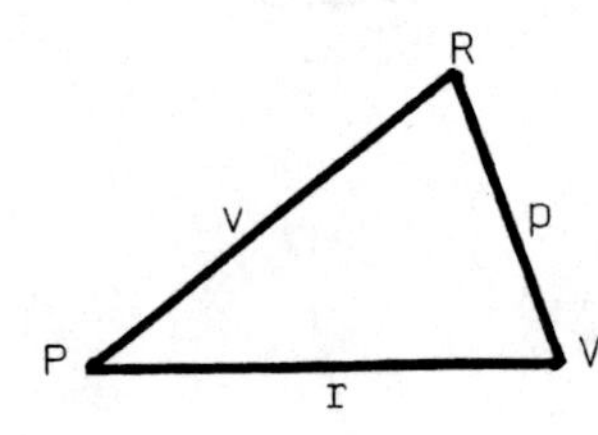

5. p^2 = ____________________

6. v^2 = ____________________

Referring to triangle ABC, complete these formulas using the Law of Cosines.

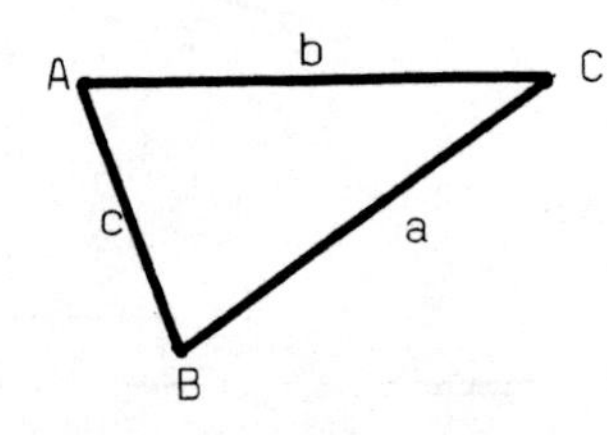

7. cos A = ____________

8. cos C = ____________

In triangle DEF, find side "f", angle D, and angle E. Round the side to tenths. Round each angle to a whole number.

9. f = ________

10. D = ________

11. E = ________

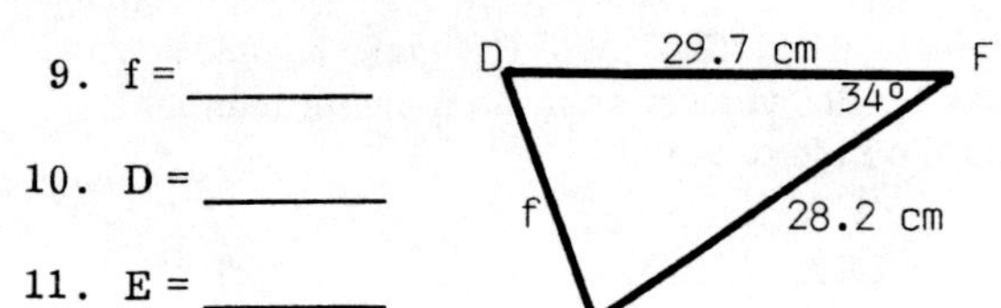

In triangle GPT, find angles G, P, and T. Round each angle to a whole number.

12. G = ________

13. P = ________

14. T = ________

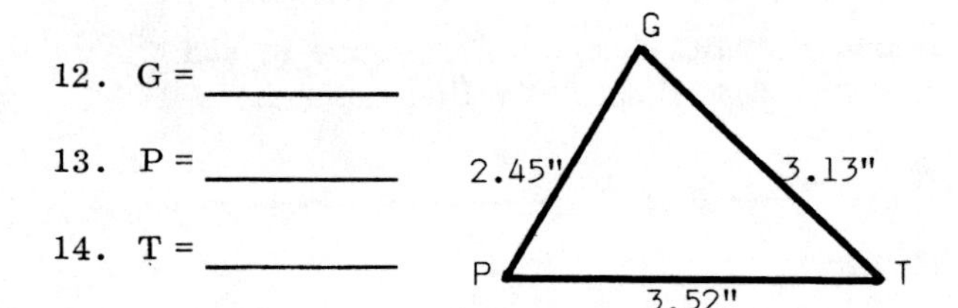

Assignment 6

Using a calculator, find each value. Round to four decimal places.

1. sin 152°   2. sin 94°   3. cos 103°   4. cos 179°

In Problems 5-7, each angle is obtuse. Find each angle. Round to a whole number.

5. If sin P = 0.3256, P = ______

6. If cos F = -0.6691, F = ______

7. If sin A = 0.9945, A = ______

8. In triangle RST, find side "s". Round to tenths.

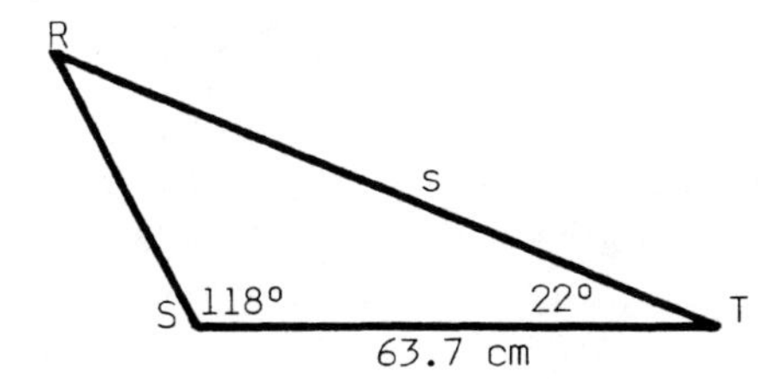

9. In triangle MNP, find angle N. Round to a whole number.

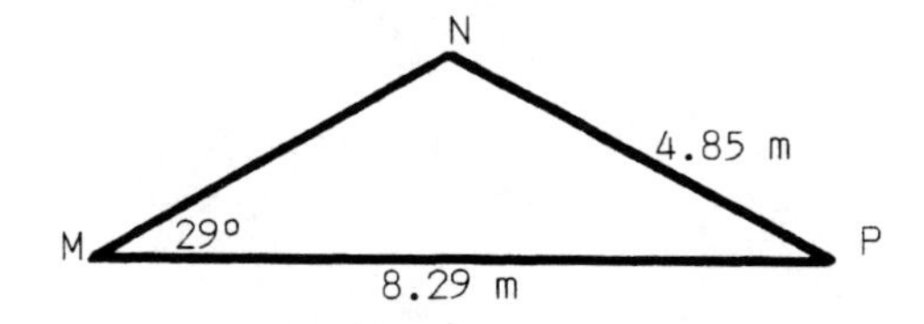

10. In triangle DEF, find side "f". Round to a whole number.

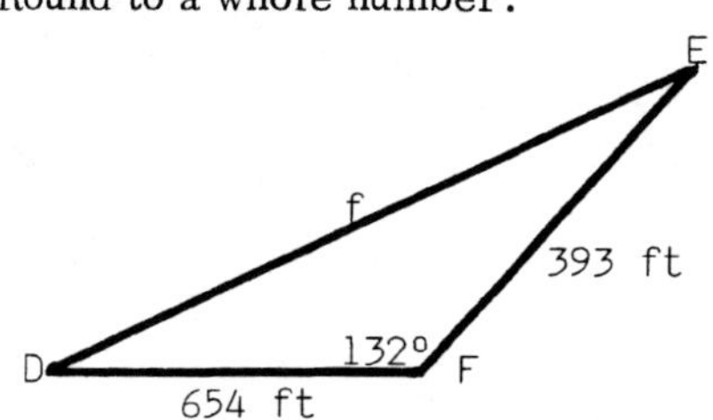

11. In triangle GHK, find angle H. Round to a whole number.

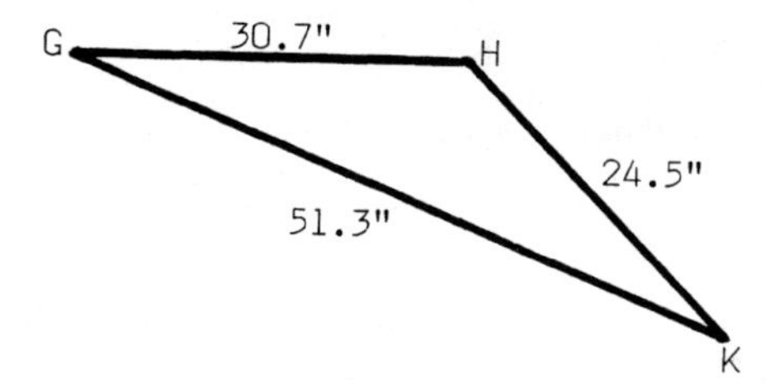

12. To find distance "d" across a swamp, a surveyor made the measurements shown. Find "d". Round to a whole number.

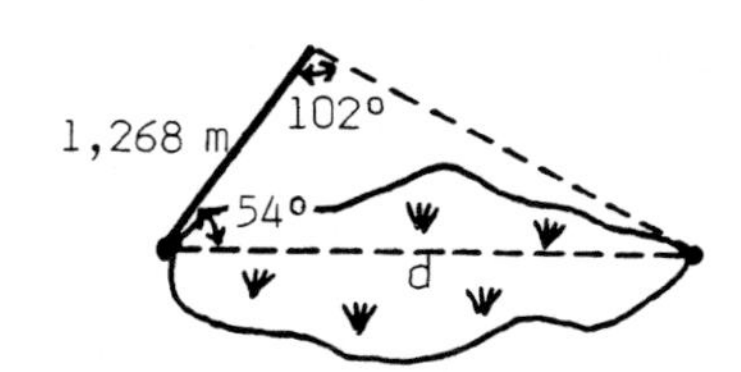

13. The center-to-center distances of three holes in a metal plate are shown. Find angle A and dimension "h". Round A to a whole number. Round "h" to hundredths.

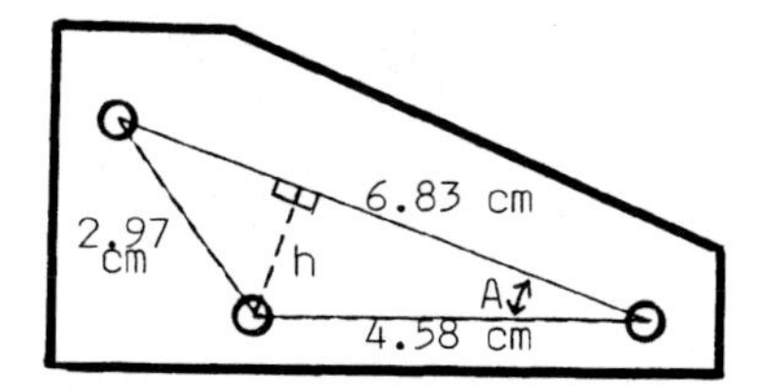

Chapter 3 TRIGONOMETRIC FUNCTIONS

In this chapter, we will show how positive and negative angles of any size can be generated by rotations on the coordinate system. Three trigonometric functions (sine, cosine, and tangent) for angles of any size are discussed and graphed. Degree and radian measures of angles are defined and conversions between degrees and radians are included.

3-1 STANDARD POSITION FOR ANGLES FROM 0° TO 360°

To define the trigonometric ratios of angles of any size, we put the angles in standard position. That is, we generate them by a rotation on the coordinate system. In this section, we will discuss the standard position of angles from 0° to 360°.

---

1. The two axes divide the coordinate system into four parts called "quadrants". The four quadrants are numbered at the right.

y
Quadrant 2 Quadrant 1
x
Quadrant 3 Quadrant 4

a) The quadrants are numbered in a ________________ (clockwise/counterclockwise) direction.

b) The x-axis is the ________________ (horizontal/vertical) axis.

---

2. A 40° angle is shown in standard position at the right. Notice these points:

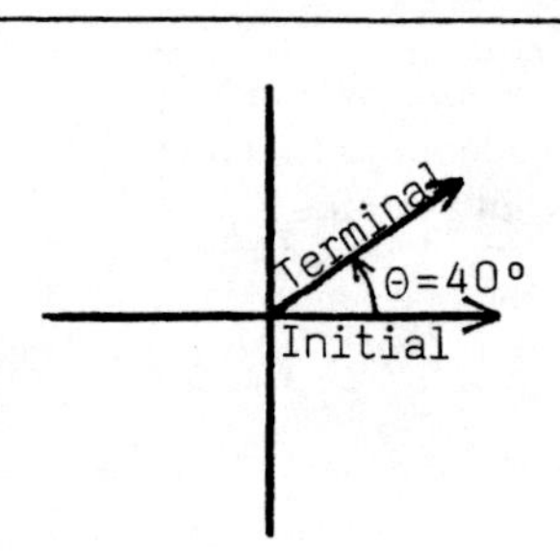

1. We use "θ" as the symbol for an angle in standard position. "θ" is the Greek letter theta. It is pronounced "thay-ta".
2. Both the initial and terminal sides of θ are arrows called "vectors".
3. The initial side of θ is the right side of the x-axis.
4. The curved arrow shows that θ was generated by rotating the initial side in a counterclockwise direction.

a) counterclockwise
b) horizontal

3. A 140° angle is shown in standard position at the right. The curved arrow shows the path of rotation.

   a) The direction of rotation is ________________ (clockwise/ counterclockwise.

   b) The terminal side lies in what quadrant? ________________

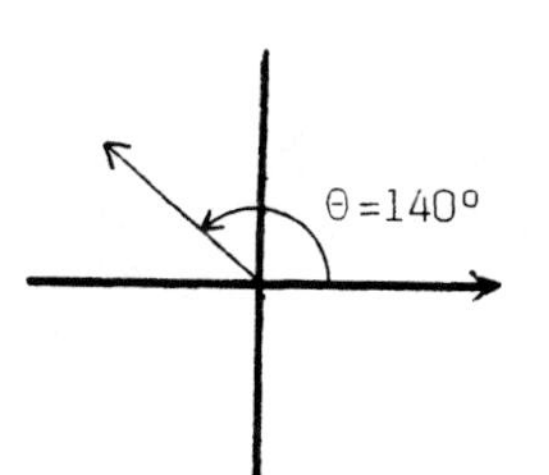

---

4. A 230° angle is shown in standard position at the right.

   a) The angle was generated by a rotation in a ________________ (clockwise/counterclockwise) direction.

   b) The terminal side lies in what quadrant? ________________

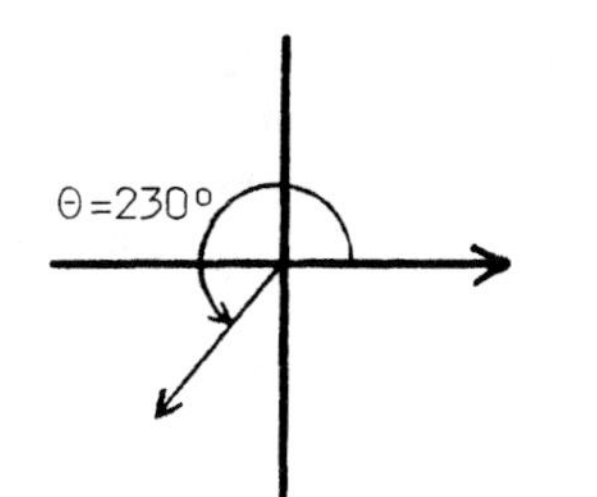

a) counterclockwise

b) Quadrant 2

---

5. A 320° angle is shown in standard position at the right.

   a) The initial side is the right side of the ____________ (x-axis/y-axis).

   b) The terminal side lies in what quadrant? ________________

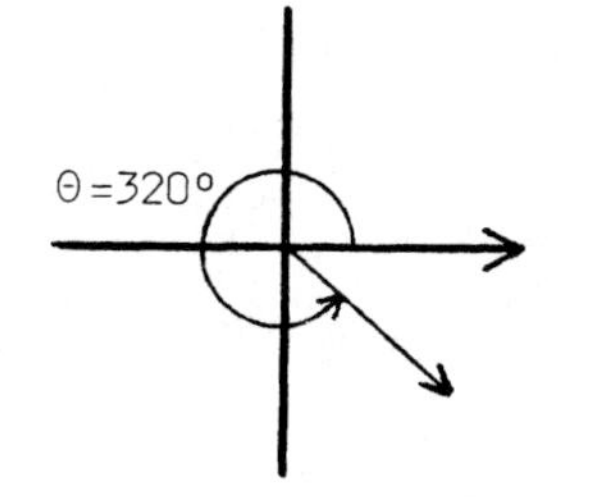

a) counterclockwise

b) Quadrant 3

---

6. These angles (90°, 180°, 270°) are shown in standard position below. The terminal side of each angle is on an axis.

θ=90°

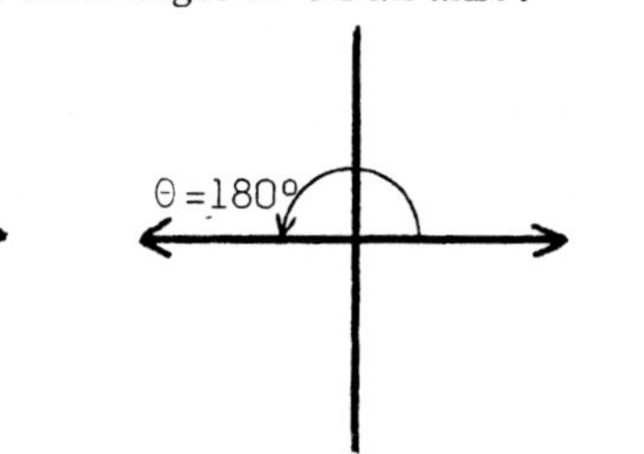

θ=270°

a) x-axis

b) Quadrant 4

Two angles (0°, 360°) are shown in standard position below. The terminal side of each angle coincides with the initial side.

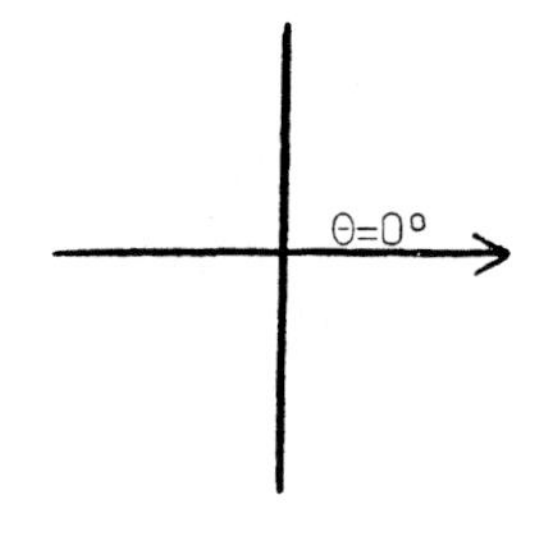

θ=360°

<u>Note</u>: 1. A 360° angle is one complete revolution.
2. A 0° angle has <u>no</u> <u>rotation</u> <u>at</u> <u>all</u>.

7. The angles in each quadrant are given below. Each quadrant contains 90°.

Quadrant 1	0° to 90°
Quadrant 2	90° to 180°
Quadrant 3	180° to 270°
Quadrant 4	270° to 360°

In which quadrant (1, 2, 3, or 4) does the terminal side of each angle lie?

a) 75° _____ b) 193° _____ c) 93° _____ d) 291° _____

---

a) 1

b) 3

c) 2

d) 4

8. An angle with its terminal side in the first quadrant is called a "first-quadrant angle". Similarly, angles with terminal sides in the second, third, or fourth quadrants are called "second-quadrant angles", "third-quadrant angles", or "fourth-quadrant angles".

Which of the following are "second-quadrant angles"? _____________

a) 317° b) 177° c) 241° d) 99° e) 88°

---

b) and d)

3-2 REFERENCE ANGLES FOR ANGLES FROM 0° TO 360°

Any angle in standard position has a reference angle. In this section, we will discuss the reference angles for angles from 0° to 360°.

---

9. A 120° angle is shown in standard position at the right. Angle α is its reference angle. Notice these points:

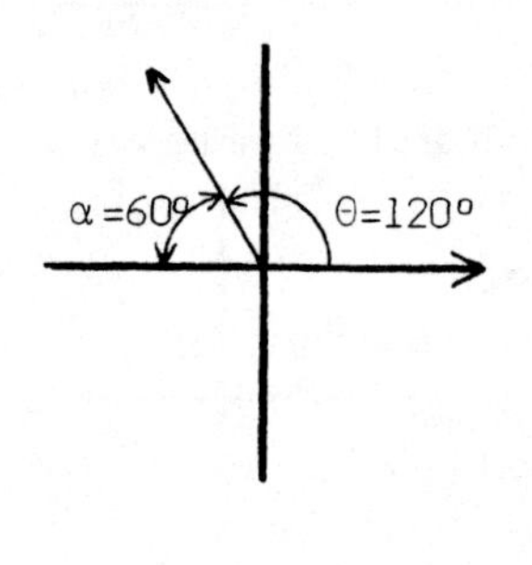

1. Angle α is the acute angle formed by the terminal side of the 120° angle and the horizontal axis.
2. We use "α" as the symbol for any reference angle. "α" is the Greek letter alpha. It is pronounced "al-fa".

a) How large is α, the reference angle? __________

b) The sum of θ and α is __________

---

a) 60° b) 180°

10. The terminal side of each angle below is in the second quadrant. α is the reference angle for each.

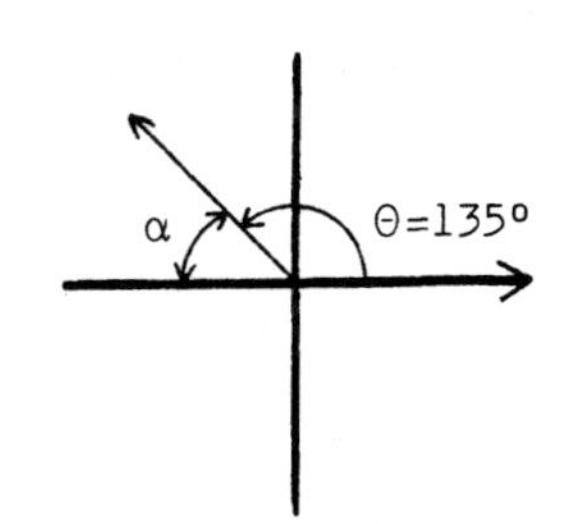

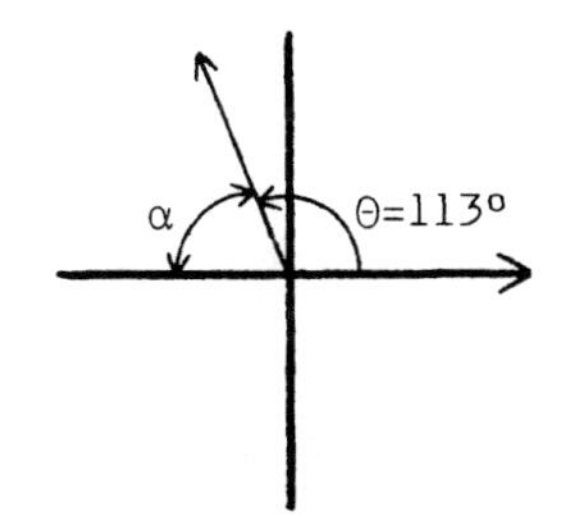

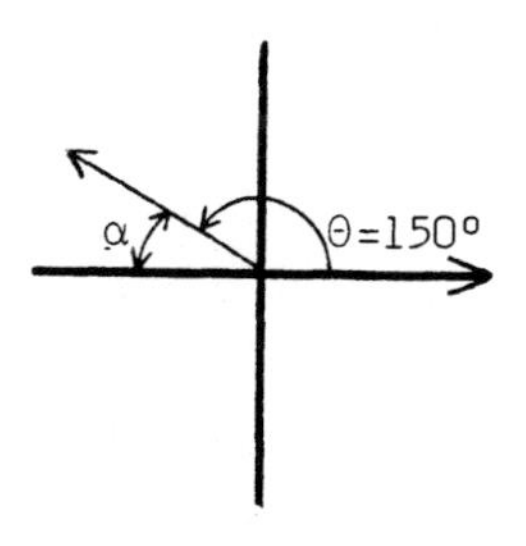

The sum of any second-quadrant angle and its reference angle is 180°. That is:

$$\theta + \alpha = 180°$$

Therefore, for any second-quadrant angle, we can find the size of α by subtracting θ from 180°. That is:

$$\alpha = 180° - \theta$$

Therefore: For $\theta = 135°$, $\alpha = 180° - 135° = 45°$

a) For $\theta = 113°$, $\alpha = 180° - 113° =$ ________

b) For $\theta = 150°$, $\alpha = 180° - 150° =$ ________

a) 67°	b) 30°

11. By subtracting θ from 180°, find the reference angle for these second-quadrant angles.

a) If $\theta = 94°$, $\alpha =$ ________

b) If $\theta = 129°$, $\alpha =$ ________

c) If $\theta = 178°$, $\alpha =$ ________

Answers (10): a) 67° b) 30°

12. A 230° angle is shown in standard position at the right. Angle α is its reference angle. Notice that α is the acute angle formed by the terminal side of the 230° angle and the horizontal axis.

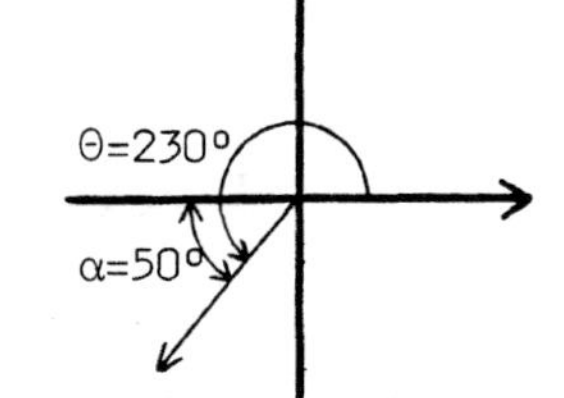

How large is α, the reference angle? ________

Answers (11): a) 86° b) 51° c) 2°

13. The terminal side of each angle below is in the third quadrant. α is the reference angle for each.

Answer (12): 50°

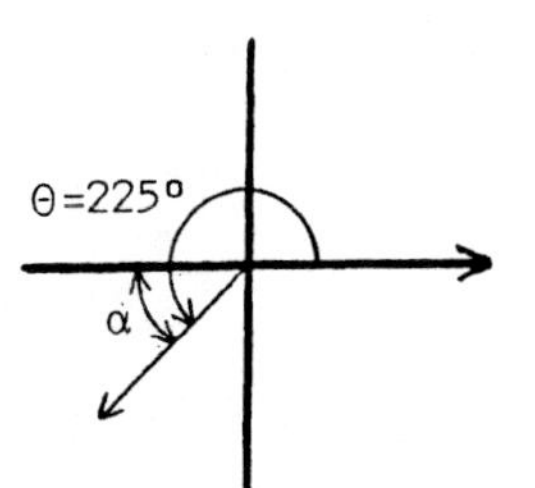

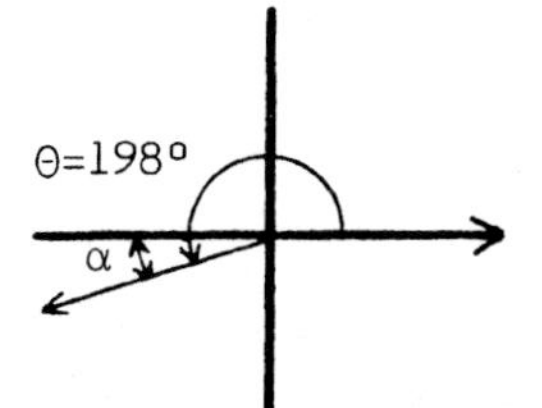

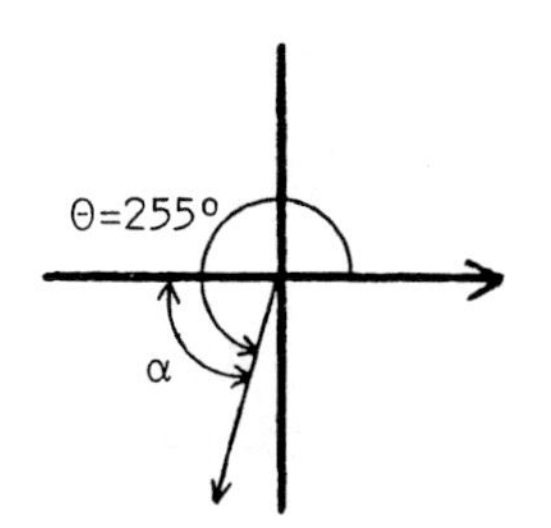

Continued on following page.

13. Continued

For any third-quadrant angle, we can find the size of α by subtracting 180° from θ. That is:

$$\alpha = \theta - 180°$$

Therefore: For $\theta = 225°$, $\alpha = 225° - 180° = 45°$

a) For $\theta = 198°$, $\alpha = 198° - 180° =$ ______

b) For $\theta = 255°$, $\alpha = 255° - 180° =$ ______

---

14. By subtracting 180° from θ, find the reference angle for these third-quadrant angles.

a) If $\theta = 185°$, $\alpha =$ ______ b) If $\theta = 268°$, $\alpha =$ ______ c) If $\theta = 237°$, $\alpha =$ ______

Answers: a) 18° b) 75°

---

15. A 310° angle is shown in standard position at the right. Angle α is its reference angle. Notice that α is the acute angle formed by the terminal side of the 310° angle and the horizontal axis.

θ=310° α=50°

a) How large is α, the reference angle? ______

b) The sum of θ and α is ______.

Answers: a) 5° b) 88° c) 57°

---

16. The terminal side of each angle below is in the fourth quadrant. α is the reference angle for each.

Answers: a) 50° b) 360°

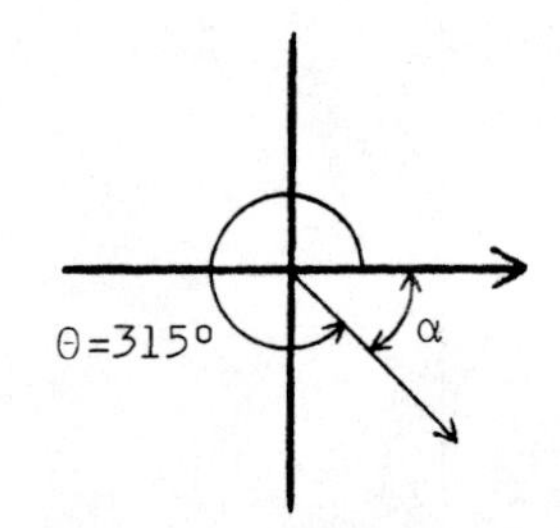

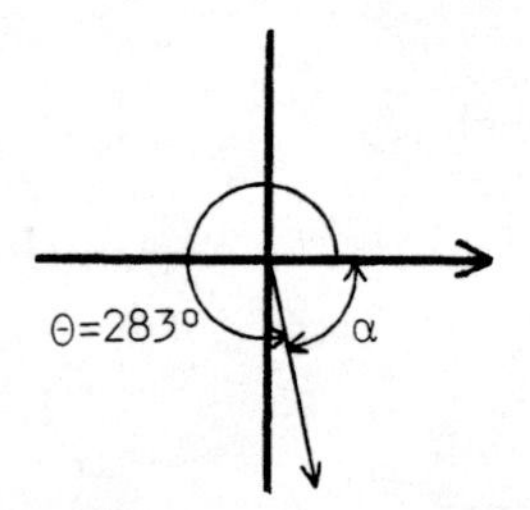

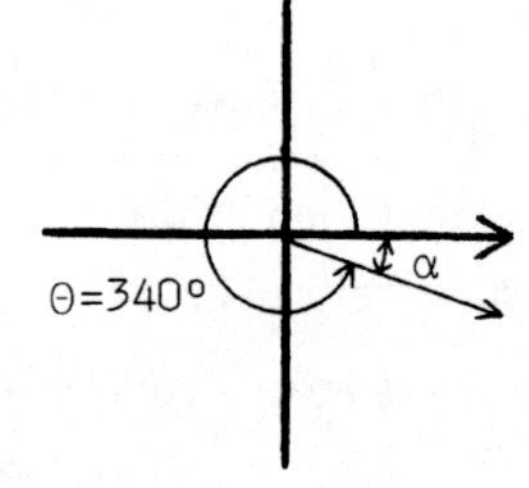

The sum of any fourth-quadrant angle and its reference angle is 360°. That is:

$$\theta + \alpha = 360°$$

Therefore, for any fourth-quadrant angle, we can find the size of α by subtracting θ from 360°. That is:

$$\alpha = 360° - \theta$$

Therefore: For $\theta = 315°$, $\alpha = 360° - 315° = 45°$

a) For $\theta = 283°$, $\alpha = 360° - 283° =$ ______

b) For $\theta = 340°$, $\alpha = 360° - 340° =$ ______

Answers: a) 77° b) 20°

17. By subtracting θ from 360°, find the reference angle for these fourth-quadrant angles.

a) For $\theta = 275°$, $\alpha =$ ________ b) For $\theta = 327°$, $\alpha =$ ________ c) For $\theta = 351°$, $\alpha =$ ________

---

a) 85°
b) 33°
c) 9°

18. We have sketched an angle in standard position in the second, third, and fourth quadrants below.

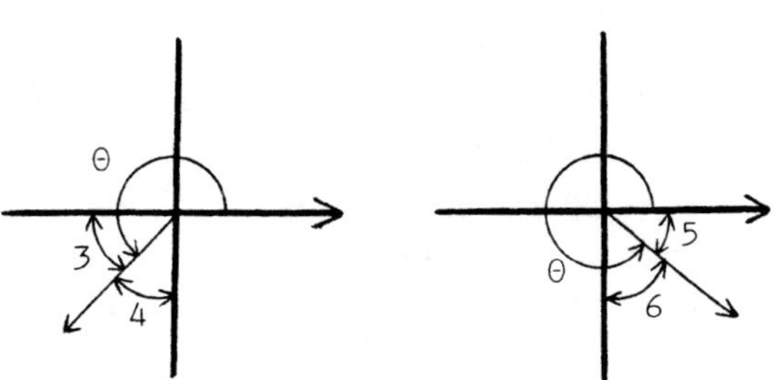

Reference angles are formed by the terminal side of θ and the <u>horizontal</u> axis, <u>not the vertical axis</u>. That is:

For the second-quadrant angle, α is angle 2 (not angle 1).

a) For the third-quadrant angle, α is angle ______.

b) For the fourth-quadrant angle, α is angle ______.

---

a) 3 (not angle 4)
b) 5 (not angle 6)

19. Three angles in standard position are sketched below.

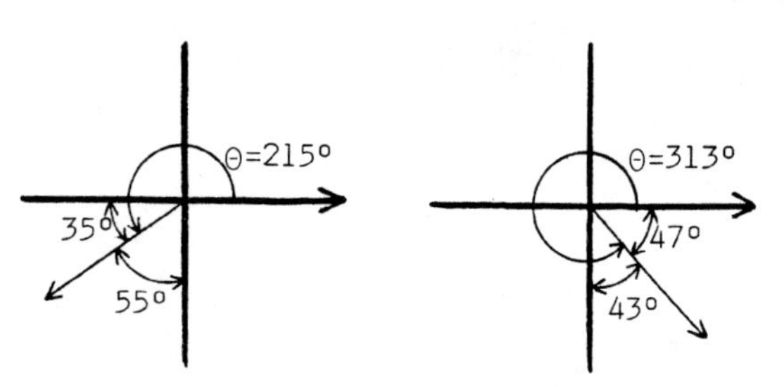

For the second-quadrant angle, $\alpha = 50°$ (not 40°).

a) For the third-quadrant angle, $\alpha =$ ________.

b) For the fourth-quadrant angle, $\alpha =$ ________.

---

a) 35° (not 55°)
b) 47° (not 43°)

20. The terminal side of θ at the right is in the first quadrant.

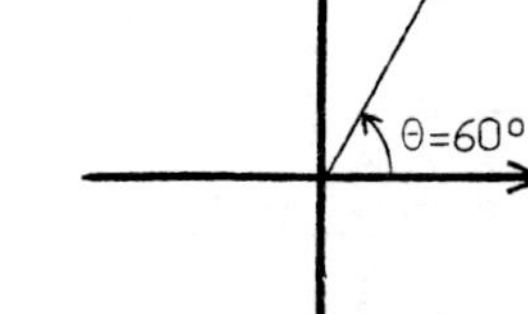

For any first-quadrant angle, α is the same as the angle. That is:

$$\alpha = \theta$$

Therefore: When $\theta = 60°$, $\alpha = 60°$

a) When $\theta = 21°$, $\alpha =$ ________

b) When $\theta = 7°$, $\alpha =$ ________

21. Reference angle α is 45° for each angle below.

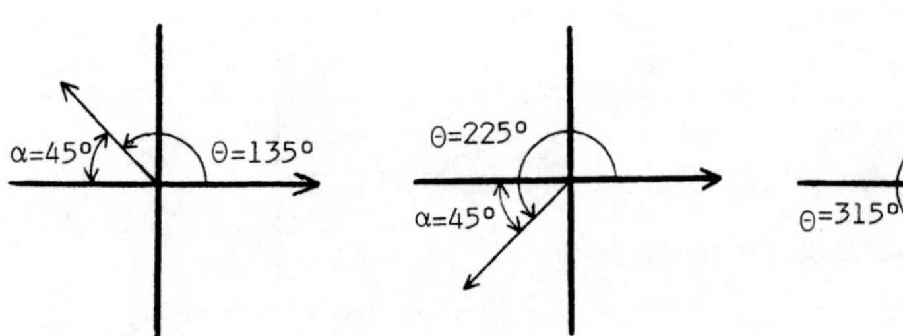

To distinguish between the quadrants when specifying reference angles, we use the following abbreviations:

Q1 for Quadrant 1
Q2 for Quadrant 2
Q3 for Quadrant 3
Q4 for Quadrant 4

Therefore: When $\theta = 135°$, $\alpha = 45°$ (Q2)

When $\theta = 225°$, $\alpha = 45°$ (Q3)

When $\theta = 315°$, $\alpha =$ ______________

a) 21°

b) 7°

---

22. We have summarized below the formulas for the reference angles in the four quadrants.

Quadrant 1:	$\alpha = \theta$
Quadrant 2:	$\alpha = 180° - \theta$
Quadrant 3:	$\alpha = \theta - 180°$
Quadrant 4:	$\alpha = 360° - \theta$

Using the formulas, we found α for each angle below. <u>Notice that we used Q1, Q2, Q3, and Q4 to specify the quadrant</u>.

For $\theta = 62°$, $\alpha = 62°$ (Q1)

For $\theta = 110°$, $\alpha = 180° - 110° = 70°$ (Q2)

For $\theta = 200°$, $\alpha = 200° - 180° = 20°$ (Q3)

For $\theta = 305°$, $\alpha = 360° - 305° = 55°$ (Q4)

Complete these. Be sure to include Q1, Q2, Q3, or Q4 to specify the quadrant.

a) If $\theta = 100°$, $\alpha =$ __________ c) If $\theta = 77°$, $\alpha =$ __________

b) If $\theta = 282°$, $\alpha =$ __________ d) If $\theta = 193°$, $\alpha =$ __________

45° (Q4)

---

a) 80° (Q2) c) 77° (Q1)

b) 78° (Q4) d) 13° (Q3)

23. We always use Q1, Q2, Q3, or Q4 to specify the quadrant for a reference angle. Complete these.

a) If $\theta = 12°$, $\alpha =$ ________ c) If $\theta = 99°$, $\alpha =$ ________

b) If $\theta = 205°$, $\alpha =$ ________ d) If $\theta = 330°$, $\alpha =$ ________

---

24. Complete these:

a) If $\theta = 260°$, $\alpha =$ ________ c) If $\theta = 199°$, $\alpha =$ ________

b) If $\theta = 175°$, $\alpha =$ ________ d) If $\theta = 275°$, $\alpha =$ ________

Answers to 23:
a) 12° (Q1)
b) 25° (Q3)
c) 81° (Q2)
d) 30° (Q4)

Answers to 24: a) 80° (Q3) b) 5° (Q2) c) 19° (Q3) d) 85° (Q4)

3-3 TRIGONOMETRIC RATIOS OF REFERENCE ANGLES

In this section, we will discuss the sine, cosine, and tangent of reference angles in the four quadrants. Before doing so, we will briefly discuss vectors on the coordinate system.

---

25. The arrows on the coordinate system at the right are called "vectors". Vectors can be horizontal, vertical, or slanted.

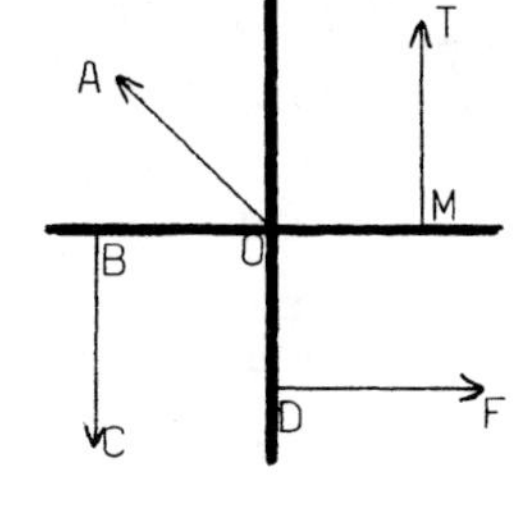

When naming a vector, we use two letters. The first letter represents where the vector begins. The second letter represents the tip or arrowhead of the vector. That is:

The name of the vertical vector in the first quadrant is "vector MT" (not vector TM).

a) The name of the slanted vector in the second quadrant is vector ______.

b) The name of the vertical vector in the third quadrant is vector ______.

c) The name of the horizontal vector in the fourth quadrant is vector ______.

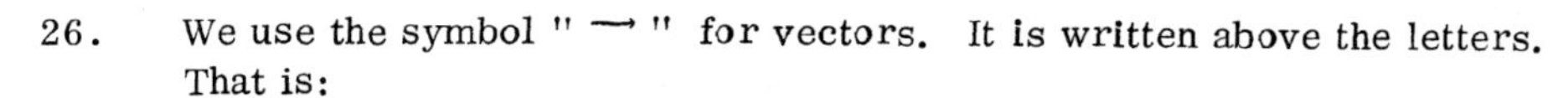

26. We use the symbol " ⟶ " for vectors. It is written above the letters. That is:

Instead of "vector AB", we write $\overrightarrow{AB}$.

The same symbol is used for slanted, horizontal, and vertical vectors. For example:

Instead of "vector OC", we write $\overrightarrow{OC}$.

a) Instead of "vector FG", we write ______.

b) Instead of "vector RT", we write ______.

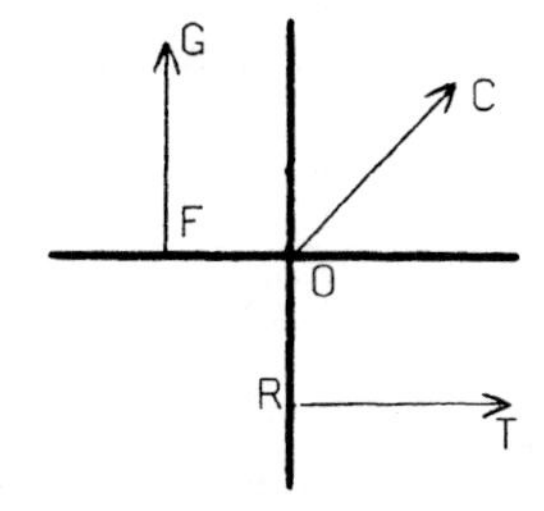

Answers to 25:
a) OA (not AO)
b) BC (not CB)
c) DF (not FD)

27. Some vertical vectors are drawn on the coordinate system at the right.

Since vectors have both length and direction, they can be represented by signed numbers.

Upward vectors are positive.
Downward vectors are negative.

Therefore: Since $\overrightarrow{AB}$ is 2 units upward, it is represented by +2.

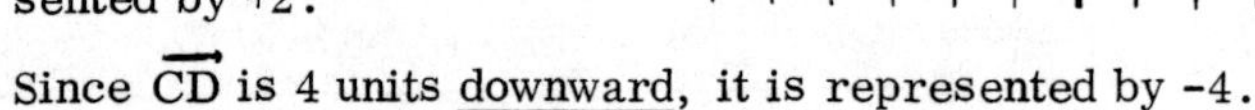

Since $\overrightarrow{CD}$ is 4 units downward, it is represented by -4.

a) Since $\overrightarrow{GH}$ is 3 units upward, it is represented by ______.

b) Since $\overrightarrow{EF}$ is 3 units downward, it is represented by ______.

a) $\overrightarrow{FG}$

b) $\overrightarrow{RT}$

---

28. Some horizontal vectors are drawn on the coordinate system at the right.

Vectors to the right are positive.
Vectors to the left are negative.

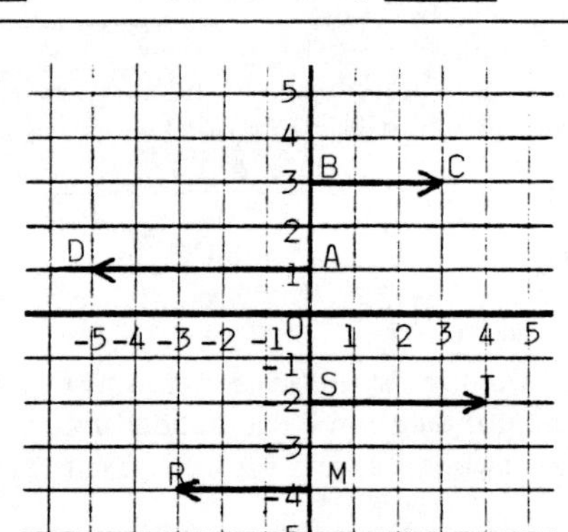

Therefore: Since $\overrightarrow{BC}$ is 3 units to the right, it is represented by +3.

Since $\overrightarrow{AD}$ is 5 units to the left, it is represented by -5.

a) Since $\overrightarrow{ST}$ is 4 units to the right, it is represented by ______.

b) Since $\overrightarrow{MR}$ is 3 units to the left, it is represented by ______.

a) +3

b) -3

---

29. Some horizontal and vertical vectors lie on the axes at the right. The same rules for signs apply. That is:

To the right and upward are positive.

To the left and downward are negative.

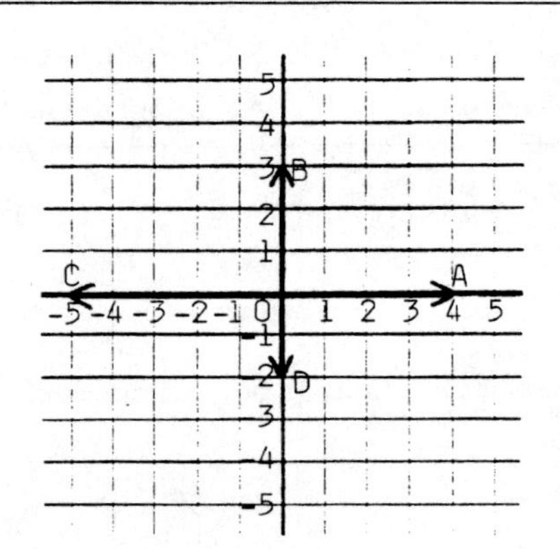

Therefore: $\overrightarrow{OA}$ = +4

a) $\overrightarrow{OB}$ = ______

b) $\overrightarrow{OC}$ = ______

c) $\overrightarrow{OD}$ = ______

a) +4

b) -3

---

a) +3

b) -5

d) -2

30. Some slanted vectors are drawn on the axes at the right. All slanted vectors are positive. Therefore:

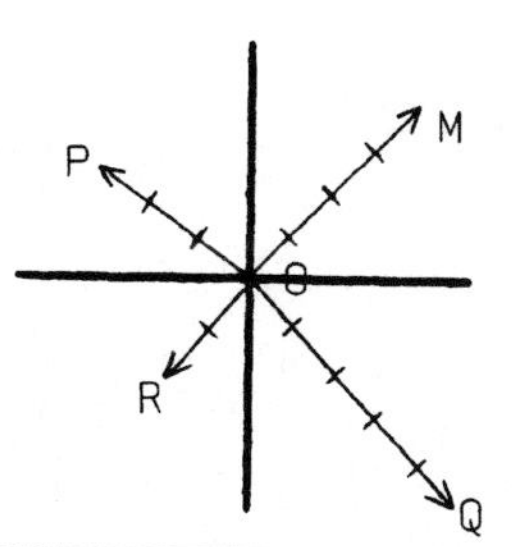

$\overrightarrow{OM} = +4$ $\overrightarrow{OP} = +3$

a) $\overrightarrow{OR}$ = _____ b) $\overrightarrow{OQ}$ = _____

---

31. A sketch of a reference angle in each quadrant is shown below.

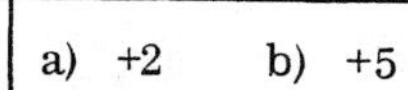

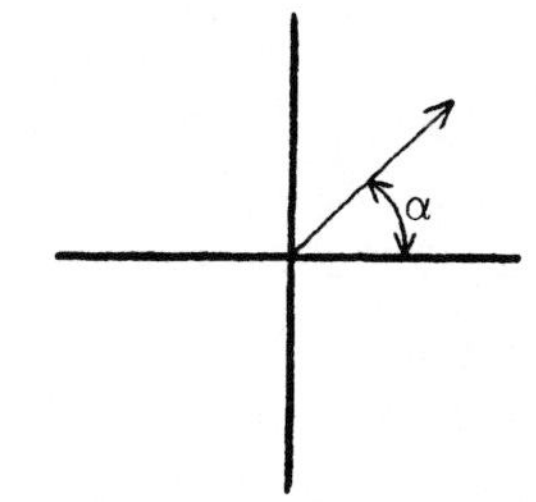

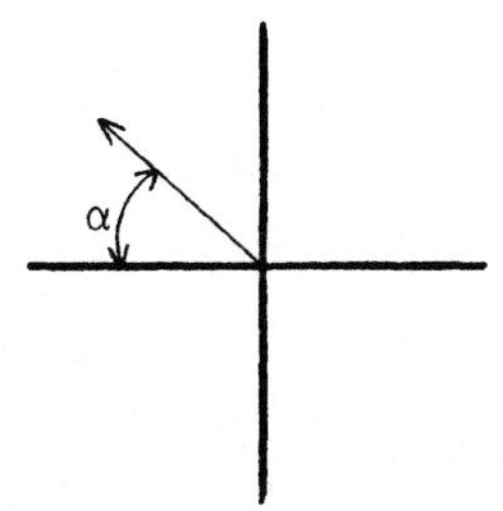

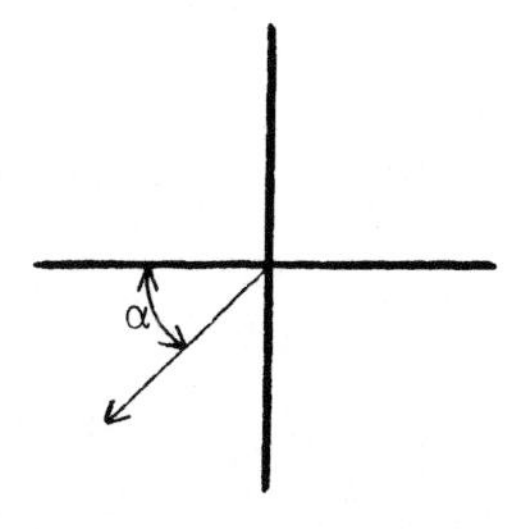

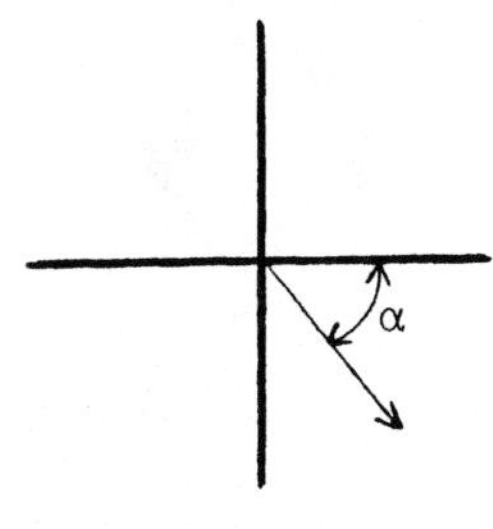

To find the sine, cosine, and tangent of reference angles, we complete right triangles as we have done below. Then we can compare the sides.

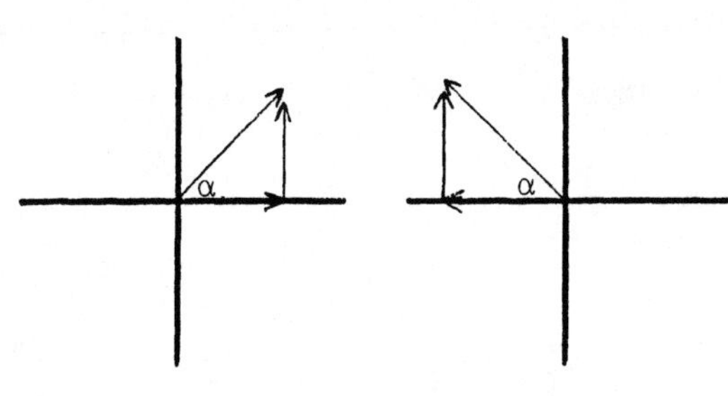

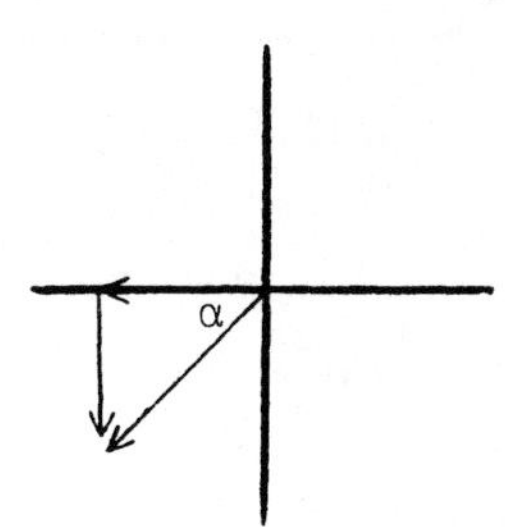

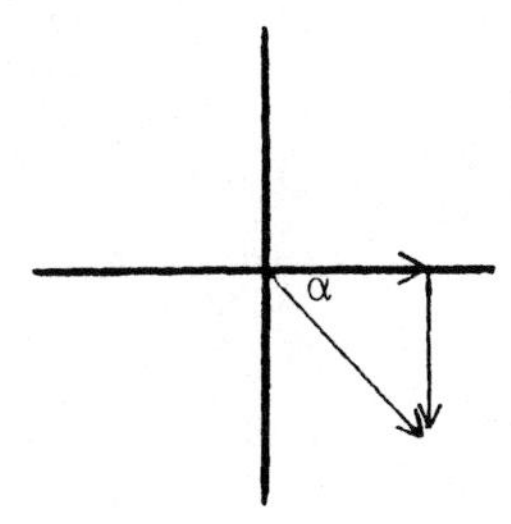

Notice these points about the right triangles:

1. α is an acute angle.
2. The three sides are vectors.

---

32. A second-quadrant reference angle is drawn at the right. Since the sides of the right triangle are vectors, the sine, cosine, and tangent of α are ratios of vectors. That is:

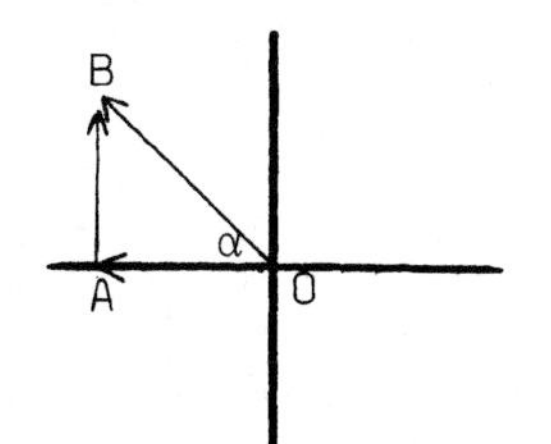

$\sin \alpha = \dfrac{\overrightarrow{AB}}{\overrightarrow{OB}}$ $\cos \alpha = \dfrac{\overrightarrow{OA}}{\overrightarrow{OB}}$ $\tan \alpha =$ _______

$\dfrac{\overrightarrow{AB}}{\overrightarrow{OA}}$

33. Since vectors can be either positive or negative, the trig ratios of a reference angle can be either positive or negative. An example is discussed below.

A second-quadrant reference angle is shown at the right.

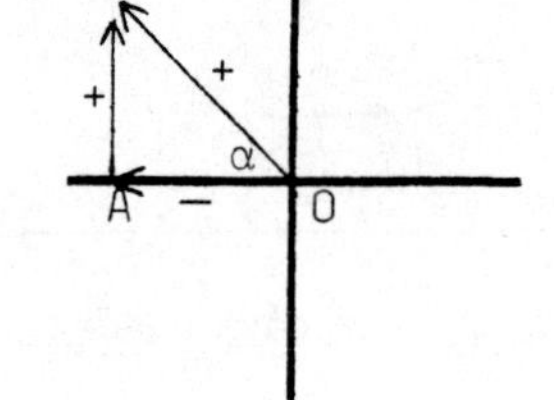

$\overrightarrow{OB}$ and $\overrightarrow{AB}$ are positive.

$\overrightarrow{OA}$ is negative.

Therefore, for a second-quadrant reference angle:

$$\sin \alpha = \frac{\overrightarrow{AB}}{\overrightarrow{OB}} = \frac{+}{+} = \text{"+"}$$

$$\cos \alpha = \frac{\overrightarrow{OA}}{\overrightarrow{OB}} = \frac{-}{+} = \text{"–"}$$

$$\tan \alpha = \frac{\overrightarrow{AB}}{\overrightarrow{OA}} = \frac{+}{-} = \text{"–"}$$

Which two trig ratios are negative for a second-quadrant reference angle? ______________ and ______________

Answer: $\cos \alpha$ and $\tan \alpha$

34. To find the trig ratios of a second quadrant reference angle, we need two steps.

1. Use a calculator to find the numerical value.
2. Then attach the proper sign.

Complete these. Round to four decimal places.

a) sin 53° (Q2) = ____________

b) cos 53° (Q2) = ____________

c) tan 53° (Q2) = ____________

Answers: a) +0.7986 b) −0.6018 c) −1.3270

35. A third-quadrant reference angle is shown at the right.

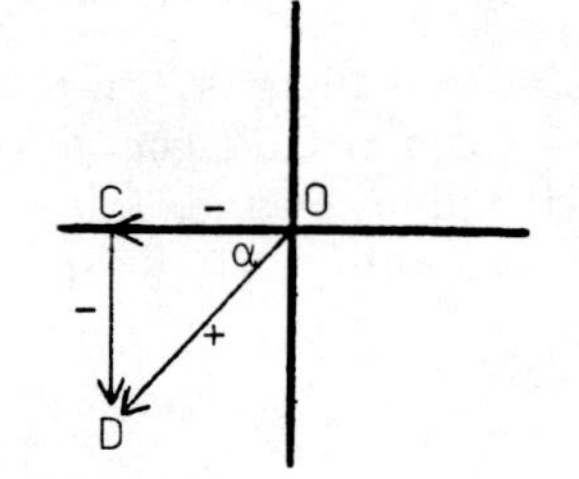

$\overrightarrow{OD}$ is positive.

$\overrightarrow{OC}$ and $\overrightarrow{CD}$ are negative.

Therefore, for a third-quadrant reference angle:

$$\sin \alpha = \frac{\overrightarrow{CD}}{\overrightarrow{OD}} = \frac{-}{+} = \text{"–"}$$

$$\cos \alpha = \frac{\overrightarrow{OC}}{\overrightarrow{OD}} = \frac{-}{+} = \text{"–"}$$

$$\tan \alpha = \frac{\overrightarrow{CD}}{\overrightarrow{OC}} = \frac{-}{-} = \text{"+"}$$

Which two trig ratios are negative for a third-quadrant reference angle? ____________ and ____________

36. To complete these, use a calculator and then attach the proper sign. Do so. Round to four decimal places. a) sin 17° (Q3) = ________ b) cos 17° (Q3) = ________ c) tan 17° (Q3) = ________	sin α and cos α
37. A fourth-quadrant reference angle is shown at the right. $\overrightarrow{OP}$ and $\overrightarrow{OT}$ are positive. $\overrightarrow{PT}$ is negative. 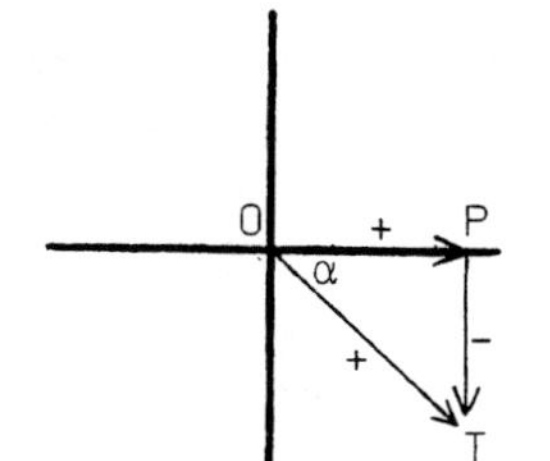 Therefore, for a fourth-quadrant reference angle: $\sin \alpha = \frac{\overrightarrow{PT}}{\overrightarrow{OT}} = \frac{-}{+} = \text{"}-\text{"}$ $\cos \alpha = \frac{\overrightarrow{OP}}{\overrightarrow{OT}} = \frac{+}{+} = \text{"}+\text{"}$ $\tan \alpha = \frac{\overrightarrow{PT}}{\overrightarrow{OP}} = \frac{-}{+} = \text{"}-\text{"}$ Which two trig ratios are negative for a fourth-quadrant reference angle? ________ and ________	a) -0.2924 b) -0.9563 c) +0.3057
38. To complete these, use a calculator and then attach the proper sign. Round to four decimal places. a) sin 34° (Q4) = ________ b) cos 34° (Q4) = ________ c) tan 34° (Q4) = ________	sin α and tan α
39. A first-quadrant reference angle is shown at the right. Since all three vectors are positive, all three ratios are positive. 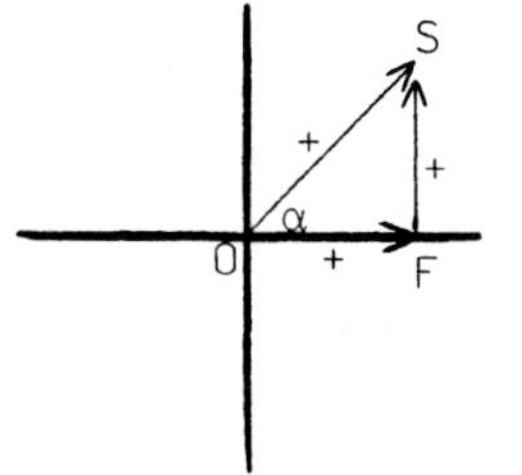 Use a calculator for these. Round to four decimal places. a) sin 80° (Q1) = ________ b) cos 80° (Q1) = ________ c) tan 80° (Q1) = ________	a) -0.5592 b) +0.8290 c) -0.6745
	a) +0.9848 b) +0.1736 c) +5.6713

40. The table below summarizes the "signs" of the trig ratios for reference angles in each quadrant.

	$\sin \alpha$	$\cos \alpha$	$\tan \alpha$
Q1	+	+	+
Q2	+	−	−
Q3	−	−	+
Q4	−	+	−

As you can see, each ratio is positive in two quadrants and negative in two quadrants.

a) $\sin \alpha$ is negative in quadrants _____ and _____.

b) $\cos \alpha$ is positive in quadrants _____ and _____.

c) $\tan \alpha$ is negative in quadrants _____ and _____.

---

41. Though the signs of the ratios in different quadrants can be memorized, it is just as easy to use sketches. An example is discussed below.

The sketch at the right can be used to get the signs for second-quadrant reference angles. Use a calculator for these. Round to four decimal places.

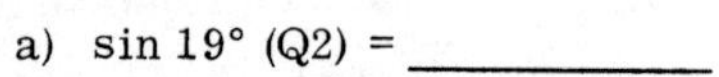

a) $\sin 19°$ (Q2) = __________

b) $\cos 58°$ (Q2) = __________

c) $\tan 75°$ (Q2) = __________

Answers to 40:

a) 3 and 4

b) 1 and 4

c) 2 and 4

---

42. The sketch at the right can be used to get the signs for third-quadrant reference angles. Do these. Round to four decimal places.

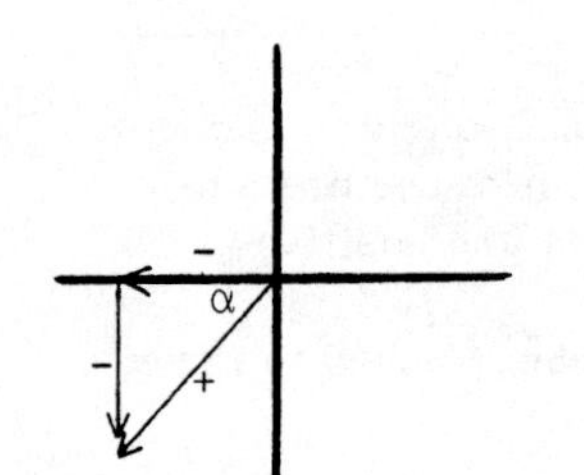

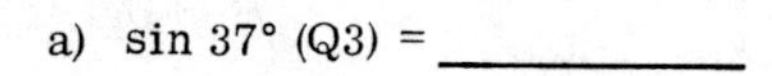

a) $\sin 37°$ (Q3) = __________

b) $\cos 9°$ (Q3) = __________

c) $\tan 22°$ (Q3) = __________

Answers to 41:

a) +0.3256

b) −0.5299

c) −3.7321

Answers to 42:

a) −0.6018

b) −0.9877

c) +0.4040

43. The sketch at the right can be used to get the signs for fourth-quadrant reference angles. Do these. Round to four decimal places.

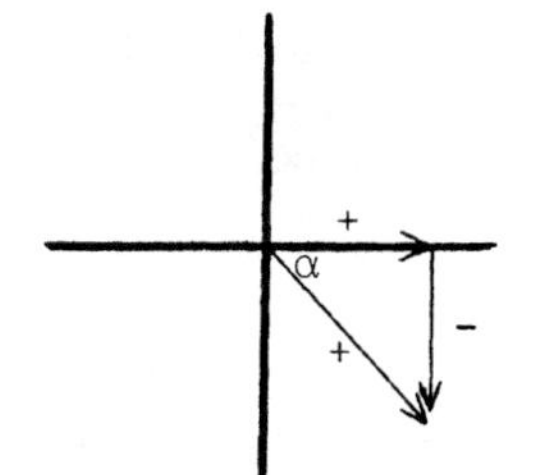

a) sin 61° (Q4) = ___________

b) cos 72° (Q4) = ___________

c) tan 81° (Q4) = ___________

---

a) -0.8746

b) +0.3090

c) -6.3138

44. Use a sketch to complete these. Round to four decimal places.

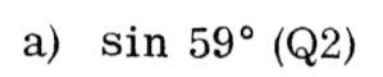

a) sin 59° (Q2) = ___________   b) cos 18° (Q3) = ___________   c) tan 77° (Q4) = ___________

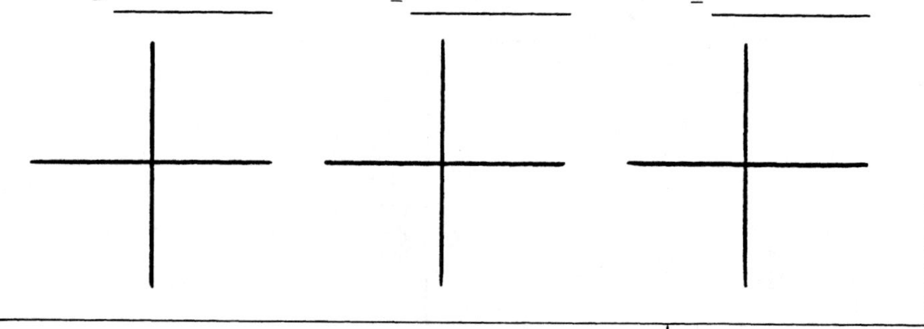

---

a) +0.8572   b) -0.9511   c) -4.3315

3-4 TRIGONOMETRIC RATIOS OF ANGLES FROM 90° TO 360°

In this section, we will discuss the sine, cosine, and tangent of angles from 90° to 360°.

45. The sine, cosine, and tangent of any standard-position angle from 90° to 360° is the same as the sine, cosine, and tangent of its reference angle. That is:

sin 100° = sin 80° (Q2)

cos 210° = cos 30° (Q3)

tan 330° = tan 30° (Q4)

Using the fact above, complete these:

a) sin 235° = ___________   c) tan 185° = ___________

b) cos 123° = ___________   d) cos 295° = ___________

---

a) sin 55° (Q3)

b) cos 57° (Q2)

c) tan 5° (Q3)

d) cos 65° (Q4)

46. A 140° angle is sketched at the right. Its reference angle α is 40° (Q2). Use your calculator for these. Round to four decimal places.

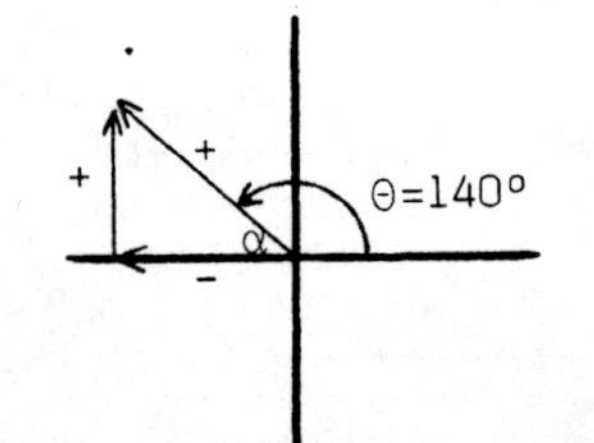

a) $\sin 140° = \sin 40°$ (Q2) = __________

b) $\cos 140° = \cos 40°$ (Q2) = __________

c) $\tan 140° = \tan 40°$ (Q2) = __________

---

47. A 235° angle is sketched at the right. Its reference angle α is 55° (Q3). Use your calculator for these. Round to four decimal places.

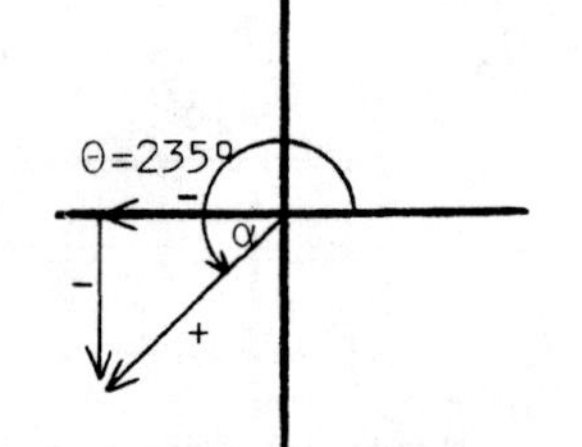

a) $\sin 235° = \sin 55°$ (Q3) = __________

b) $\cos 235° = \cos 55°$ (Q3) = __________

c) $\tan 235° = \tan 55°$ (Q3) = __________

a) +0.6428

b) −0.7660

c) −0.8391

---

48. A 325° angle is sketched at the right. Its reference angle is 35° (Q4). Use your calculator for these. Round to four decimal places.

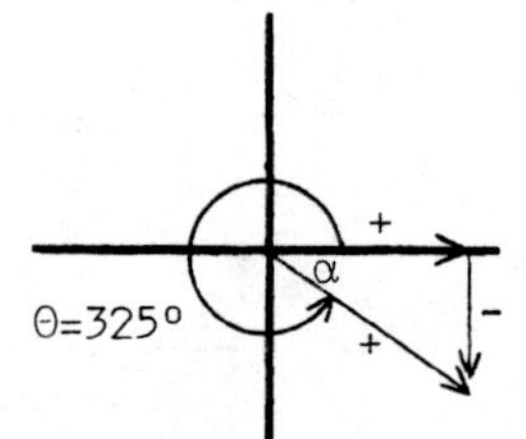

a) $\sin 325° = \sin 35°$ (Q4) = __________

b) $\cos 325° = \cos 35°$ (Q4) = __________

c) $\tan 325° = \tan 35°$ (Q4) = __________

a) −0.8192

b) −0.5736

c) +1.4281

---

49. The table below summarizes the "signs" of the trig ratios for standard-position angles whose terminal sides lie in each of the four quadrants.

	$\sin \theta$	$\cos \theta$	$\tan \theta$
0° to 90°	+	+	+
90° to 180°	+	−	−
180° to 270°	−	−	+
270° to 360°	−	+	−

Therefore: For angles from 90° to 180°, both $\cos \theta$ and $\tan \theta$ are negative.

a) For angles from 270° to 360°, both __________ and __________ are negative.

b) For angles from 180° to 270°, only __________ is positive.

a) −0.5736

b) +0.8192

c) −0.7002

50. Fortunately, we do not have to worry about getting the proper signs for the trig ratios of angles from 90° to 360° because the calculator does it for us.

a) Find sin 123° by entering 123 and pressing [sin].

sin 123° = ________

b) Find cos 237° by entering 237 and pressing [cos].

cos 237° = ________

c) Find tan 309° by entering 309 and pressing [tan].

tan 309° = ________

> a) sin θ and tan θ
>
> b) tan θ

---

51. Use a calculator for these. Round to four decimal places.

a) sin 200° = ________ c) tan 99° = ________

b) cos 349° = ________ d) tan 187° = ________

> a) +0.8386706
>
> b) -0.544639
>
> c) -1.2348972

---

52. Use a calculator for these. Round to four decimal places.

a) cos 201° = ________ c) tan 101° = ________

b) sin 347° = ________ d) cos 298° = ________

> a) -0.3420
>
> b) +0.9816
>
> c) -6.3138
>
> d) +0.1228

> a) -0.9336 b) -0.2250 c) -5.1446 d) +0.4695

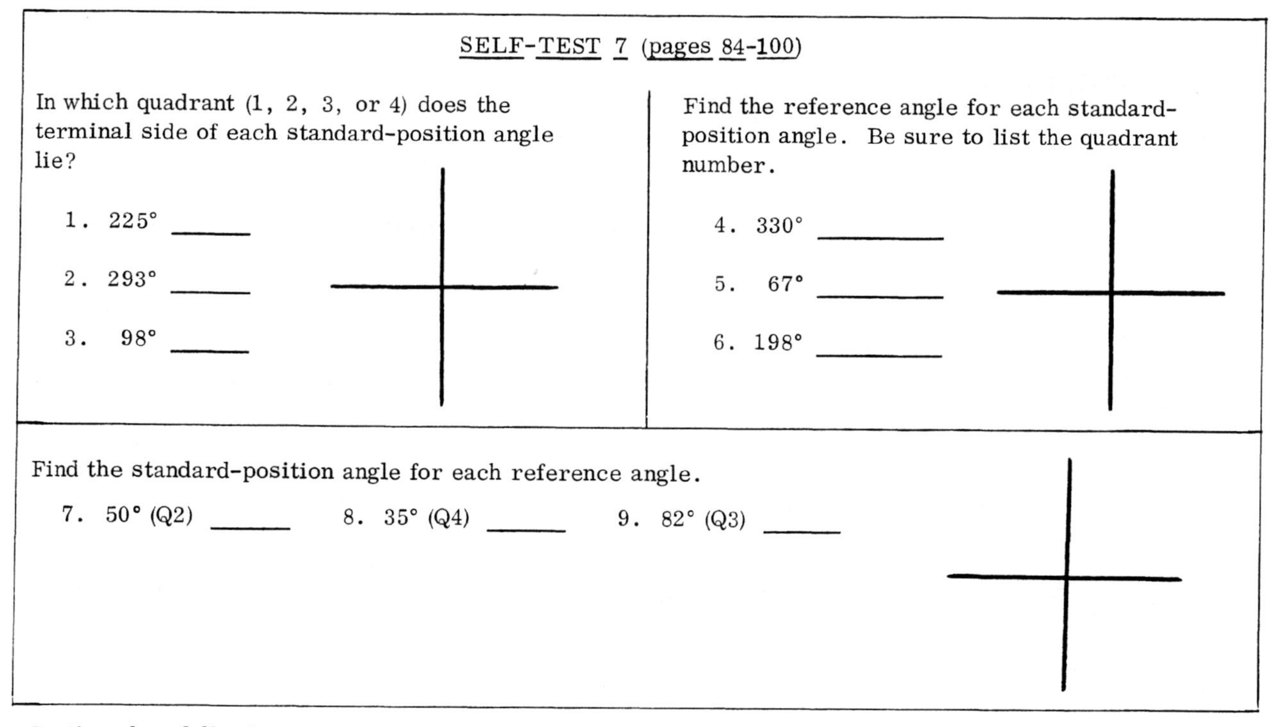

SELF-TEST 7 (pages 84-100)

In which quadrant (1, 2, 3, or 4) does the terminal side of each standard-position angle lie?

1. 225° ____
2. 293° ____
3. 98° ____

Find the reference angle for each standard-position angle. Be sure to list the quadrant number.

4. 330° ________
5. 67° ________
6. 198° ________

Find the standard-position angle for each reference angle.

7. 50° (Q2) ____ 8. 35° (Q4) ____ 9. 82° (Q3) ____

Continued on following page.

SELF-TEST 7 (pages 84-100) - Continued

10. In which quadrants is the sine ratio negative? ____________

11. In which quadrants is the cosine ratio negative? ____________

12. In which quadrants is the tangent ratio positive? ____________

Using a calculator, find the numerical value of each of the following. Round to four decimal places.

13. tan 71° (Q3) = ____________

14. sin 14° (Q4) = ____________

15. cos 49° (Q2) = ____________

16. cos 293° = ____________

17. tan 319° = ____________

18. sin 105° = ____________

ANSWERS:

1. Q3	4. 30° (Q4)	7. 130°	10. Q3 and Q4	13. 2.9042	16. 0.3907
2. Q4	5. 67° (Q1)	8. 325°	11. Q2 and Q3	14. -0.2419	17. -0.8693
3. Q2	6. 18° (Q3)	9. 262°	12. Q1 and Q3	15. -0.6561	18. 0.9659

3-5 TRIGONOMETRIC RATIOS OF 0°, 90°, 180°, 270°, 360°

In this section, we will discuss the sine, cosine, and tangent of five angles (0°, 90°, 180°, 270°, and 360°) whose terminal sides lie on an axis.

53. A calculator gives the following value.

sin 90° = 1

We can use the figures below to show how the value "1" is obtained. In each figure, $\overrightarrow{OB}$ is 5 units.

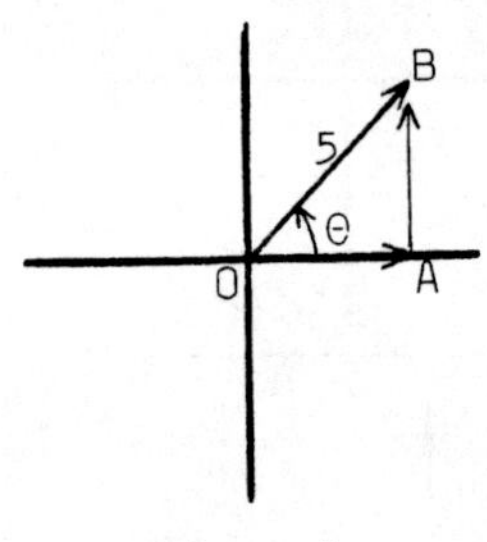

Figure 1

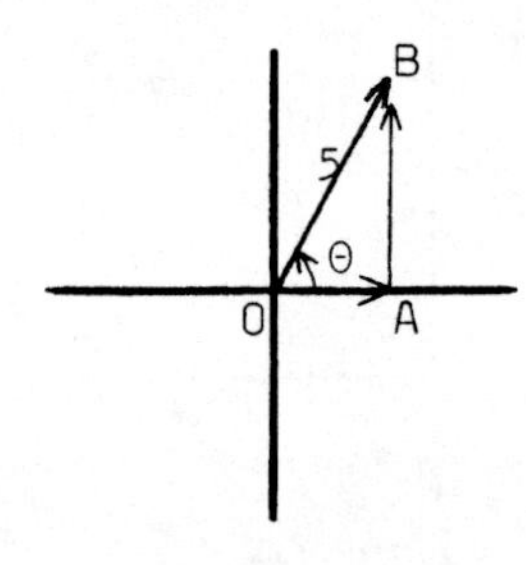

Figure 2

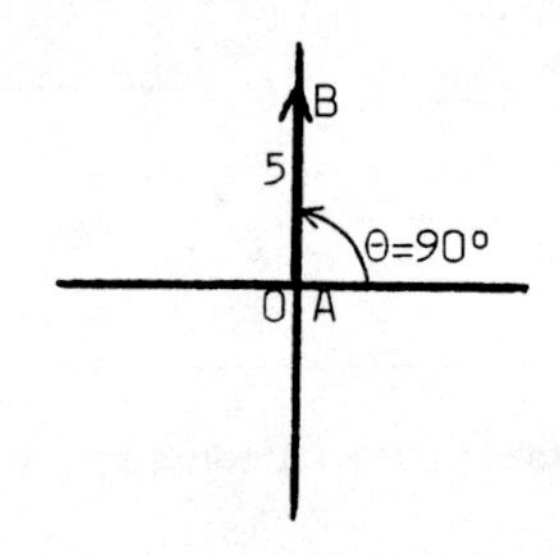

Figure 3

Continued on following page.

53. Continued

From Figure 1 to Figure 3, angle θ increases in size until it becomes 90° in Figure 3. There is no triangle in Figure 3 because $\overrightarrow{OB}$ coincides with the vertical axis. Notice these points:

1. In each figure, $\sin\theta = \frac{\overrightarrow{AB}}{\overrightarrow{OB}} = \frac{\overrightarrow{AB}}{5}$

2. From Figure 1 to Figure 3, the length of $\overrightarrow{AB}$ increases and therefore $\sin\theta$ increases.

3. In Figure 3, $\theta = 90°$ and $\overrightarrow{AB} = \overrightarrow{OB} = 5$ units.

   Therefore: $\sin 90° = \frac{\overrightarrow{AB}}{\overrightarrow{OB}} = \frac{5}{5} =$ ______

---

54. A calculator gives the following value. | 1

$$\boxed{\cos 90° = 0}$$

We can use the figures in the last frame to show how the value "0" is obtained. Notice these points:

1. In each figure, $\cos\theta = \frac{\overrightarrow{OA}}{\overrightarrow{OB}} = \frac{\overrightarrow{OA}}{5}$

2. From Figure 1 to Figure 3, the length of $\overrightarrow{OA}$ decreases and therefore $\cos\theta$ decreases.

3. In Figure 3, $\theta = 90°$ and $\overrightarrow{OA} = 0$ units.

   Therefore: $\cos 90° = \frac{\overrightarrow{OA}}{\overrightarrow{OB}} = \frac{0}{5} =$ ______

---

55. A calculator gives the following value. | 0

$$\boxed{\tan 90° = \text{Error}}$$

The word "error" really means "undefined". We can use the figures in Frame 53 to see that "undefined" makes sense for tan 90°. Notice these points:

1. In each figure, $\tan\theta = \frac{\overrightarrow{AB}}{\overrightarrow{OA}}$.

2. In figure 3 where $\theta = 90°$, $\overrightarrow{AB} = 5$ units and $\overrightarrow{OA} = 0$ units.

   Therefore: $\tan 90° = \frac{\overrightarrow{AB}}{\overrightarrow{OA}} = \frac{5}{0} =$ undefined

Note: $\frac{5}{0}$ means: divide 5 by 0. But division by "0" is impossible.
Therefore, tan 90° is "undefined".

56. A calculator gives the following values for a 0° angle.

$\sin 0° = 0$	$\cos 0° = 1$	$\tan 0° = 0$

We can use the series of figures below to show how the values above are obtained.

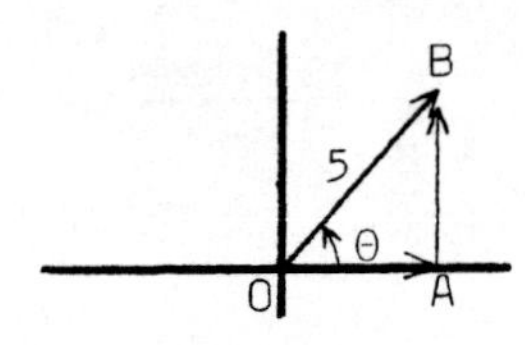

Figure 1

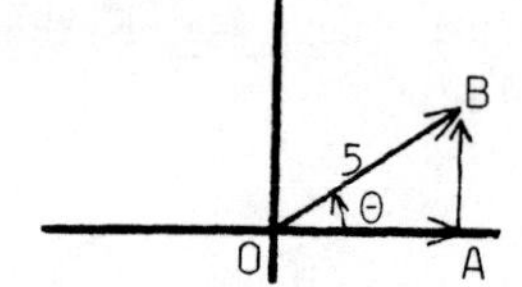

Figure 2

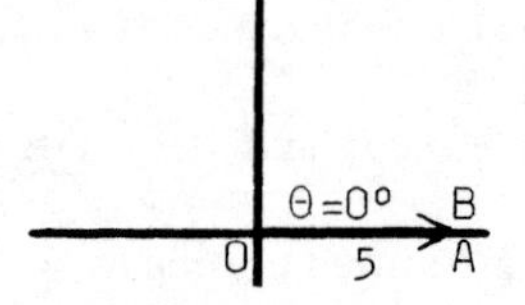

Figure 3

From Figure 1 to Figure 3, angle θ decreases in size until it becomes 0° in Figure 3. There is no triangle in Figure 3 because $\overrightarrow{OB}$ coincides with $\overrightarrow{OA}$ on the horizontal axis. In Figure 3, where $\theta = 0°$:

$\overrightarrow{OB} = 5$ units $\qquad \overrightarrow{OA} = 5$ units $\qquad \overrightarrow{AB} = 0$ units

Therefore: a) $\sin 0° = \dfrac{\overrightarrow{AB}}{\overrightarrow{OB}} = \dfrac{0}{5} =$ ______

b) $\cos 0° = \dfrac{\overrightarrow{OA}}{\overrightarrow{OB}} = \dfrac{5}{5} =$ ______

c) $\tan 0° = \dfrac{\overrightarrow{AB}}{\overrightarrow{OA}} = \dfrac{0}{5} =$ ______

---

a) 0

b) 1

c) 0

57. Since 0° is the reference angle for both 180° and 360°, and 90° is the reference angle for 270°, it is necessary that you know the values of the three ratios for 0° and 90°. They are:

θ	0°	90°
sin θ	0	1
cos θ	1	0
tan θ	0	Undef.

As θ increases from 0° to 90°:

a) sin θ increases from ______ to ______.

b) cos θ decreases from ______ to ______.

---

a) 0 to 1

b) 1 to 0

58. A calculator gives these values for a 180° angle.

$\sin 180° = 0$	$\cos 180° = -1$	$\tan 180° = 0$

To understand the values, we have diagrammed a second-quadrant angle at the right. As θ increases to 180°, α decreases to 0°. Therefore, the reference angle for 180° is 0° (Q2).

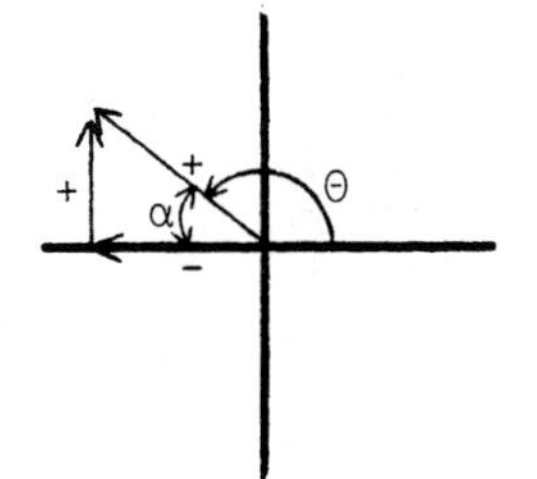

Using the correct signs for second-quadrant angles (and remembering that both "+0" and "-0" equal "0"), we get:

$\sin 180° = \sin 0°$ (Q2) $= +0$ or 0

$\cos 180° = \cos 0°$ (Q2) $= -1$

$\tan 180° = \tan 0°$ (Q2) $= -0$ or ______

0

59. A calculator gives these values for a 270° angle.

$\sin 270° = -1$	$\cos 270° = 0$	$\tan 270° =$ Error

To understand the values, we have diagrammed a third-quadrant angle at the right. As θ increases to 270°, α increases to 90°. Therefore, the reference angle for 270° is 90° (Q3).

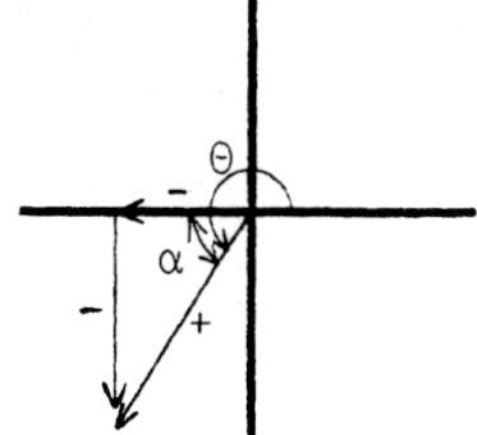

Using the correct signs for third-quadrant angles, we get:

$\sin 270° = \sin 90°$ (Q3) $= -1$

$\cos 270° = \cos 90°$ (Q3) $= -0$ or 0

$\tan 270° = \tan 90°$ (Q3) $=$ ______________

undefined

60. A calculator gives these values for 360°.

$\sin 360° = 0$	$\cos 360° = 1$	$\tan 360° = 0$

To understand the values, we have diagrammed a fourth-quadrant angle at the right. As θ increases to 360°, α decreases to 0°. Therefore, the reference angle for 360° is 0° (Q4).

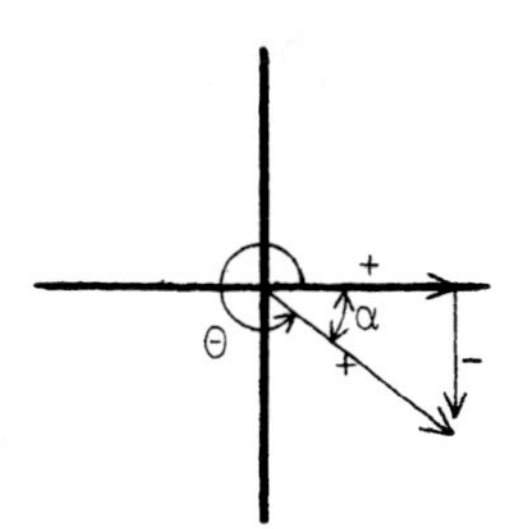

Using the correct signs for fourth-quadrant angles, we get:

$\sin 360° = \sin 0°$ (Q4) $= -0$ or 0

$\cos 360° = \cos 0°$ (Q4) $= +1$

$\tan 360° = \tan 0°$ (Q4) $= -0$ or ______

61. The values of the trig ratios for 0°, 90°, 180°, 270°, and 360° are given in the table below.

θ	0°	90°	180°	270°	360°
sin θ	0	+1	0	-1	0
cos θ	+1	0	-1	0	+1
tan θ	0	Undef.	0	Undef.	0

Because their terminal sides are in the same positions, two angles have identical trig ratios. The angles are ________ and ________.

Answer: 0

---

62. Try to do these without using the table or your calculator.

a) sin 0° = ______ b) cos 180° = ______ c) tan 360° = ______

Answer: 0° and 360°

---

63. Complete these:

a) cos 90° = ______ b) sin 270° = ______ c) tan 90° = ______

Answer: a) 0 b) -1 c) 0

---

64. Complete these:

a) tan 180° = ______ b) sin 360° = ______ c) cos 0° = ______

Answer: a) 0 b) -1 c) undefined

---

65. Of the five angles (0°, 90°, 180°, 270°, 360°):

a) There is one whose <u>sine</u> is +1. It is __________.

b) There is one whose <u>sine</u> is -1. It is __________.

c) There is one whose <u>cosine</u> is -1. It is __________.

Answer: a) 0 b) 0 c) 1

---

66. Of the five angles (0°, 90°, 180°, 270°, 360°):

a) There are two whose <u>cosines</u> are "0". They are __________ and __________.

b) There are two whose <u>cosines</u> are +1. They are __________ and __________.

c) There are two whose <u>tangents</u> are "undefined". They are __________ and __________.

Answer: a) 90° b) 270° c) 180°

Answer: a) 90° and 270° b) 0° and 360° c) 90° and 270°

3-6 TRIG RATIOS OF ANGLES GREATER THAN 360°

In this section, we will define angles greater than 360°. Then we will discuss the sine, cosine, and tangent of angles of that size.

---

67. A 360° angle is shown at the left below. It is one complete revolution. To get an angle greater than 360°, we must rotate the vector beyond one complete revolution. A 410° angle is shown at the right below.

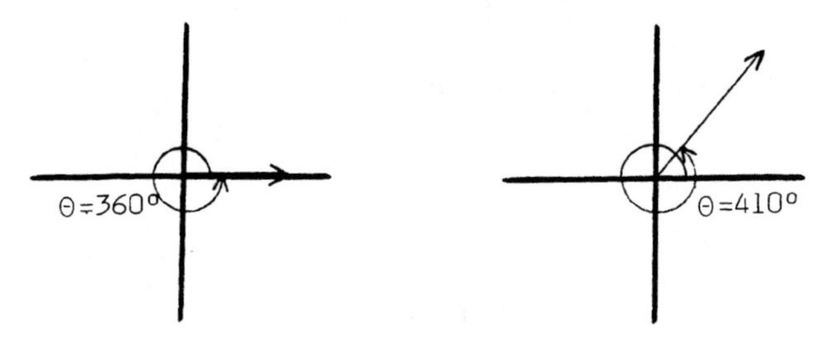

A 410° angle is 50° more than one complete revolution, since 410° - 360° = 50°. Therefore, a 410° angle has the same terminal side as a ______° angle.

Answer: 50°

---

68. A 480° angle and a 587° angle are shown below.

θ=480°

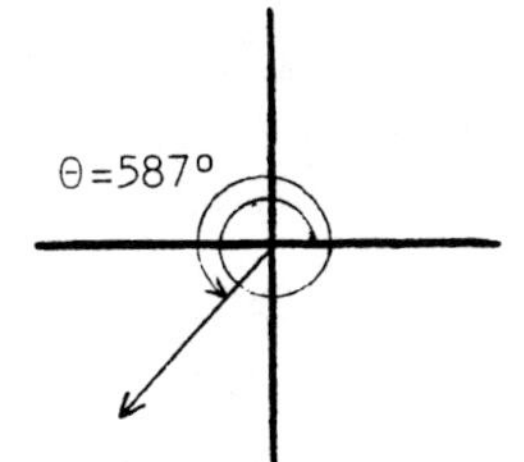

By subtracting 360° from the angle, complete these:

a) A 480° angle has the same terminal side as a ________° angle.

b) A 587° angle has the same terminal side as a ________° angle.

Answers:

a) 120°

b) 227°

69. A 686° angle and a 720° angle are shown below. Notice that 720° is exactly <u>two complete revolutions</u>.

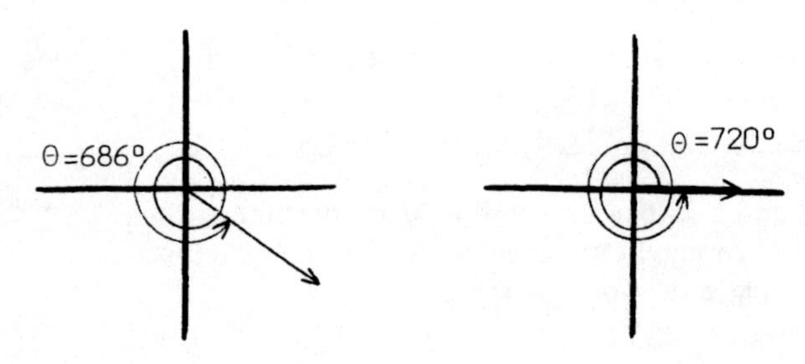

a) A 686° angle has the same terminal side as one smaller angle. It is ________°.

b) A 720° angle has the same terminal side as two smaller angles. They are ________° and ________°.

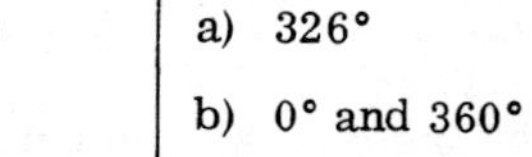
a) 326°

b) 0° and 360°

70. Angles can extend beyond two revolutions. For example, a 850° angle and a 1,080° angle are shown below. Notice that 1,080° is exactly <u>three complete revolutions</u>.

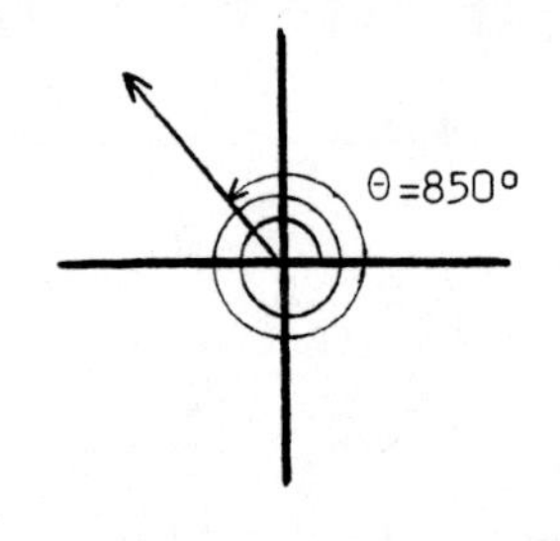

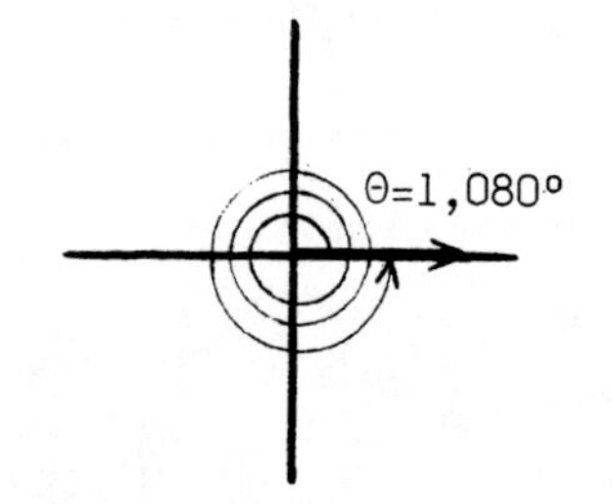

a) A 850° angle has the same terminal side as two smaller angles. They are ________° and ________°.

b) A 1,080° angle has the same terminal side as three smaller angles. They are ________°, ________°, and ________°.

---

a) 130° and 490°, from:

850° - 720° = 130°
850° - 360° = 490°

b) 0°, 360°, 720°

71. For any angle between 0° and 360°, larger angles having the same terminal side can be found by adding 360°, 720°, 1080°, etc., to that angle. For example, all of the angles below have the same terminal side as a 60° angle.

420°	=	60° + 360°	(1 revolution)
780°	=	60° + 720°	(2 revolutions)
1140°	=	60° + 1,080°	(3 revolutions)

Name three angles with the same terminal side as a 240° angle.

__________, __________, and __________

600°, 960°, 1320°

72. By adding 360° and 720° to each angle, name two angles with the same terminal side as each of these.

a) 0° ________ and ________ c) 180° ________ and ________

b) 90° ________ and ________ d) 270° ________ and ________

---

73. We can use a calculator to find the sine, cosine, and tangent of any angle greater than 360°. For example, we found sin 427°, cos 641°, and tan 811° below.

Enter	Press	Display
427	sin	0.9205049
641	cos	0.190809
811	tan	-57.289958

Answers to 72:

a) 360° and 720°

b) 450° and 810°

c) 540° and 900°

d) 630° and 990°

---

74. <u>All angles with the same terminal side have the same sine, cosine, and tangent</u>. For example:

sin 30° = sin 390° = sin 750° = <u>0.5</u>

Use a calculator for these. Round to four decimal places.

a) sin 125° = sin 485° = sin 845° = ____________

b) cos 260° = cos 620° = cos 980° = ____________

c) tan 310° = tan 670° = tan 1030° = ____________

---

75. Complete these:

a) sin 90° = sin 450° = sin 810° = ____________

b) cos 180° = cos 540° = cos 900° = ____________

c) tan 270° = tan 630° = tan 990° = ____________

Answers to 74:

a) 0.8192

b) -0.1736

c) -1.1918

Answers to 75:

a) 1 b) -1 c) undefined

3-7 TRIG RATIOS OF NEGATIVE ANGLES

In this section, we will define negative angles and discuss their trig ratios.

76. Negative angles are generated by rotating a vector in a clockwise direction. For example, $\theta = -120°$ and $\theta = -240°$ are shown below.

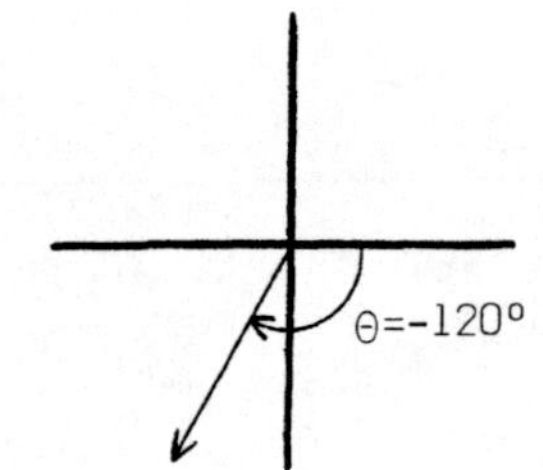

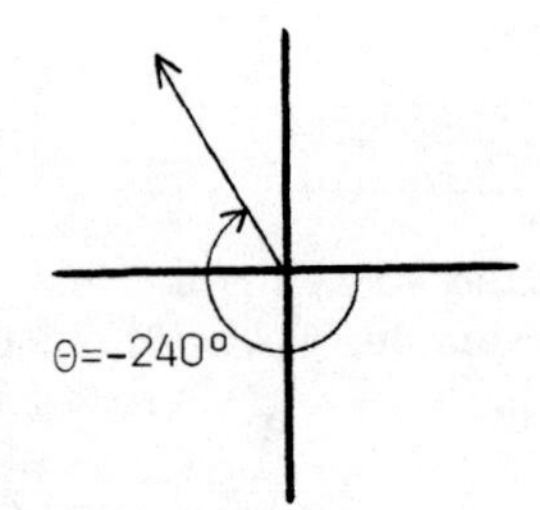

a) A -120° angle has its terminal side in quadrant ______.

b) A -240° angle has its terminal side in quadrant ______.

---

77. Negative angles beyond -360° involve more than one complete revolution in a clockwise direction. Two examples are shown.

Θ-390°

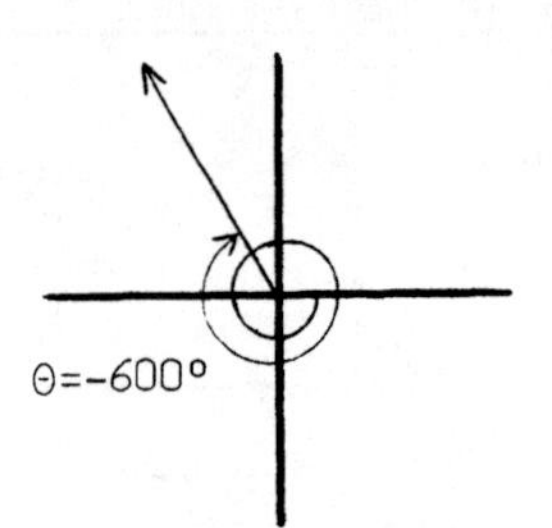

A -390° angle has the same terminal side as a -30° angle.

A -600° angle has the same terminal side as a ________° angle.

a) 3

b) 2

---

78. For any negative angle between 0° and -360°, other negative angles having the same terminal side can be found by adding -360°, -720°, -1080°, etc. For example, all of the angles on the left below have the same terminal side as a -40° angle.

$$-400° = (-40°) + (-360°)$$

$$-760° = (-40°) + (-720°)$$

$$-1120° = (-40°) + (-1080°)$$

Name three negative angles with the same terminal side as a -100° angle.

__________, __________, and __________

-240°

-460°, -820°, and -1180°

79. In each figure below, a negative angle between 0° and -360° and a positive angle between 0° and 360° have the same terminal side.

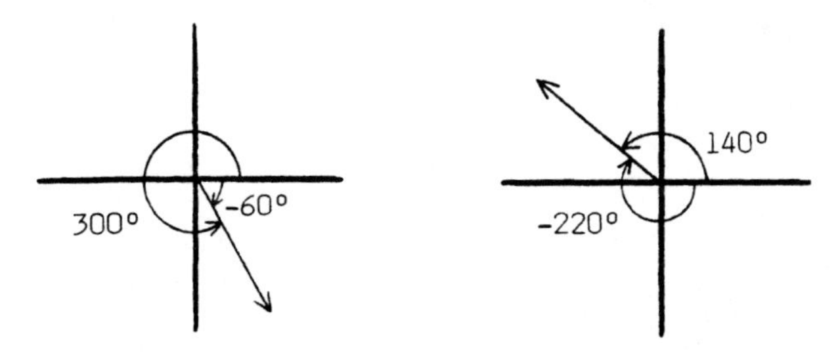

You can see that the sum of the absolute values of the two angles is 360°. That is:

$$60° + 300° = 360°$$

$$220° + 140° = 360°$$

Using the fact above, name a positive angle between 0° and 360° with the same terminal side as each of these.

a) -20° b) -135° c) -270° d) -330°

________ ________ ________ ________

Answers:
a) 340°
b) 225°
c) 90°
d) 30°

80. Any negative angle between 0° and -360° has the same terminal side as many other negative and positive angles. For example:

-120° has the same terminal side as:

-480°, -840°, and -1200°.

-120° has the same terminal side as:

240°, 600°, and 960°,

Name two negative and two positive angles with the same terminal side as -200°.

________ and ________, ________ and ________

Answers:
-560° and -920°
160° and 520°

81. We can use a calculator to find the sine, cosine, and tangent of any negative angle. For example, we found sin (-75°), cos (-284°), and tan (-644°) below.

Enter	Press	Display
75	[+/-] [sin]	-0.9659258
284	[+/-] [cos]	0.2419219
644	[+/-] [tan]	4.0107809

82. All angles with the same terminal side have the same sine, cosine, and tangent. For example:

$$\cos(-100°) = \cos(-460°) = -0.1736482$$

Use a calculator for these. Round to four decimal places.

a) $\sin(-50°) = \sin(-410°) =$ ________

b) $\cos(-210°) = \cos(-570°) =$ ________

c) $\tan(-340°) = \tan(-700°) =$ ________

---

83. A negative angle and a positive angle with the same terminal side have the same sine, cosine, and tangent. For example:

$$\tan(-130°) = \tan 230° = 1.1917536$$

Use a calculator for these. Round to four decimal places.

a) $\sin(-80°) = \sin 280° =$ ________

b) $\cos(-155°) = \cos 205° =$ ________

c) $\tan(-263°) = \tan 97° =$ ________

Answers to 82: a) -0.7660 b) -0.8660 c) 0.3640

---

84. Complete these:

a) $\sin(-90°) = \sin(-450°) = \sin 270° =$ ________

b) $\cos(-270°) = \cos(-630°) = \cos 90° =$ ________

c) $\tan(-180°) = \tan(-540°) = \tan 180° =$ ________

Answers to 83: a) -0.9848 b) -0.9063 c) -8.1443

Answers to 84: a) -1 b) 0 c) 0

3-8 GRAPHS OF TRIGONOMETRIC FUNCTIONS

In this section, we will discuss the graphs of the sine, cosine, and tangent functions.

---

85. The equation of the "sine function" is given below. The angle "θ" is the independent variable; "y" is the dependent variable.

$$\boxed{y = \sin\theta}$$

By substituting values for "θ", we can find the corresponding values for "y". For example:

If $\theta = 0°$, $y = \sin 0° = 0$

If $\theta = 30°$, $y = \sin 30° = 0.5$

If $\theta = 90°$, $y = \sin 90° =$ ______

Answer to 85: 1

86. Some pairs of values for θ and y for the sine function are given in the table below.

$$y = \sin \theta$$

θ	y	θ	y	θ	y	θ	y
0°	0.00	90°	1.00	180°	0.00	270°	-1.00
30°	0.50	120°	0.87	210°	-0.50	300°	-0.87
45°	0.71	135°	0.71	225°	-0.71	315°	-0.71
60°	0.87	150°	0.50	240°	-0.87	330°	-0.50
90°	1.00	180°	0.00	270°	-1.00	360°	0.00

By examining the table, you can see these facts:

As θ increases from 0° to 90°, y increases from 0 to +1.

a) As θ increases from 90° to 180°, y decreases from _____ to _____.

b) As θ increases from 180° to 270°, y decreases from _____ to _____.

c) As θ increases from 270° to 360°, y increases from _____ to _____.

---

a) +1 to 0

b) 0 to -1

c) -1 to 0

---

87. Using the pairs of values from the table in the last frame, we graphed $y = \sin \theta$ from $\theta = 0°$ to $\theta = 360°$ below. The graph from 0° to 360° is the basic cycle (or period) of the sine function.

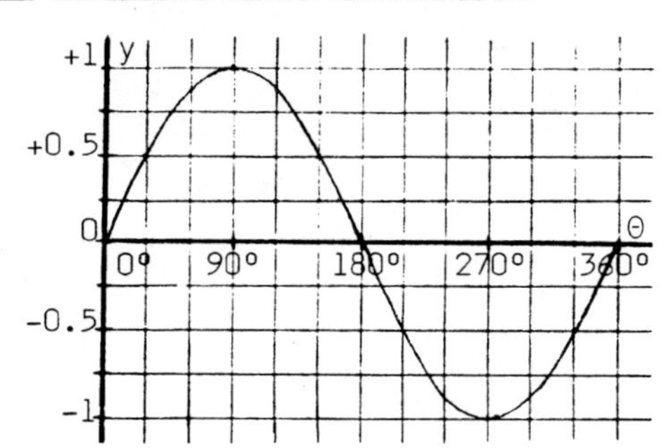

Examine the graph to complete these:

a) For angles between 0° and 180°, is y positive or negative? _______________

b) For angles between 180° and 360°, is y positive or negative? _______________

c) For what angle θ does $y = +1$? ___________

d) For what angle θ does $y = -1$? ___________

---

a) positive

b) negative

c) 90°

d) 270°

88. The graph of $\boxed{y = \sin \theta}$ from $\theta = -720°$ to $\theta = 720°$ is given below.

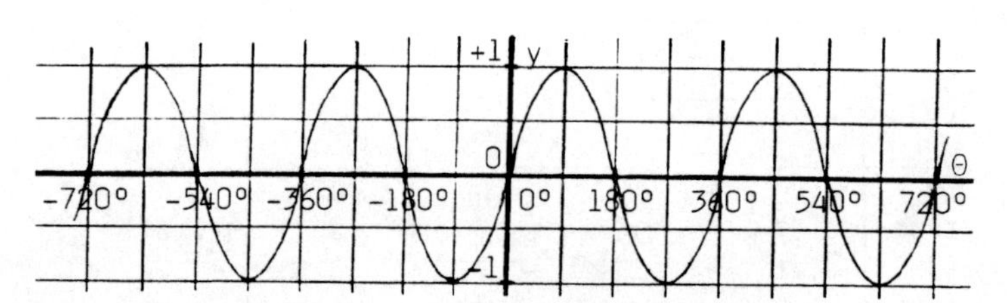

The basic cycle (or period) of the sine function is from 0° to 360°. You can see that that cycle (or period) repeats itself from -720° to -360°, from -360° to 0°, and from 360° to 720°. The basic cycle will continue to repeat itself if the graph is extended to the left of -720° and to the right of 720°. Therefore, $y = \sin \theta$ is called a "periodic" function.

Examine the graph to complete these:

a) y = +1 for what four angles? __________, __________, __________, and __________

b) y = -1 for what four angles? __________, __________, __________, and __________

c) Is the value of y ever greater than +1? __________

d) Is the value of y ever less than -1? __________

---

a) -630°, -270°, 90°, 450°

b) -450°, -90°, 270°, 630°

c) No

d) No

89. The equation of the "cosine function" is given below.

$$\boxed{y = \cos \theta}$$

Some pairs of values for θ and y for the cosine function are given in the table below.

θ	y	θ	y	θ	y	θ	y
0°	1.00	90°	0.00	180°	-1.00	270°	0.00
30°	0.87	120°	-0.50	210°	-0.87	300°	0.50
45°	0.71	135°	-0.71	225°	-0.71	315°	0.71
60°	0.50	150°	-0.87	240°	-0.50	330°	0.87
90°	0.00	180°	-1.00	270°	0.00	360°	1.00

By examining the table, you can see these facts:

As θ increases from 0° to 90°, y decreases from +1 to 0.

a) As θ increases from 90° to 180°, y decreases from ______ to ______.

b) As θ increases from 180° to 270°, y increases from ______ to ______.

c) As θ increases from 270° to 360°, y increases from ______ to ______.

90. Using the pairs of values from the table in the last frame, we graphed $\boxed{y = \cos\theta}$ from $\theta = 0°$ to $\theta = 360°$ below. The graph from 0° to 360° is the basic cycle (or period) of the cosine function.

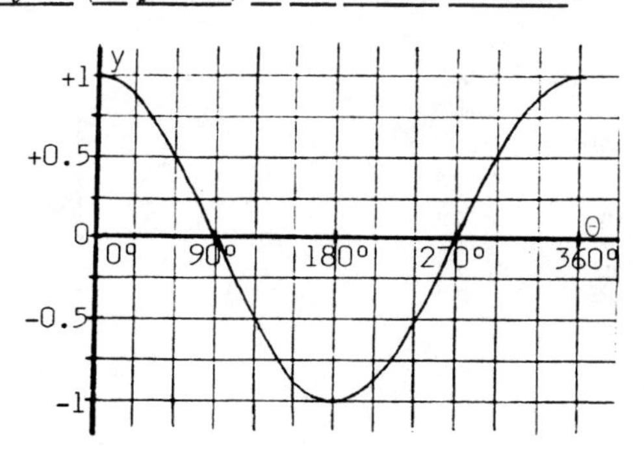

Examine the graph to complete these:

a) For what angles does $y = +1$? ________ and ________

b) For what angle does $y = -1$? ________

c) From 90° to 180°, y is ____________ (positive/negative).

d) From 270° to 360°, y is ____________ (positive/negative).

a) 0 to -1

b) -1 to 0

c) 0 to +1

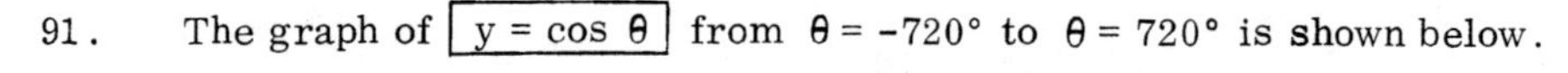

91. The graph of $\boxed{y = \cos\theta}$ from $\theta = -720°$ to $\theta = 720°$ is shown below.

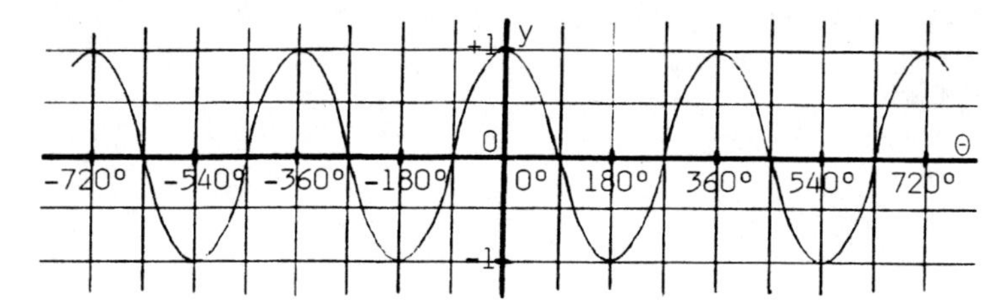

The basic cycle (or period) of the cosine function is from 0° to 360°. You can see that that cycle (or period) repeats itself from -720° to -360°, from -360° to 0°, and from 360° to 720°. That cycle will continue to repeat itself if the graph is extended to the left of -720° and to the right of 720°.

Since the cycle (or period) repeats itself, the cosine function is called a ____________ function.

a) 0° and 360°

b) 180°

c) negative

d) positive

---

92. The equation of the "tangent function" is given below:

$$\boxed{y = \tan\theta}$$

periodic

Continued on following page.

92. Continued

Some pairs of values for θ and y for the tangent function are given in the table below.

θ	y	θ	y
-90°	Undef.	0°	0.00
-89°	-57.29	30°	0.57
-85°	-11.43	45°	1.00
-75°	-3.73	60°	1.73
-60°	-1.73	75°	3.73
-45°	-1.00	85°	11.43
-30°	-0.57	89°	57.29
0°	0.00	90°	Undef.

By examining the table, you can see these facts:

a) As θ increases from -89° to 0°, y increases from ________ to ________.

b) As θ increases from 0° to 89°, y increases from ________ to ________.

c) y is undefined for two angles, ________ and ________.

---

93. Using the pairs of values from the table in the last frame, we graphed $y = \tan\theta$ from $\theta = -90°$ to $\theta = 90°$ below. The graph from -90° to 90° is the basic period of the tangent function.

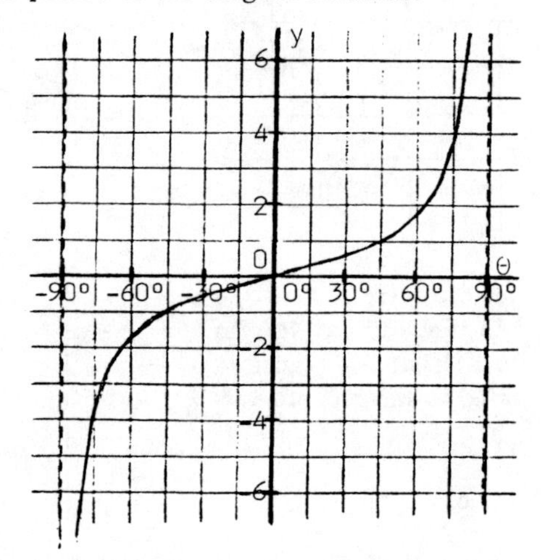

The two dashed lines at $\theta = -90°$ and $\theta = 90°$ are not part of the graph. They are called "asymptotes". Though the graph approaches the asymptotes, it never touches them because tan (-90°) and tan 90° are undefined.

Examine the graph to complete these:

a) From -90° to 0°, y is ____________ (positive/negative).

b) From 0° to 90°, y is ____________ (positive/negative).

Answers to frame 92:

a) -57.29 to 0

b) 0 to +57.29

c) -90° and 90°

94. The graph of $y = \tan \theta$ from -270° to 270° is shown below. The dashed lines are the asymptotes.

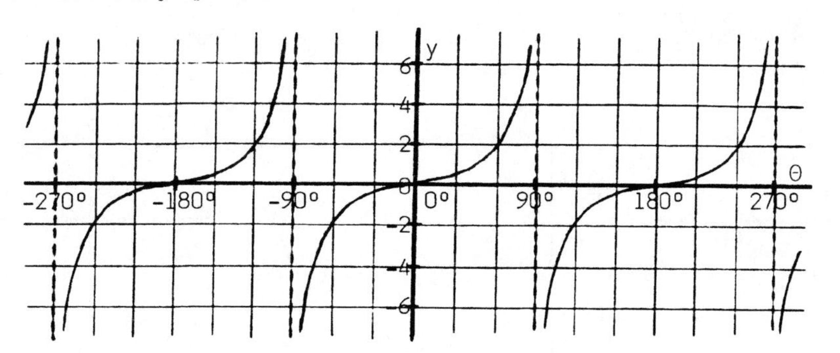

The basic period of the tangent function is from -90° to 90°. That period repeats itself from -270° to -90° and from 90° to 270°. It will continue to repeat itself if the graph is extended to the left of -270° and to the right of 270°.

Since the period repeats itself, the tangent function is also called a ______________ function.

a) negative

b) positive

---

periodic

SELF-TEST 8 (pages 100-116)

Given these five angles: 0° 90° 180° 270° 360°

1. For which angle or angles does $\cos \theta = 1$? ______________
2. For which angle or angles does $\sin \theta = -1$? ______________
3. For which angle or angles does $\tan \theta = 0$? ______________

For each angle below, find an angle between 0° and 360° with the same terminal side.

4. 610° ________
5. 875° ________
6. -290° ________
7. -435° ________

Continued on following page.

SELF-TEST 8 (pages 100-116) - Continued

Find two positive angles between 360° and 1080° with the same terminal side as a 290° angle.

8. ________

9. ________

Find two negative angles between 0° and -720° with the same terminal side as a -920° angle.

10. ________

11. ________

Fill each blank with a positive angle between 0° and 360°.

12. sin 740° = sin ________

13. cos 1200° = cos ________

14. tan (-500°) = tan ________

15. sin (-90°) = sin ________

Find the numerical value of each. Round to four decimal places.

16. tan 472° = ______

17. sin 1298° = ______

18. cos (-418°) = ______

19. tan (-270°) = ______

By referring to the graphs on pages 111 and 115, complete the following.

20. For the function $y = \sin \theta$, the largest value of y is ______.

21. For the function $y = \cos \theta$, as θ increases from 0° to 90°, y ____________ (increases/decreases).

22. For the function $y = \tan \theta$, as θ increases from 90° to 180°, y ____________ (increases/decreases).

ANSWERS:

1. 0° and 360°
2. 270°
3. 0°, 180°, and 360°
4. 250°
5. 155°
6. 70°
7. 285°
8. 650°
9. 1010°
10. -200°
11. -560°
12. sin 20°
13. cos 120°
14. tan 220°
15. sin 270°
16. -2.4751
17. -0.6157
18. 0.5299
19. undefined
20. +1
21. decreases
22. increases

3-9 DEFINITION OF A "DEGREE"

A "degree" is one unit used to measure the size of angles. We will define a "degree" in this section.

95. Before defining a "degree", we must review some facts about circles.

A central angle of a circle is an angle formed by two radii. Angle DOE is a central angle of this circle.

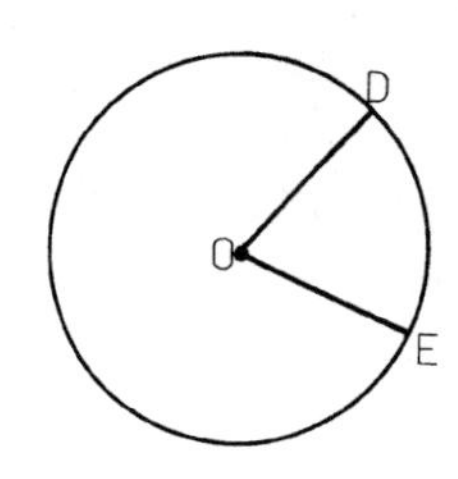

An arc of a circle is the curved line between two points on the circle. The curved line between points D and E is an arc of this circle. We call it "arc DE" or $\overset{\frown}{DE}$. (Note: The curved line over DE means "arc".)

Any central angle cuts off an arc on its circle. For example:

There are three central angles in the circle at the right. They are angles MOP, MOQ, and POQ. Name the arc cut off by:

a) angle MOP ________

b) angle MOQ ________

c) angle POQ ________

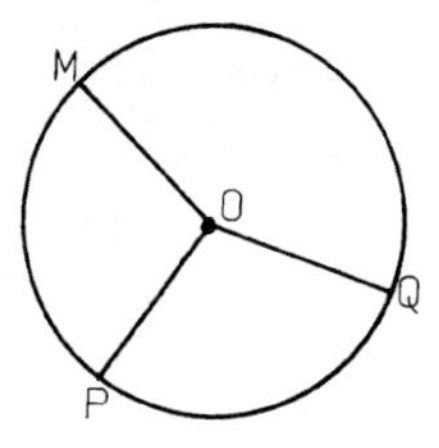

96. In any circle, equal arcs have equal central angles.

In this circle, the three arcs ($\overset{\frown}{AB}$, $\overset{\frown}{AC}$, and $\overset{\frown}{BC}$) are equal. Name the three equal central angles.

Angles ________, ________, ________

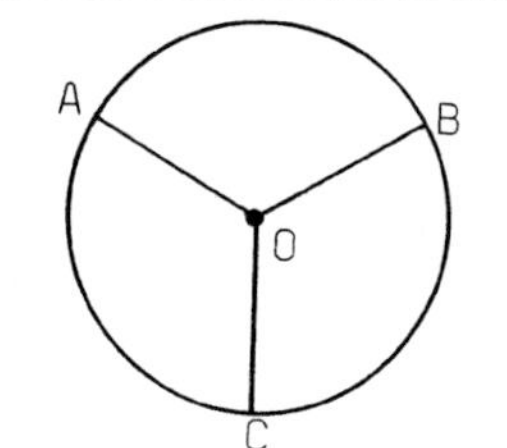

a) $\overset{\frown}{MP}$

b) $\overset{\frown}{MQ}$

c) $\overset{\frown}{PQ}$

97. The circumference of a circle is its boundary or the length of its boundary. We can compare the length of an arc ("arc-length") with the length of the circumference. For example, since there are four equal arcs in Figure 1 below, $\overset{\frown}{AB}$ is 1/4 of the circumference of the circle.

A
B

Figure 1

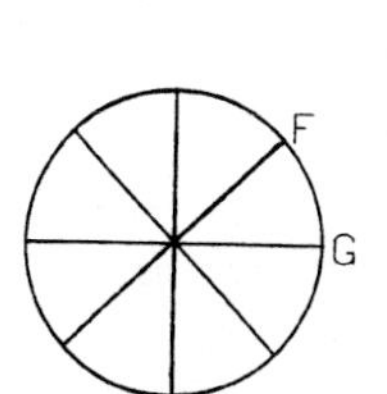

Figure 2

F
G

Figure 3

Angles AOB, AOC, BOC

Continued on following page.

97. Continued

a) There are three equal arcs in Figure 2. Therefore, $\overset{\frown}{CD}$ is what fractional part of the circumference? ______

b) There are eight equal arcs in Figure 3. Therefore, $\overset{\frown}{FG}$ is what fractional part of the circumference? ______

Answers to 97: a) $\frac{1}{3}$ b) $\frac{1}{8}$

98. A "degree" is defined in terms of a central angle of a circle which cuts off a definite arc-length. Its definition is:

AN ANGLE OF 1 DEGREE IS A CENTRAL ANGLE WHOSE ARC-LENGTH IS $\frac{1}{360}$ OF THE CIRCUMFERENCE OF THE CIRCLE.

The central angle in the diagram is 1° because its arc-length is $\frac{1}{360}$ of the circumference (C) of the circle.

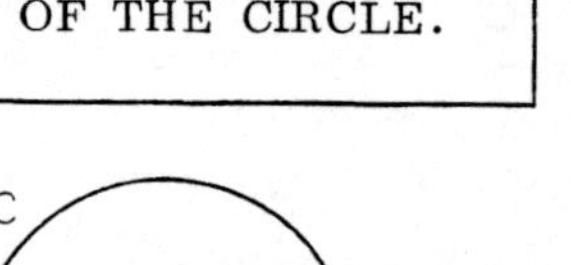
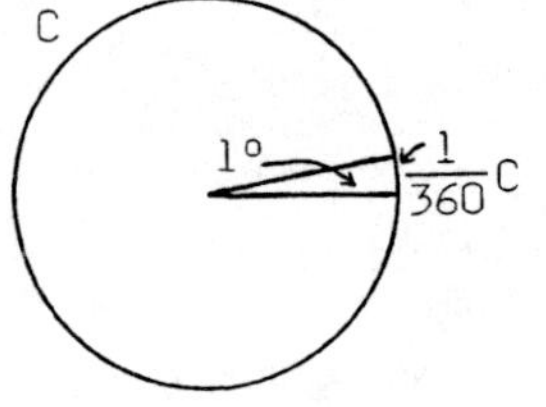

a) How many 1° central angles can be drawn in the circle? ______

b) How many degrees are there in a complete circle? ______

Answers to 98: a) 360 b) 360°

99. We can use the fact that there are 360° in a complete circle to find the size of each central angle below.

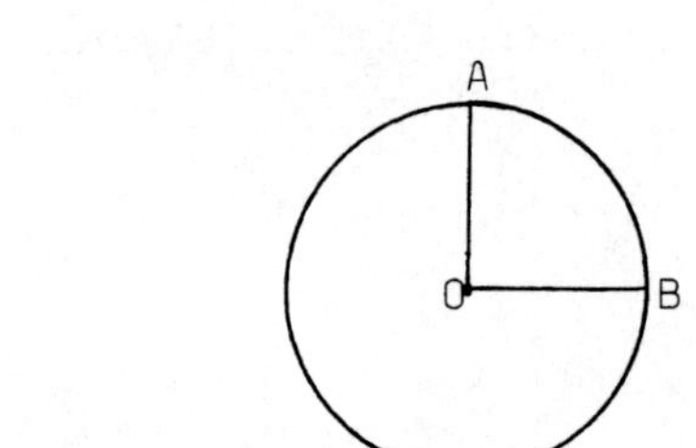

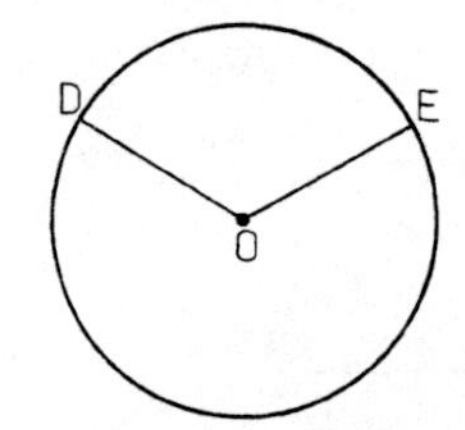

In the circle at the left, $\overset{\frown}{AB}$ is $\frac{1}{4}$ of the circumference of the circle. Therefore:

$$\text{Angle AOB} = \frac{1}{4} \times 360° = \frac{360°}{4} = 90°$$

In the circle at the right, $\overset{\frown}{DE}$ is $\frac{1}{3}$ of the circumference of the circle. Therefore:

$$\text{Angle DOE} = \frac{1}{3} \times 360° = \_\_\_\_\_\_$$

Answer to 99: 120°

100. We can think of standard-position angles as central angles. To do so, we include a circle whose center is at the origin on the coordinate system. Two examples are discussed.

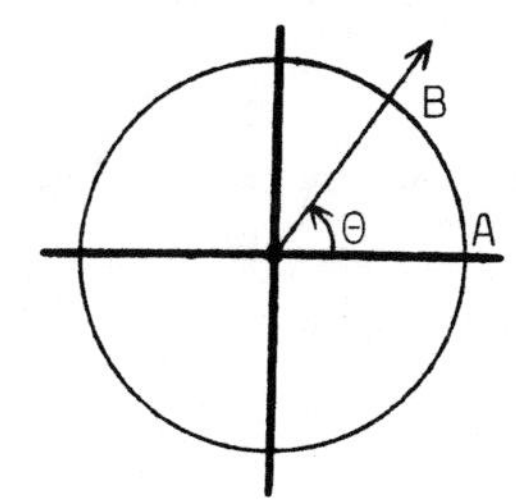

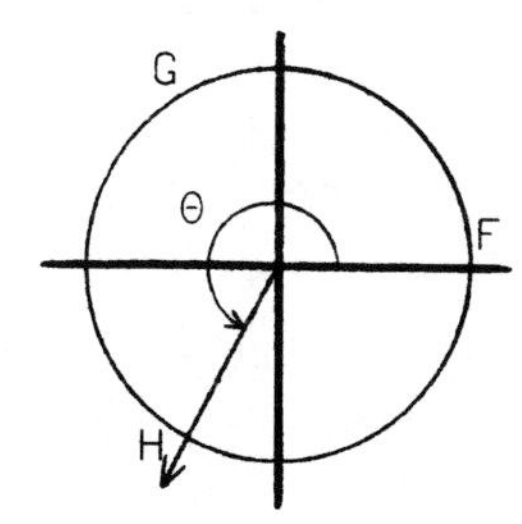

In the circle at the left, $\widehat{AB}$ is $\frac{1}{6}$ of the circumference of the circle. Therefore:

$$\theta = \frac{1}{6} \times 360° = \frac{360°}{6} = 60°$$

In the circle at the right, $\widehat{FGH}$ is $\frac{2}{3}$ of the circumference of the circle. Therefore:

$$\theta = \frac{2}{3} \times 360° = \_\_\_\_\_\_$$

Answer: 240°

101. The angle at the right is greater than 360°. Its arc-length is the circumference (one complete revolution) plus $\widehat{MT}$. Since $\widehat{MT}$ is $\frac{1}{6}$ of the circumference, the total arc-length is $1 + \frac{1}{6}$ or $\frac{7}{6}$ of the circumference. Therefore:

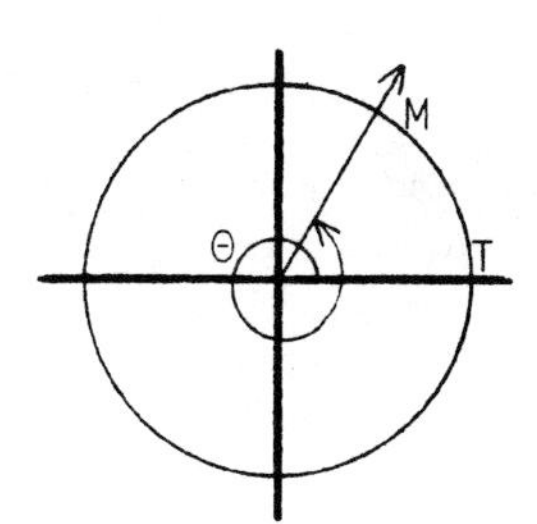

$$\theta = \frac{7}{6} \times 360° = \_\_\_\_\_\_$$

Answer: 420°

102. Negative angles can also be thought of as central angles with definite arc-lengths. Two examples are discussed.

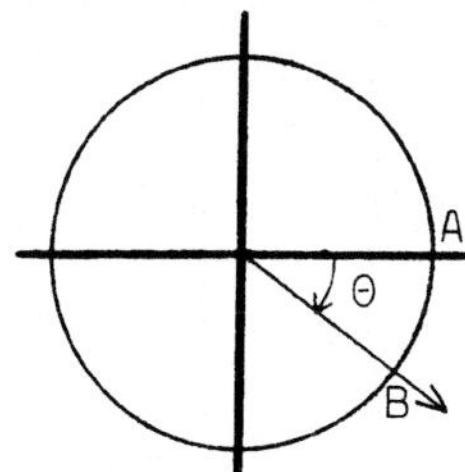

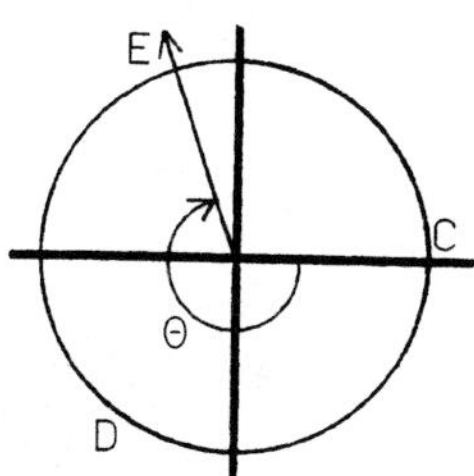

At the left, $\widehat{AB}$ is $\frac{1}{10}$ of the circumference. Therefore:

$$\theta = \frac{1}{10} \times (-360°) = \frac{-360°}{10} = -36°$$

At the right, $\widehat{CDE}$ is $\frac{7}{10}$ of the circumference. Therefore:

$$\theta = \frac{7}{10} \times (-360°) = \_\_\_\_\_\_$$

Answer: -252°

3-10 DEFINITION OF A "RADIAN"

Angles are sometimes measured with a unit called a "radian". We will define a "radian" in this section.

---

103. Units of angular measurement are defined in terms of central angles which cut off definite arc-lengths. When defining one degree, the arc-length chosen was $\frac{1}{360}$ of the circumference of the circle. We could have chosen any arc-length.

A "radian" is another measuring unit for angles. It is defined as a central angle which cuts off an arc-length equal to the radius of the circle. That is:

AN ANGLE OF 1 RADIAN IS A CENTRAL ANGLE WHOSE ARC-LENGTH EQUALS THE RADIUS OF THE CIRCLE.

The word "radian" is derived from the word "radius".

A diagram of an angle of 1 radian is shown at the right. Note that the radius of the circle is "r" and the length of $\overset{\frown}{GH}$ is "r".

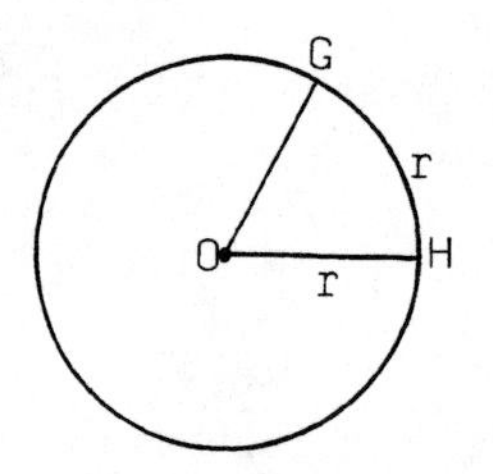

---

104. Given the radius of the circle and the arc-length of a central angle, we can use the following formula to find the number of radians in a central angle.

$$\boxed{\theta = \frac{s}{r}}$$

θ = angle (in radians)
s = arc-length
r = radius

Note: When $s = r$, $\theta = \frac{s}{r} = \frac{r}{r} = 1$ radian

We can use the above formula to determine the number of radians in each central angle below.

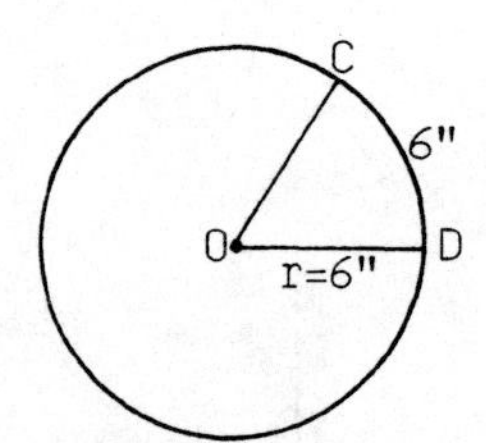

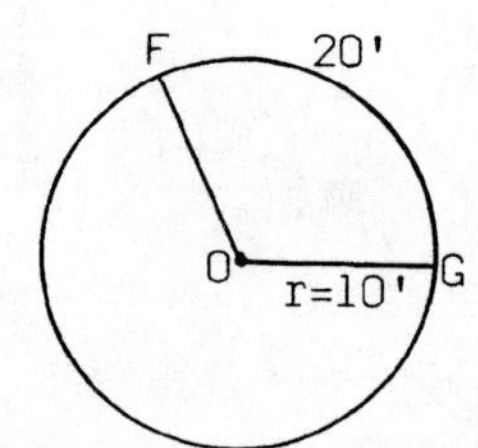

Continued on following page.

104. Continued

At the left, $r = 6''$ and $\widehat{CD} = 6''$. Therefore:

$$\text{Angle COD} = \frac{s}{r} = \frac{6''}{6''} = 1 \text{ radian}$$

At the right, $r = 10'$ and $\widehat{FG} = 20'$. Therefore:

$$\text{Angle FOG} = \frac{s}{r} = \frac{20'}{10'} = \underline{\quad\quad} \text{ radians}$$

---

105. The number of radians in an angle is not always a whole number. Here are two examples:

2 radians

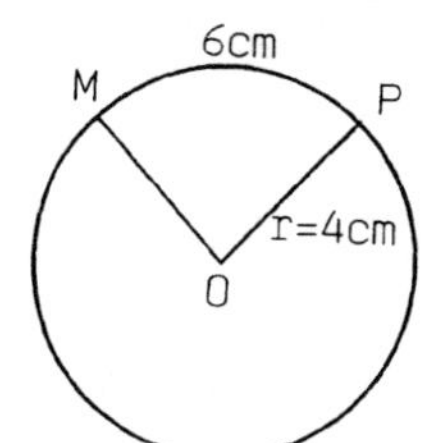

At the left, $r = 10m$ and $\widehat{AB} = 7m$. Therefore:

$$\text{Angle AOB} = \frac{s}{r} = \frac{7m}{10m} = 0.7 \text{ radian}$$

At the right, $r = 4$ cm and $\widehat{MP} = 6$ cm. Therefore:

$$\text{Angle MOP} = \frac{s}{r} = \frac{6 \text{ cm}}{4 \text{ cm}} = \underline{\quad\quad\quad} \text{ radians}$$

---

106. Angles in standard position can be measured in radians. To do so, we include a circle whose center is at the origin on the coordinate system. Two examples are discussed.

1.5 radians

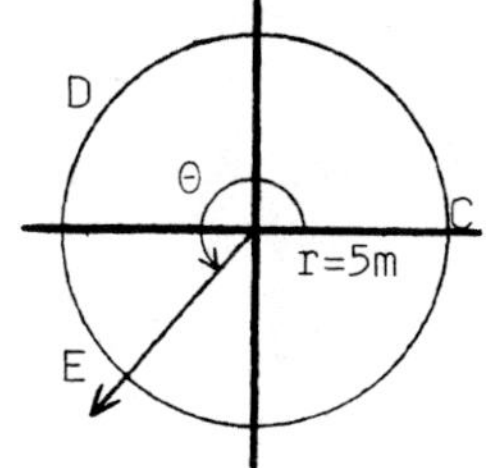

At the left, $r = 10''$ and $\widehat{AB} = 23''$. Therefore:

$$\theta = \frac{s}{r} = \frac{23''}{10''} = 2.3 \text{ radians}$$

At the right, $r = 5m$ and $\widehat{CDE} = 20m$. Therefore:

$$\theta = \frac{s}{r} = \frac{20m}{5m} = \underline{\quad\quad} \text{ radians}$$

4 radians

107. The angle at the right is greater than 360°. Its arc-length is the circumference (one revolution) plus $\widehat{AB}$. $\widehat{AB}$ is 7.00". Since the radius is 5.00", the circumference is:

$$C = 2\pi r = 2(3.14)(5.00") = 31.4"$$

The arc-length is: 31.4" + 7.00" = 38.4"

Therefore: $\theta = \frac{s}{r} = \frac{38.4"}{5.00"} = $ ________ radians

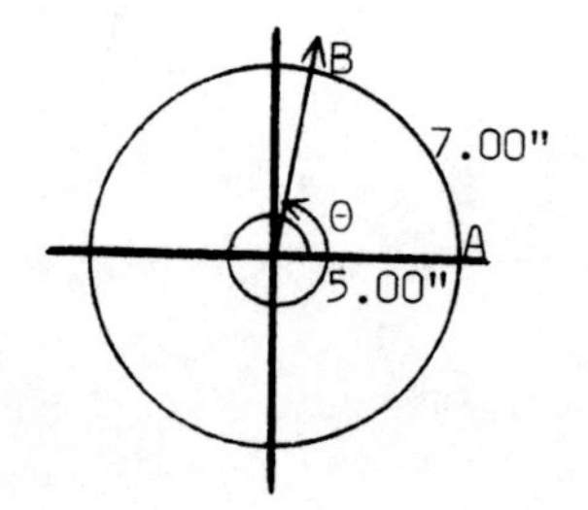

7.68 radians

108. Negative angles can be measured in radians. For example, the radius at the right is 4.0 cm. $\widehat{PR}$ is 10.0 cm. Therefore:

$\theta = $ ________ radians

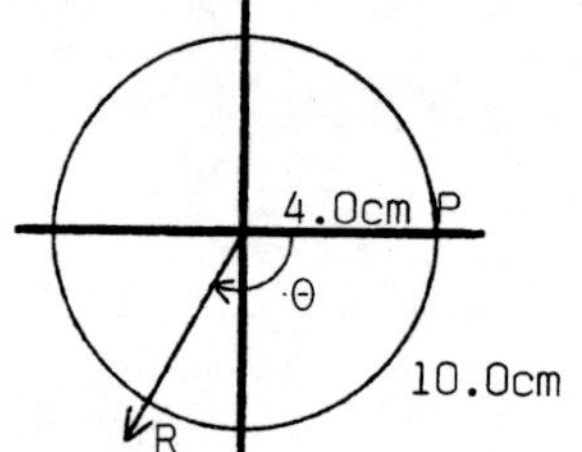

-2.5 radians

3-11 RADIANS EXPRESSED IN TERMS OF "π"

For some specific angles (like 30°, 45°, 60°, 90°, and so on), radian measures are often expressed in terms of "π". We will discuss radian measures of that type in this section.

109. Angles of 180° and 360° are expressed in terms of "π" radians in the box below.

180° =	π radians
360° =	2π radians

We can show that the relations above are true. To do so, a 360° angle and a 180° angle are shown in standard position below.

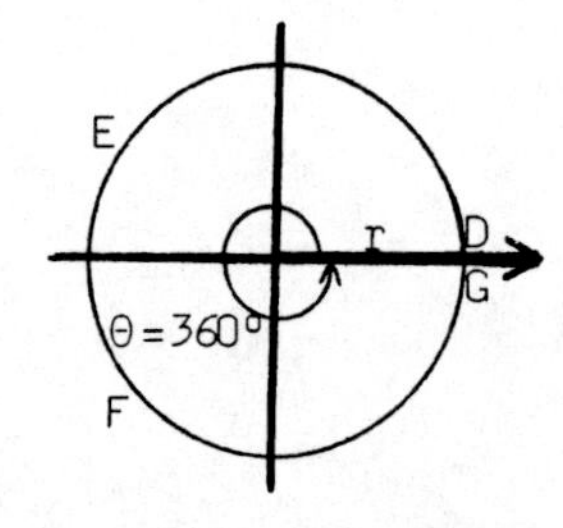

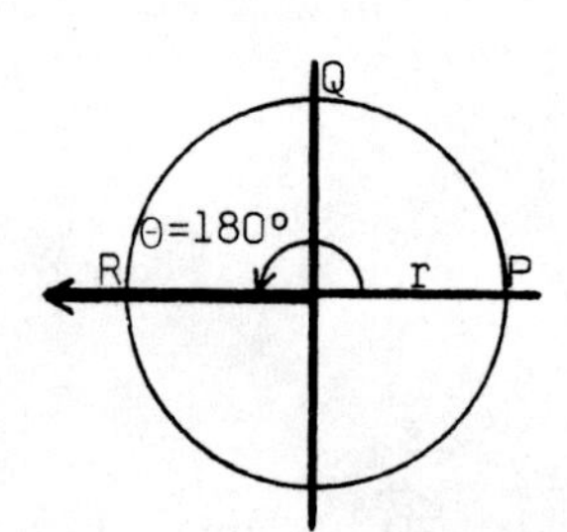

Continued on following page.

109. Continued

At the left, the arc of the 360° angle is $\widehat{DEFG}$ which equals the circumference of the circle. The circumference is $2\pi r$. Therefore:

$$360° \text{ (in radians)} = \frac{s}{r} = \frac{2\pi r}{r} = 2\pi\left(\frac{r}{r}\right) = \underline{2\pi \text{ radians}}$$

At the right, the arc of the 180° angle is $\widehat{PQR}$ which equals $\frac{1}{2}$ of the circumference of the circle. Since the circumference is $2\pi r$, $\frac{1}{2}$ the circumference is $\frac{1}{2}(2\pi r)$ or πr. Therefore:

$$180° \text{ (in radians)} = \frac{s}{r} = \frac{\pi r}{r} = \pi\left(\frac{r}{r}\right) = \underline{\pi \text{ radians}}$$

---

110. To convert from degrees to radians expressed in terms of "π", we use this relation:

$$\boxed{180° = \pi \text{ radians}}$$

Substituting π radians for 180°, complete these conversions.

$$90° = \frac{1}{2}(180°) = \frac{1}{2}(\pi \text{ radians}) = \frac{\pi}{2} \text{ radians.}$$

a) $60° = \frac{1}{3}(180°) = \frac{1}{3}(\pi \text{ radians}) = \_\_\_\_ \text{ radians.}$

b) $45° = \frac{1}{4}(180°) = \frac{1}{4}(\pi \text{ radians}) = \_\_\_\_ \text{ radians.}$

c) $30° = \frac{1}{6}(180°) = \frac{1}{6}(\pi \text{ radians}) = \_\_\_\_ \text{ radians.}$

---

111. Substituting "π radians" for 180°, complete these conversions.

a) $\frac{\pi}{3}$ radians

b) $\frac{\pi}{4}$ radians

c) $\frac{\pi}{6}$ radians

$$120° = \frac{2}{3}(180°) = \frac{2}{3}(\pi \text{ radians}) = \frac{2\pi}{3} \text{ radians.}$$

a) $135° = \frac{3}{4}(180°) = \frac{3}{4}(\pi \text{ radians}) = \_\_\_\_ \text{ radians.}$

b) $150° = \frac{5}{6}(180°) = \frac{5}{6}(\pi \text{ radians}) = \_\_\_\_ \text{ radians.}$

c) $270° = \frac{3}{2}(180°) = \frac{3}{2}(\pi \text{ radians}) = \_\_\_\_ \text{ radians.}$

---

a) $\frac{3\pi}{4}$ radians

b) $\frac{5\pi}{6}$ radians

c) $\frac{3\pi}{2}$ radians

112. To convert from radians expressed in terms of "π" to degrees, we substitute 180° for "π radians". For example:

$$\frac{\pi}{2} \text{ radians} = \frac{1}{2}(\pi \text{ radians}) = \frac{1}{2}(180°) = 90°$$

a) $\frac{\pi}{3}$ radians $= \frac{1}{3}(\pi$ radians$) = \frac{1}{3}(180°) =$ ______

b) $\frac{\pi}{4}$ radians $= \frac{1}{4}(\pi$ radians$) = \frac{1}{4}(180°) =$ ______

c) $\frac{\pi}{6}$ radians $= \frac{1}{6}(\pi$ radians$) = \frac{1}{6}(180°) =$ ______

---

113. Substituting 180° for "π radians", complete these conversions.

$$\frac{2\pi}{3} \text{ radians} = \frac{2}{3}(\pi \text{ radians}) = \frac{2}{3}(180°) = 120°$$

a) $\frac{3\pi}{4}$ radians $= \frac{3}{4}(\pi$ radians$) = \frac{3}{4}(180°) =$ ______

b) $\frac{5\pi}{6}$ radians $= \frac{5}{6}(\pi$ radians$) = \frac{5}{6}(180°) =$ ______

c) $\frac{3\pi}{2}$ radians $= \frac{3}{2}(\pi$ radians$) = \frac{3}{2}(180°) =$ ______

Answers to 112: a) 60° b) 45° c) 30°

---

114. It is helpful to remember that expressions like $\frac{2\pi}{3}$, $\frac{5\pi}{6}$, and $\frac{3\pi}{2}$ are really "π" with a fractional coefficient. That is:

$$\frac{2\pi}{3} = \frac{2}{3}\pi \qquad \frac{5\pi}{6} = \frac{5}{6}\pi \qquad \frac{3\pi}{2} = \text{\_\_\_\_\_\_}$$

Answers to 113: a) 135° b) 150° c) 270°

---

115. In the table below, we summarized the degree-radian conversions for some specific angles. <u>They should be memorized</u>.

Degrees	0°	30°	45°	60°	90°	120°	135°	150°	180°	270°	360°
Radians	0 or 0π	$\frac{\pi}{6}$ or $\frac{1}{6}\pi$	$\frac{\pi}{4}$ or $\frac{1}{4}\pi$	$\frac{\pi}{3}$ or $\frac{1}{3}\pi$	$\frac{\pi}{2}$ or $\frac{1}{2}\pi$	$\frac{2\pi}{3}$ or $\frac{2}{3}\pi$	$\frac{3\pi}{4}$ or $\frac{3}{4}\pi$	$\frac{5\pi}{6}$ or $\frac{5}{6}\pi$	π or 1π	$\frac{3\pi}{2}$ or $\frac{3}{2}\pi$	2π or 2π

Using the table, complete these:

a) 30° = ______ radians

b) 135° = ______ radians

c) $\frac{\pi}{2}$ radians = ______°

d) $\frac{5\pi}{6}$ radians = ______°

Answer to 114: $\frac{3}{2}\pi$

116. Complete these:

a) $0° =$ ____ radians

b) $360° =$ ______ radians

c) π radians $=$ ________$°$

d) $\frac{3\pi}{2}$ radians $=$ ________$°$

a) $\frac{\pi}{6}$ b) $\frac{3\pi}{4}$ c) $90°$ d) $150°$

---

117. Complete these:

a) $90° =$ ______ radians

b) $120° =$ ______ radians

c) $\frac{\pi}{3}$ radians $=$ ________$°$

d) $\frac{\pi}{4}$ radians $=$ ________$°$

a) 0 b) 2π c) $180°$ d) $270°$

---

a) $\frac{\pi}{2}$ b) $\frac{2\pi}{3}$ c) $60°$ d) $45°$

3-12 DEGREE-RADIAN CONVERSIONS

In this section, we will show a simple method for converting from degrees to radians and from radians to degrees.

---

118. Using the fact that $180° = \pi$ radians, we can set up two fractions that equal "1". Because the fractions equal "1", they are called "<u>unity</u>" fractions.

$$\frac{180°}{\pi \text{ radians}} = 1 \qquad \frac{\pi \text{ radians}}{180°} = 1$$

Using the unity fraction at the left, we can convert 1 radian to degrees. Notice that the "radians" cancel.

$$1 \text{ radian} = 1\,\cancel{\text{radian}}\left(\frac{180°}{\pi\,\cancel{\text{radians}}}\right) = \frac{180°}{\pi}$$

We can use a caluclator to complete the conversion. If your calculator has a [π] key, use it. If it has "π" in the upper register, use [INV] [π] or whatever gets you to the upper register. The steps are:

Enter	Press	Display
180	[÷]	180
	or [÷] [INV]	
π	[=]	57.29578

Rounding to the nearest tenth, we get: 1 radian = __________$°$

57.3°

119. Using the unity-fraction method, we can convert 2.5 radians to degrees. We get:

$$2.5 \text{ radians} = 2.5\,\cancel{\text{radians}}\left(\frac{180°}{\pi\,\cancel{\text{radians}}}\right) = \frac{2.5(180°)}{\pi}$$

We can use a calculator to complete the conversion. The steps are:

Enter	Press	Display
2.5	[x]	2.5
180	[÷]	450
	or [÷] [INV]	
π	[=]	143.23945

Rounding to a whole number, we get: 2.5 radians = ________°

Answer: 143°

120. Using the same method, complete each conversion below. Round to tenths.

a) 0.7 radian = ________° b) 1.6 radians = ________°

Answers: a) 40.1°, from: $\frac{0.7(180°)}{\pi}$ b) 91.7°, from: $\frac{1.6(180°)}{\pi}$

121. Using the same method, complete each conversion. Round to a whole number.

a) -2 radians = ________° b) 8.5 radians = ________°

Answers: a) -115°, from: $\frac{-2(180°)}{\pi}$ b) 487°, from: $\frac{8.5(180°)}{\pi}$

122. Notice in this conversion that both the "π's" and the "radians" cancel.

$$\frac{3\pi}{4} \text{ radians} = \frac{3\cancel{\pi}}{4}\,\cancel{\text{radians}}\left(\frac{180°}{\cancel{\pi}\,\cancel{\text{radians}}}\right) = \frac{3(180°)}{4} = \frac{540°}{4} = 135°$$

Using the same steps, complete this conversion.

$$\frac{\pi}{3} \text{ radians} = \frac{\pi}{3} \text{ radian}\left(\frac{180°}{\pi \text{ radians}}\right) = \underline{\hspace{3cm}}$$

Answer: 60°, from: $\frac{180°}{3}$

123. To convert degrees to radians, we use the following unity fraction:

$$\frac{\pi \text{ radians}}{180°} = 1$$

For example, to convert 80° to radians, we get:

$$80° = 80\not{°}\left(\frac{\pi \text{ radians}}{180\not{°}}\right) = \frac{80(\pi)}{180}\text{ radians}$$

We can use a calculator to complete the conversion. We get:

Enter	Press	Display
80	[x]	80
	or [x] [INV]	
π	[÷]	251.32741
180	[=]	1.3962634

Rounding to hundredths, we get: 80° = ________ radians

1.40 radians

124. Use a calculator to complete these. Round to hundredths.

a) $37° = 37\not{°}\left(\frac{\pi \text{ radians}}{180\not{°}}\right) = \frac{37(\pi)}{180}$ radians = ________ radians

b) $149° = 149\not{°}\left(\frac{\pi \text{ radians}}{180\not{°}}\right) = \frac{149(\pi)}{180}$ radians = ________ radians

---

a) 0.65 radian

b) 2.60 radians

125. Complete these. Round to hundredths.

a) -13° = ________ radians

c) 90° = ________ radians

b) 548° = ________ radians

d) 120° = ________ radians

---

a) -0.23 radians, from: $\frac{-13\not{°}(\pi)}{180\not{°}}$

c) 1.57 radians, from: $\frac{90\not{°}\pi}{180\not{°}}$

b) 9.56 radians, from: $\frac{548\not{°}(\pi)}{180\not{°}}$

d) 2.09 radians, from: $\frac{120\not{°}\pi}{180\not{°}}$

126. In the table below, we have given some conversions of degrees to radians. Notice that radians are expressed in terms of "π" and also as decimal numbers rounded to hundredths.

Degrees	-360°	-270°	-180°	-90°	0°	90°	180°	270°	360°
Radians	-2π or -6.28	$-\frac{3\pi}{2}$ or -4.71	$-\pi$ or -3.14	$-\frac{\pi}{2}$ or -1.57	0π or 0.00	$\frac{\pi}{2}$ or 1.57	π or 3.14	$\frac{3\pi}{2}$ or 4.71	2π or 6.28

In the next table, we have given some conversions of radians to degrees. The degrees are rounded to tenths.

Radians	-4	-3	-2	-1	0	1	2	3	4
Degrees	-229.2°	-171.9°	-114.6°	-57.3°	0.0°	57.3°	114.6°	171.9°	229.2°

You can see from the table that 1 radian = 57.3°. To make quick estimates of conversions, we can use this basic fact:

1 radian is approximately 60°

Using the above fact, approximately how many degrees are contained in an angle of:

a) 2 radians? ________°   b) 5 radians? ________°   c) 0.5 or $\frac{1}{2}$ radians? ________°

---

127. Similarly, we can estimate the number of radians in an angle given in degrees by dividing by 60°. For example:

180° is approximately 3 radians $\left(\text{from } \frac{180°}{60°}\right)$

a) 420° is approximately ______ radians.

b) 90° is approximately ______ radians.

c) -120° is approximately ______ radians.

Answers: a) 120° b) 300° c) 30°

---

128. Complete: a) 900° is approximately _______ radians.

b) 10 radians is approximately ________°.

c) -90° is approximately _______ radians.

d) 2.5 radians is approximately ________°.

Answers: a) 7 radians b) $1\frac{1}{2}$ or 1.5 radians c) -2 radians

---

a) 15 radians   b) 600°   c) $-1\frac{1}{2}$ or -1.5 radians   d) 150°

3-13 TRIG RATIOS OF ANGLES EXPRESSED IN RADIANS

In this section, we will show how a calculator can be used to find the sine, cosine, or tangent of angles expressed in radians.

---

129. The symbol ° is an abbreviation for the word "degree". The symbol is always used for denoting angles measured in degrees. For example:

tan 7° means "tan 7 degrees"

No similar symbol exists for denoting angles measured in radians. We simply write the numerical value of the angle. For example:

tan 7 means "tan 7 radians"

$\cos \frac{\pi}{4}$ means "$\cos \frac{\pi}{4}$ radians"

Complete the following:

a) sin 3 = sin 3 ________

b) cos 3° = cos 3 ________

c) $\tan \frac{\pi}{3} = \tan \frac{\pi}{3}$ ________

d) sin 1.64 = sin 1.64 ________

---

130. To use a calculator to find the sine, cosine, or tangent of an angle given in degrees, the calculator must be in the "degree mode". That is, we press the [DRG] key until "DEG" appears on the display. When a calculator is turned on, it will be in the "degree mode".

To use a calculator to find the sine, cosine, or tangent of an angle given in radians, the calculator must be in the "radian mode". That is, we push the [DRG] key until "RAD" appears on the display.

Put your calculator in the "radian mode" and then press [sin], [cos], or [tan] to complete these. Round to four decimal places.

a) sin 1 = ________

b) cos 2 = ________

c) tan 4.5 = ________

Answers to 129:
a) radians
b) degrees
c) radians
d) radians

---

131. We use the following steps to find $\sin \frac{3\pi}{4}$ on a calculator.

Enter	Press	Display
3	[x]	3
	or [x] [INV]	
π	[÷]	9.424778
4	[=] [sin]	0.7071068

Using the same steps, complete these. Round to hundredths when necessary.

a) $\cos \frac{\pi}{3}$ = ________

b) $\tan \frac{2\pi}{3}$ = ________

Answers to 130:
a) 0.8415
b) -0.4161
c) 4.6373

132. Complete these:

a) $\sin\frac{\pi}{2}$ = ______ b) $\cos \pi$ = ______ c) $\tan\frac{5\pi}{4}$ = ______

a) 0.5
b) -1.73

---

133. Complete these. Round to four decimal places.

a) $\sin(-0.86)$ = ______ b) $\cos(-5)$ = ______

a) 1
b) -1
c) 1

---

134. In graphing the sine function $y = \sin\theta$, we sometimes calibrate the horizontal axis in radians rather than degrees. The radians are expressed in terms of "π". The word "radian" does not appear on the graph.

a) -0.7578
b) 0.2837

The graph of $y = \sin\theta$ below has its horizontal axis calibrated in radians. For reference, corresponding degree values are shown.

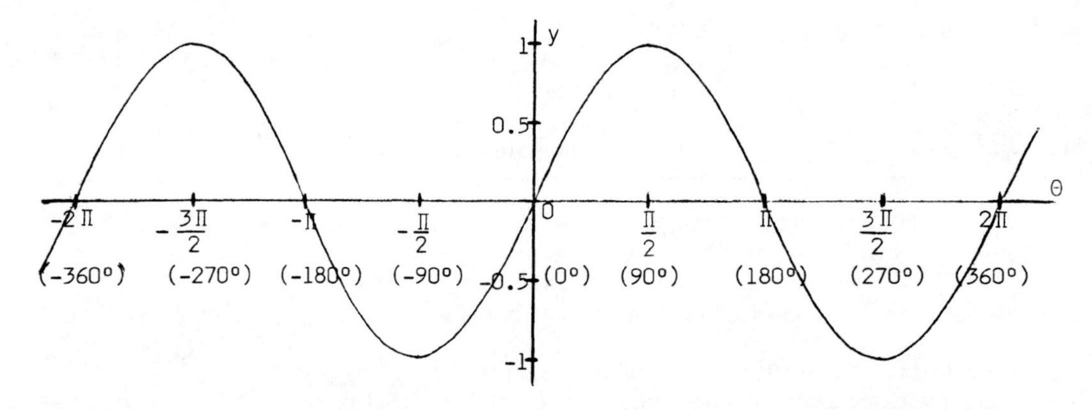

Similarly, the cosine function and the tangent function are sometimes graphed with the horizontal axis calibrated in radians.

SELF-TEST 9 (pages 116-131)

1. The circumference of a circle is divided into 18 equal arcs. Find the number of degrees in each central angle. ______

2. If $\overset{\frown}{ABC}$ is two-fifths of the circumference, find angle θ in degrees.

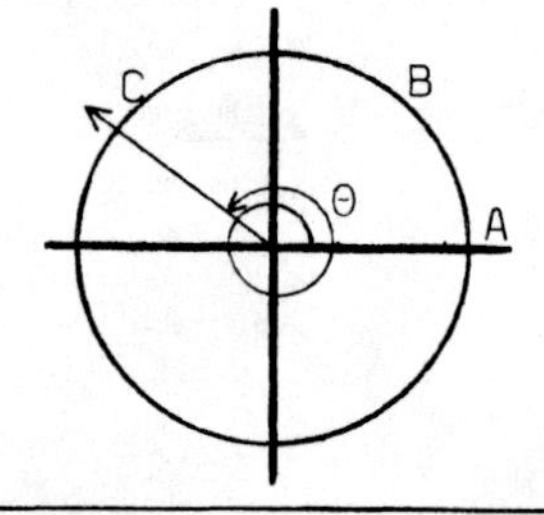

θ = ______

3. In a circle of radius 20 cm, what length of arc is cut off by an angle of 1 radian? ______

Continued on following page.

SELF-TEST 9 (pages 116-131) - Continued

4. In a circle of radius 28.9" a central angle cuts off a 17.4" arc-length. Find the central angle in radians. Round to hundredths.

________ radian

5. If radius OF = 15 cm and $\overset{\frown}{FGH}$ = 62 cm, find angle θ in radians. Round to tenths.

θ = ______ radians

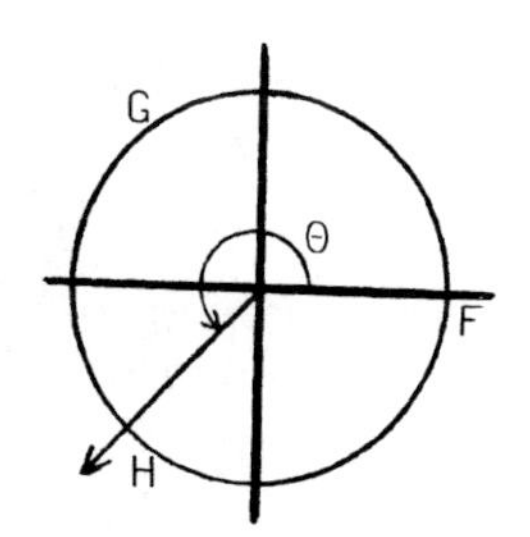

Convert each angle to degrees.

6. $\frac{\pi}{2}$ radians = ________

7. $\frac{3\pi}{4}$ radians = ________

8. $\frac{7\pi}{6}$ radians = ________

Convert each angle to radians expressed in terms of "π".

9. 180° = ________ radians

10. 270° = ________ radians

11. 300° = ________ radians

Use the "unity-fraction" method for these conversions.

Round each to tenths.

12. 1 radian = __________°

13. 2.71 radians = __________°

14. $\frac{8\pi}{3}$ radians = __________°

Round each to hundredths.

15. 152° = __________ radians

16. 90° = __________ radians

17. -28° = __________ radians

Using the calculator, find the value of each. Round to four decimal places.

18. $\sin\frac{\pi}{3}$ = ______

19. $\cos\frac{3\pi}{4}$ = ______

20. tan 2 = ______

21. sin(-4.35) = ______

ANSWERS:

1. 20°
2. 504°
3. 20 cm
4. 0.60 radian
5. 4.1 radian
6. 90°
7. 135°
8. 210°
9. π radians
10. $\frac{3\pi}{2}$ radians
11. $\frac{5\pi}{3}$ radians
12. 57.3°
13. 155.3°
14. 480°
15. 2.65 radians
16. 1.57 radians
17. -0.49 radian
18. 0.8660
19. -0.7071
20. -2.1850
21. 0.9351

3-14 THE UNIT CIRCLE AND TRIG FUNCTIONS

In this section, we will discuss the sine, cosine, and tangent of angles in terms of the coordinates of a point on a unit circle.

---

135. The circle at the left below is called a "unit circle" because its radius is 1 unit. A "unit circle" is also shown on the coordinate system at the right below.

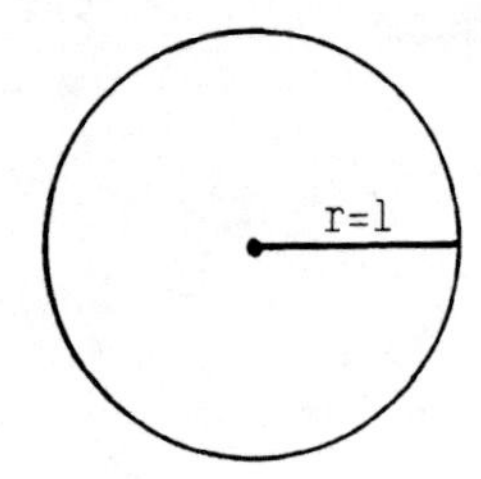

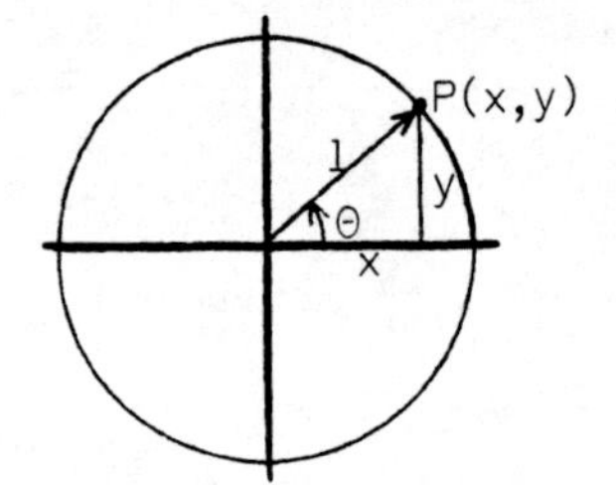

At the right above, angle θ is associated with point P on the circle. The hypotenuse of the reference triangle is "1" because it is a radius of the circle. Using the reference triangle, we can see that both $\sin\theta$ and $\cos\theta$ are one of the coordinates of P. That is:

$$\sin\theta = \frac{y}{1} = y$$

$$\cos\theta = \frac{x}{1} = x$$

If we know θ, we can find the coordinates of P. For example, if $\theta = 53°$:

$$\sin\theta = y = 0.7986$$

$$\cos\theta = x = 0.6018$$

Therefore, if $\theta = 53°$, the coordinates of P are (,).

(0.6018, 0.7986)

136. The circle at the right is a unit circle because its radius is "1". Since P is in the second quadrant, "x" is negative and "y" is positive. By finding sin 130° and cos 130°, we can find the coordinates of P.

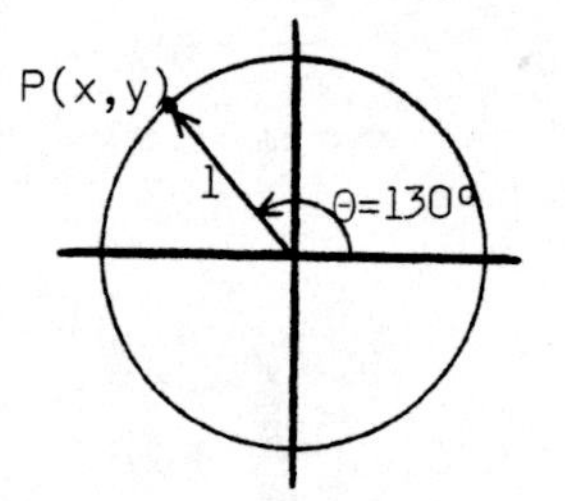

$$\sin 130° = y = 0.7660$$

$$\cos 130° = x = -0.6428$$

Therefore, the coordinates of P are (,).

(-0.6428, 0.7660)

137. The circle at the right is a unit circle. Since P is in the third quadrant, both "x" and "y" are negative. By finding sin 234° and cos 234°, we can find the coordinates of P.

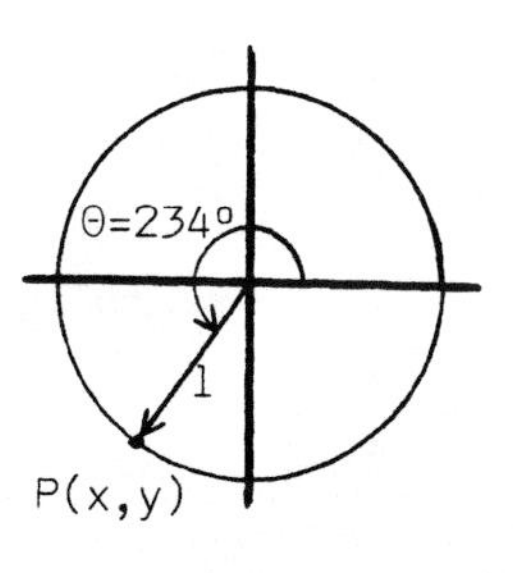

$\sin 234° = y = -0.8090$

$\cos 234° = x = -0.5878$

Therefore, the coordinates of P are (,).

---

138. The circle at the right is a unit circle. Since P is in the fourth quadrant, "x" is positive and "y" is negative. By finding sin 331° and cos 331°, we can find the coordinates of P.

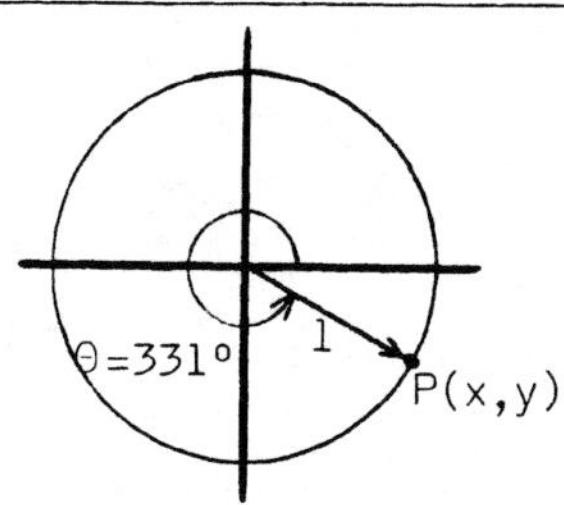

$\sin 331° = y = -0.4848$

$\cos 331° = x = 0.8746$

Therefore, the coordinates of P are (,).

Answer: (-0.5878, -0.8090)

---

139. The circle at the right is a unit circle. Using the reference triangle, we can see this fact:

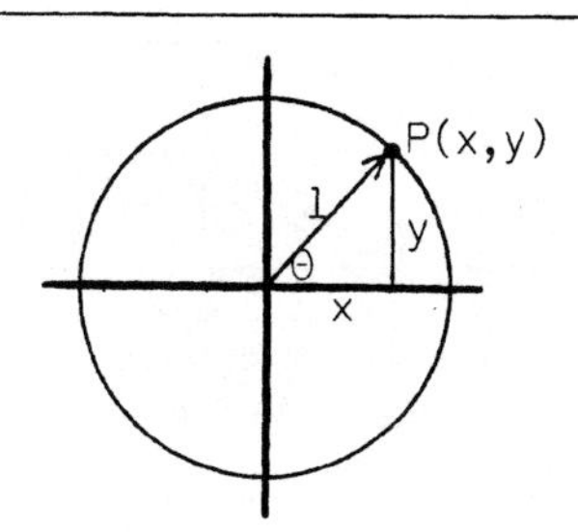

$$\tan \theta = \frac{y}{x}$$

But since $y = \sin \theta$ and $x = \cos \theta$, we can also express $\tan \theta$ in terms of $\sin \theta$ and $\cos \theta$. We get:

$$\tan \theta = \frac{\sin \theta}{\cos \theta} \quad \left(\text{from } \frac{y}{x}\right)$$

Let's use $\theta = 33°$ to confirm the fact that we can find $\tan \theta$ by dividing $\sin \theta$ by $\cos \theta$. The relationship and calculator steps are shown.

$$\tan 33° = \frac{\sin 33°}{\cos 33°}$$

Enter	Press	Display
33	[sin] [÷]	0.544639
33	[cos] [=]	0.6494076

By dividing sin 33° by cos 33°, we get tan 33° = 0.6494076. Confirm the procedure by finding tan 33° directly on a calculator. Do you get tan 33° = 0.6494076? ________

Answer: (0.8746, -0.4848)

Answer: Yes

140. In the last frame, we used $\theta = 33°$ to confirm the following relationship.

$$\tan\theta = \frac{\sin\theta}{\cos\theta}$$

Let's use $\theta = 255°$ to confirm the relationship again. We get:

$$\tan 255° = \frac{\sin 255°}{\cos 255°}$$

Using a calculator, do these:

a) Find tan 255° by dividing sin 255° by cos 255°.

tan 255° = ____________

b) Find tan 255° directly.

tan 255° = ____________

c) Is the relationship confirmed? ________

> Note: The sine, cosine, and tangent functions are sometimes called circular functions because they can be defined on the unit circle.

a) tan 255° = 3.7320508

b) tan 255° = 3.7320508

c) Yes

3-15 THE UNIT CIRCLE AND ANGLES ON THE AXES

In this section, we will use a unit circle to confirm the values of the sine, cosine, and tangent of angles on the axes (like 0°, 90°, 180°, 270°, 360°, and so on).

141. Angles of 0° and 360° are shown in standard position below. A unit circle is drawn; the coordinates of P are (1, 0) because the radius is "1".

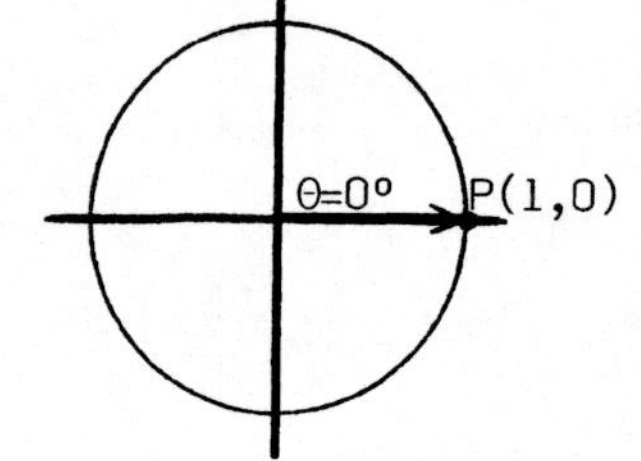

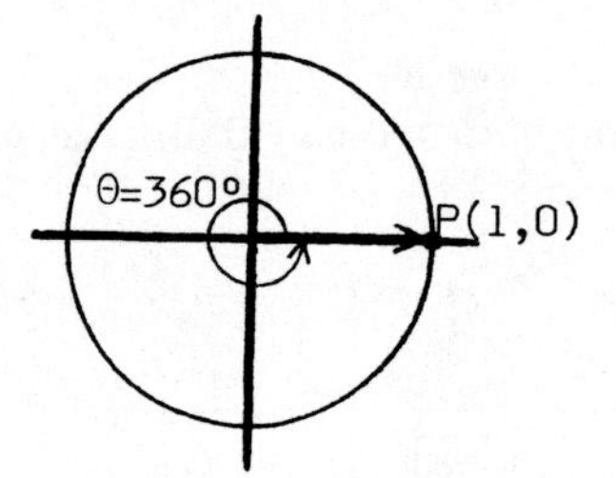

In the last section, we saw the following relationships for angles related to points on the unit circle.

$\sin\theta = y$ $\cos\theta = x$

Using (1, 0), the coordinates of P, we get:

$\sin 0° = \sin 360° = 0$

$\cos 0° = \cos 360° =$ ____

1

142. Angles of 90°, 180°, and 270° are shown in standard position below. A unit circle is drawn; the coordinates of P are given in each case.

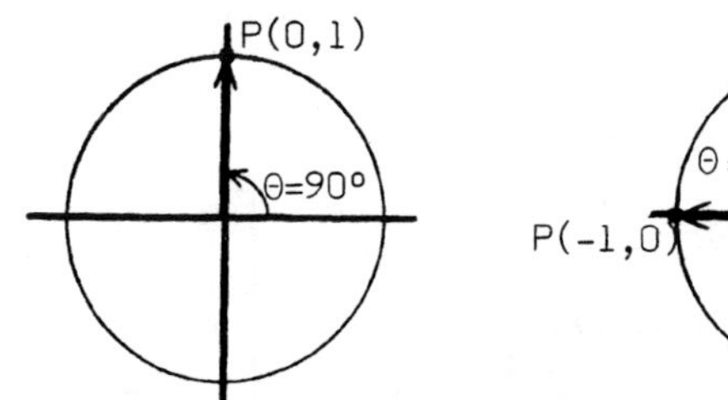

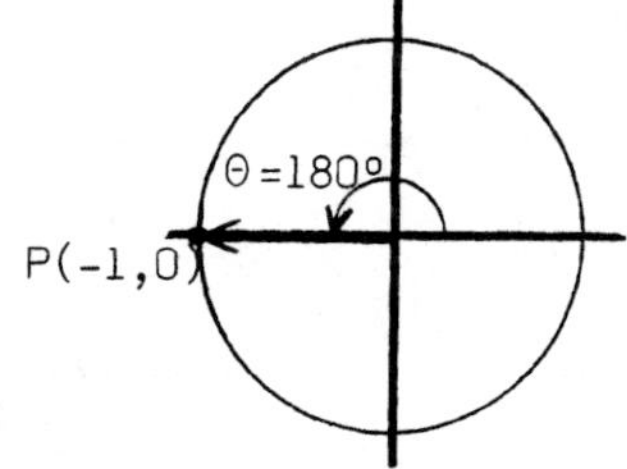

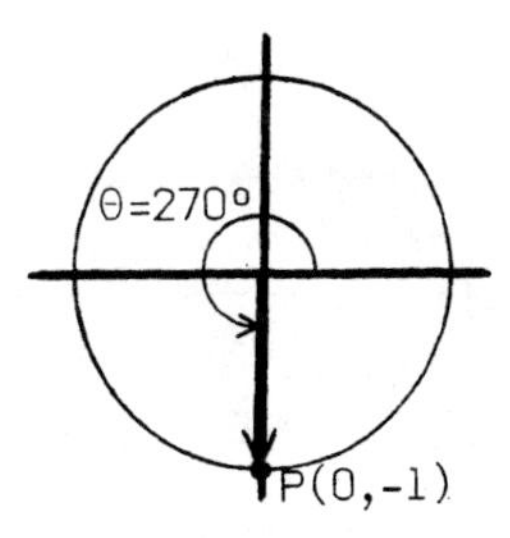

The following relationships apply to the above angles.

$\sin \theta = y$ $\cos \theta = x$

Using the coordinates of P in each case, we get:

$\sin 90° = 1$ $\cos 90° = 0$

$\sin 180° = 0$ $\cos 180° = -1$

$\sin 270° =$ ______ $\cos 270° =$ ______

$\sin 270° = -1$

$\cos 270° = 0$

143. The values of $\sin \theta$ and $\cos \theta$ for 0°, 90°, 180°, 270°, and 360° are given below.

θ	0°	90°	180°	270°	360°
$\sin \theta$	0	+1	0	-1	0
$\cos \theta$	+1	0	-1	0	+1

In the last section, we saw the following relationship for $\tan \theta$.

$$\tan \theta = \frac{\sin \theta}{\cos \theta}$$

Let's use the relationship to find tan 0° and tan 90°.

$$\tan 0° = \frac{0}{+1} = 0 \qquad \tan 90° = \frac{+1}{0} = \text{undefined}$$

Using the same relationship, find these:

a) $\tan 180° =$ ______ b) $\tan 270° =$ ______ c) $\tan 360° =$ ______

---

a) 0

b) undefined

c) 0

144. The table below contains sin θ, cos θ, and tan θ for 0°, 90°, 180°, 270°, and 360°.

θ	0°	90°	180°	270°	360°
sin θ	0	+1	0	-1	0
cos θ	+1	0	-1	0	+1
tan θ	0	Undef.	0	Undef.	0

As we saw earlier, the values above also apply to angles greater than 360° with the same terminal sides. For example, 450° has the same terminal side as 90°. Therefore:

$\sin 450° = 1$ $\cos 450° = 0$ $\tan 450° =$ ______

---

145. The same values also apply to negative angles with the same terminal sides. For example, -90° has the same terminal side as 270°. Therefore:

Answer (144): Undefined

$\sin(-90°) = -1$ $\cos(-90°) = 0$ $\tan(-90°) =$ ______

Answer (145): Undefined

3-16 THE UNIT CIRCLE AND ANGLES EXPRESSED IN RADIANS

In this section, we will show how the unit circle can be used to express angles in radians. The discussion is limited to radians expressed in terms of "π".

---

146. The circle at the right is a unit circle. Like all circles, the formula for its circumference is $C = 2\pi r$. Since its radius is 1 unit, the length of its circumference is:

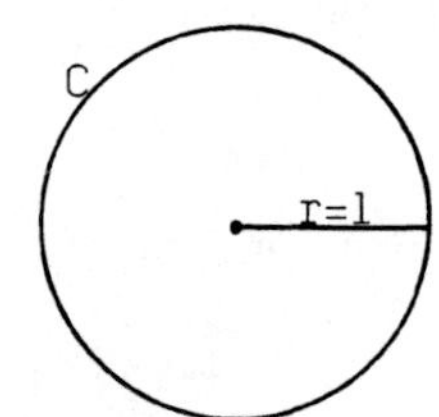

$$C = 2\pi r = 2\pi(1) = 2\pi$$

We can easily find fractional parts of the circumference. For example:

$$\frac{1}{2}C = \frac{1}{2}(2\pi) = \frac{\not{2}\pi}{\not{2}} = \pi$$

$$\frac{1}{3}C = \frac{1}{3}(2\pi) = \frac{2\pi}{3}$$

$$\frac{1}{4}C = \frac{1}{4}(2\pi) = \frac{2\pi}{4} = \_\_\_\_$$

Answer (146): $\frac{\pi}{2}$

147. A 360° angle and a 180° angle are shown in standard position below. The circle on each diagram is a unit circle.

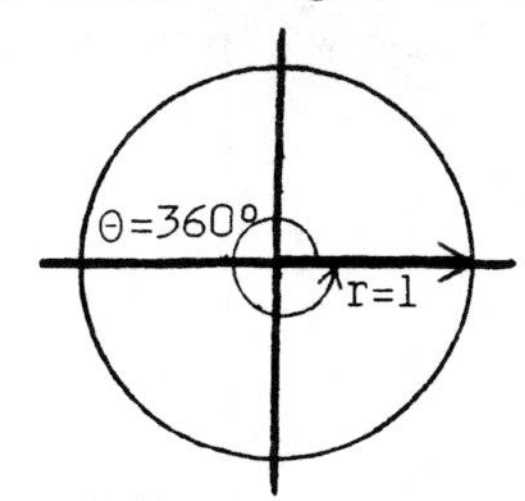

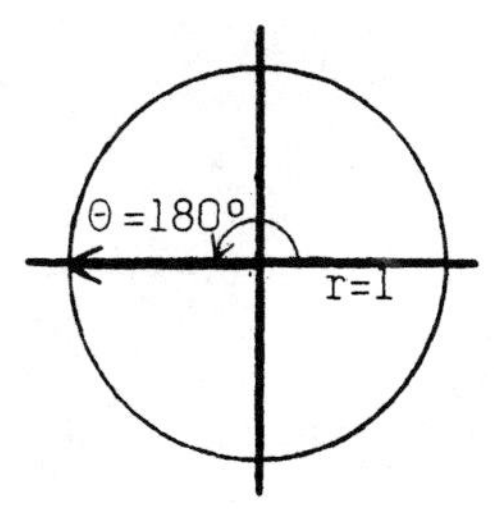

To express the angles above in radians, we use the following formula, where s = arc-length and r = radius.

$$\theta \text{ (in radians)} = \frac{s}{r}$$

However, since $r = 1$ for a unit circle, the formula simplifies to:

$$\theta = s \qquad (\text{since } \frac{s}{1} = s)$$

That is, when a unit circle is used, <u>the number of radians in an angle equals the arc-length of the angle</u>.

For a 360° angle, the arc-length is the circumference or <u>2π</u>. Therefore:

$$360° = 2\pi \text{ radians}$$

For a 180° angle, the arc-length is one-half the circumference or <u>π</u>. Therefore:

$$180° = \_\_\_\_\_\_ \text{ radians}$$

π radians

148. A 90° angle and a 270° angle are shown in standard position below. The circle in each diagram is a unit circle.

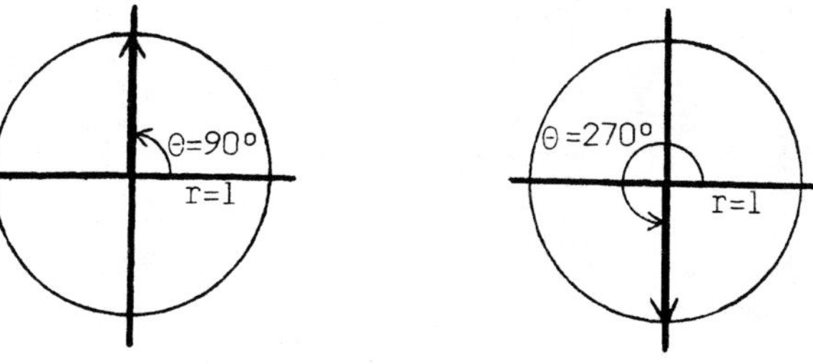

Since $r = 1$, the number of radians in each angle equals its arc-length. That is:

$$\theta \text{ (in radians)} = s$$

For a 90° angle, the arc-length is one-fourth the circumference of the circle. Therefore:

$$90° = \frac{1}{4}(2\pi) = \frac{2\pi}{4} = \frac{\pi}{2} \text{ radians}$$

For a 270° angle, the arc-length is three-fourths the circumference of the circle. Therefore:

$$270° = \frac{3}{4}(2\pi) = \frac{6\pi}{4} = \_\_\_\_\_\_ \text{ radians}$$

149. A 120° angle and a 45° angle are shown in standard position below. Each circle is a unit circle.

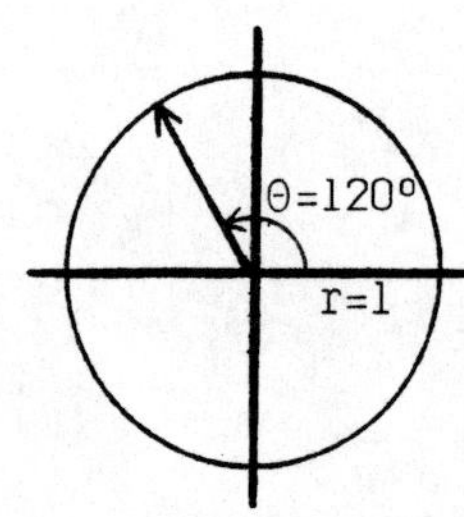

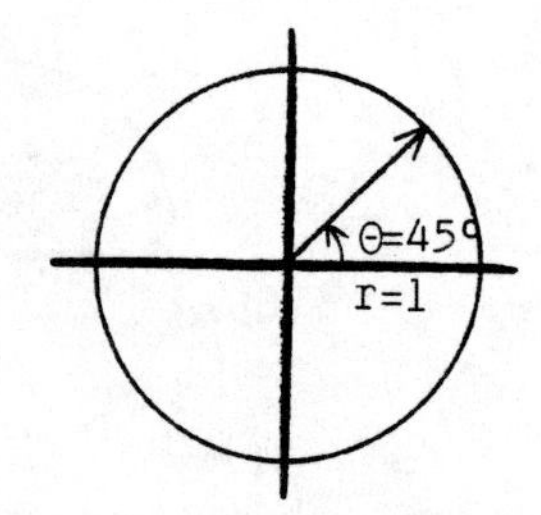

For a 120° angle, the arc-length is one-third the circumference of the circle. Therefore:

$$120° = \frac{1}{3}(2\pi) = \frac{2\pi}{3} \text{ radians}$$

For a 45° angle, the arc-length is one-eighth the circumference of the circle. Therefore:

$$45° = \frac{1}{8}(2\pi) = \underline{\qquad} \text{ radians}$$

$\frac{3\pi}{2}$ radians

$\frac{\pi}{4}$ radians

3-17 TRIG RATIOS OF 45° AND RELATED ANGLES

In this section, we will discuss the sine, cosine, and tangent of 45° and related angles.

150. The length of each side of the square below is "1". If we take half the square cut off by the diagonal, we get the triangle at the right. It is an isosceles right triangle with two 45° angles and legs of length "1".

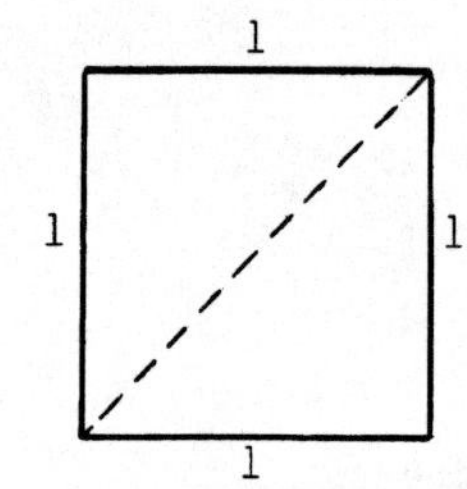

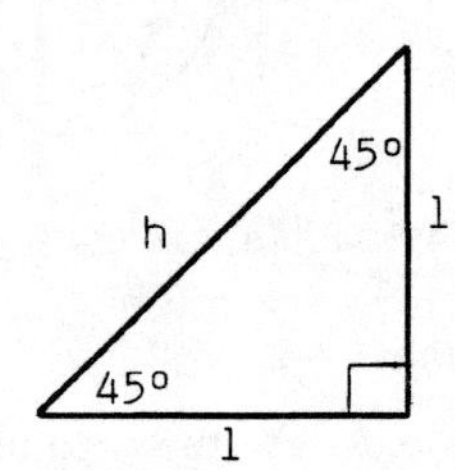

Let's use the Pythagorean Theorem to find the length of the hypotenuse. We get:

$$h^2 = (1)^2 + (1)^2$$
$$h^2 = 1 + 1$$
$$h^2 = 2$$
$$h = \sqrt{2}$$

151. Here is the triangle from the last frame. Using the sides, we get:

$$\sin 45° = \frac{1}{\sqrt{2}}$$

$$\cos 45° = \frac{1}{\sqrt{2}}$$

$$\tan 45° = \frac{1}{1} = 1$$

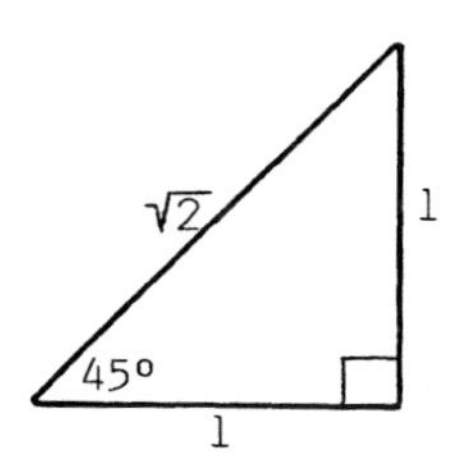

Usually we rationalize denominators. That is, we try to get expressions with the radical in the numerator rather than the denominator. The steps for rationalizing sin 45° and cos 45° are:

$$\sin 45° = \frac{1}{\sqrt{2}} = \frac{1}{\sqrt{2}}\left(\frac{\sqrt{2}}{\sqrt{2}}\right) = \frac{\sqrt{2}}{2}$$

$$\cos 45° = \frac{1}{\sqrt{2}} = \frac{1}{\sqrt{2}}\left(\frac{\sqrt{2}}{\sqrt{2}}\right) = \_\_\_\_\_\_$$

Answer: $\frac{\sqrt{2}}{2}$

152. We found that the sine, cosine, and tangent of a 45° angle are:

$$\sin 45° = \frac{\sqrt{2}}{2} \qquad \cos 45° = \frac{\sqrt{2}}{2} \qquad \tan 45° = 1$$

Let's use a calculator to confirm these values. Using a calculator, we can evaluate $\frac{\sqrt{2}}{2}$. We get:

Enter	Press	Display
2	[INV] [√x] [÷]	1.4142136
2	[=]	0.7071068

a) If you find sin 45° and cos 45° directly on a calculator, do you get 0.7071068? ______

b) If you find tan 45° directly on a calculator, do you get "1"? ______

Answers: a) Yes b) Yes

153. Using the values from the last frame, we can find the sine, cosine, and tangent of a 135° angle because its reference angle is 45° (Q2). A sketch is shown.

Using the sketch, you can see that sin α is positive, but cos α and tan α are negative. Therefore:

θ=135°

$$\sin 135° = \frac{\sqrt{2}}{2}$$

a) cos 135° = ______ b) tan 135° = ______

154. We can use the trig ratios of a 45° angle to find the sine, cosine, and tangent of a 225° angle because its reference angle is 45° (Q3). A diagram is shown.

Using the sketch, you can see that sin α and cos α are <u>negative</u>, but tan α is <u>positive</u>. Therefore:

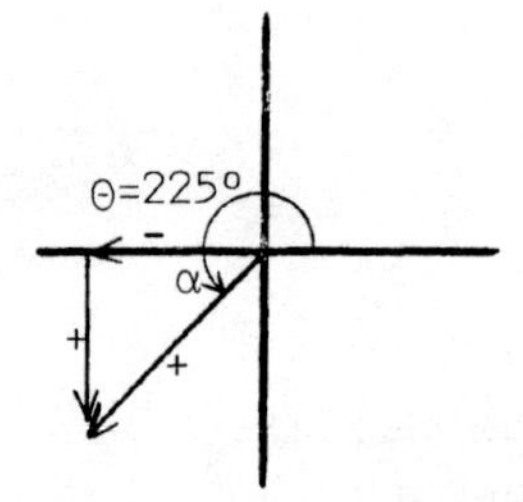

$$\sin 225° = -\frac{\sqrt{2}}{2}$$

a) cos 225° = ______

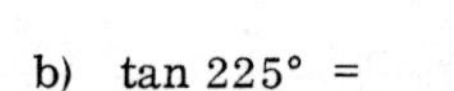

b) tan 225° = ______

a) $\cos 135° = -\frac{\sqrt{2}}{2}$

b) tan 135° = -1

---

155. We can use the trig ratios of a 45° angle to find the sine, cosine, and tangent of -45° because its reference angle is 45° (Q4). A diagram is shown.

Using the sketch, you can see that sin α and tan α are <u>negative</u>, but cos α is <u>positive</u>. Therefore:

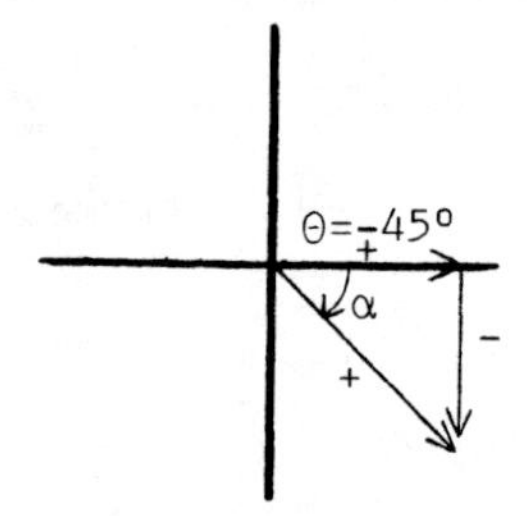

$$\sin(-45°) = -\frac{\sqrt{2}}{2}$$

a) cos(-45°) = ______

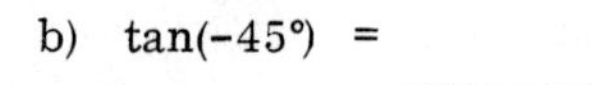

b) tan(-45°) = ______

a) $-\frac{\sqrt{2}}{2}$

b) 1

---

a) $\frac{\sqrt{2}}{2}$ b) -1

3-18 TRIG RATIOS OF 30°, 60°, AND RELATED ANGLES

In this section, we will discuss the sine, cosine, and tangent of 30°, 60°, and related angles.

---

156. The length of each side of the equilateral triangle at the left below is "2". Each angle is 60°. If we take half the triangle cut off by the perpendicular, we get the triangle at the right. It is a right triangle with 30° and 60° angles. Its hypotenuse is "2"; its right side is "1".

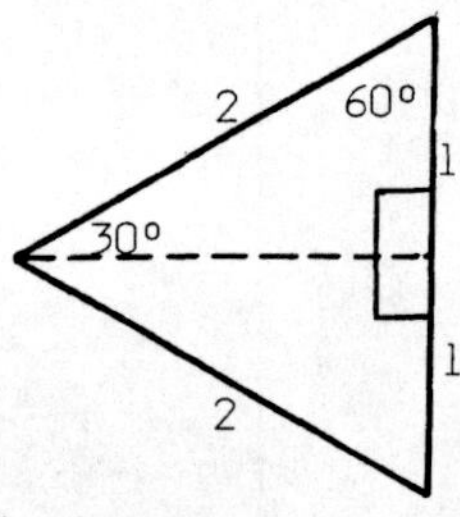

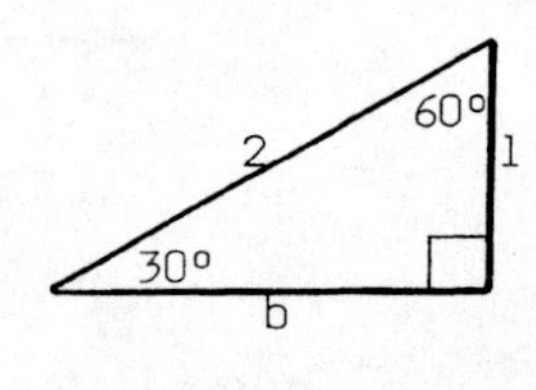

Continued on following page.

156. Continued

Using the Pythagorean Theorem, we can find the bottom side "b". We get:

$$b^2 = (2)^2 - (1)^2$$
$$b^2 = 4 - 1$$
$$b^2 = 3$$
$$b = \sqrt{3}$$

---

157. Substituting $\sqrt{3}$ for b, we get the values at the right. Using those values, we can state the trig ratios for 30°. We get:

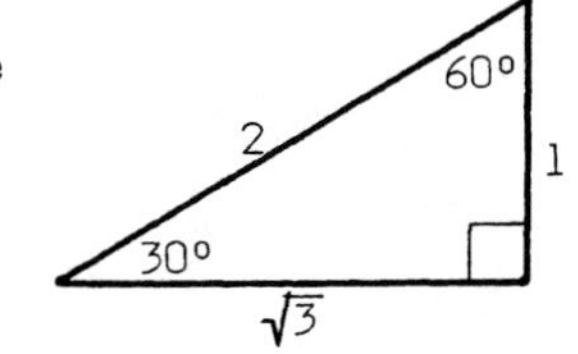

$$\sin 30° = \frac{1}{2} \text{ or } 0.5$$

$$\cos 30° = \frac{\sqrt{3}}{2}$$

$$\tan 30° = \frac{1}{\sqrt{3}}$$

Rationalizing the denominator for $\frac{1}{\sqrt{3}}$, we get:

$$\tan 30° = \frac{1}{\sqrt{3}} = \frac{1}{\sqrt{3}}\left(\frac{\sqrt{3}}{\sqrt{3}}\right) = \_\_\_\_\_\_$$

Answer: $\frac{\sqrt{3}}{3}$

158. We found that the sine, cosine, and tangent of a 30° angle are:

$$\sin 30° = \frac{1}{2} \qquad \cos 30° = \frac{\sqrt{3}}{2} \qquad \tan 30° = \frac{\sqrt{3}}{3}$$

Using a calculator, we can convert the values above to decimal numbers. We get:

$$\sin 30° = \frac{1}{2} = 0.5$$

$$\cos 30° = \frac{\sqrt{3}}{2} = 0.8660254$$

$$\tan 30° = \frac{\sqrt{3}}{3} = 0.5773503$$

Are these the same values we would get if we find sin 30°, cos 30°, and tan 30° directly on a calculator? ______

Answer: Yes

159. Using the values from the last frame, we can find the sine, cosine, and tangent of a 150° angle because its reference angle is 30° (Q2). A sketch is shown.

Complete these. Be sure to attach the correct sign.

$\sin 150° = \frac{1}{2}$

a) $\cos 150° =$ ______

b) $\tan 150° =$ ______

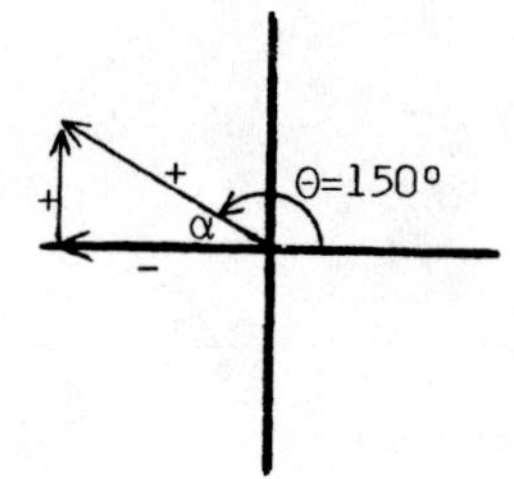

---

160. Using the same values, we can find the sine, cosine, and tangent of a -150° angle because its reference angle is 30° (Q3). A sketch is shown.

Complete these. Be sure to attach the correct sign.

a) $\sin(-150°) =$ ______

b) $\cos(-150°) =$ ______

c) $\tan(-150°) =$ ______

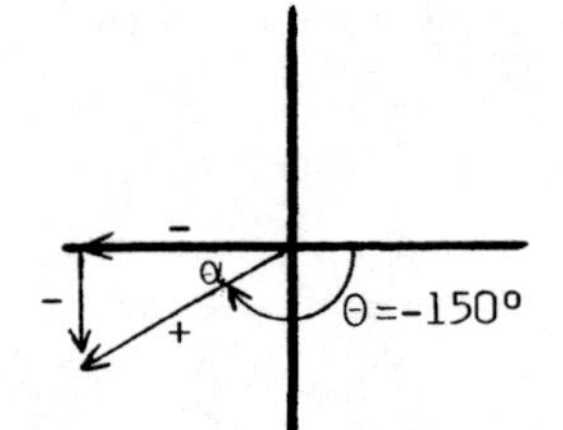

Answers to 159:

a) $-\frac{\sqrt{3}}{2}$

b) $-\frac{\sqrt{3}}{3}$

---

161. Using the triangle at the right, we can also state the trig ratios for 60°. We get:

$\sin 60° = \frac{\sqrt{3}}{2}$

$\cos 60° = \frac{1}{2}$

$\tan 60° = \frac{\sqrt{3}}{1} = \sqrt{3}$

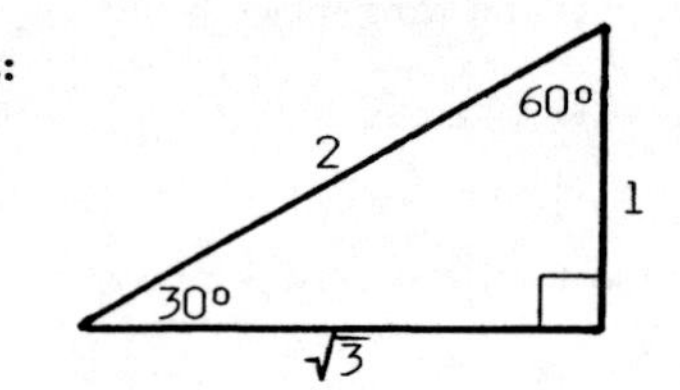

Using a calculator, we can convert the values above to decimal numbers. We get:

$\sin 60° = \frac{\sqrt{3}}{2} = 0.8660254$

$\cos 60° = \frac{1}{2} = 0.5$

$\tan 60° = \sqrt{3} = 1.7320508$

Are these the same values we get if we find sin 60°, cos 60°, and tan 60° directly on a calculator? ______

Answers to 160:

a) $-\frac{1}{2}$

b) $-\frac{\sqrt{3}}{2}$

c) $\frac{\sqrt{3}}{3}$

Answer to 161: Yes

162. Using the values from the last frame, we can find the sine, cosine, and tangent of a 120° angle because its reference angle is 60° (Q2). A sketch is shown.

Complete these. Be sure to attach the correct sign.

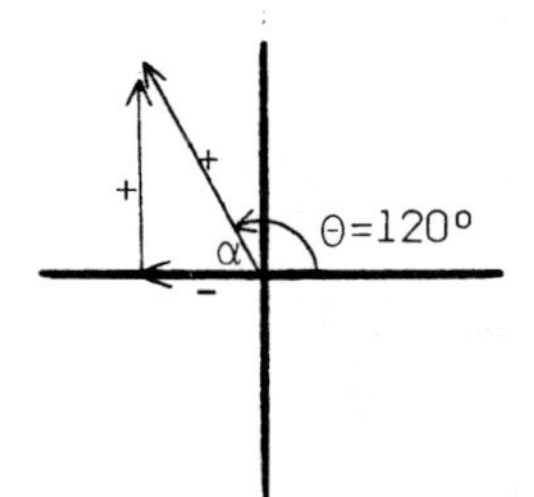

$\sin 120° = \frac{\sqrt{3}}{2}$

a) $\cos 120° =$ ______

b) $\tan 120° =$ ______

---

163. Using the same values, we can find the sine, cosine, and tangent of a -60° angle because its reference angle is 60° (Q4). A sketch is shown.

Complete these. Attach the correct signs.

$\sin(-60°) = -\frac{\sqrt{3}}{2}$

a) $\cos(-60°) =$ ______

b) $\tan(-60°) =$ ______

Answers to 162:

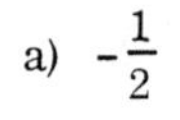

a) $-\frac{1}{2}$

b) $-\sqrt{3}$

---

164. A summary of the trig ratios for 30°, 45°, and 60°, is given in the table below. The values for 0° and 90° are included.

θ	0°	30°	45°	60°	90°
sin θ	0	$\frac{1}{2}$	$\frac{\sqrt{2}}{2}$	$\frac{\sqrt{3}}{2}$	1
cos θ	1	$\frac{\sqrt{3}}{2}$	$\frac{\sqrt{2}}{2}$	$\frac{1}{2}$	0
tan θ	0	$\frac{\sqrt{3}}{3}$	1	$\sqrt{3}$	Undef.

Answers to 163:

a) $\frac{1}{2}$

b) $-\sqrt{3}$

SELF-TEST 10 (pages 132-144)

In terms of the coordinates of point P on the unit circle below, define these circular functions:

1. $\sin \theta =$ ______
2. $\cos \theta =$ ______
3. $\tan \theta =$ ______

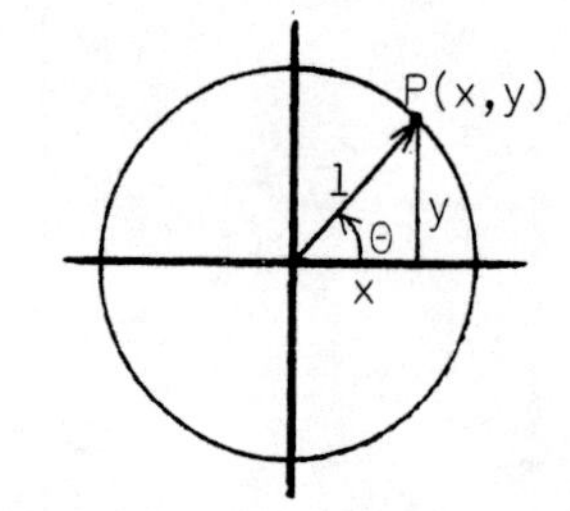

Point P with coordinates (x, y) lies on the unit circle. For each angle below, find the coordinates of point P. Round to four decimal places.

4. $\theta = 157°$ ______
5. $\theta = 292°$ ______

On the unit circle at the right, the coordinates of point P are (-1, 0). Using these coordinates and the definitions of the circular functions, find:

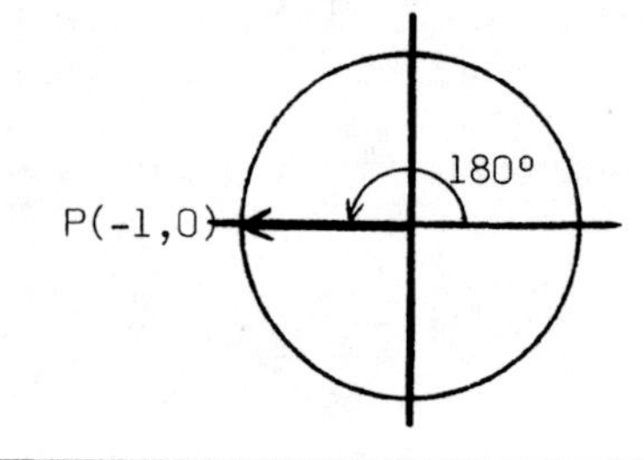

6. $\sin 180° =$ ____
7. $\cos 180° =$ ____
8. $\tan 180° =$ ____

9. For the unit circle, the radian formula $\theta = \frac{s}{r}$ simplifies to: ________

10. On the unit circle, the arc-length cut off by angle θ equals angle θ measured in ____________.

For a 240° angle: 11. Expressed in terms of "π", what arc-length does the 240° angle cut off on the unit circle? ________

12. Expressed in terms of "π", how many radians are in 240°? ________ radians

The length of each side of the equilateral triangle at the right is 2 units. Leaving your answers in radical form, find the value of each of the following.

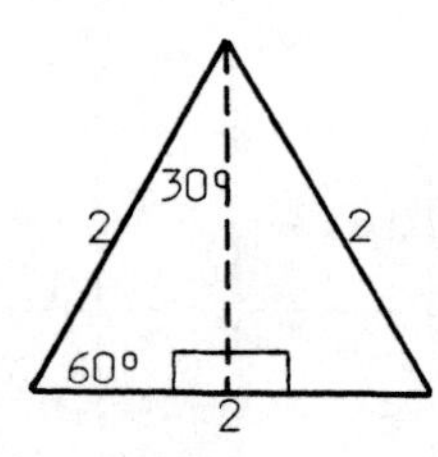

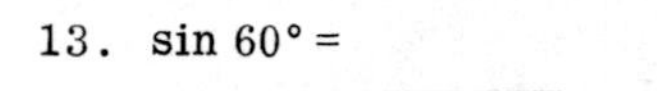

13. $\sin 60° =$ ______
14. $\tan 30° =$ ______

ANSWERS:

1. y
2. x
3. $\frac{y}{x}$
4. (-0.9205, 0.3907)
5. (0.3746, -0.9272)
6. 0
7. -1
8. 0
9. $\theta = s$
10. radians
11. $\frac{4\pi}{3}$
12. $\frac{4\pi}{3}$
13. $\frac{\sqrt{3}}{2}$
14. $\frac{1}{\sqrt{3}}$ or $\frac{\sqrt{3}}{3}$

SUPPLEMENTARY PROBLEMS - CHAPTER 3

Assignment 7

Given these standard-position angles:

312°	94°	51°	127°	275°	230°	89°	186°

Which of these angles have their terminal sides in:

1. Quadrant 1? 2. Quadrant 2? 3. Quadrant 3? 4. Quadrant 4?

Find the reference angle for each standard-position angle. Be sure to include the quadrant number.

5. 280° 6. 95° 7. 247° 8. 72° 9. 336° 10. 198°

Find the standard-position angle for each reference angle.

11. 22° (Q3) 12. 86° (Q2) 13. 49° (Q4) 14. 14° (Q1) 15. 75° (Q4) 16. 31° (Q2)

Complete: 17. The cosine ratio is positive in quadrants ______ and ______.

18. The sine ratio is positive in quadrants ______ and ______.

19. The tangent ratio is negative in quadrants ______ and ______.

20. Which of the following are negative?

a) sin 20° (Q3) b) cos 50° (Q4) c) tan 72° (Q2) d) cos 10° (Q3) e) sin 8° (Q1)

Write the correct reference angle in each blank.

21. sin 153° = sin ______ 22. cos 318° = cos ______ 23. tan 255° = tan ______

Find the numerical value of each of the following. Round to four decimal places.

24. cos 42° (Q2) = ________	27. sin 287° = ________	30. tan 199° = ________
25. tan 65° (Q3) = ________	28. cos 245° = ________	31. sin 170° = ________
26. sin 17° (Q4) = ________	29. tan 98° = ________	32. cos 351° = ________

Assignment 8

Without using a calculator, find the numerical value of:

1. sin 180° 2. cos 360° 3. tan 0° 4. sin 270°

Given these five angles: | 0° 90° 180° 270° 360° |

5. For which angle or angles does $\cos\theta = -1$?
6. For which angle or angles does $\sin\theta = 1$?
7. For which angle or angles is $\tan\theta$ undefined?

For each angle below, find an angle between 0° and 360° with the same terminal side.

8. 638° 9. 915° 10. -330° 11. -514°

Find the reference angle for each standard-position angle. Be sure to include the quadrant number.

12. 517° 13. 1,000° 14. -135° 15. -705°

Using the example below as a pattern, complete the following. Note that (a) is an angle between 0° and 360°, (b) is a reference angle, and (c) is a numerical value rounded to four decimal places.

			(a)		(b)		(c)
Example:	tan 475°	= tan	115°	= tan	65° (Q2)	=	-2.1445
16.	sin 954°	= sin	______	sin	______	=	______
17.	cos 519°	= cos	______	cos	______	=	______
18.	tan (-290°)	= tan	______	tan	______	=	______
19.	cos (-445°)	= cos	______	cos	______	=	______

Find the numerical value of each. Round to four decimal places.

20. cos 862° 21. sin 1,250° 22. tan (-578°) 23. sin (-925°)

By referring to the graphs on pages 111-115, answer the following.

For the function $y = \sin\theta$:

24. What is the smallest value of y?
25. As θ increases from 270° to 360°, does y increase or decrease?

For the function $y = \cos\theta$:

26. What is the largest value of y?
27. As θ increases from 90° to 180°, does y increase or decrease?

For the function $y = \tan\theta$:

28. As θ increases from -90° to 90°, does y increase or decrease?
29. For what three values of θ does y equal 0?

Assignment 9

1. Find the number of degrees in the central angle of a circle whose arc-length is two-thirds of the circumference.

2. Find the number of degrees in each central angle of a circle whose circumference is divided into 8 equal arcs.

Complete these definitions of "degree" and "radian":

3. An angle of 1 degree is a central angle whose arc-length is __________ of the circumference of the circle.

4. An angle of 1 radian is a central angle whose arc-length equals the __________ of the circle.

In a circle of radius 16 cm, what arc-length is cut off by an angle of:

5. 1 radian? 6. 2 radians? 7. 0.5 radian?

8. Write the formula relating angle "θ" (in radians), radius "r", and arc-length "s".

Find central angle "θ" of each circle in radians. Round to hundredths.

9. Radius = 12.0"
Arc-Length = 29.6"

10. Radius = 5.75 cm
Arc-Length = 3.49 cm

11. Radius = 200 mm
Arc-Length = 753 mm

Convert each angle to degrees.

12. $\frac{\pi}{3}$ radians 13. $\frac{3\pi}{2}$ radians 14. $\frac{7\pi}{4}$ radians 15. 4π radians

Convert each angle to radians expressed in terms of "π".

16. 30° 17. 90° 18. 120° 19. -360°

20. Rounded to thousandths, how many degrees are there in 1 radian?

Using the unity-fraction method, convert to degrees. Round to tenths where necessary.

21. 0.372 radian 22. -1.84 radians 23. $\frac{5\pi}{6}$ radians 24. $-\frac{9\pi}{4}$ radians

Using the unity-fraction method, convert to radians. Round to hundredths.

25. 83° 26. 35° 27. 258° 28. -309°

Using the calculator, find each value. Round to four decimal places.

29. $\sin \frac{2\pi}{5}$ 30. $\cos \frac{11\pi}{6}$ 31. sin 3.5 32. tan (-0.76)

Using "1 radian is approximately 60°", roughly convert radians to degrees, and degrees to radians, in the following problems.

33. 3 radians 34. 1.5 radians 35. 30° 36. 150°

Assignment 10

Referring to the unit-circle diagram below, what trig function (sin θ, cos θ, or tan θ) is represented by:

1. "x"
2. 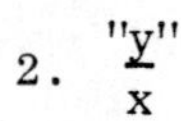 $\frac{\text{"y"}}{\text{x}}$
3. "y"

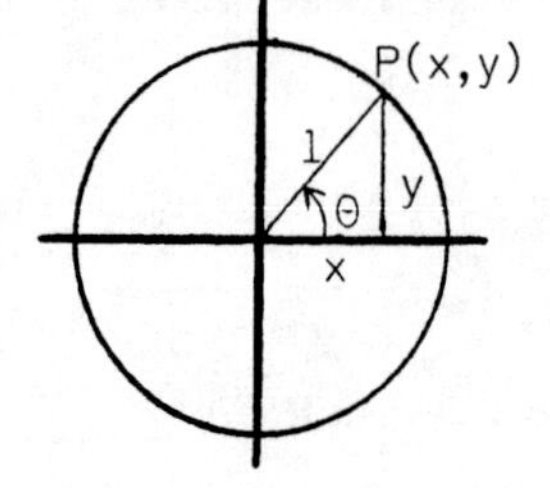

On the unit circle below, point P has coordinates (0, -1). Using these coordinates and the definitions of the trig functions, find:

4. sin 270°
5. cos 270°
6. tan 270°

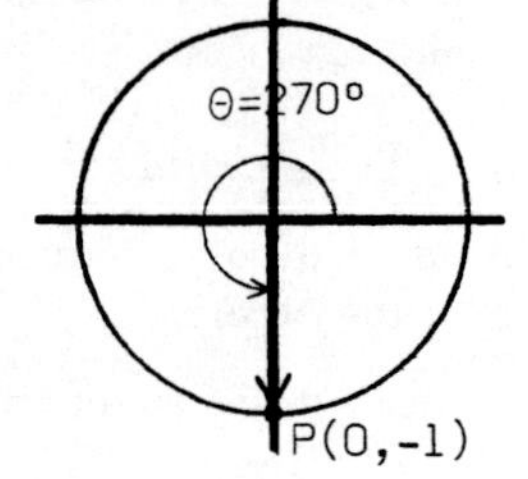

Point P with coordinates (x, y) lies on the unit circle. Find the coordinates of point P for each angle below. Round to four decimal places.

7. θ = 225° 8. θ = 19° 9. θ = 303° 10. θ = 98°

11. If central angle θ on the unit circle is 1 radian, what arc-length does it cut off on the circumference?

Expressed in terms of "π", what arc-length does each angle below cut off on the unit circle?

12. 60° 13. 180° 14. 210° 15. 300°

Referring to your answers to Problems 12-15, how many radians are there in each angle below? Express each answer in terms of "π".

16. 60° 17. 180° 18. 210° 19. 300°

Referring to the right triangle below, find the numerical value of each of the following. Leave answers in radical form.

20. sin 45°
21. cos 45°
22. tan 45°

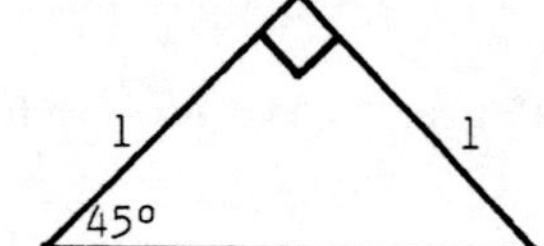

Referring to the equilateral triangle below, find the numerical value of each of the following. Leave answers in radical form.

23. sin 60°
24. tan 60°
25. sin 30°
26. cos 30°

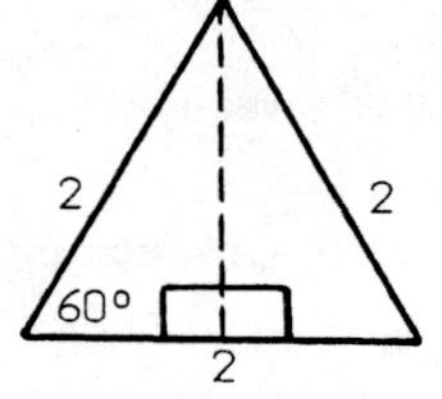

Chapter 4 VECTORS

Vector mathematics is used in science and technology to represent concepts like force, voltage, current, velocity, and acceleration. In this chapter, we will discuss vectors, the components of vectors, methods of adding vectors, and vectors in a state of equilibrium.

4-1 VECTORS

In this section, we will define vectors and discuss some examples of vectors.

1. Some quantities (like distance, time, temperature, area, volume, and others) have only magnitude. Quantities of that type are called "scalar quantities" or "scalars". Some examples are:

   A woman worked 10 hours.
   A man jogged 5 miles.
   The area of a circle is 25.6 cm^2.
   The high temperature for the day was 85°F.

   Other quantities (like force, displacement, velocity, acceleration, voltage, current, and others) have both magnitude and direction. Quantities of that type are called "vector quantities" or "vectors". Some examples are:

   A woman takes 3 steps to the right.
   An airplane travels 450 miles/hour to the northeast.
   A hoist exerts an upward force of 3,000 pounds to lift a car.
   A voltage has a 20-volt amplitude with a 60° phase angle.

   Which of the following is a vector? __________

   a) The volume of a cylinder is 476 in^3.

   b) A train travels 200 kilometers in a 225° direction.

Only (b)

2. In mathematics, vectors are represented by arrows that have both length and direction. The length is chosen, according to some scale, to represent the magnitude of the vector. The direction of the arrow represents the direction of the vector.

The arrows below represent the vectors "3 steps to the right" and "4 steps to the left". A scale of approximately 1 centimeter per step was used.

Using the same scale, the arrow below would represent the vector "______ steps to the __________ (right/left)".

---

3. Arrows A and B at the right represent the vectors "a 10-pound force upward" and "a 15-pound force upward". A scale of approximately 1 centimeter per 5 pounds was used. Therefore:

A is 2 cm, since $2 \times 5 = 10$

B is 3 cm, since $3 \times 5 = 15$

Using the same scale, arrow C represents "a ______-pound force upward".

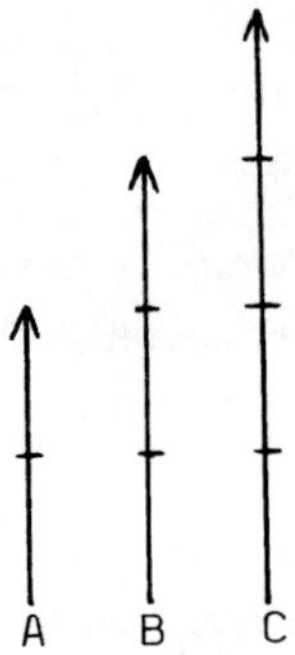

A B C

Answer: 5 steps to the left

---

4. Arrow A at the right represents the vector "300 kilometers/hour to the northwest". A scale of approximately 1 centimeter per 100 kilometers/hour was used. Therefore, A is 3 cm, since $3 \times 100 = 300$.

Using the same scale, arrow B represents the vector "______ kilometers/hour to the northwest".

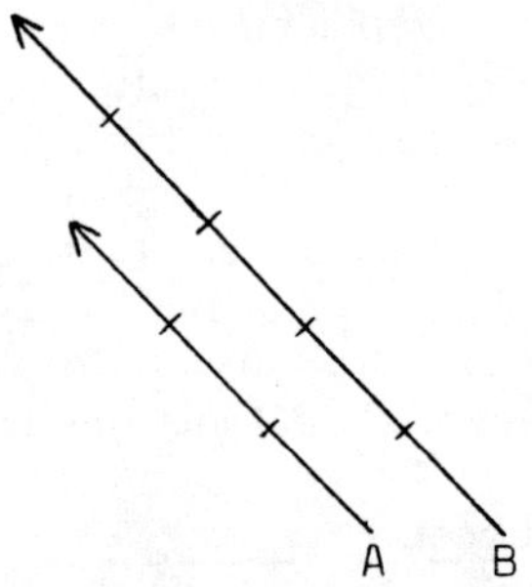

A B

Answer: 20-pound force upward

---

5. Generally, the direction of a vector is given by an angle (like 50° or 225°). When an angle is used, the arrow representing the vector is frequently drawn on the coordinate system. An example is discussed below.

$\vec{OA}$ at the right represents "150 miles at 120°". Since the scale is 1 unit for each 50 miles, $\vec{OA}$ is 3 units long.

Specify α, the reference angle.

α = __________

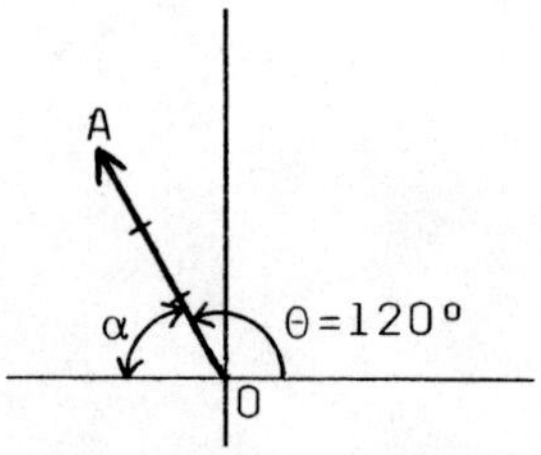

A α $\theta = 120°$ O

Answer: 500 kilometers/hour to the northwest

Answer: 60° (Q2)

6. Sometimes the direction of a vector is given as a negative angle. An example is given.

$\overrightarrow{OP}$ at the right represents "a force of 100 kilograms at -130°". Since the scale is 1 unit for each 25 kilograms, $\overrightarrow{OP}$ is 4 units long.

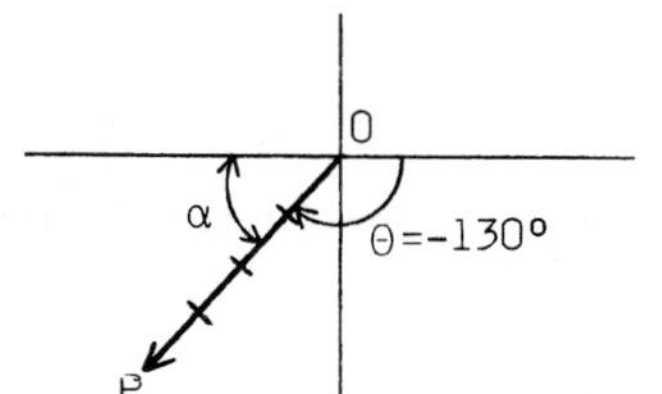

Specify α, the reference angle.

α = ___________

---

7. "Vectors" is a topic in mathematics. In mathematics, vectors are usually drawn on the coordinate system. The length of the vector is given in units. The direction of the vector is given by a standard-position angle or its reference angle.

50° (Q3)

$\overrightarrow{OB}$ at the right is in the third quadrant. Its length is 5 units. Its direction is 45° (Q3).

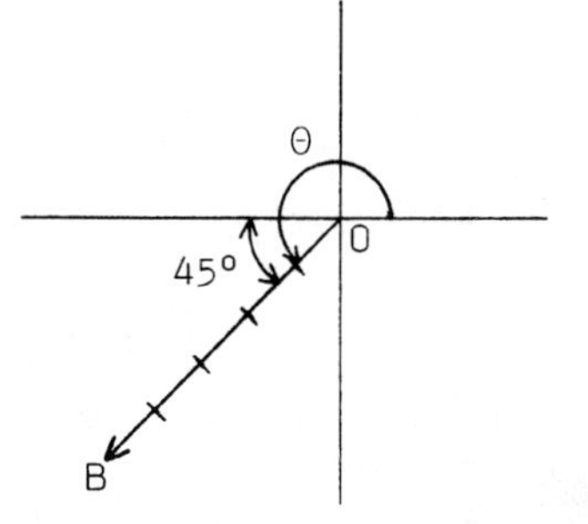

Specify θ, the standard-position angle.

θ = ___________

θ = 225°

4-2 VECTOR ADDITION

In this section, we will use some applied problems to show what is meant by an addition of two vectors. The tail-to-head method and the parallelogram (rectangle) method are shown.

---

8. The displacement problem below can be solved by an addition of two vectors.

A woman takes 4 steps east and then 3 steps north. How far is she and what direction is she from the starting point?

The tail-to-head model for the vector addition is shown at the right.

$\overrightarrow{AB}$ represents "4 steps east".

$\overrightarrow{BC}$ represents "3 steps north".

$\overrightarrow{AC}$ is the sum of $\overrightarrow{AB}$ and $\overrightarrow{BC}$. It is called the "resultant vector" or simply the "resultant".

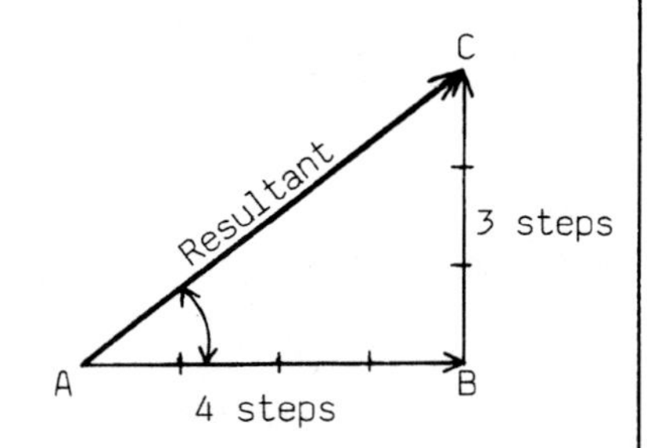

Continued on following page.

8. Continued

To complete the solution, we must find the length of the resultant $\overrightarrow{AC}$ and its direction (angle A). Since ABC is a right triangle:

1. We can use the Pythagorean Theorem to find the <u>length</u> of $\overrightarrow{AC}$.

$$(AC)^2 = (AB)^2 + (BC)^2$$
$$(AC)^2 = (4)^2 + (3)^2$$
$$(AC)^2 = 16 + 9$$
$$(AC)^2 = 25$$
$$AC = \sqrt{25} = 5 \text{ steps}$$

2. We can use the tangent ratio to find angle A.

$$\tan A = \frac{\overrightarrow{BC}}{\overrightarrow{AB}} = \frac{3}{4} = 0.75$$

$$A = 37° \text{ (rounded to a whole number)}$$

Therefore: a) The woman is ______ steps from the starting point.

b) Her final direction is ______° north of east.

Answers: a) 5 b) 37°

9. This displacement problem can also be solved by an addition of two vectors.

A plane flies 300 miles east. After landing, it flies 250 miles northwest. How far is it and what direction is it from the starting point?

The tail-to-head model for the vector addition is shown at the right.

$\overrightarrow{CD}$ represents "300 miles east".

$\overrightarrow{DE}$ represents "250 miles northwest".

$\overrightarrow{CE}$ is the resultant of $\overrightarrow{CD}$ and $\overrightarrow{DE}$.

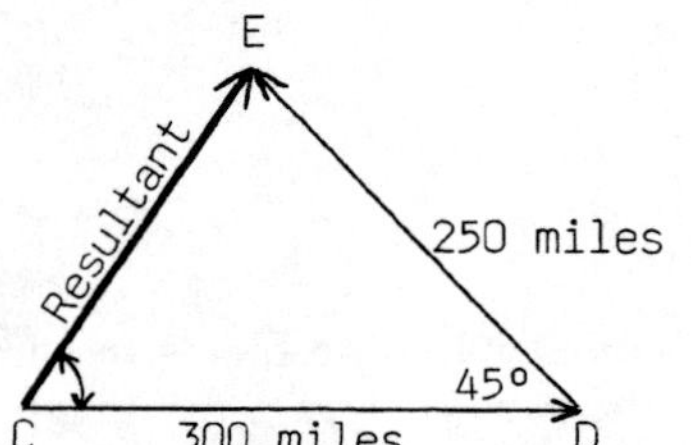

To complete the solution, we must find the length of $\overrightarrow{CE}$ and its direction (angle C). Since CDE is an oblique triangle:

1. We can use the Law of Cosines to find the <u>length</u> of $\overrightarrow{CE}$.

$$(CE)^2 = (CD)^2 + (DE)^2 - 2(CD)(DE)(\cos D)$$
$$(CE)^2 = (300)^2 + (250)^2 - 2(300)(250)(\cos 45°)$$
$$CE = \sqrt{(300)^2 + (250)^2 - 2(300)(250)(\cos 45°)}$$
$$CE = 215 \text{ miles (rounded to a whole number)}$$

Continued on following page.

9. Continued

2. We can now use the Law of Sines to find angle C.

$$\frac{\sin C}{DE} = \frac{\sin D}{CE}$$

$$\frac{\sin C}{250} = \frac{\sin 45°}{215}$$

$$\sin C = \frac{(250)(\sin 45°)}{215}$$

$$C = 55° \text{ (rounded to a whole number)}$$

Therefore: a) The plane is __________ miles from the starting point.

b) Its final direction is ______° north of east.

---

a) 215

b) 55°

10. The force problem below can be solved by an addition of vectors.

A 50 kg force and a 35 kg force act on an object at right angles to each other. Find their sum, or resultant, and the angle the resultant makes with the 50 kg force.

Two diagrams are shown below. At the left, we show the two forces acting on object O at right angles to each other. At the right, we completed a rectangle. $\overrightarrow{AR}$ is equivalent to $\overrightarrow{OB}$. Therefore, $\overrightarrow{OR}$ is the resultant of $\overrightarrow{OA}$ and $\overrightarrow{OB}$.

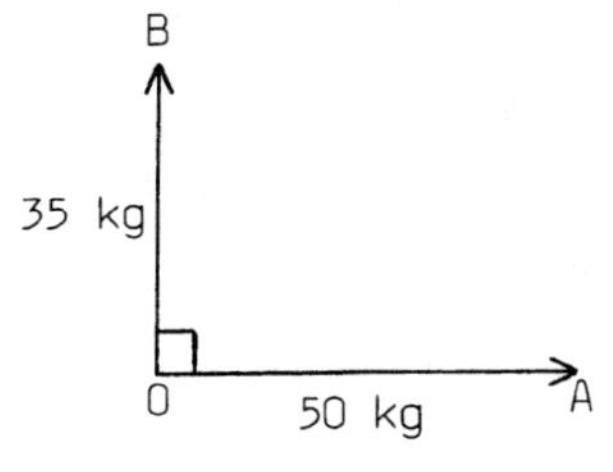

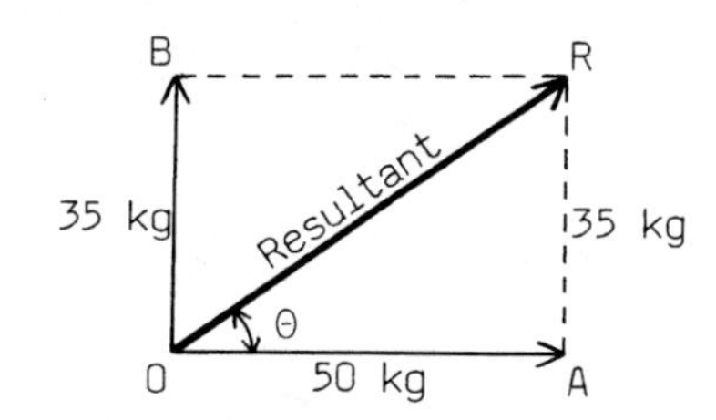

To complete the solution, we must find the length of the resultant $\overrightarrow{OR}$ and the angle (θ) it makes with the 50 kg force. Since OAR is a right triangle:

1. We can use the Pythagorean Theorem to find the length of $\overrightarrow{OR}$.

$$(OR)^2 = (OA)^2 + (AR)^2$$

$$(OR)^2 = (50)^2 + (35)^2$$

$$(OR)^2 = 2,500 + 1,225$$

$$(OR)^2 = 3,725$$

$$OR = 61 \text{ kg (rounded to a whole number)}$$

Continued on following page.

10. Continued

2. We can use the tangent ratio to find θ.

$$\tan \theta = \frac{\overrightarrow{AR}}{\overrightarrow{OA}} = \frac{35}{50} = 0.7$$

$\theta = 35°$ (rounded to a whole number)

Therefore: a) The resultant force is ______ kg.

b) The angle between the resultant force and the 50 kg force is ______°.

---

11. The force problem below can be solved by an addition of vectors.

A 200 lb force and a 150 lb force act on an object at a 70° angle to each other. Find the resultant force and the angle it makes with the 200 lb force.

Two diagrams are shown below. At the left, we show the two forces acting on object O at a 70° angle with each other. At the right, we completed a parallelogram. $\overrightarrow{CR}$ is equivalent to $\overrightarrow{OD}$. Therefore, $\overrightarrow{OR}$ is the resultant of $\overrightarrow{OD}$ and $\overrightarrow{OC}$.

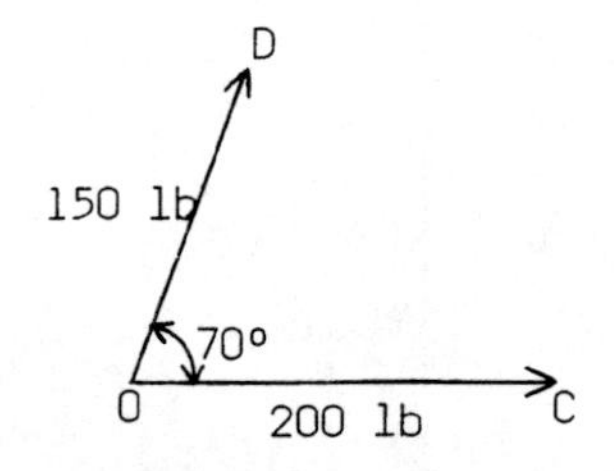

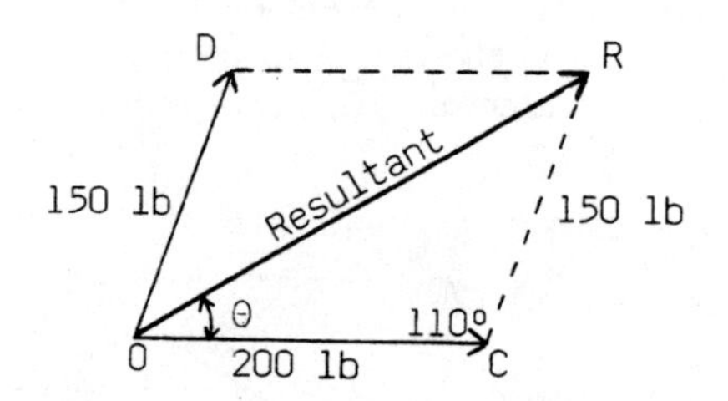

To complete the solution, we must find the length of $\overrightarrow{OR}$ and the angle (θ) it makes with the 200 lb force. Since OCR is an oblique triangle:

1. We can use the Law of Cosines to find the length of $\overrightarrow{OR}$. (Note: Since angle O is 70°, angle C is 110°.)

$$(OR)^2 = (OC)^2 + (CR)^2 - 2(OC)(CR)(\cos C)$$

$$(OR)^2 = (200)^2 + (150)^2 - 2(200)(150)(\cos 110°)$$

$$OR = \sqrt{(200)^2 + (150)^2 - 2(200)(150)(\cos 110°)}$$

$OR = 288$ lb (rounded to a whole number)

2. We can now use the Law of Sines to find angle θ.

$$\frac{\sin \theta}{CR} = \frac{\sin C}{OR}$$

$$\frac{\sin \theta}{150} = \frac{\sin 110°}{288}$$

$$\sin \theta = \frac{(150)(\sin 110°)}{288}$$

$\theta = 29°$ (rounded to a whole number)

Continued on following page.

a) 61 kg

b) 35°

11. Continued

Therefore: a) The resultant force is _________ lb.

b) The angle between the resultant force and the 200 lb force is _____°.

a) 288 lb b) 29°

4-3 PARALLELOGRAMS

In this section, we will review some basic facts about parallelograms. These facts are needed in the next section where we discuss the oblique-triangle method for vector addition.

12. ABCD at the right is a parallelogram. From the figure, you can see these facts about parallelograms:

1. Opposite sides are equal.
2. Opposite angles are equal.

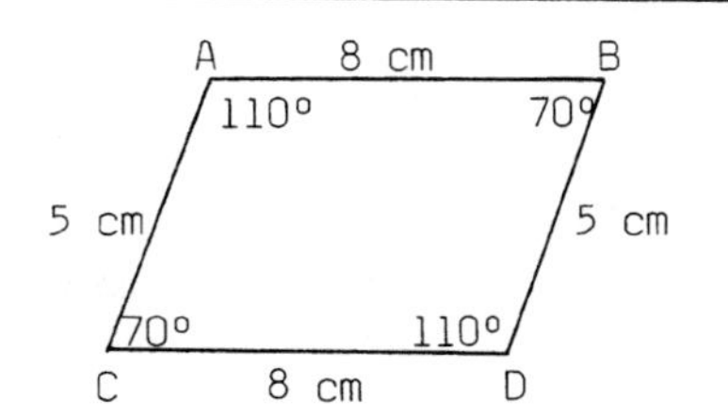

That is: a) Side AB is the same length as side _______.

b) Angle C is the same size as angle _______.

13. From the parallelogram at the right, you can see these two facts:

1. The sum of the four angles of a parallelogram is 360°.
2. The sum of any two successive angles in a parallelogram is 180°. That is:

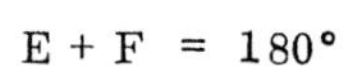

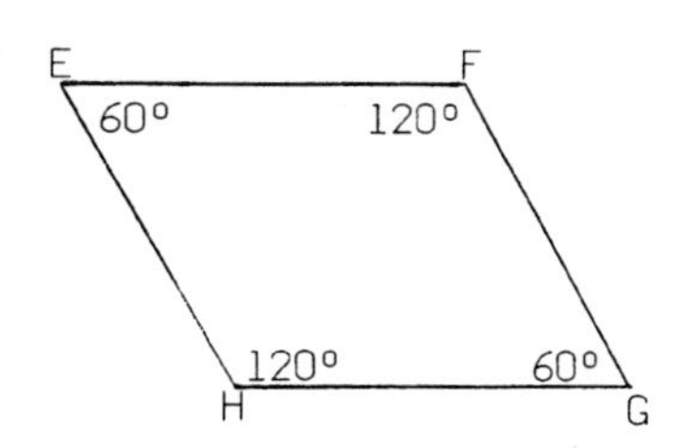

$$E + F = 180°$$
$$F + G = 180°$$
$$G + H = 180°$$
$$H + E = 180°$$

When the sum of two angles is 180°, we call them "supplementary angles". How many pairs of supplementary angles are there in parallelogram EFGH? _________

a) CD

b) B

four

14. The length of two sides and the size of one angle are given in the parallelogram at the right.

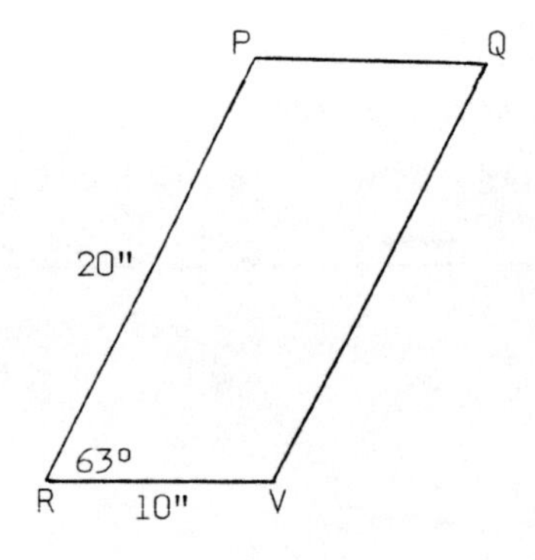

a) How long is PQ? __________

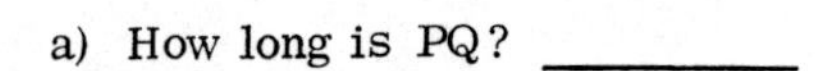

b) How long is QV? __________

c) How large is angle P? __________

d) How large is angle Q? __________

e) How large is angle V? __________

---

a) 10"
b) 20"
c) 117°
d) 63°
e) 117°

15. One diagonal of parallelogram FPQT is drawn at the right. You can see this fact:

<u>The diagonal of a parallelogram does not necessarily divide the angles into two equal angles.</u>

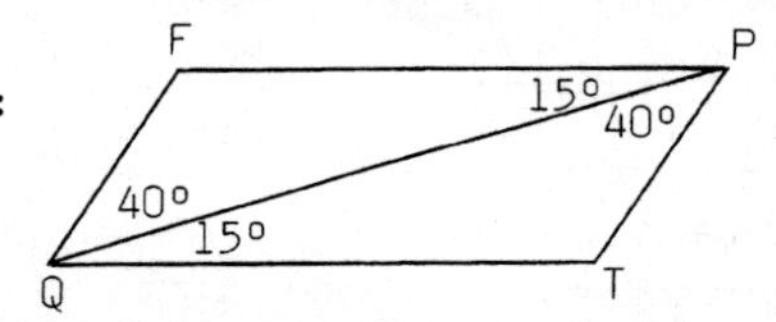

<u>Note</u>: Angles P and Q are divided into two <u>unequal</u> angles, a 15° angle and a 40° angle.

One diagonal of parallelogram ABCD is drawn at the right. It divides angle D into angles #1 and #2.

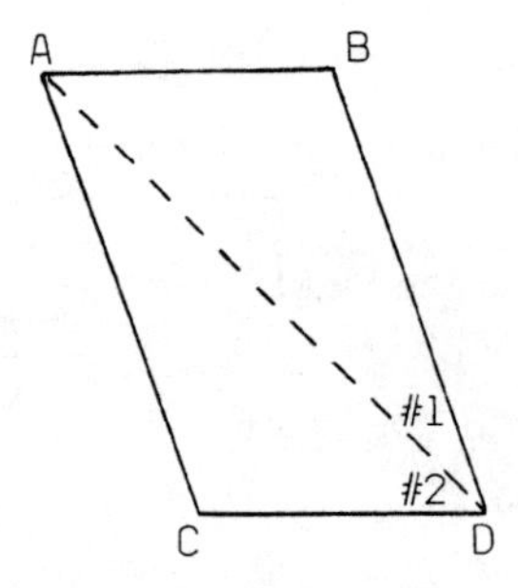

If angle D is 70°:

a) How large is angle C? __________

b) Can you tell the size of angles #1 and #2? __________

---

a) 110°
b) No

16. To add the two vectors at the left below, we must complete the parallelogram and draw the resultant as we have done at the right. Notice how the dashed lines are extended.

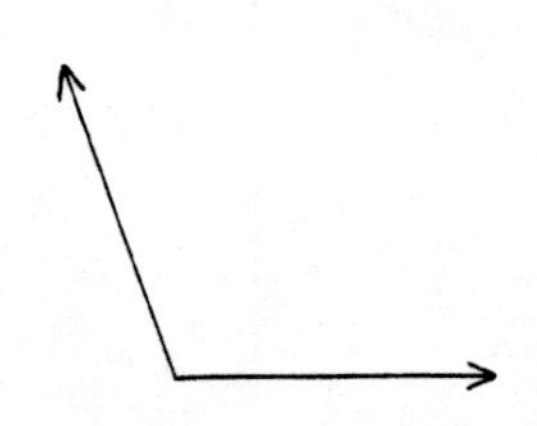

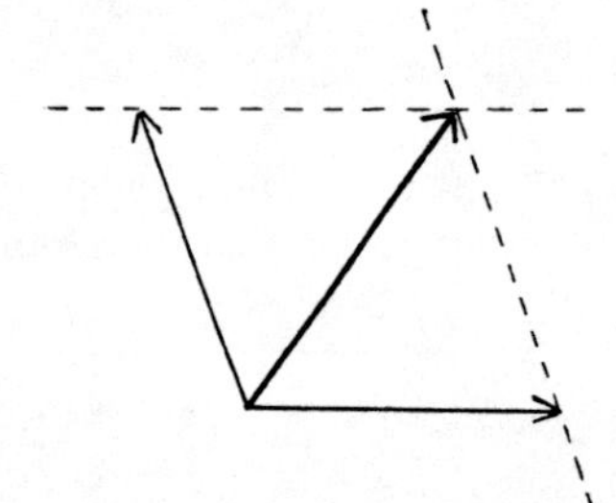

Continued on following page.

16. Continued

For each pair of vectors, complete the parallelogram and draw the resultant.

a) 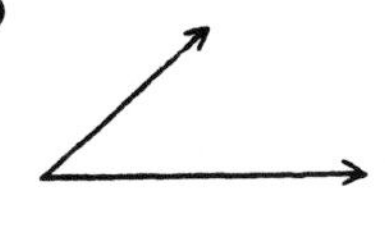 b) 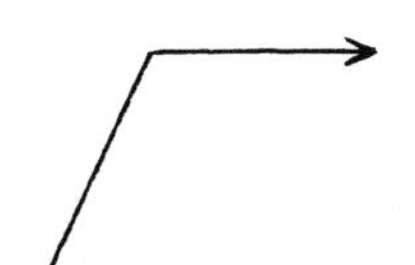 c) 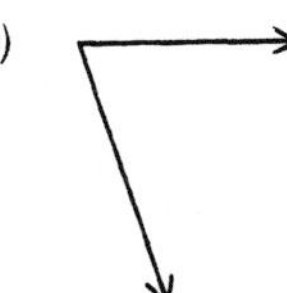

---

17. We completed two parallelograms below. In each one, $\overrightarrow{OR}$ is the resultant. Notice that $\overrightarrow{OR}$ is smaller than both of the two added vectors in its parallelogram.

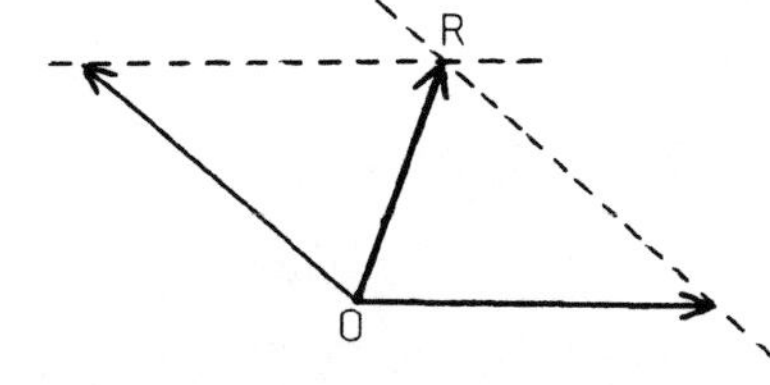

When computing the length of a resultant, don't be surprised if it is smaller than both of the two added vectors.

Answers to Frame 16:

a) 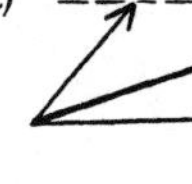

b)

c)

4-4 THE OBLIQUE-TRIANGLE METHOD FOR VECTOR ADDITION

In this section, we will discuss the oblique-triangle method for adding two vectors.

---

18. In Figure 1 below, we have drawn $\overrightarrow{OA}$ and $\overrightarrow{OB}$. The angle between them is 38°. Their lengths are 25 units and 30 units. In Figure 2, we have completed the parallelogram and drawn the resultant $\overrightarrow{OR}$. In Figure 3, we have drawn oblique triangle OBR alone.

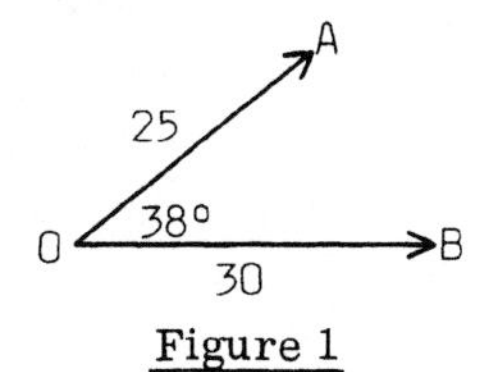

Figure 1

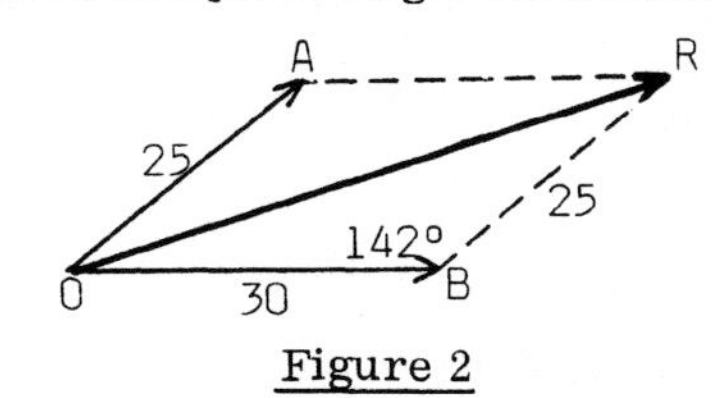

Figure 2

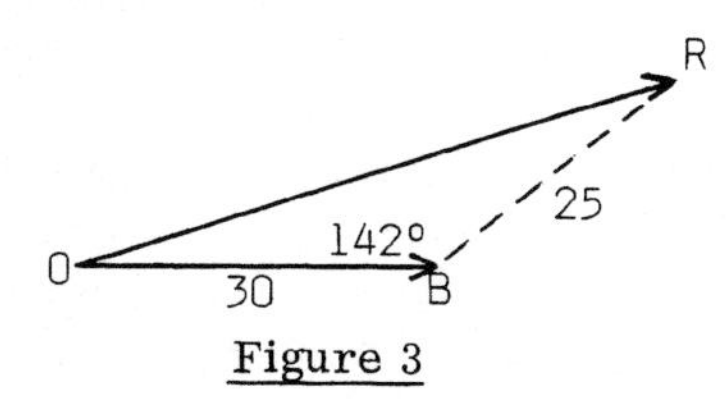

Figure 3

In triangle OBR, we know two sides and the included angle. Therefore, we can use the Law of Cosines to find the length of resultant $\overrightarrow{OR}$. What equation would we use to find the length of $\overrightarrow{OR}$?

____

$$(OR)^2 = (OB)^2 + (BR)^2 - 2(OB)(BR)(\cos B)$$

or

$$(OR)^2 = 30^2 + 25^2 - 2(30)(25)(\cos 142°)$$

19. In Figure 1 below, we have drawn $\overrightarrow{OC}$ and $\overrightarrow{OD}$. The angle between them is 120°. Their lengths are 40 units and 50 units. In Figure 2, we completed the parallelogram and drew the resultant $\overrightarrow{OR}$.

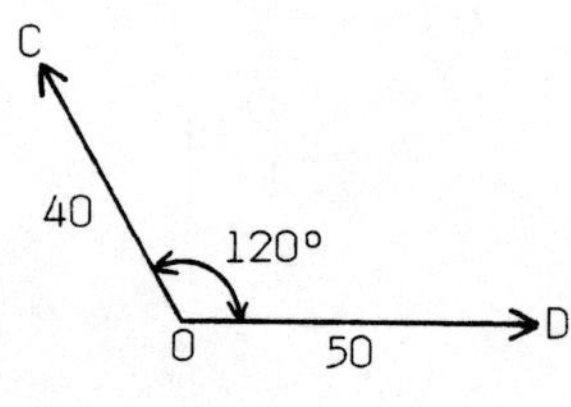

Figure 1

C R 40 120° O 50 D

Figure 2

a) If angle O is 120°, angle D is ________ °.

b) How long is $\overrightarrow{DR}$? ________ units

c) Using triangle ODR, what equation would be used to find the length of $\overrightarrow{OR}$?

____

20. We want to add $\overrightarrow{OF}$ and $\overrightarrow{OT}$. The resultant is $\overrightarrow{OR}$.

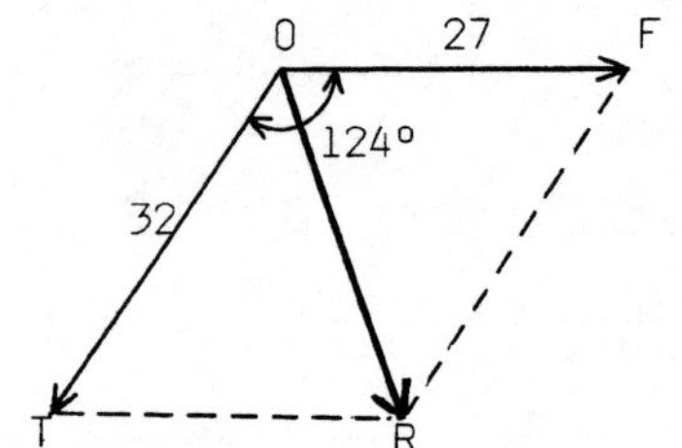

a) Angle O is 124°. How large is angle F? ________ °

b) How long is $\overrightarrow{FR}$? ________ units

c) Using triangle OFR, what equation would be used to find the length of $\overrightarrow{OR}$.

____

Answers to 19:

a) 60° b) 40 units

c) $(OR)^2 = (OD)^2 + (DR)^2 - 2(OD)(DR)(\cos D)$

or

$(OR)^2 = (50)^2 + (40)^2 - 2(50)(40)(\cos 60°)$

21. In Figure 1 below, the angle between $\overrightarrow{OM}$ and $\overrightarrow{OP}$ is 112°. In Figure 2, we have completed the parallelogram and drawn the resultant $\overrightarrow{OR}$. Using the Law of Cosines, we found that $\overrightarrow{OR}$ is 38.4 units long. We want to find the size of angles #1 and #2, the angles formed by the resultant and each original vector.

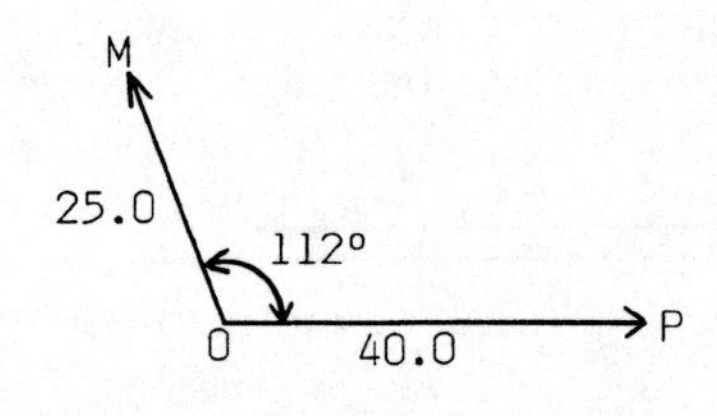

Figure 1

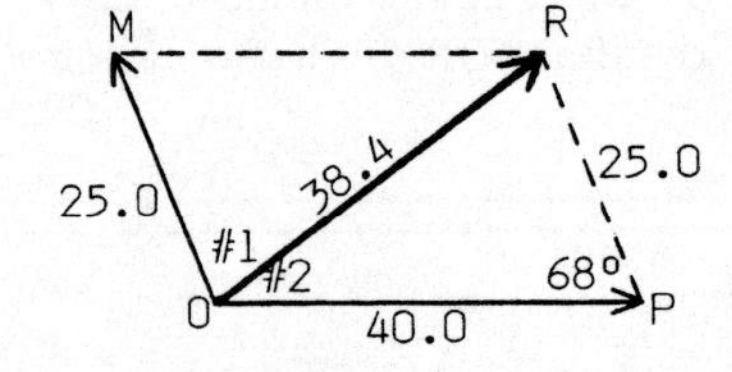

Figure 2

Continued on following page.

Answers to 20:

a) 56° b) 32 units

c) $(OR)^2 = (OF)^2 + (FR)^2 - 2(OF)(FR)(\cos F)$

or

$(OR)^2 = (27)^2 + (32)^2 - 2(27)(32)(\cos 56°)$

21. Continued

Since angle #1 + angle #2 equals 112°, we can find the size of both angles if we can find the size of one of them.

Triangle OPR is drawn at the right. We can use the Law of Sines to find angle #2.

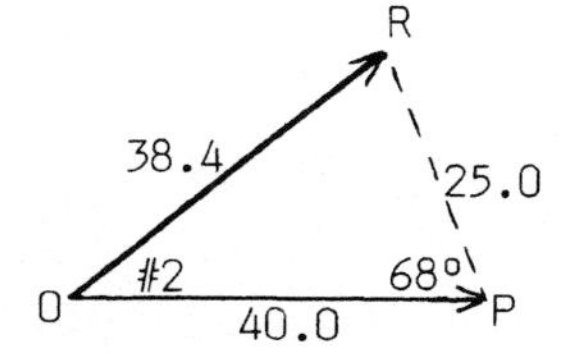

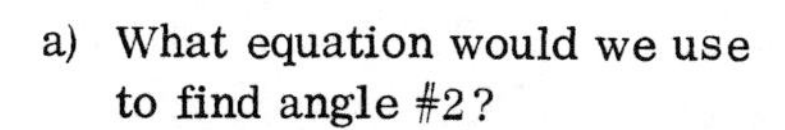

a) What equation would we use to find angle #2?

____

b) Angle #2 contains ________°. Round to a whole number.

c) Angle #1 (in Figure 2) contains ________°.

Answers to 21:

a) $\frac{\sin \#2}{PR} = \frac{\sin P}{OR}$ or $\frac{\sin \#2}{25.0} = \frac{\sin 68°}{38.4}$

b) 37°

c) 75°, from: 112° - 37°

---

22. We want to add $\overrightarrow{OA}$ and $\overrightarrow{OB}$ at the left below. The angle between them is 118°. Their lengths are 20.0 units and 15.0 units. We completed the parallelogram and drew the resultant at the right below.

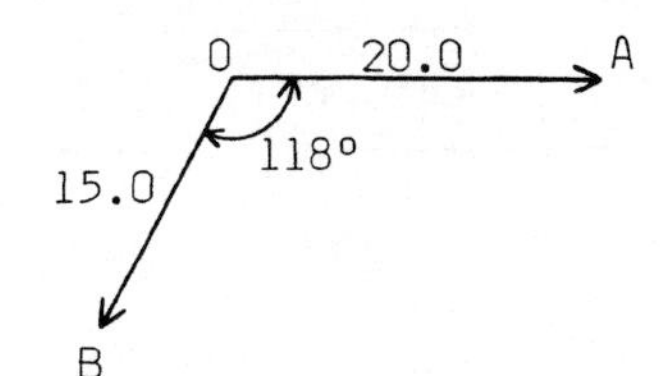

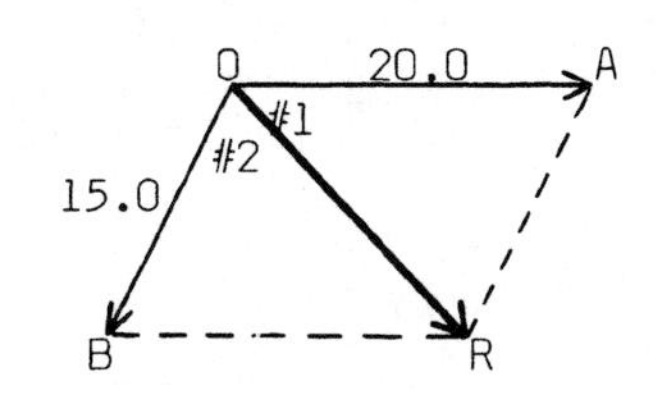

a) How large is angle A? ________°

b) How long is $\overrightarrow{OR}$? Round to tenths. ________ units

c) How large is angle #1? Round to a whole number. ________°

d) How large is angle #2? ________°

Answers to 22:

a) 62°

b) 18.5 units

c) 46°

d) 72°

23. The two vectors at the right are 25.0 units and 45.0 units. They act at an angle of 55°.

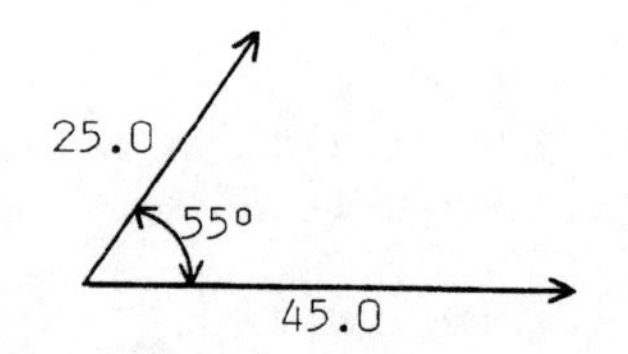

a) Find the length of their resultant. Round to tenths.

____

b) Find the angle between the resultant and the 45.0 unit vector. Round to a whole number.

____

a) 62.8 units b) 19°

4-5 THE OBLIQUE-TRIANGLE METHOD WITH VECTORS ON THE COORDINATE SYSTEM

The oblique-triangle method can be used to add two vectors on the coordinate system. We will discuss vector-additions of that type in this section.

---

24. When two vectors on the coordinate system are added by the oblique-triangle method, the direction of the vectors is given by standard-position angles. We must begin by finding the angle formed by the two vectors. Two examples are given below.

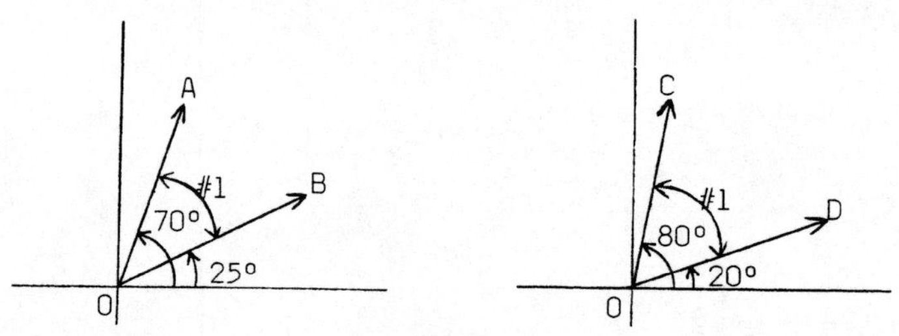

At the left above, the direction of $\overrightarrow{OA}$ is 70°; the direction of $\overrightarrow{OB}$ is 25°. We must find angle #1, the angle formed by the two vectors. To do so, we can subtract the smaller angle from the larger angle. We get:

$$\text{Angle \#1} = 70° - 25° = 45°$$

At the right above, the angle formed by $\overrightarrow{OC}$ and $\overrightarrow{OD}$ is also labeled angle #1. How large is that angle? _______

60°, from 80° - 20°

25. In the diagram at the right, $\overrightarrow{OF}$ is in the second quadrant. The directions of the two vectors are:

138° for $\overrightarrow{OF}$

22° for $\overrightarrow{OH}$

Angle #1 is the angle formed by $\overrightarrow{OF}$ and $\overrightarrow{OH}$. How large is angle #1? ________

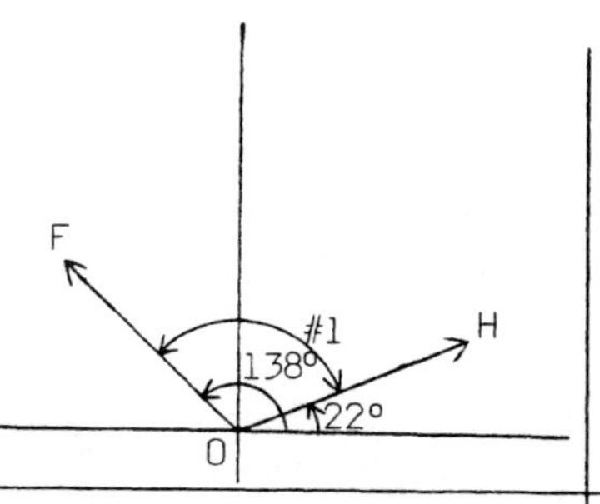

116°, from: 138° - 22°

26. The same vector addition is shown twice below. At the left, the direction of $\overrightarrow{OB}$ is given as the positive angle 315°. At the right, the direction of $\overrightarrow{OB}$ is given as the negative angle -45°. We must find angle #1, the angle between $\overrightarrow{OA}$ and $\overrightarrow{OB}$.

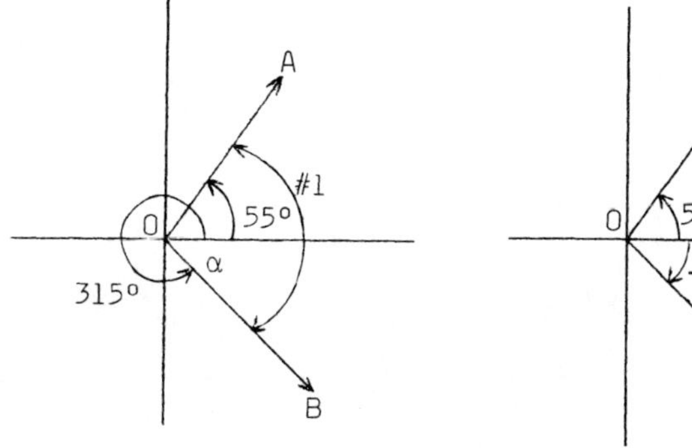

To find angle #1 at the left, we add 55° and α. Since the reference angle is 45°, we get: angle #1 = 55° + 45° = 100°

To find angle #1 at the right, we add 55° and the absolute value of -45°. We get: angle #1 = 55° + 45° = 100°

Did we get the same value for angle #1 in both figures? ________

Yes

27. We want to add $\overrightarrow{OA}$ which is 5.0 units at 120° and $\overrightarrow{OB}$ which is 10.0 units at 40°. The two vectors are shown twice below. At the left, their directions (120° and 40°) and the angle between them (80°) are given. At the right, the parallelogram is completed and the resultant $\overrightarrow{OR}$ is drawn.

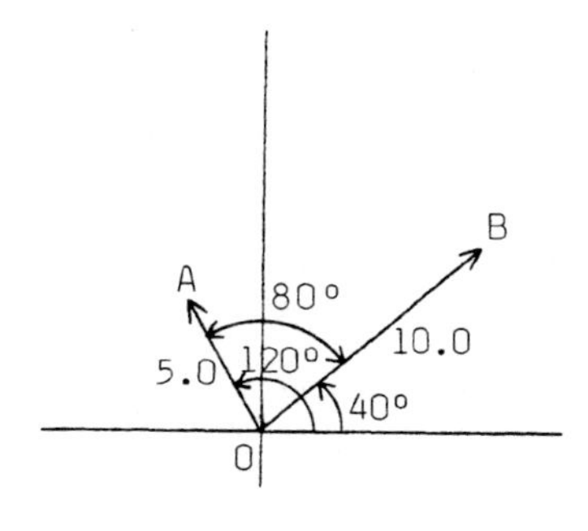

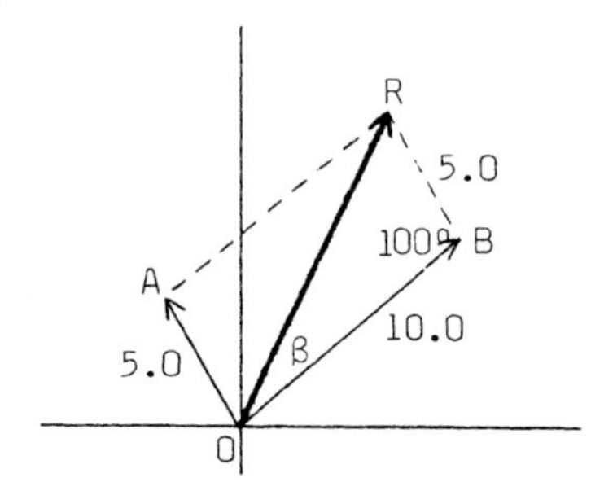

Continued on following page.

27. Continued

We want to find the length and direction of $\overrightarrow{OR}$. We know these facts:

1. The direction of $\overrightarrow{OA}$ is 120°.
   The direction of $\overrightarrow{OB}$ is 40°.
2. Since the angle formed by $\overrightarrow{OA}$ and $\overrightarrow{OB}$ is 120° - 40° = 80°, angle B contains 100°.
3. OB = 10.0 units and BR = 5.0 units.
4. Angle β is the angle formed by $\overrightarrow{OB}$ and $\overrightarrow{OR}$.

a) Based on triangle OBR, what Law-of-Cosines equation would we use to find the length of $\overrightarrow{OR}$?

____

b) Find the length of $\overrightarrow{OR}$. Round to tenths.

__________ units

c) Based on triangle OBR, what Law-of-Sines equation would we use to find angle β ?

____

d) Find angle β. Round to a whole number.

β = ______

e) The standard-position angle of the resultant $\overrightarrow{OR}$ is the sum of angle β and the standard-position angle of $\overrightarrow{OB}$. Therefore: the direction of $\overrightarrow{OR}$ is ________°.

---

a) $(OR)^2 = (OB)^2 + (BR)^2 - 2(OB)(BR)(\cos B)$

or

$(OR)^2 = (10.0)^2 + (5.0)^2 - 2(10.0)(5.0)(\cos 100°)$

b) 11.9 units

c) $\dfrac{\sin \beta}{BR} = \dfrac{\sin B}{OR}$

or

$\dfrac{\sin \beta}{5.0} = \dfrac{\sin 100°}{11.9}$

d) 24°

e) 64°, from: 24° + 40°

28. At the right, $\overrightarrow{OR}$ is the resultant of $\overrightarrow{OC}$ and $\overrightarrow{OD}$. We want to find the length and direction of $\overrightarrow{OR}$. We know these facts:

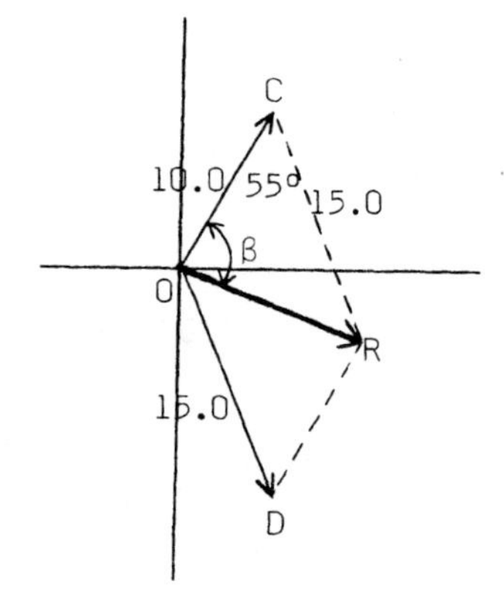

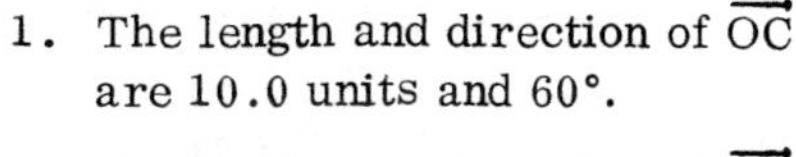

1. The length and direction of $\overrightarrow{OC}$ are 10.0 units and 60°.

   The length and direction of $\overrightarrow{OD}$ are 15.0 units and -65°.

2. The angle formed by $\overrightarrow{OC}$ and $\overrightarrow{OD}$ is $60° + 65° = 125°$. Therefore, angle C is 55°.

3. Angle β is the angle formed by $\overrightarrow{OC}$ and $\overrightarrow{OR}$.

a) Using triangle OCR and the Law of Cosines, find the length of $\overrightarrow{OR}$. Round to tenths.

_________ units

b) Using triangle OCR and the Law of Sines, find angle β. Round to a whole number.

β = _________

c) The resultant $\overrightarrow{OR}$ is in the fourth quadrant. To find its reference angle, we subtract the standard-position angle of $\overrightarrow{OC}$ (60°) from angle β (82°). We get: $82° - 60° = 22°$ (Q4)

State the direction of $\overrightarrow{OR}$ as both a negative and positive standard-position angle. _______ and _______

---

29. When using the oblique-triangle method to add two vectors, we must complete the parallelogram and draw the resultant vector. Do so for each pair of vectors below and identify the quadrant of the resultant.

a) 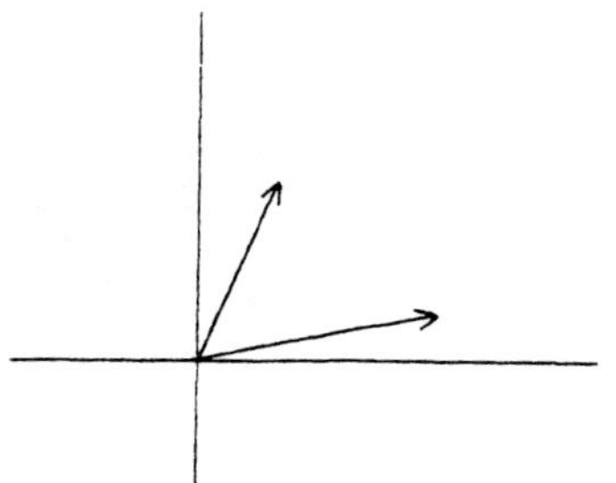

Quadrant ____

b) 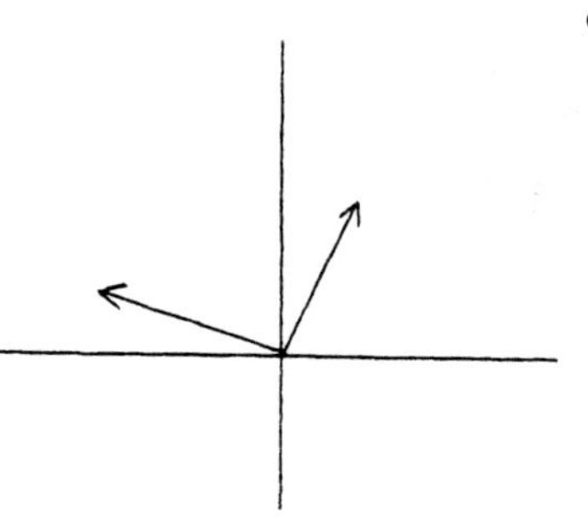

Quadrant ____

c) 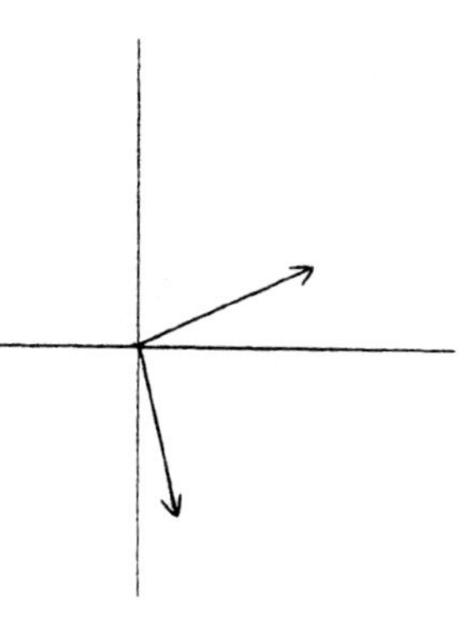

Quadrant ____

Answers to 28:

a) 12.4 units

b) 82°

c) -22° and 338°

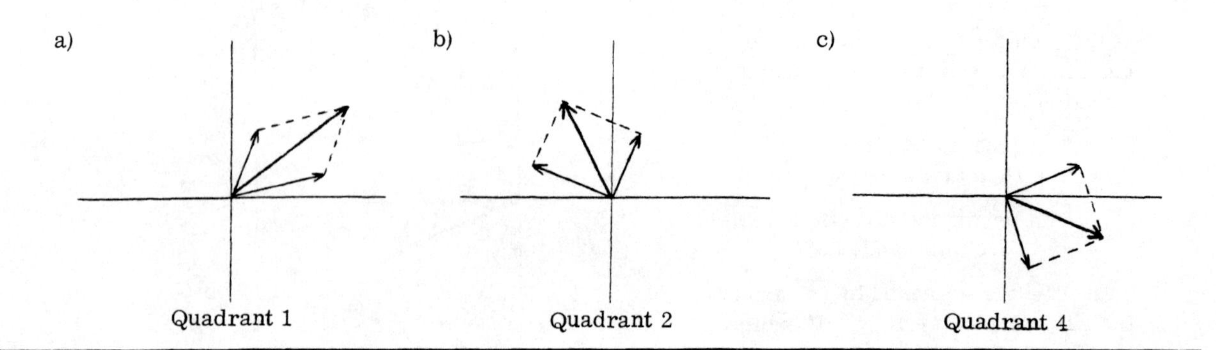

30. Ordinarily, we are only given the length and direction of two vectors to add. We must sketch them on the coordinate system, complete the parallelogram, and draw the resultant. An example is given at the left below. Sketch the addition at the right below.

Two vectors have these lengths and directions:

5.0 units at 50°
10.0 units at 150°

The vector addition is sketched below.

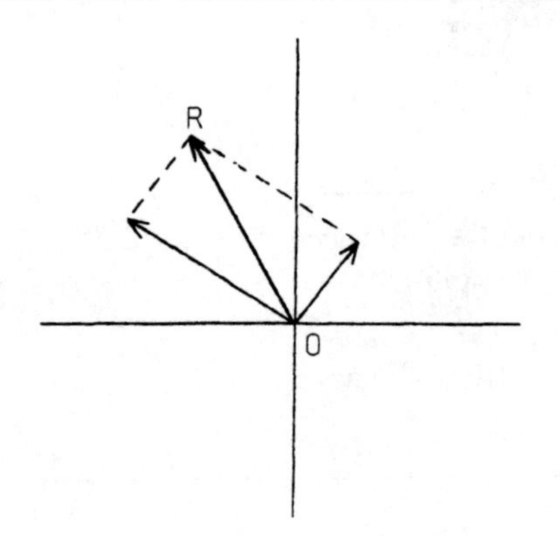

Two vectors have these lengths and directions:

10.0 units at 35°
15.0 units at -70°

Sketch the vector addition.

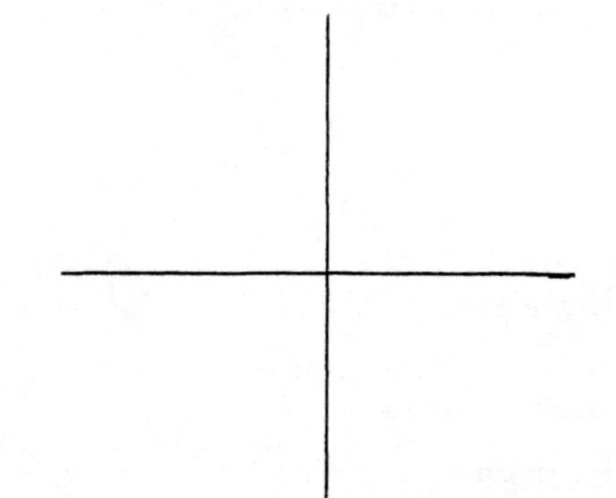

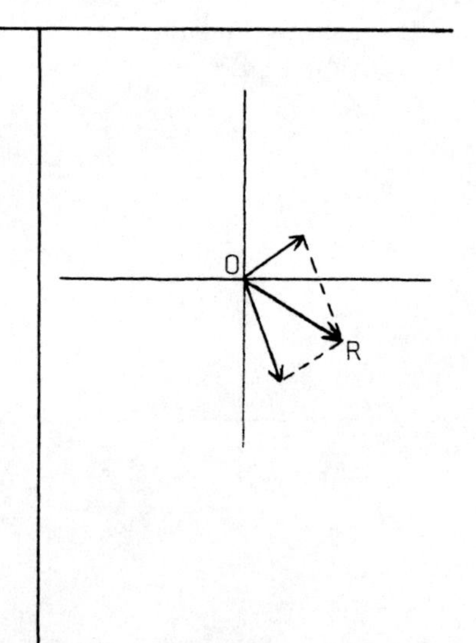

31. Two vectors have these lengths and directions:

10.0 units at 60°
20.0 units at 140°

a) Sketch the vector addition at the right.

b) Find the length of the resultant $\overrightarrow{OR}$. Round to tenths.

c) Find the standard-position angle of the resultant $\overrightarrow{OR}$. Round to a whole number.

---

32. The addition of $\overrightarrow{OA}$ and $\overrightarrow{OB}$ is sketched at the right.

$\overrightarrow{OA}$ is 12.0 units at 0°
$\overrightarrow{OB}$ is 8.0 units at 130°

Since angle BOA is 130°, angle A is 50°.

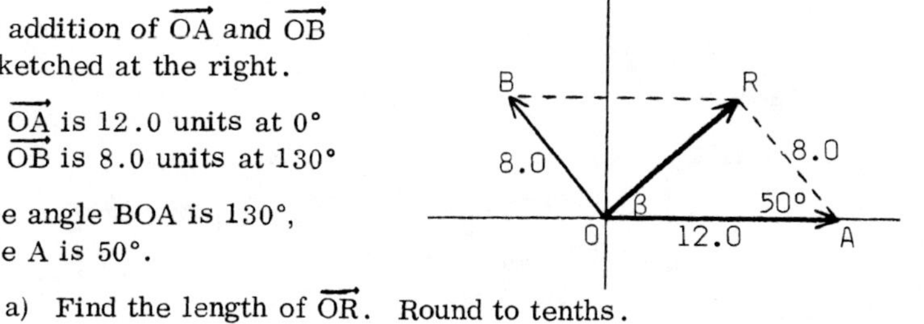

a) Find the length of $\overrightarrow{OR}$. Round to tenths.

b) Angle β is the direction of $\overrightarrow{OR}$. Find the direction of $\overrightarrow{OR}$. Round to a whole number.

β = ________

a) R O

b) 23.9 units

c) 115°

---

a) 9.2 units

b) 42°

SELF-TEST 11 (pages 149-166)

1. Which of the following are vector quantities:

   a) A 36°C temperature.

   b) A 100 lb force downward.

   c) A 15 mph northwest wind.

   d) A 280 ft^2 room area.

   e) A 60 km/hr westward velocity.

   f) A 24v voltage at 30° phase angle.

   g) A 15 second duration time.

2. A 25 kg force and a 15 kg force act on an object at 90° to each other.

   a) Find the resultant force. Round to a whole number.

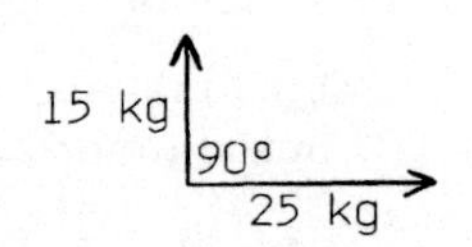

   b) Find the angle between the resultant and the 25 kg force. Round to a whole number.

Sketch the resultant of each pair of vectors by completing the parallelogram. Label the resultant R.

3.

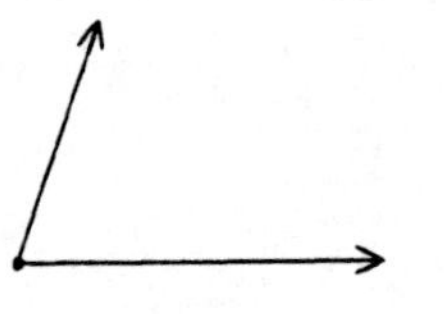

4.

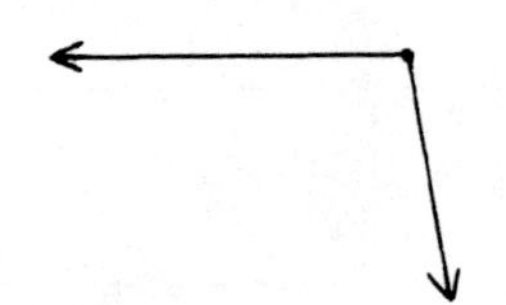

5.

6. Two vectors of 73.8 units and 31.4 units act at an angle of 62°.

   a) Find the length of their resultant. Round to tenths.

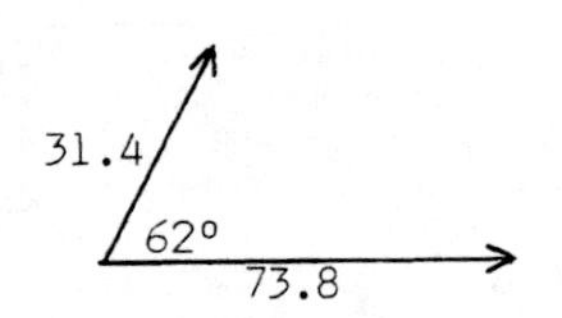

   b) Find the angle between the resultant and the 73.8 unit vector.

7. Two vectors on the coordinate system are:

   $\overrightarrow{OA}$ is 450 units at 40°
   $\overrightarrow{OB}$ is 680 units at 150°

   a) Find the length of their resultant $\overrightarrow{OR}$. Round to a whole number.

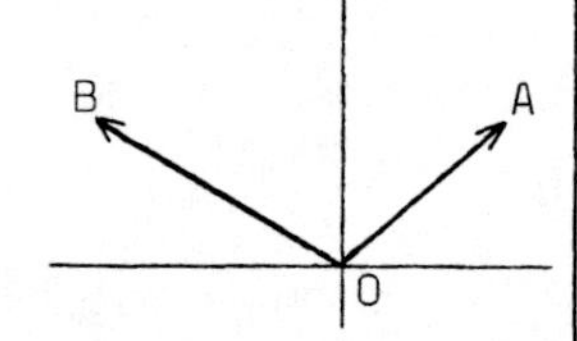

   b) Find the standard-position angle of the resultant. Round to a whole number.

ANSWERS:

1. b, c, e, f

2. a) 29 kg
   b) 31°

3. 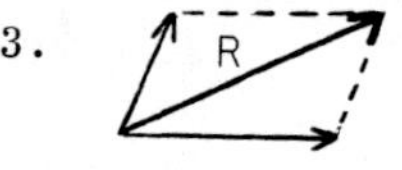

4. R

5. 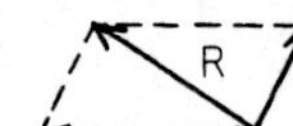

6. a) 92.8 units
   b) 17°

7. a) 675 units
   b) 111°

4-6 COMPONENTS OF VECTORS

In this section, we will discuss the components of vectors. The discussion is limited to the components of slanted vectors.

---

33. $\overrightarrow{OA}$ is a vector in the first quadrant. The other two vectors ($\overrightarrow{OB}$ and $\overrightarrow{BA}$) are called the "components" of $\overrightarrow{OA}$.

    $\overrightarrow{OB}$ is the horizontal component of $\overrightarrow{OA}$.

    $\overrightarrow{BA}$ is the vertical component of $\overrightarrow{OA}$.

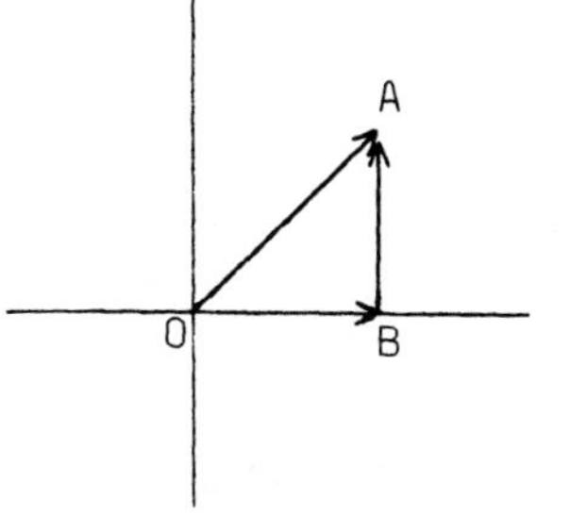

In the last chapter, we discussed the "signs" of component vectors like $\overrightarrow{OB}$ and $\overrightarrow{BA}$. We saw that:

$\overrightarrow{OB}$ is a positive vector because it is "to the right".

$\overrightarrow{BA}$ is a positive vector because it is "upward".

In the last chapter, we saw that all slanted vectors like $\overrightarrow{OA}$ are ______________ (positive/negative).

---

34. $\overrightarrow{OD}$ is a vector in the second quadrant. Its components are $\overrightarrow{OF}$ and $\overrightarrow{FD}$.

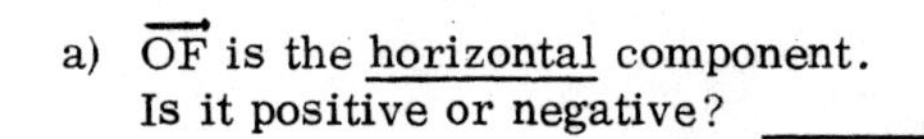

a) $\overrightarrow{OF}$ is the horizontal component. Is it positive or negative? ___________

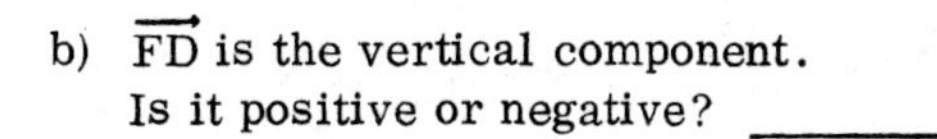

b) $\overrightarrow{FD}$ is the vertical component. Is it positive or negative? ___________

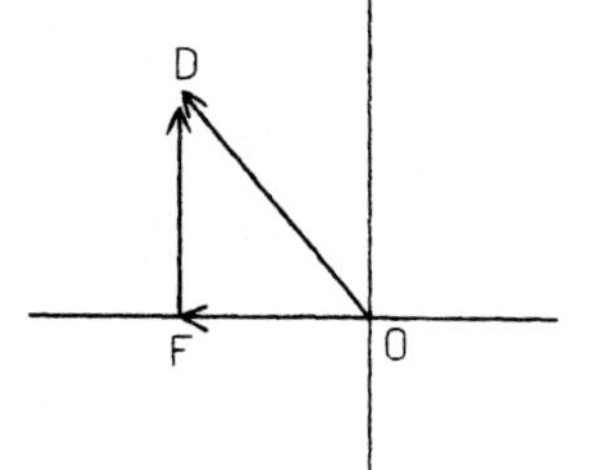

Answer (33): positive

---

35. $\overrightarrow{OT}$ is a vector in the third quadrant. Its components are $\overrightarrow{OP}$ and $\overrightarrow{PT}$.

a) $\overrightarrow{OP}$ is the ______________ component.

b) $\overrightarrow{PT}$ is the ______________ component.

c) Is $\overrightarrow{OP}$ positive or negative? ___________

d) Is $\overrightarrow{PT}$ positive or negative? ___________

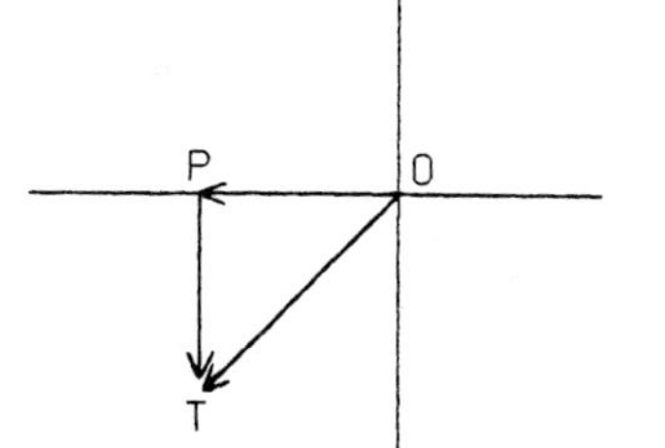

Answers (34):
a) negative
b) positive

Answers (35):
a) horizontal
b) vertical
c) negative
d) negative

36. $\overrightarrow{OM}$ is a vector in the fourth quadrant. Its components are $\overrightarrow{OC}$ and $\overrightarrow{CM}$.

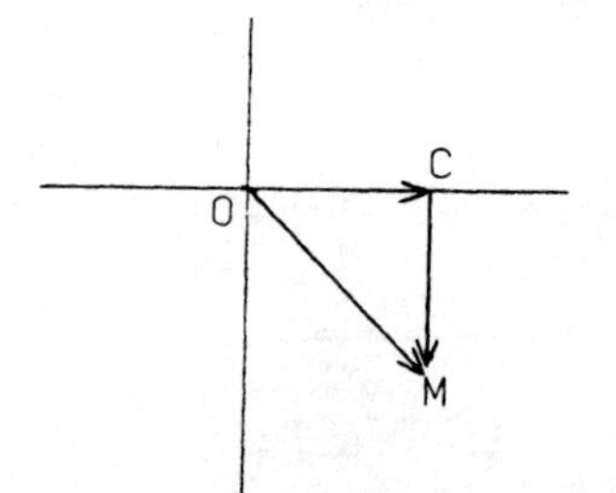

a) The horizontal component is ________.

b) The vertical component is ________.

c) The sign of the horizontal component is ____________.

d) The sign of the vertical component is ____________.

---

37. The coordinates of the tip of a slanted vector give the signed-values of the two components. An example is given below.

a) $\overrightarrow{OC}$

b) $\overrightarrow{CM}$

c) + (or positive)

d) - (or negative)

The coordinates of point P, the tip of $\overrightarrow{OP}$, are (6.4, 4.5). The coordinates give the signed-values of the components, $\overrightarrow{OR}$ and $\overrightarrow{RP}$.

6.4 is the x-coordinate. It is the signed-value of the horizontal component. Therefore:

$\overrightarrow{OR}$ = +6.4 or 6.4

4.5 is the y-coordinate. It is the signed value of the vertical component. Therefore:

$\overrightarrow{RP}$ = ________

---

38. The coordinates of point A at the right are (-2.6, 3.7). Therefore:

$\overrightarrow{RP}$ = +4.5 or 4.5

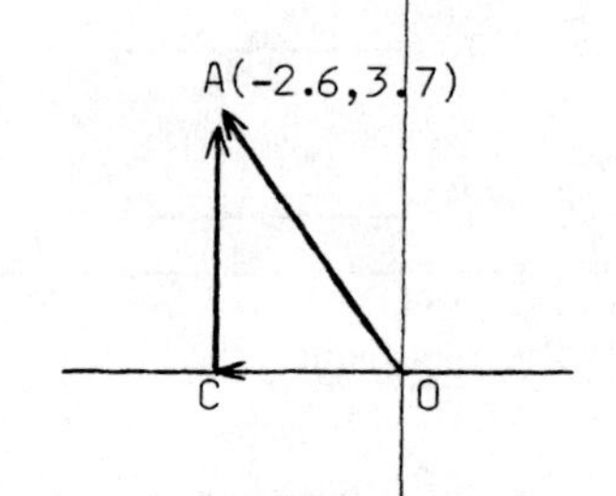

a) What signed number represents $\overrightarrow{OC}$? ________

b) What signed number represents $\overrightarrow{CA}$? ________

---

39. If $\overrightarrow{OD}$ is a -3 vector and $\overrightarrow{DT}$ is a -4 vector, what are the coordinates of point T?

a) -2.6

b) +3.7 or 3.7

(,)

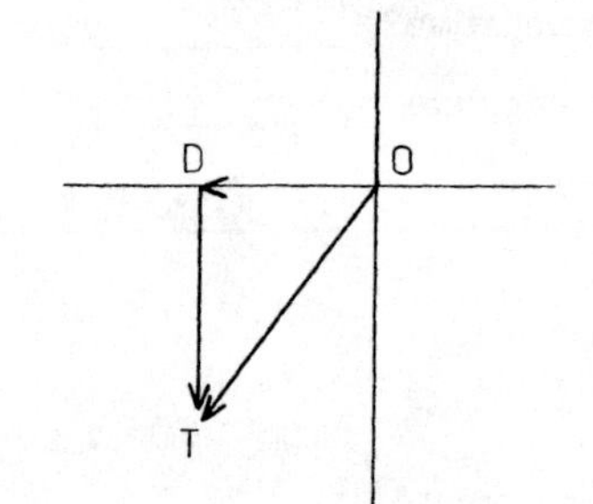

(-3, -4)

40. If $\overrightarrow{OC}$ is a +7.5 vector and $\overrightarrow{CG}$ is a -5.3 vector, what are the coordinates of point G?

(,)

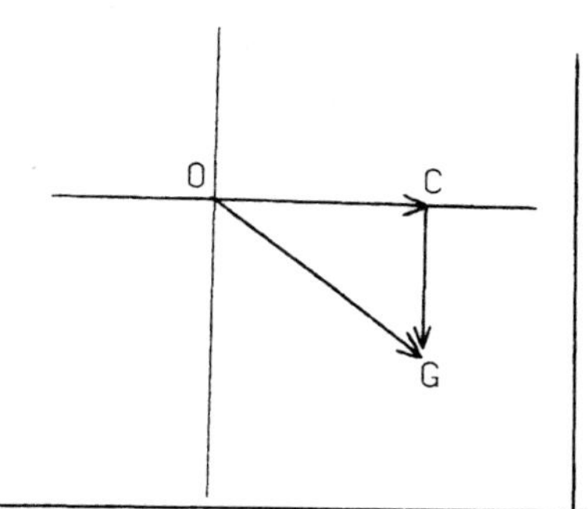

(7.5, -5.3)

4-7 FINDING THE COMPONENTS OF VECTORS

If the length and direction of a vector are known, we can find its components. We will discuss the method in this section.

41. $\overrightarrow{OA}$ is a vector in the first quadrant. Its length is 3.75 units. Its direction is 25°. We want to find the signed-lengths of its components, $\overrightarrow{OB}$ and $\overrightarrow{BA}$.

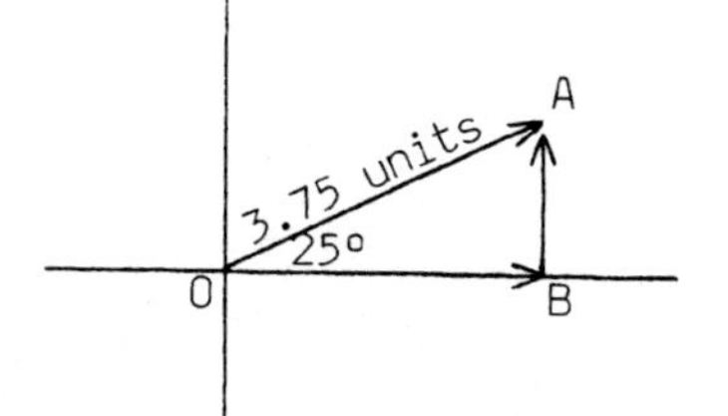

<u>To find the signed-length of $\overrightarrow{OB}$, we use cos 25°.</u>

Since $\cos 25° = \dfrac{\overrightarrow{OB}}{\overrightarrow{OA}}$, $\overrightarrow{OB} = (\overrightarrow{OA})(\cos 25°)$

$= 3.75(\cos 25°)$

$= 3.3986542$

<u>To find the signed-length of $\overrightarrow{BA}$, we use sin 25°.</u>

Since $\sin 25° = \dfrac{\overrightarrow{BA}}{\overrightarrow{OA}}$, $\overrightarrow{BA} = (\overrightarrow{OA})(\sin 25°)$

$= 3.75(\sin 25°)$

$= 1.5848185$

Rounding to hundredths, we get: $\overrightarrow{OB} = 3.40$, $\overrightarrow{BA} = 1.58$

Therefore, the coordinates of point A are (,)

(3.40, 1.58)

42. $\overrightarrow{OP}$ is a vector in the second quadrant. Its length is 18.6 units. Its direction is 41° (Q2) or 139° (since 180° - 41° = 139°). We want to find the signed-lengths of its components, $\overrightarrow{OQ}$ and $\overrightarrow{QP}$.

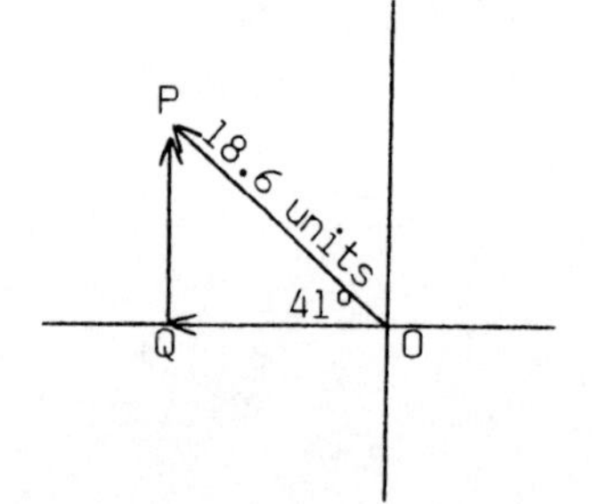

We know these facts about signs:

$\overrightarrow{OQ}$ is <u>negative</u>.

$\overrightarrow{QP}$ is <u>positive</u>.

We also know that the standard-position angle and reference angle have the same sine and cosine. That is:

$$\sin 41° \text{ (Q2)} = \sin 139°$$
$$\cos 41° \text{ (Q2)} = \cos 139°$$

To avoid sign errors, it is easier to use sin 139° and cos 139° rather than sin 41° (Q2) and cos 41° (Q2) <u>because the sign is attached by the calculator automatically</u>. It does not have to be attached separately. Therefore:

<u>To find the signed-length of $\overrightarrow{OQ}$, we use cos 139°.</u>

$$\text{Since } \cos 139° = \frac{\overrightarrow{OQ}}{\overrightarrow{OP}}, \quad \overrightarrow{OQ} = (\overrightarrow{OP})(\cos 139°)$$
$$= 18.6(\cos 139°)$$
$$= -14.037598$$

<u>To find the signed-length of $\overrightarrow{QP}$, we use sin 139°.</u>

$$\text{Since } \sin 139° = \frac{\overrightarrow{QP}}{\overrightarrow{OP}}, \quad \overrightarrow{QP} = (\overrightarrow{OP})(\sin 139°)$$
$$= 18.6(\sin 139°)$$
$$= 12.202698$$

Rounding to tenths, we get: $\overrightarrow{OQ} = -14.0$, $\overrightarrow{QP} = 12.2$.

Therefore, the coordinates of point P are (,).

(-14.0, 12.2)

---

43. $\overrightarrow{OF}$ is a vector in the third quadrant. Its length is 27.5 units. Its direction is 52 (Q3) or 232° (since 180° + 52° = 232°). We want to find the signed-lengths of its components, $\overrightarrow{OB}$ and $\overrightarrow{BF}$.

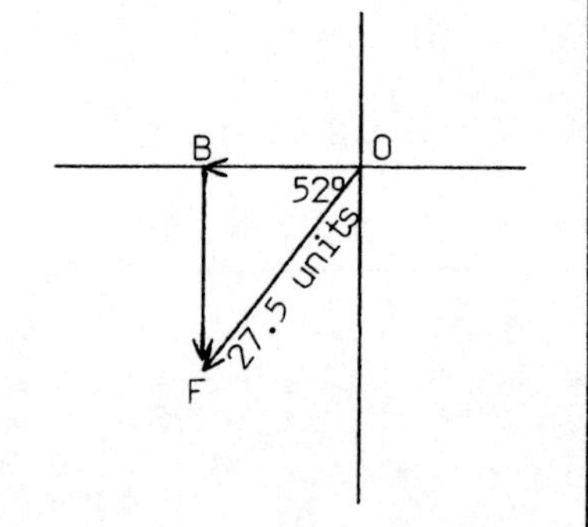

We know these facts about signs.

$\overrightarrow{OB}$ is <u>negative</u>.

$\overrightarrow{BF}$ is <u>negative</u>.

Continued on following page.

43. Continued

To avoid sign errors, it is easier to use sin 232° and cos 232° rather than sin 52° (Q3) and cos 52° (Q3). Therefore:

a) <u>To find the signed-length of $\overrightarrow{OB}$, we use cos 232°.</u>

$$\text{Since } \cos 232° = \frac{\overrightarrow{OB}}{\overrightarrow{OF}}, \quad \overrightarrow{OB} = (\overrightarrow{OF})(\cos 232°)$$

$$= 27.5(\cos 232°)$$

$$= \underline{\hspace{3cm}}$$

(Round to tenths.)

b) <u>To find the signed-length of $\overrightarrow{BF}$, we use sin 232°.</u>

$$\text{Since } \sin 232° = \frac{\overrightarrow{BF}}{\overrightarrow{OF}}, \quad \overrightarrow{BF} = (\overrightarrow{OF})(\sin 232°)$$

$$= 27.5(\sin 232°)$$

$$= \underline{\hspace{3cm}}$$

(Round to tenths.)

---

a) -16.9 units

b) -21.7 units

44. 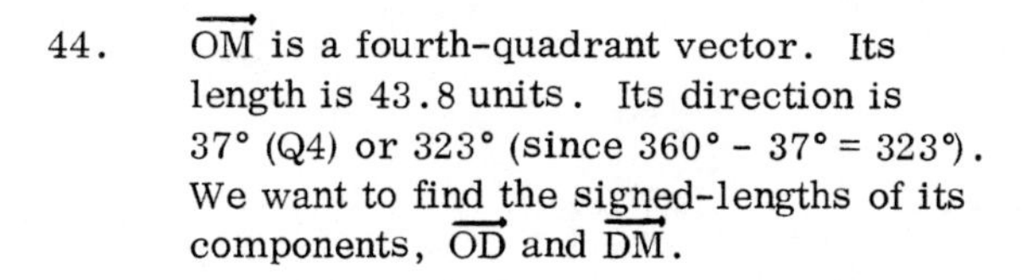
$\overrightarrow{OM}$ is a fourth-quadrant vector. Its length is 43.8 units. Its direction is 37° (Q4) or 323° (since 360° - 37° = 323°). We want to find the signed-lengths of its components, $\overrightarrow{OD}$ and $\overrightarrow{DM}$.

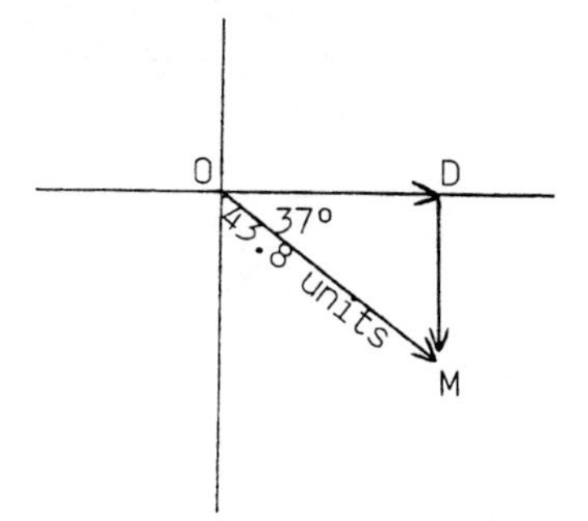

We know these facts about signs:

$\overrightarrow{OD}$ is <u>positive</u>.

$\overrightarrow{DM}$ is <u>negative</u>.

To avoid sign errors, it is easier to use sin 323° and cos 323° rather than sin 37° (Q4) and cos 37° (Q4).

a) Using cos 323°, find the signed-length of $\overrightarrow{OD}$. Round to tenths.

$\overrightarrow{OD}$ = ____________

b) Using sin 323°, find the signed-length of $\overrightarrow{DM}$. Round to tenths.

$\overrightarrow{DM}$ = ____________

---

a) 35.0 units

b) -26.4 units

45. Sometimes the direction of a vector is given as a negative angle. Typically, the vector is in the third or fourth quadrant. Two examples are discussed.

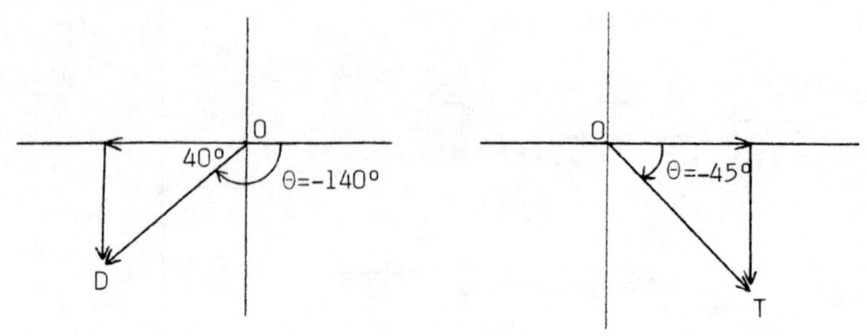

At the left, the direction of $\overrightarrow{OD}$ is -140°. Its reference angle is 40° (Q3). 40° (Q3) and -140° have the same sine and cosine because they have the same terminal side. That is:

$$\sin 40° \text{ (Q3)} = \sin(-140°) = -0.6427876$$

$$\cos 40° \text{ (Q3)} = \cos(-140°) = -0.7660444$$

At the right, the direction of $\overrightarrow{OT}$ is -45°. Its reference angle is 45° (Q4). 45° (Q4) and -45° also have the same sine and cosine. that is:

a) $\sin 45° \text{ (Q4)} = \sin(-45°) =$ __________

b) $\cos 45° \text{ (Q4)} = \cos(-45°) =$ __________

---

a) -0.7071068

b) 0.7071068

46. $\overrightarrow{OC}$ is a fourth-quadrant vector whose direction is -48°. Its reference angle is 48° (Q4). Its length is 18.6 units. We want to find the signed-lengths of $\overrightarrow{OA}$ and $\overrightarrow{AC}$.

To avoid sign errors when finding $\overrightarrow{OA}$ and $\overrightarrow{AC}$, it is easier to use sin(-48°) and cos(-48°) rather than sin 48° (Q4) and cos 48° (Q4).

a) Using cos(-48°), find the signed-length of $\overrightarrow{OA}$. Round to tenths.

$\overrightarrow{OA} =$ __________

b) Using sin(-48°), find the signed-length of $\overrightarrow{AC}$. Round to tenths.

$\overrightarrow{AC} =$ __________

---

a) 12.4 units

b) -13.8 units

47. $\overrightarrow{OR}$ is a third-quadrant vector whose direction is -145°. Its reference angle is 35° (Q3). Its length is 6.83 units. We want to find the signed-lengths of $\overrightarrow{OF}$ and $\overrightarrow{FR}$.

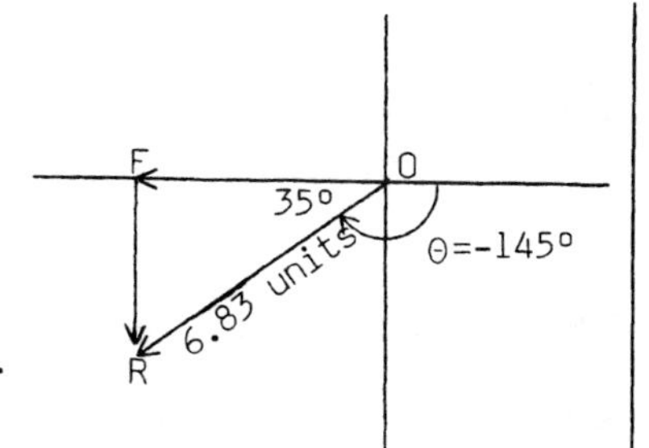

a) Using cos(-145°), find the signed-length of $\overrightarrow{OF}$. Round to hundredths.

$\overrightarrow{OF}$ = ______________

b) Using sin(-145°), find the signed-length of $\overrightarrow{FR}$. Round to hundredths.

$\overrightarrow{FR}$ = ______________

c) The coordinates of point R are (,).

a) -5.59 units b) -3.92 units c) (-5.59, -3.92)

4-8 FINDING ANGLES WITH KNOWN TANGENTS

In this section, we will discuss the procedure for finding angles with known tangents. The angles are limited to positive angles from 0° to 360° and negative angles from 0° to -180°.

48. A 50° angle and a 230° angle are shown in standard position below.

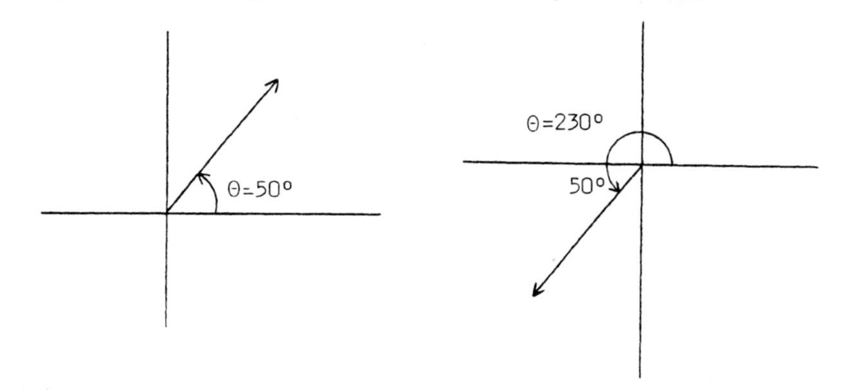

The tangents of angles are positive in both the first and third quadrants. Since the reference angle for both angles above is 50°, both have the same tangent. Using a calculator, we get:

$$\tan 50° = \tan 230° = 1.1917536$$

Continued on following page.

48. Continued

When using a calculator to find the angle whose tangent is 1.1917536, we enter that number and press [INV] [tan]. We get:

Enter	Press	Display
1.1917536	[INV] [tan]	50

The calculator gives us the first-quadrant angle. To get the third-quadrant angle, we must add 180°. We get:

$$50° + 180° = 230°$$

Use a calculator and then add 180° for these. Round to a whole number.

a) Find the third-quadrant angle whose tangent is 0.324669 .

θ = ________

b) Find the third-quadrant angle whose tangent is 2.465124 .

θ = ________

---

49. In the last frame, we saw that tan 230° = tan 50°. However, as you can see at the right, a -130° angle also has the same tangent because it has the same terminal side as a 230° angle. Therefore:

$$\tan 50° = \tan 230° = \tan(-130°)$$

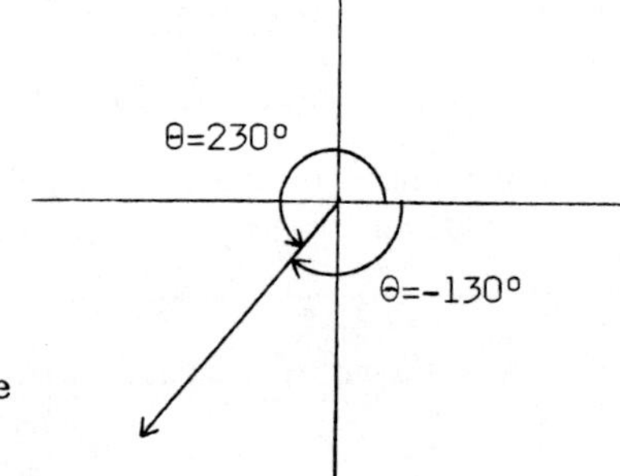

Notice that we can find the absolute value of -130° by subtracting 230° from 360°. Use that method for these.

Find the negative angle with the same tangent as each of these third-quadrant angles.

a) 200° ________ b) 185° ________ c) 256° ________

Answers to 48:
a) 198°, from: 18° + 180°
b) 248°, from: 68° + 180°

---

50. To use a calculator to find the negative angle whose tangent is 0.81099, we use these steps.

1. Find the first-quadrant angle whose tangent is 0.81099. Round to a whole number. — It is 39°.
2. Add 180° to get the positive third-quadrant angle. — It is 219°, from: 39° + 180°.
3. Subtract 219° from 360° to get the absolute value of the negative angle. — The absolute value is 141°. The negative angle is -141°.

Answers to 49:
a) -160°
b) -175°
c) -104°

Continued on following page.

50. Continued

Using the steps above, complete these. Round to a whole number.

a) Find the negative angle whose tangent is 0.20879 . θ = ________

b) Find the negative angle whose tangent is 14.226 . θ = ________

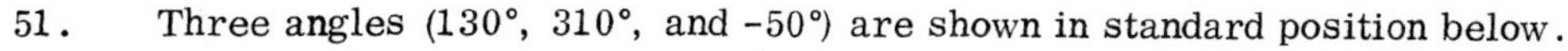

a) -168° b) -94°

51. Three angles (130°, 310°, and -50°) are shown in standard position below.

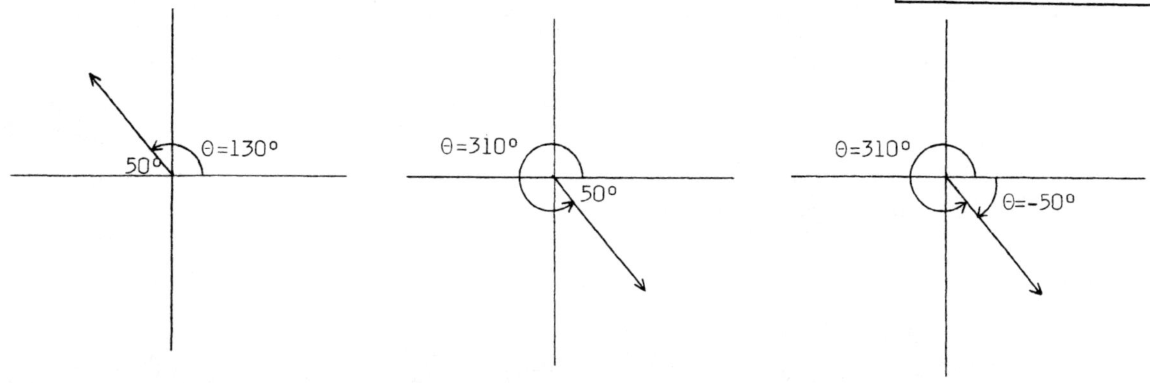

The tangents of angles are negative in both the second and fourth quadrants. Since 50° is the reference angle for both 130° and 310°, those two angles have the same tangent. And since -50° has the same terminal side as 310°, it also has the same tangent. Using a calculator, we get:

tan 130° = tan 310° = tan(-50°) = -1.1917536

When using a calculator to find the angle whose tangent is -1.1917536, we enter that number and press [INV] [tan] . We get:

Enter	Press	Display
1.1917536	[+/-] [INV] [tan]	-50

The calculator gives us -50°, the negative fourth-quadrant angle. We can get the reference angle (50°) by simply dropping the negative sign. Then we can subtract that value from 360° and 180° to find the positive fourth-quadrant and second-quadrant angles with the same tangent.

360° - 50° = 310°
180° - 50° = 130°

Name the positive fourth-quadrant and second-quadrant angles with the same tangent as -71°.

________ and ________

289° and 109°

52. Rounding to a whole number, find the following angles with a tangent of -0.25336.

a) a negative fourth-quadrant angle ________

b) a positive fourth-quadrant angle ________

c) a positive second-quadrant angle ________

53. Round each of these to a whole number.

a) Find the <u>negative</u> fourth-quadrant angle whose tangent is -13.9707 . ________

b) Find the <u>positive</u> fourth-quadrant angle whose tangent is -2.295668 . ________

c) Find the <u>positive</u> second-quadrant angle whose tangent is -0.40267 . ________

a) -14°

b) 346°

c) 166°

---

54. Round each of these to a whole number.

a) Find the <u>positive</u> third-quadrant angle whose tangent is 0.205791 . ________

b) Find the <u>positive</u> second-quadrant angle whose tangent is -1.53804 . ________

a) -86°

b) 294°, from: 360° - 66°

c) 158°, from: 180° - 22°

---

55. Round each of these to a whole number.

a) Find the <u>negative</u> fourth-quadrant angle whose tangent is -0.953302 . ________

b) Find the <u>negative</u> third-quadrant angle whose tangent is 0.664419 .

a) 192°, from: 180° + 12°

b) 123°, from: 180° - 57°

---

a) -44° b) -146°, from: 360° - 214°

4-9 FINDING THE LENGTH AND DIRECTION OF VECTORS

If the components of a slanted vector are known, we can find its length and direction. The method is shown in this section.

---

56. $\overrightarrow{OA}$ is a vector in the first quadrant. The coordinates of point A are (6.0, 5.0). Therefore:

$\overrightarrow{OC}$ is 6.0 units.

$\overrightarrow{CA}$ is 5.0 units.

We want to find: 1. The <u>direction of</u> $\overrightarrow{OA}$ (the size of α).
2. The <u>length</u> of $\overrightarrow{OA}$.

a) <u>To find the size of α, we use the tangent ratio.</u>

$$\tan \alpha = \frac{\overrightarrow{CA}}{\overrightarrow{OC}} = \frac{5.0}{6.0}$$

α contains ________° (Round to a whole number.)

Continued on following page.

56. Continued

b) <u>To find the length of $\overrightarrow{OA}$, we use the Pythagorean Theorem</u>.

$$(OA)^2 = (OC)^2 + (CA)^2$$

$$OA = \sqrt{(OC)^2 + (CA)^2}$$

$$= \sqrt{(6.0)^2 + (5.0)^2}$$

$$= \sqrt{61} = \_\_\_\_\_\_\_\_ \text{ (Round to tenths.)}$$

---

57. $\overrightarrow{OM}$ is in the third quadrant. The coordinates of point M are (-4.0, -5.0). Therefore:

$\overrightarrow{OD}$ is -4.0 units.

$\overrightarrow{DM}$ is -5.0 units.

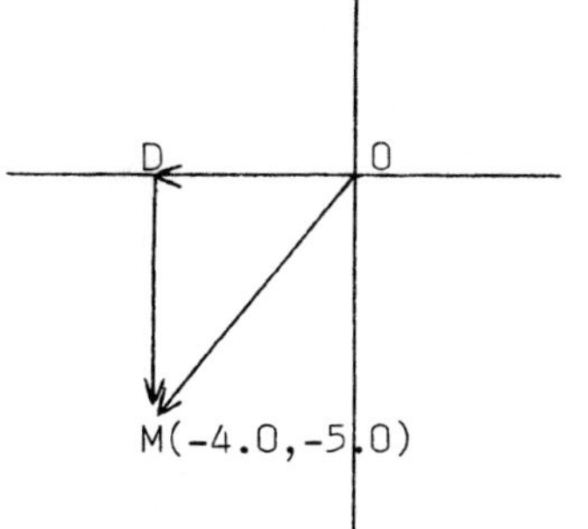

To find the length of $\overrightarrow{OM}$, we use the Pythagorean Theorem. <u>Remember</u>: squaring a negative number gives a positive number.

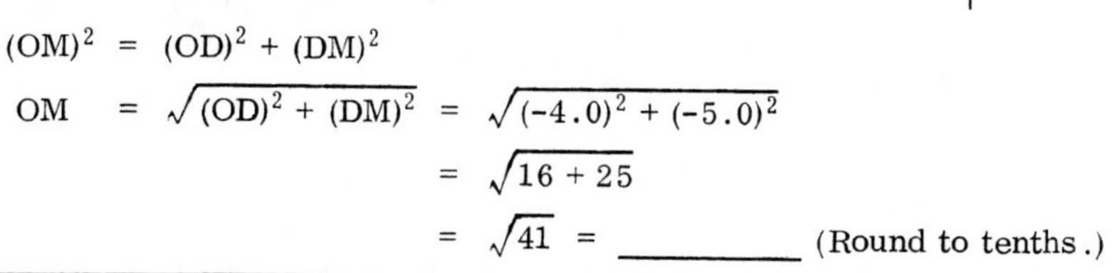

$$(OM)^2 = (OD)^2 + (DM)^2$$

$$OM = \sqrt{(OD)^2 + (DM)^2} = \sqrt{(-4.0)^2 + (-5.0)^2}$$

$$= \sqrt{16 + 25}$$

$$= \sqrt{41} = \_\_\_\_\_\_\_\_ \text{ (Round to tenths.)}$$

Answers (to frame 56):

a) 40°

b) 7.8 units

---

58. $\overrightarrow{OP}$ is in the second quadrant. The coordinates of point P are given.

Use the Pythagorean Theorem to find the length of $\overrightarrow{OP}$. Round to hundredths.

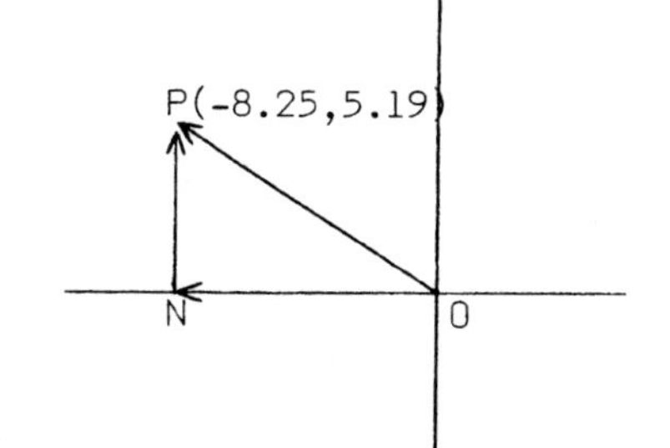

The length of $\overrightarrow{OP}$ is ________ units.

Answer (to frame 57): 6.4 units

---

59. $\overrightarrow{OF}$ is in the third quadrant. The coordinates of point F are (-3.5, -4.5). Therefore:

$\overrightarrow{OB}$ is -3.5 units.

$\overrightarrow{BF}$ is -4.5 units.

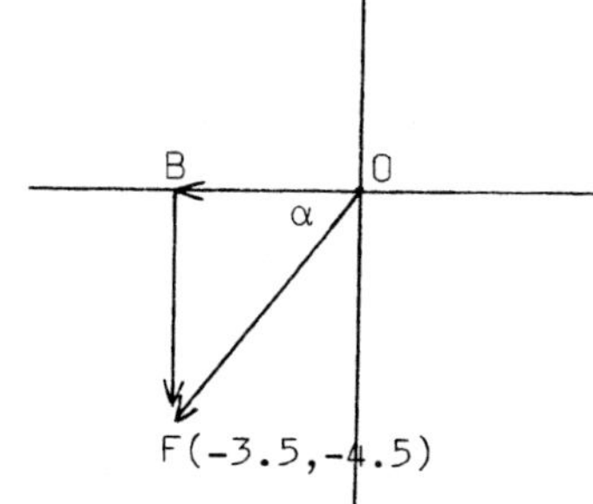

Answer (to frame 58): 9.75 units

Continued on following page.

59. Continued

To find the direction of $\overrightarrow{OF}$, we use $\tan\alpha$. Notice that $\tan\alpha$ is positive even though both components are negative.

$$\tan\alpha = \frac{-4.5}{-3.5} = 1.2857143$$

a) State the direction of $\overrightarrow{OF}$ as a <u>positive</u> standard-position angle. Round to a whole number. θ = ________

b) State the direction of $\overrightarrow{OF}$ as a <u>negative</u> standard-position angle. θ = ________

---

60. $\overrightarrow{OH}$ is in the fourth quadrant. The coordinates of point H are given. Therefore:

$\overrightarrow{OF}$ is 4.51 units.

$\overrightarrow{FH}$ is -6.14 units.

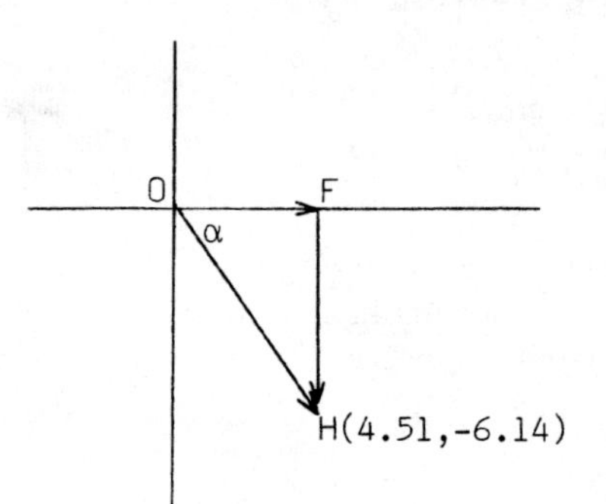

To find the direction of $\overrightarrow{OH}$, we use the tangent ratio.

$$\tan\alpha = \frac{\overrightarrow{FH}}{\overrightarrow{OF}} = \frac{-6.14}{4.51} = -1.3614191$$

Report each of these as a whole number.

a) The reference angle α is ________.

b) The <u>negative</u> standard-position angle is ________.

c) The <u>positive</u> standard-position angle is ________.

Answers (59):
a) 232°, from: 52° + 180°
b) -128°, from: 360° - 232°

---

61. $\overrightarrow{OJ}$ is in the second quadrant. The coordinates of point J are given. Therefore:

$\overrightarrow{OA}$ is -6.5 units.

$\overrightarrow{AJ}$ is 5.7 units.

J(-6.5,5.7)

α

A O

To find the direction of $\overrightarrow{OJ}$, we use the tangent ratio.

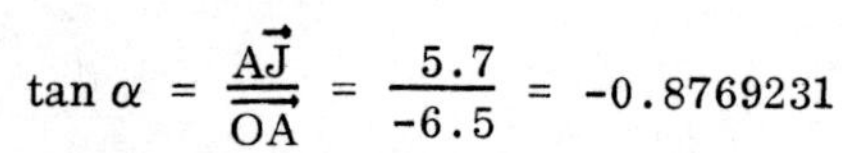

$$\tan\alpha = \frac{\overrightarrow{AJ}}{\overrightarrow{OA}} = \frac{5.7}{-6.5} = -0.8769231$$

Report each of these as a whole number.

a) The reference angle α is ________.

b) The positive standard-position angle is ________.

Answers (60):

a) 54° (Q4)
b) -54°
c) 306°

Answers (61):
a) 41° (Q2)
b) 139°

62. $\overrightarrow{OT}$ is in the third quadrant. The coordinates of point T are given.

a) Find the length of $\overrightarrow{OT}$. Round to tenths.

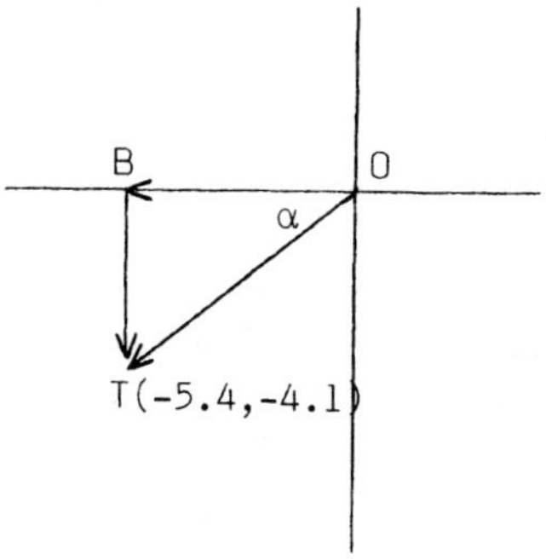

B O α T(-5.4,-4.1)

The length of $\overrightarrow{OT}$ is ________ units.

b) Find the direction of $\overrightarrow{OT}$. State the direction as both a positive and negative standard-position angle. Round to a whole number.

θ= __________ and __________

63. Referring to the diagram:

a) Find the length of $\overrightarrow{OD}$. Round to hundredths.

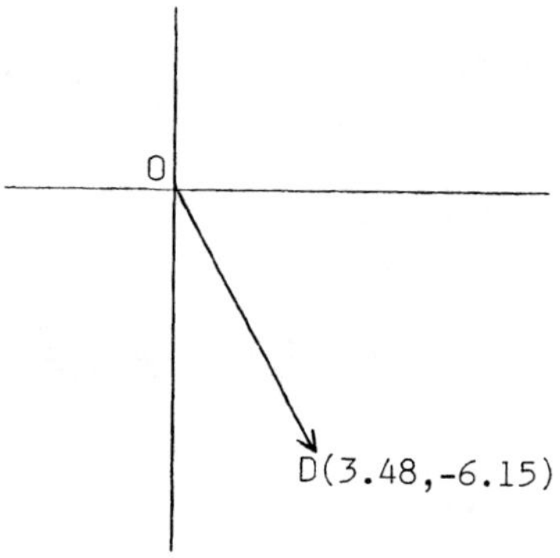

O D(3.48,-6.15)

The length of $\overrightarrow{OD}$ is ________ units.

b) Find the direction of $\overrightarrow{OD}$. State the direction as both a positive and negative standard-position angle. Round to a whole number.

θ = __________ and __________

a) 6.8 units

b) 217° and -143°

64. The coordinates of a second-quadrant vector are (-685, 427).

a) Sketch the vector at the right.

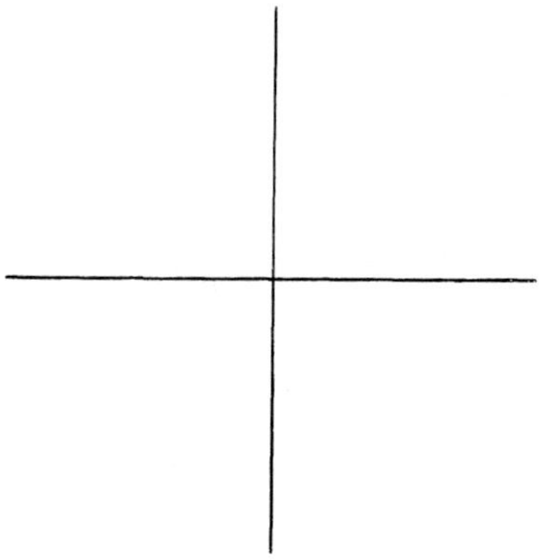

b) Find the length of the vector. Round to a whole number.

____

c) State the direction of the vector as a positive standard-position angle. Round to a whole number.

____

a) 7.07 units

b) 300° and -60°

Answers to Frame 64:

a)

(-685,427)

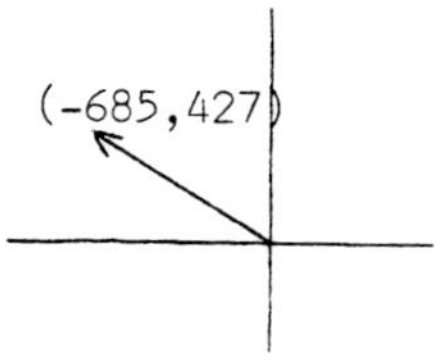

b) 807 units

c) 148°

4-10 VECTORS ON THE AXES

In this section, we will discuss the length, direction, and components of vectors on the axes.

65. Two vectors on the axes are shown below. The coordinates of the tip of each vector give the signed-values of its components.

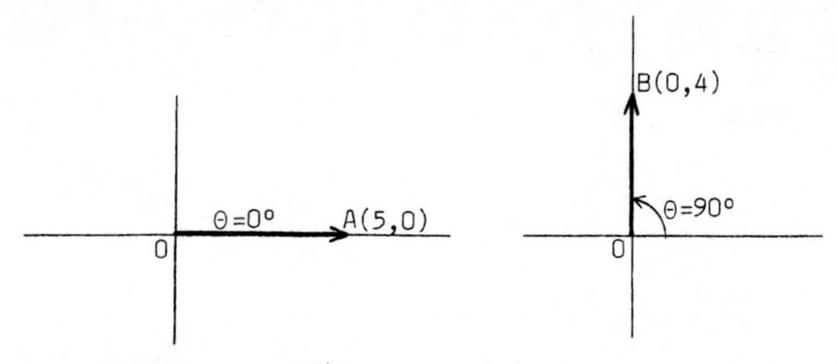

Notice these points about $\overrightarrow{OA}$ on the horizontal axis.

1. Its length is 5 units.
2. Its direction is 0°.
3. Its horizontal component is 5.
4. Its vertical component is 0.

$\overrightarrow{OB}$ is on the vertical axis. From the sketch, you can see that:

a) Its length is ______ units.

b) Its direction is ________°.

c) Its horizontal component is ______.

d) Its vertical component is ______.

66. Two more vectors on the axes are shown below.

a) 4 units

b) 90°

c) 0

d) 4

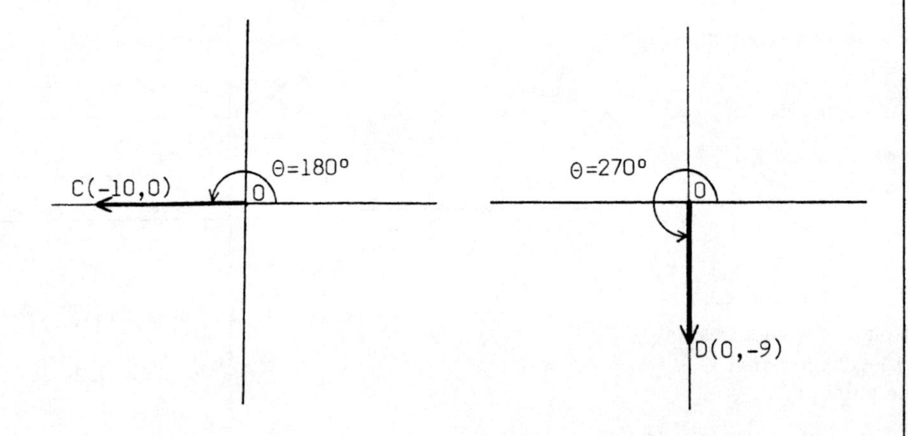

Continued on following page.

66. Continued

Even when one of its components is negative, <u>the length of a vector on an axis is always positive</u>. Therefore, for $\overrightarrow{OC}$ on the horizontal axis:

1. Its length is +10 or 10 units.
2. Its direction is 180°.
3. Its horizontal component is -10.
4. Its vertical component is 0.

$\overrightarrow{OD}$ is on the vertical axis. From the sketch, you can see that:

a) Its length is ________ units.

b) Its direction is ________°.

c) Its horizontal component is ______.

d) Its vertical component is ________.

---

67. Each vector below is on an axis. The direction for each is given as both a positive and a negative angle.

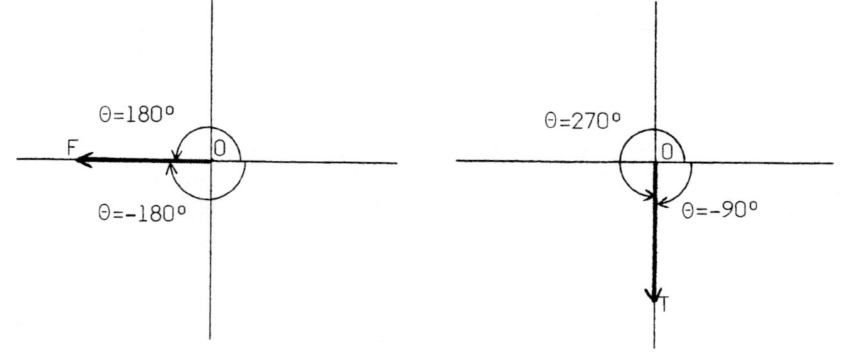

As you can see from the sketches:

The direction of $\overrightarrow{OF}$ is either 180° or -180°.

The direction of $\overrightarrow{OT}$ is either ________ or ________.

Answers to 66:

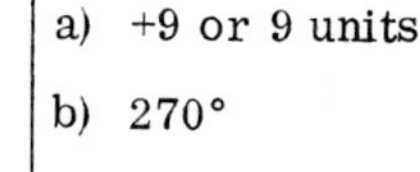
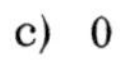

a) +9 or 9 units

b) 270°

c) 0

d) -9

---

68. $\overrightarrow{OM}$ at the right is on the vertical axis.

a) Its horizontal component is ______.

b) Its vertical component is __________.

c) Its length is __________ units.

d) Its direction is either __________ or __________.

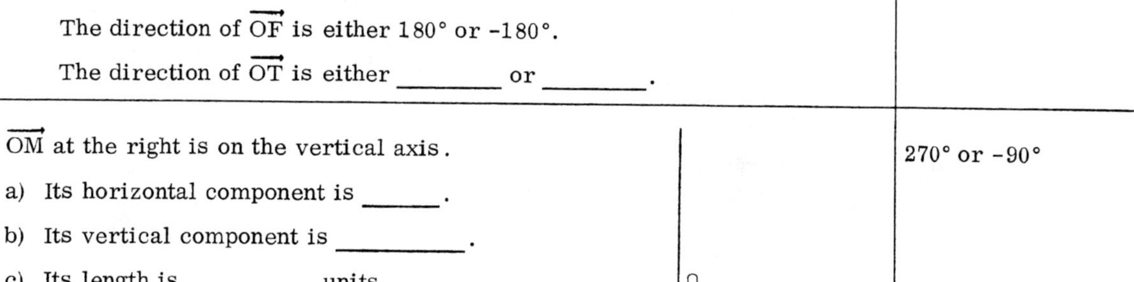
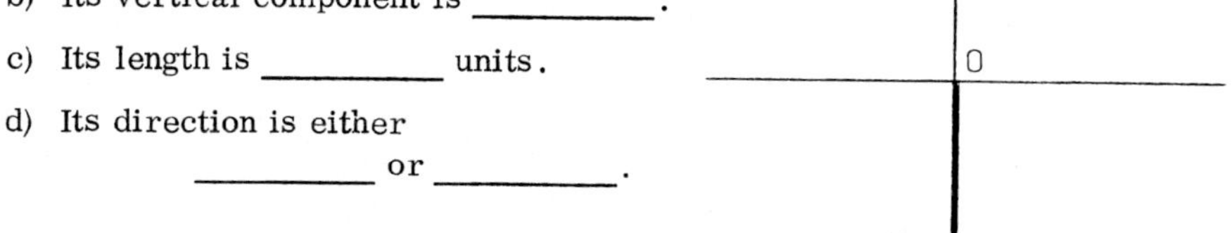

Answer to 67: 270° or -90°

Answers to 68: a) 0 b) -3.75 c) +3.75 or 3.75 d) 270° or -90°

SELF-TEST 12 (pages 167-182)

1. The coordinates of the tip of a vector are (18, -27).

a) Its horizontal component is ______. b) Its vertical component is ______.

2. A vector $\overrightarrow{OG}$ has a length of 52.7 units and a direction of 157°.

a) Find its horizontal component. Round to tenths.

b) Find its vertical component. Round to tenths.

3. A vector has a length of 415 units and a direction of -73°.

a) Find its horizontal component. Round to a whole number.

b) Find its vertical component. Round to a whole number.

4. Rounded to a whole number, find the positive second-quadrant angle whose tangent is -4.3129 .

5. Rounded to a whole number, find the negative third-quadrant angle whose tangent is 0.1394 .

6. Vector $\overrightarrow{OM}$ has a horizontal component of -4.39 units and a vertical component of -7.95 units.

a) Find its length. Round to hundredths.

b) Find its direction. Round to a whole number.

7. Vector $\overrightarrow{OC}$ has a horizontal component of -48.3 units and a vertical component of 41.9 units.

a) Find its length. Round to tenths.

b) Find its direction. Round to a whole number.

8. The coordinates of the tip of vector $\overrightarrow{OF}$ are (0, -72).

a) Find its horizontal component.

b) Find its vertical component.

c) Find its length.

d) Find its direction.

ANSWERS:

1. a) 18
   b) -27

2. a) -48.5 units
   b) 20.6 units

3. a) 121 units
   b) -397 units

4. 103°

5. -172°

6. a) 9.08 units
   b) 241° or -119°

7. a) 63.9 units
   b) 139°

8. a) 0
   b) -72
   c) 72
   d) -90° or 270°

4-11 FINDING THE COMPONENTS OF A RESULTANT

When adding vectors, we can find the components of the resultant by adding the components of the vector-addends. We will confirm that fact in this section.

69. To add $\overrightarrow{OA}$ and $\overrightarrow{OB}$ at the right, we completed the parallelogram and drew the resultant $\overrightarrow{OR}$. The components of the three vectors are given. Notice these points:

1. The <u>horizontal</u> component of $\overrightarrow{OR}$ is the sum of the horizontal components of $\overrightarrow{OA}$ and $\overrightarrow{OB}$. That is:

   $6 = 1 + 5$

2. The <u>vertical</u> component of $\overrightarrow{OR}$ is the sum of the vertical components of $\overrightarrow{OA}$ and $\overrightarrow{OB}$. That is:

   $5 = 4 + 1$

The relationships among the components are summarized in the table below.

	Horizontal Component	Vertical Component
$\overrightarrow{OA}$	1	4
$\overrightarrow{OB}$	5	1
$\overrightarrow{OR}$	6	5

70. To add $\overrightarrow{OC}$ and $\overrightarrow{OD}$ at the right, we completed the parallelogram and drew the resultant $\overrightarrow{OR}$. The components of the vectors are given.

Notice again that the components of $\overrightarrow{OR}$ are the sums of the components of $\overrightarrow{OC}$ and $\overrightarrow{OD}$. To confirm that fact, complete the table below.

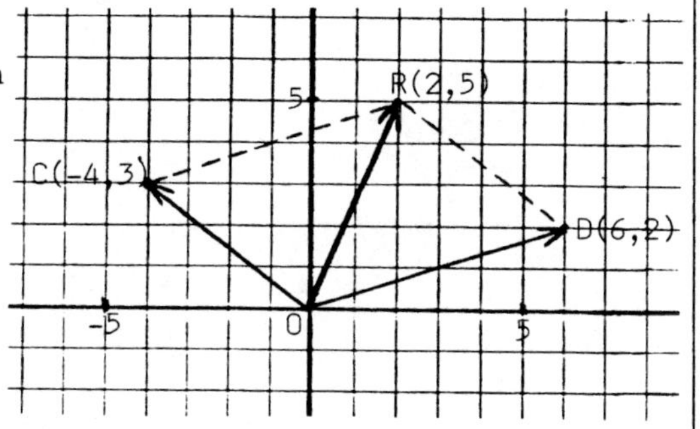

	Horizontal Component	Vertical Component
$\overrightarrow{OC}$		
$\overrightarrow{OD}$		
$\overrightarrow{OR}$		

71. Another vector addition is sketched at the right. $\overrightarrow{OR}$ is the resultant. The components of the three vectors are given.

To confirm the fact that the components of $\overrightarrow{OR}$ can be obtained by adding the components of $\overrightarrow{OF}$ and $\overrightarrow{OH}$, complete the table below.

	Horizontal Component	Vertical Component
$\overrightarrow{OF}$		
$\overrightarrow{OH}$		
$\overrightarrow{OR}$		

	Horizontal Component	Vertical Component
$\overrightarrow{OC}$	-4	3
$\overrightarrow{OD}$	6	2
$\overrightarrow{OR}$	2	5

72. The components of $\overrightarrow{F_1}$ and $\overrightarrow{F_2}$ are given. By completing the table below, find the components of their resultant $\overrightarrow{R}$.

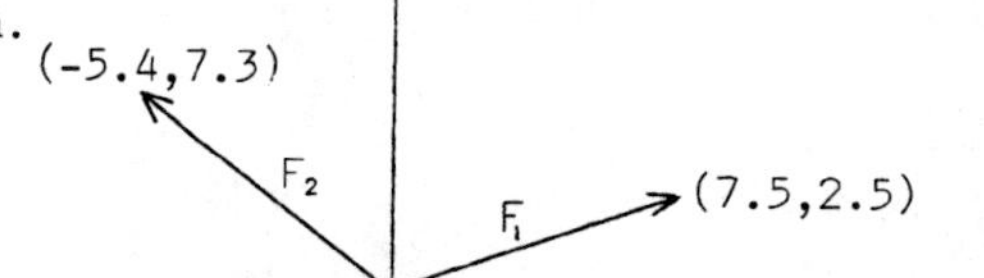

	Horizontal Component	Vertical Component
$\overrightarrow{F_1}$		
$\overrightarrow{F_2}$		
$\overrightarrow{R}$		

The components of resultant $\overrightarrow{R}$ show that it lies in Quadrant ____.

	Horizontal Component	Vertical Component
$\overrightarrow{OF}$	5	2
$\overrightarrow{OH}$	-3	-4
$\overrightarrow{OR}$	2	-2

	Horizontal Component	Vertical Component
$\overrightarrow{F_1}$	7.5	2.5
$\overrightarrow{F_2}$	-5.4	7.3
$\overrightarrow{R}$	2.1	9.8

Quadrant 1

73. The components of $\vec{F_1}$ and $\vec{F_2}$ are given. By completing the table below, find the components of their resultant $\vec{R}$.

	Horizontal Component	Vertical Component
$\vec{F_1}$		
$\vec{F_2}$		
$\vec{R}$		

The components of resultant $\vec{R}$ show that it lies in Quadrant _____.

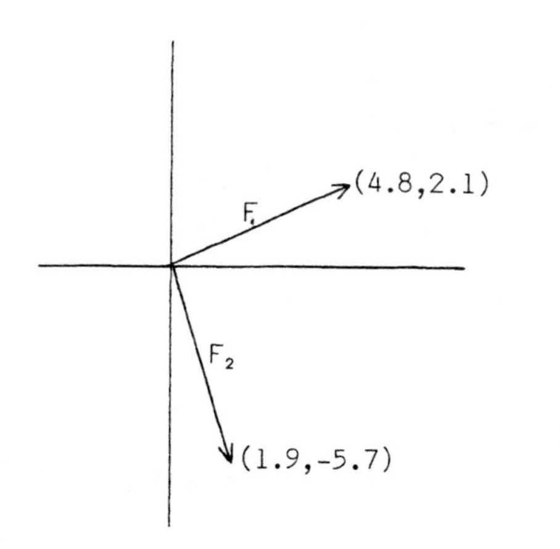

	Horizontal Component	Vertical Component
$\vec{F_1}$	4.8	2.1
$\vec{F_2}$	1.9	-5.7
$\vec{R}$	6.7	-3.6

Quadrant 4

4-12 THE COMPONENT METHOD FOR VECTOR ADDITION

In this section, we will discuss the component method for adding two vectors.

74. In the steps below, we will add $\overrightarrow{OA}$ and $\overrightarrow{OB}$ and then find the length and direction of their resultant $\overrightarrow{OR}$.

Step 1: Find the components of the resultant.

	Horizontal Component	Vertical Component
$\overrightarrow{OA}$	4.0	6.0
$\overrightarrow{OB}$	12.0	3.0
$\overrightarrow{OR}$	16.0	9.0

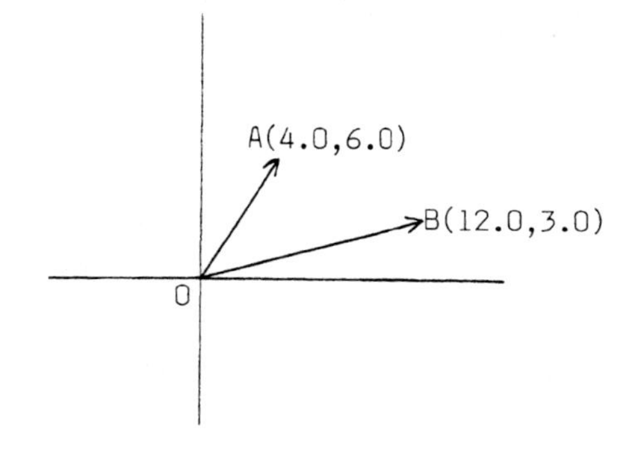

Continued on following page.

74. Continued

Step 2: Sketch the parallelogram and the resultant $\overrightarrow{OR}$.

We do so to check the sensibleness of the completed components and to identify the quadrant of the resultant.

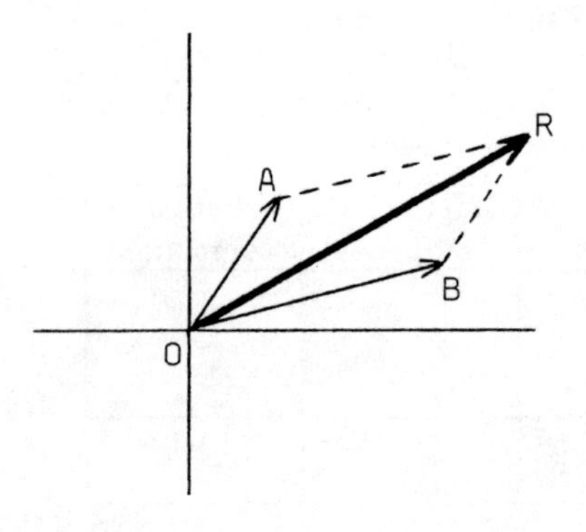

Step 3: Find the length and direction of the resultant $\overrightarrow{OR}$.

a) Using the Pythagorean Theorem, find the length of $\overrightarrow{OR}$. Round to tenths.

_______ units

b) $\overrightarrow{OR}$ is in what quadrant? _____

c) Using the tangent ratio, find the direction of $\overrightarrow{OR}$. Round to a whole number.

_______°

---

a) 18.4 units

b) Quadrant 1

c) 29°

---

75. We want to add $\overrightarrow{OC}$ and $\overrightarrow{OD}$ and find the length and direction of their resultant $\overrightarrow{OR}$.

a) Complete the table to find the components of $\overrightarrow{OR}$. Round to tenths.

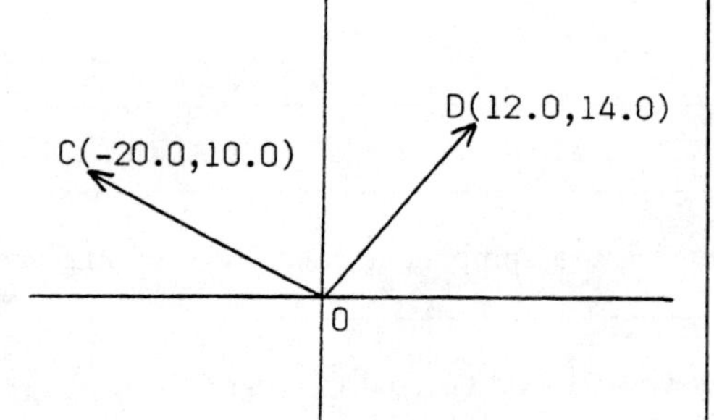

	Horizontal Component	Vertical Component
$\overrightarrow{OC}$		
$\overrightarrow{OD}$		
$\overrightarrow{OR}$		

b) Sketch the parallelogram and the resultant $\overrightarrow{OR}$ on the figure.

c) Find the length of $\overrightarrow{OR}$. Round to tenths.

_______ units

d) $\overrightarrow{OR}$ is in what quadrant? _____

e) Find the direction of $\overrightarrow{OR}$. Round to a whole number.

_______°

76. We want to add $\overrightarrow{OF}$ and $\overrightarrow{OT}$ and find the length and direction of their resultant $\overrightarrow{OR}$.

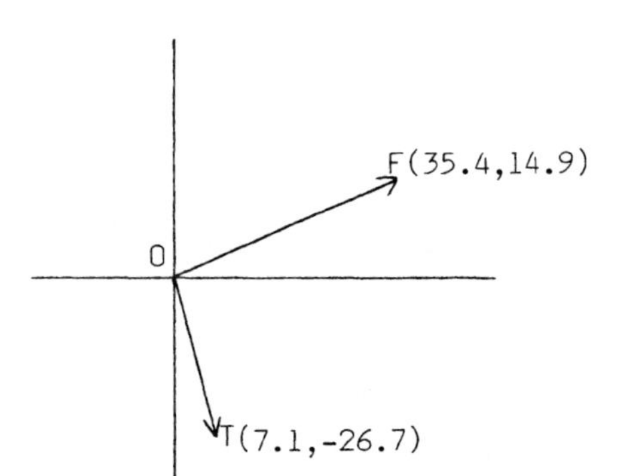

a) Complete the table to find the components of $\overrightarrow{OR}$. Round to tenths.

	Horizontal Component	Vertical Component
$\overrightarrow{OF}$		
$\overrightarrow{OT}$		
$\overrightarrow{OR}$		

b) Sketch the parallelogram and the resultant $\overrightarrow{OR}$ on the figure.

c) Find the length of $\overrightarrow{OR}$. Round to tenths.

_______ units

d) $\overrightarrow{OR}$ is in what quadrant? _______

e) Find the direction of $\overrightarrow{OR}$. Round to a whole number.

_______°

a)

	Horizontal Component	Vertical Component
$\overrightarrow{OC}$	-20.0	10.0
$\overrightarrow{OD}$	12.0	14.0
$\overrightarrow{OR}$	-8.0	24.0

b)

c) 25.3 units

d) Quadrant 2

e) 108°

77. Ordinarily we are not given the components of the vector-addends. We must find those components first. An example is discussed below.

We want to add $\overrightarrow{OA}$ and $\overrightarrow{OB}$ and find the length and direction of their resultant $\overrightarrow{OR}$. We are given these facts:

$\overrightarrow{OA}$ is 25.0 units at 130°

$\overrightarrow{OB}$ is 12.0 units at 30°

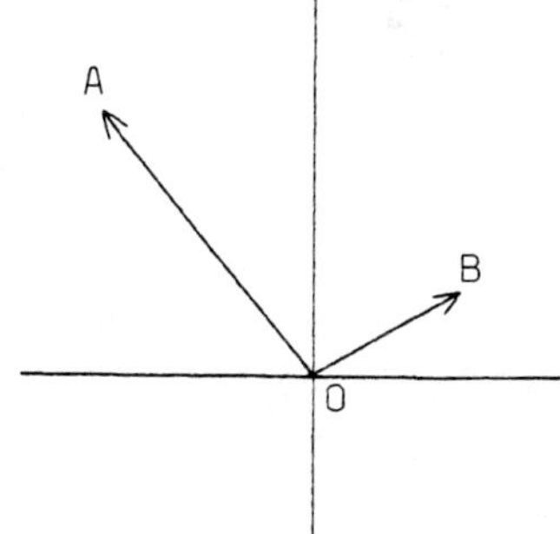

Continued on following page.

a)

	Horizontal Component	Vertical Component
$\overrightarrow{OF}$	35.4	14.9
$\overrightarrow{OT}$	7.1	-26.7
$\overrightarrow{OR}$	42.5	-11.8

b)

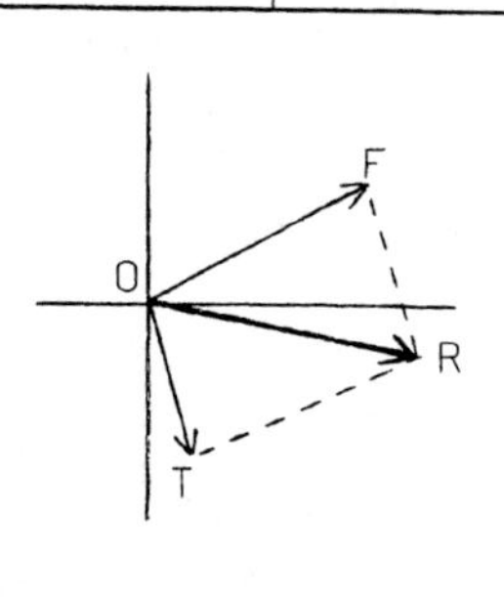

c) 44.1 units

d) Quadrant 4

e) -16° (or 344°)

77. Continued

We can use the cosine and sine ratios to find the components of $\overrightarrow{OA}$ and $\overrightarrow{OB}$. We get:

Horizontal component of $\overrightarrow{OA}$ = 25.0 (cos 130°) = -16.1 units
Vertical component of $\overrightarrow{OA}$ = 25.0 (sin 130°) = 19.2 units

Horizontal component of $\overrightarrow{OB}$ = 12.0 (cos 30°) = 10.4 units
Vertical component of $\overrightarrow{OB}$ = 12.0 (sin 30°) = 6.0 units

Now we can complete the addition in the usual way.

a) Complete the table to find the components of $\overrightarrow{OR}$. Round to tenths.

	Horizontal Component	Vertical Component
$\overrightarrow{OA}$		
$\overrightarrow{OB}$		
$\overrightarrow{OR}$		

b) Sketch the parallelogram and the resultant $\overrightarrow{OR}$ on the figure.

c) Find the length of $\overrightarrow{OR}$. Round to tenths.

_______ units

d) $\overrightarrow{OR}$ is in what quadrant? _________

e) Find the direction of $\overrightarrow{OR}$. Round to a whole number.

_______°

a)

	Horizontal Component	Vertical Component
$\overrightarrow{OA}$	-16.1	19.2
$\overrightarrow{OB}$	10.4	6.0
$\overrightarrow{OR}$	-5.7	25.2

b)

c) 25.8 units

d) Quadrant 2

e) 103°

78. We want to add $\overrightarrow{OC}$ and $\overrightarrow{OD}$ and find the length and direction of their resultant $\overrightarrow{OR}$. We know these facts:

$\overrightarrow{OC}$ is 448 units at -145°.
$\overrightarrow{OD}$ is 207 units at -70°.

a) Complete the table to find the components of $\overrightarrow{OR}$. Round to whole numbers.

	Horizontal Component	Vertical Component
$\overrightarrow{OC}$		
$\overrightarrow{OD}$		
$\overrightarrow{OR}$		

Continued on following page.

78. Continued

b) Sketch the parallelogram and the resultant $\overrightarrow{OR}$ on the figure.

c) Find the length of $\overrightarrow{OR}$. Round to a whole number.

_______ units

d) $\overrightarrow{OR}$ is in what quadrant? _______

e) Find the direction of $\overrightarrow{OR}$. Round to a whole number.

_______°

a)

	Horizontal Component	Vertical Component
$\overrightarrow{OC}$	-367	-257
$\overrightarrow{OD}$	71	-195
$\overrightarrow{OR}$	-296	-452

b)

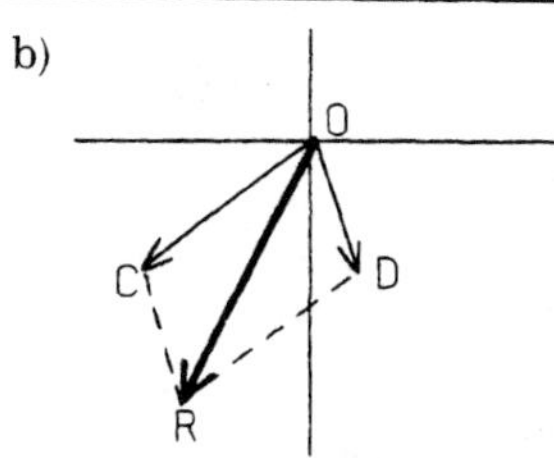

c) 540 units

d) Quadrant 3

e) -123° (or 237°)

4-13 ADDING VECTORS THAT LIE ON AN AXIS

In this section, we will discuss vector additions in which one or both of the vectors lie on an axis.

79. We want to add $\overrightarrow{OA}$ and $\overrightarrow{OB}$. Both vectors are on the horizontal axis. We know these facts:

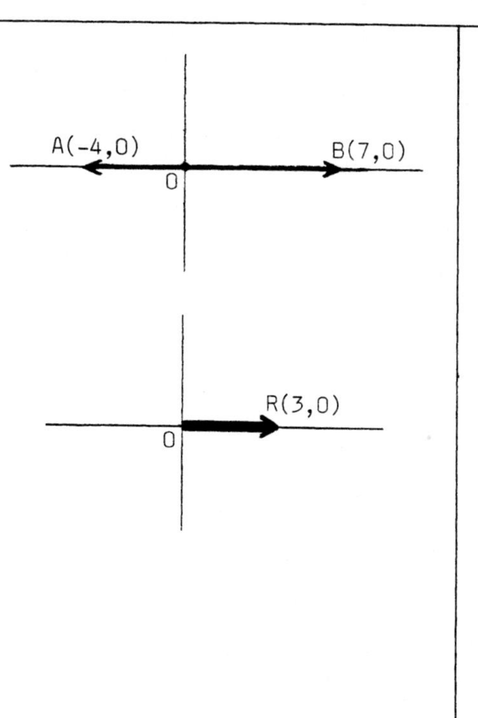

For $\overrightarrow{OA}$: Length is 4 units.
Direction is 180°.
Components are (-4, 0).

For $\overrightarrow{OB}$: Length is 7 units.
Direction is 0°.
Components are (7, 0).

We used the table below to find the components of the resultant $\overrightarrow{OR}$ which is sketched at the right.

	Horizontal Component	Vertical Component
$\overrightarrow{OA}$	-4	0
$\overrightarrow{OB}$	7	0
$\overrightarrow{OR}$	3	0

Continued on following page.

79. Continued

From the sketch of $\overrightarrow{OR}$, you can see that:

a) The components of $\overrightarrow{OR}$ are (,).

b) The length of $\overrightarrow{OR}$ is ______ units.

c) The direction of $\overrightarrow{OR}$ is ______°.

---

80. We want to add $\overrightarrow{OC}$ and $\overrightarrow{OD}$. Both vectors are on the vertical axis. We know these facts:

For $\overrightarrow{OC}$: Length is 3 units.
Direction is 90°.
Components are (0, 3).

For $\overrightarrow{OD}$: Length is 5 units.
Direction is -90° (or 270°).
Components are (0, -5).

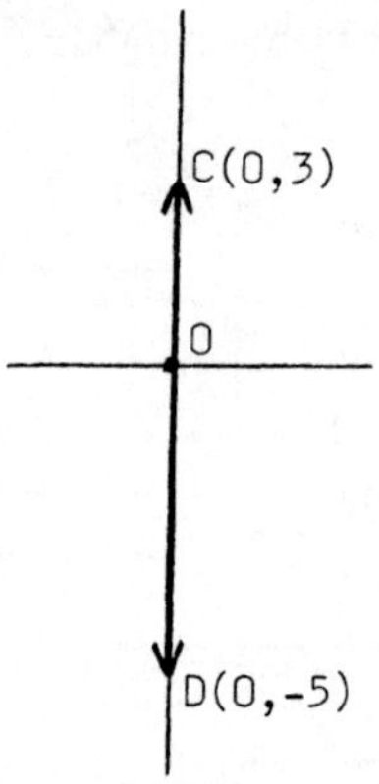

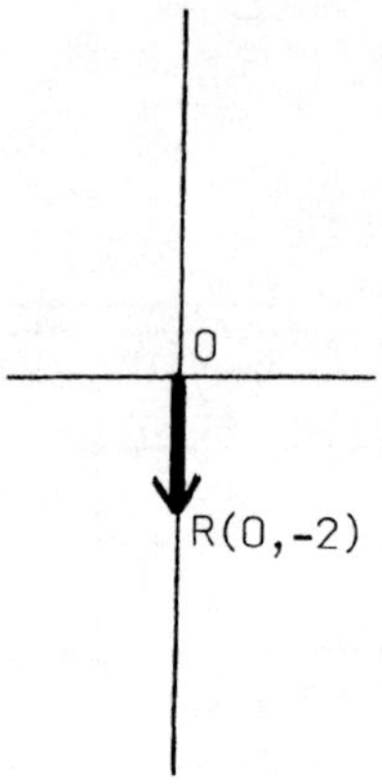

We used the table below to find the components of the resultant $\overrightarrow{OR}$ which is also sketched.

	Horizontal Component	Vertical Component
$\overrightarrow{OC}$	0	3
$\overrightarrow{OD}$	0	-5
$\overrightarrow{OR}$	0	-2

From the sketch of $\overrightarrow{OR}$, you can see these facts:

a) The components of $\overrightarrow{OR}$ are (,).

b) The length of $\overrightarrow{OR}$ is ______ units.

c) The direction of $\overrightarrow{OR}$ is ______°.

Answers (79):

a) (3, 0)

b) 3 units

c) 0°

Answers (80):

a) (0, -2)

b) 2 units

c) -90° (or 270°)

81. We want to add $\overrightarrow{OF}$ and $\overrightarrow{OM}$ which are shown at the right. Both vectors have a direction of 0°. We used the table below to find the components of their resultant $\overrightarrow{OR}$ which is also sketched.

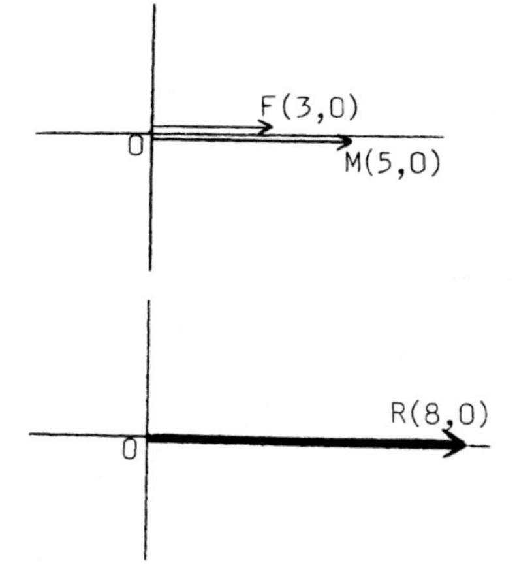

	Horizontal Component	Vertical Component
$\overrightarrow{OF}$	3	0
$\overrightarrow{OM}$	5	0
$\overrightarrow{OR}$	8	0

From the sketch of $\overrightarrow{OR}$, you can see these facts:

a) The components of $\overrightarrow{OR}$ are (,).

b) The length of $\overrightarrow{OR}$ is ________ units.

c) The direction of $\overrightarrow{OR}$ is ________°.

---

a) (8, 0)

b) 8 units

c) 0°

---

82. We want to add $\overrightarrow{OA}$ and $\overrightarrow{OB}$. Their components are given at the right. We know these facts:

$\overrightarrow{OA}$ is 10.0 units at 90°.
$\overrightarrow{OB}$ is 16.0 units at 180°.

We completed the parallelogram and drew the resultant $\overrightarrow{OR}$. Notice that the parallelogram is really a rectangle.

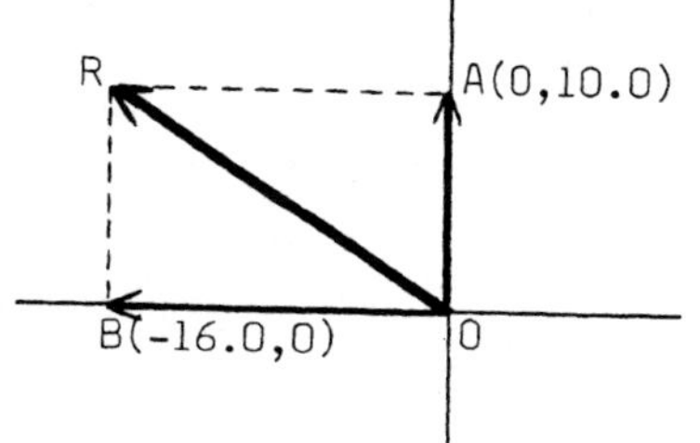

a) Using the table below, find the components of $\overrightarrow{OR}$.

	Horizontal Component	Vertical Component
$\overrightarrow{OA}$		
$\overrightarrow{OB}$		
$\overrightarrow{OR}$		

b) Using the Pythagorean Theorem, find the length of $\overrightarrow{OR}$. Round to tenths. ________ units

c) Using the tangent ratio, find the direction of $\overrightarrow{OR}$. Round to a whole number. ________°

83. We want to add $\overrightarrow{OC}$ and $\overrightarrow{OD}$. We are given these facts:

$\overrightarrow{OC}$ is 5.1 units at 0°.
$\overrightarrow{OD}$ is 7.5 units at -90°.

a) Using the table below, find the components of the resultant $\overrightarrow{OR}$.

	Horizontal Component	Vertical Component
$\overrightarrow{OC}$		
$\overrightarrow{OD}$		
$\overrightarrow{OR}$		

b) Complete the rectangle and draw the resultant $\overrightarrow{OR}$ on the figure.

c) Find the length of $\overrightarrow{OR}$. Round to tenths. ________ units

d) Find the direction of $\overrightarrow{OR}$. Round to a whole number. ________°

a)

	Horizontal Component	Vertical Component
$\overrightarrow{OA}$	0	10.0
$\overrightarrow{OB}$	-16.0	0
$\overrightarrow{OR}$	-16.0	10.0

b) 18.9 units

c) 148°

84. We want to add $\overrightarrow{OF}$ and $\overrightarrow{OH}$ at the right. We know these facts:

$\overrightarrow{OF}$ is 22.4 units at 0°.
$\overrightarrow{OH}$ is 16.7 units at 65°.

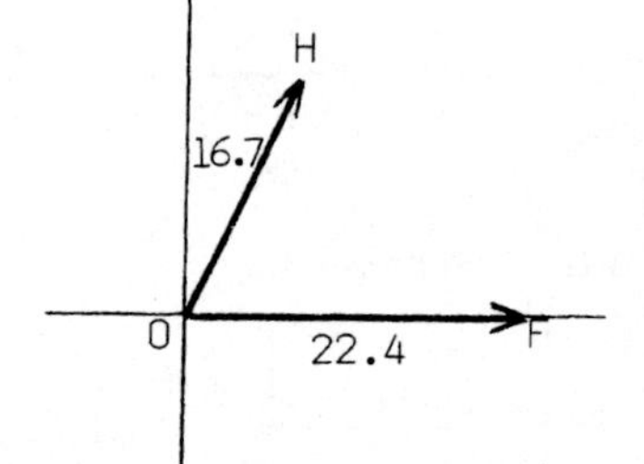

a) Complete the parallelogram and draw the resultant $\overrightarrow{OR}$ on the figure.

b) Find the components of $\overrightarrow{OF}$ and $\overrightarrow{OH}$ and then compute the components of $\overrightarrow{OR}$. Round to tenths.

	Horizontal Component	Vertical Component
$\overrightarrow{OF}$		
$\overrightarrow{OH}$		
$\overrightarrow{OR}$		

c) Find the length of $\overrightarrow{OR}$. Round to tenths.

________ units

d) Find the direction of $\overrightarrow{OR}$. Round to a whole number.

________°

a)

	Horizontal Component	Vertical Component
$\overrightarrow{OC}$	5.1	0
$\overrightarrow{OD}$	0	-7.5
$\overrightarrow{OR}$	5.1	-7.5

b)

c) 9.1 units

d) -56° (or 304°)

85. We are given these facts about $\overrightarrow{OM}$ and $\overrightarrow{OT}$ at the right.

$\overrightarrow{OM}$ is 156 units at -90°.
$\overrightarrow{OT}$ is 109 units at -153°.

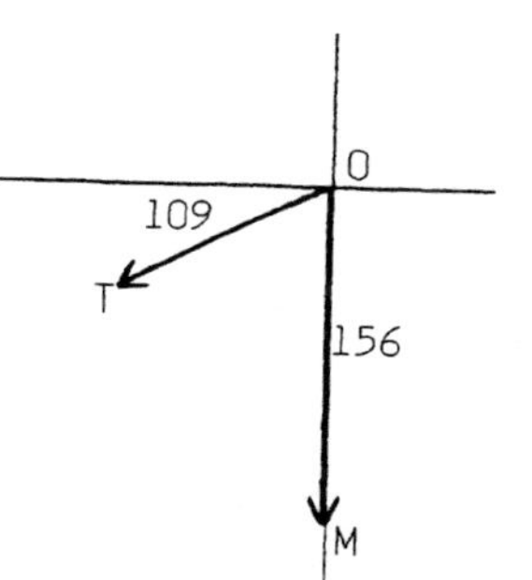

Let's use the oblique-triangle method to find the length and direction of the resultant $\overrightarrow{OR}$.

a) Complete the parallelogram and draw $\overrightarrow{OR}$ on the figure.

b) How large is angle TOM? ________°

c) How large is angle T? ________°

d) Using triangle TOR and the Law of Cosines, find the length of $\overrightarrow{OR}$. Round to a whole number.

________ units

e) Using triangle TOR and the Law of Sines, find the size of angle TOR. Round to a whole number.

________°

f) The direction of $\overrightarrow{OR}$ is ________°.

a)

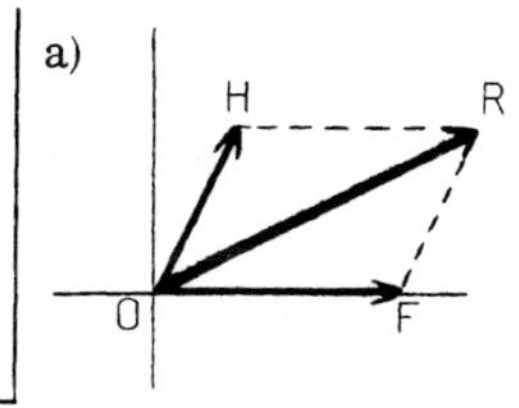

b)

	Horizontal Component	Vertical Component
$\overrightarrow{OF}$	22.4	0
$\overrightarrow{OH}$	7.1	15.1
$\overrightarrow{OR}$	29.5	15.1

c) 33.1 units

d) 27°

a)

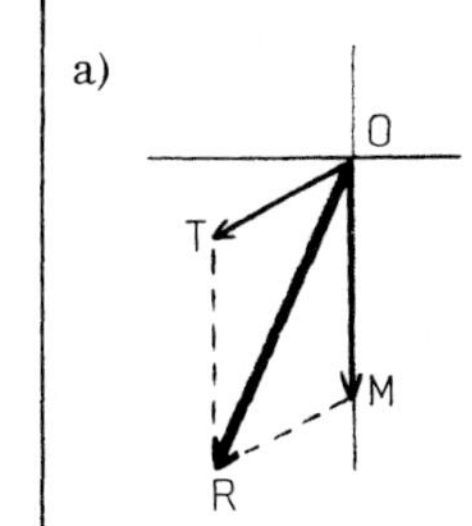

b) 63°

c) 117° (from 180° - 63°)

d) 227 units

e) 38°

f) -115° (or 245°)

4-14 FINDING A VECTOR-ADDEND

In a vector addition, the added vectors can be called "vector-addends"; the sum is called the "resultant". In this section, we will discuss a method to find the second vector-addend when one vector-addend and the resultant are given.

86. In a vector-addition, the two added vectors can be called "vector-addends". At the left below, we are given one vector-addend $\overrightarrow{OC}$ and the resultant $\overrightarrow{OR}$. At the right below, we completed the parallelogram to get a rough idea of the length and direction of the second vector-addend $\overrightarrow{OD}$.

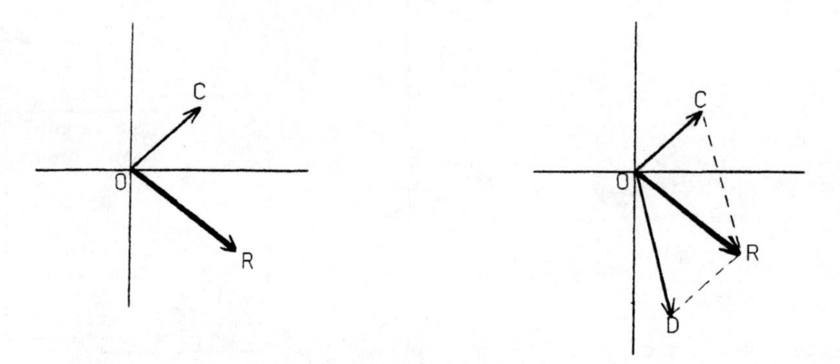

As a help in completing parallelograms like the one above:

1. Connect the tips of the two known vectors first. Above, it is dashed line CR.
2. Starting at the origin, draw a line parallel to the dashed line. Above, it is line OD.
3. Complete the parallelogram. Remember that the resultant is the diagonal of the parallelogram.

In each figure below, one vector-addend $\overrightarrow{OF}$ and the resultant $\overrightarrow{OR}$ are given. Complete the parallelogram to find the second vector-addend $\overrightarrow{OM}$.

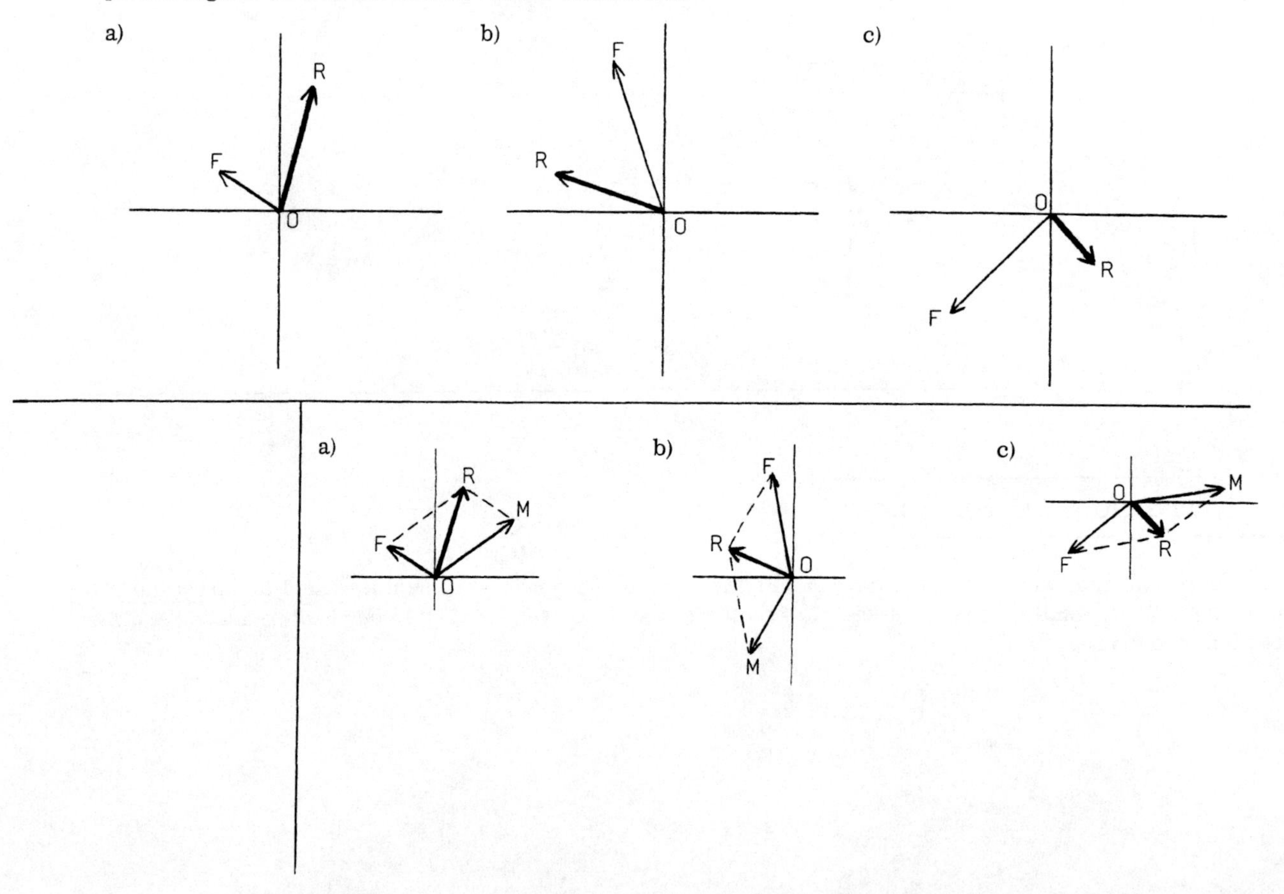

87. At the left below, we are given the components of a vector-addend $\overrightarrow{OA}$ and the resultant $\overrightarrow{OR}$. At the right below, we completed the parallelogram to find the second vector-addend $\overrightarrow{OB}$. We call its components (h, v).

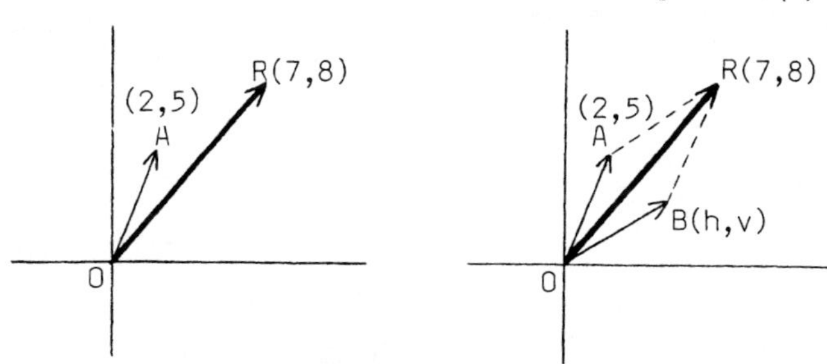

Since we add the components of $\overrightarrow{OA}$ and $\overrightarrow{OB}$ to find the components of $\overrightarrow{OR}$, we can set up the following equations.

$$2 + h = 7 \qquad (h = 5)$$

$$5 + v = 8 \qquad (v = 3)$$

Therefore, the components of $\overrightarrow{OR}$ are (,).

(5, 3)

---

88. On the graph at the right, $\overrightarrow{OF}$ is a vector-addend and $\overrightarrow{OR}$ is a resultant. We want to find the components of the second vector-addend $\overrightarrow{OH}$.

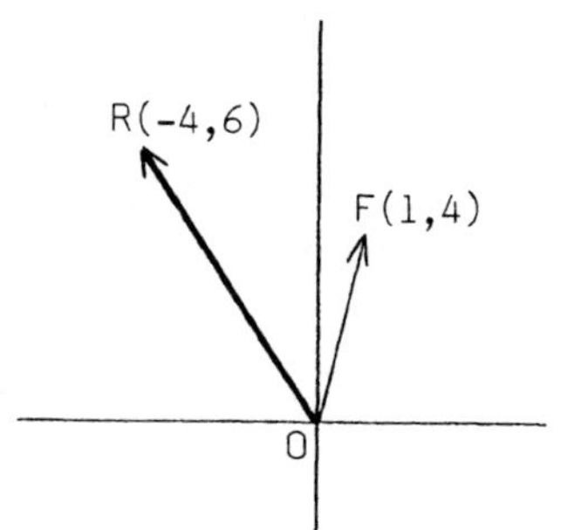

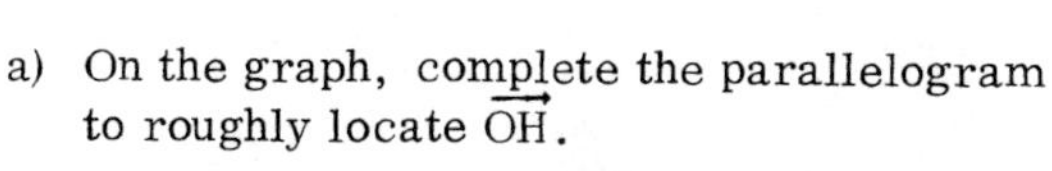

a) On the graph, complete the parallelogram to roughly locate $\overrightarrow{OH}$.

b) If (h, v) are the components of $\overrightarrow{OH}$, set up equations to find "h" and "v".

____

____

c) The components of $\overrightarrow{OH}$ are (,).

a)

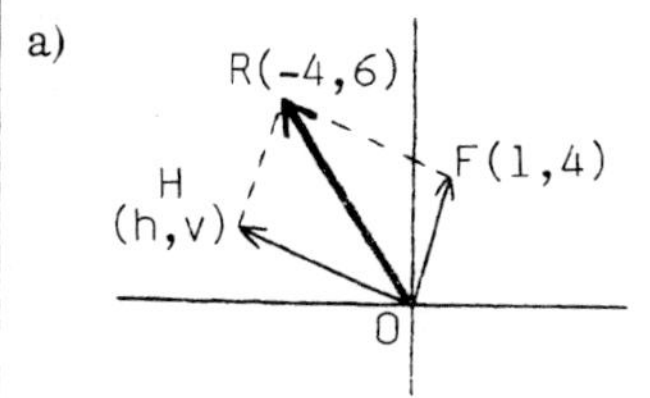

b) $1 + h = -4$
$4 + v = 6$

c) (-5, 2)

---

89. On the graph at the right, $\overrightarrow{OA}$ is a vector-addend and $\overrightarrow{OR}$ is a resultant. We want to find the components of the second vector-addend $\overrightarrow{OD}$.

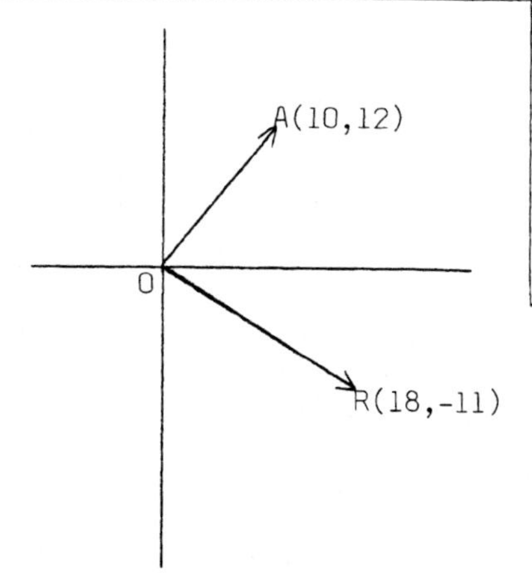

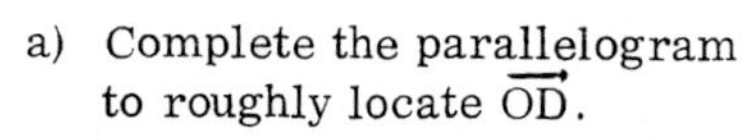

a) Complete the parallelogram to roughly locate $\overrightarrow{OD}$.

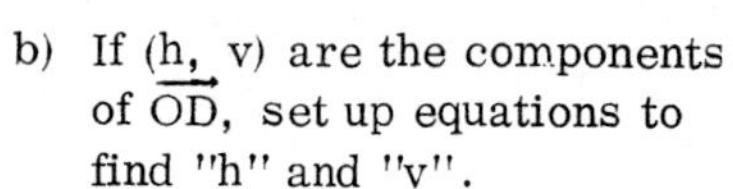

b) If (h, v) are the components of $\overrightarrow{OD}$, set up equations to find "h" and "v".

____

____

c) The components of $\overrightarrow{OD}$ are (,).

90. At the right, $\overrightarrow{OM}$ is a vector-addend and $\overrightarrow{OR}$ is the resultant. We are given these facts:

$\overrightarrow{OM}$ is 9.0 units at 147°.
$\overrightarrow{OR}$ is 16.9 units at -154°.

We want to find the length and direction of the other vector-addend $\overrightarrow{OT}$.

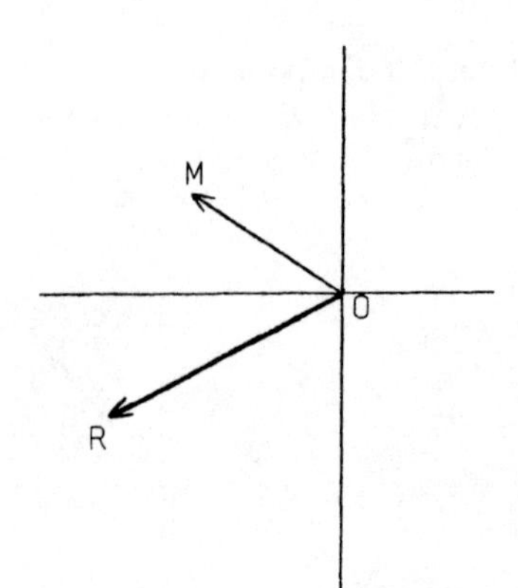

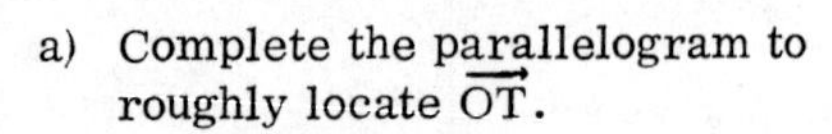

a) Complete the parallelogram to roughly locate $\overrightarrow{OT}$.

b) Find the components of $\overrightarrow{OM}$. Round to tenths.

(,)

c) Find the components of $\overrightarrow{OR}$. Round to tenths.

(,)

d) Using (h, v) as the components of $\overrightarrow{OT}$, set up equations to find "h" and "v".

____

____

e) The components of $\overrightarrow{OT}$ are (,).

f) The length of $\overrightarrow{OT}$ is ______ units. Round to tenths.

g) The direction of $\overrightarrow{OT}$ is ______°. Round to a whole number.

a)

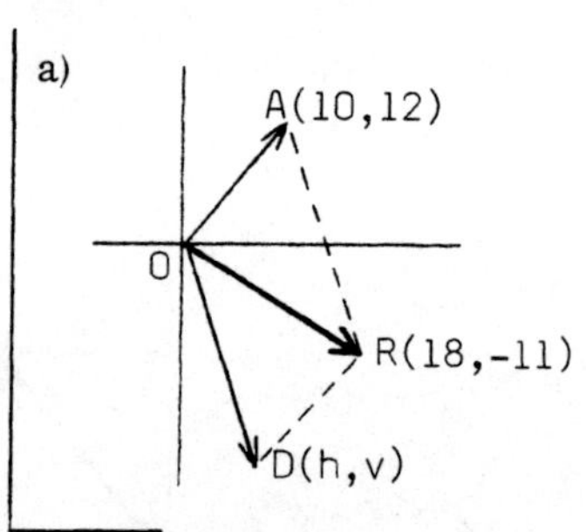

b) $10 + h = 18$
$12 + v = -11$

c) (8, -23)

a)

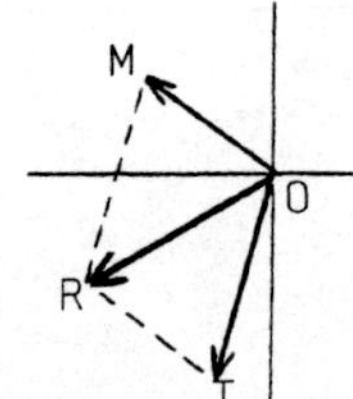

b) (-7.5, 4.9)

c) (-15.2, -7.4)

d) $-7.5 + h = -15.2$
$4.9 + v = -7.4$

e) (-7.7, -12.3)

f) 14.5 units

g) -122° (or 238°)

SELF-TEST 13 (pages 183-198)

1. Vectors $\overrightarrow{OA}$ and $\overrightarrow{OB}$ have these components:

   $\overrightarrow{OA}$: (50, 10)
   $\overrightarrow{OB}$: (-10, 20)

   a) Find the components of their resultant $\overrightarrow{OR}$.

   b) Find the length of $\overrightarrow{OR}$.

   c) Find the direction of $\overrightarrow{OR}$. Round to a whole number.

2. Vectors $\overrightarrow{OG}$ and $\overrightarrow{OH}$ have these components:

   $\overrightarrow{OG}$: (300, 200)
   $\overrightarrow{OH}$: (0, -400)

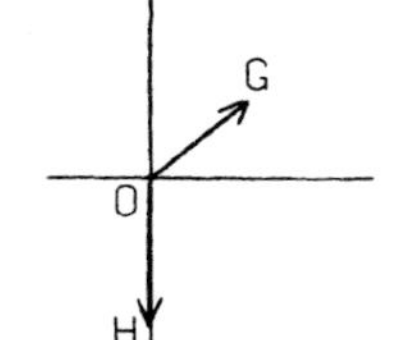

   a) Find the components of their resultant $\overrightarrow{OR}$.

   b) Find the length of $\overrightarrow{OR}$. Round to a whole number.

   c) Find the direction of $\overrightarrow{OR}$. Round to a whole number.

3. Vectors $\overrightarrow{OM}$ and $\overrightarrow{OT}$ have these lengths and directions:

   $\overrightarrow{OM}$: 5.83 units at 90°
   $\overrightarrow{OT}$: 4.59 units at 160°

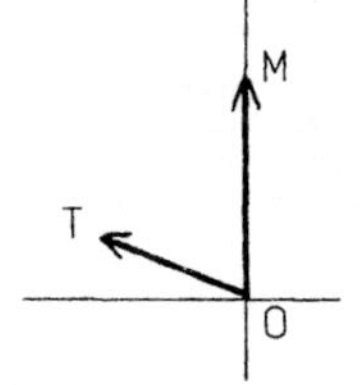

   a) Find the components of their resultant $\overrightarrow{OR}$. Round to hundredths.

   b) Find the length of $\overrightarrow{OR}$. Round to hundredths.

   c) Find the direction of $\overrightarrow{OR}$. Round to a whole number.

4. Vectors $\overrightarrow{OC}$ and $\overrightarrow{OD}$ have these lengths and directions:

   $\overrightarrow{OC}$: 33.7 units at 68°
   $\overrightarrow{OD}$: 60.5 units at -42°

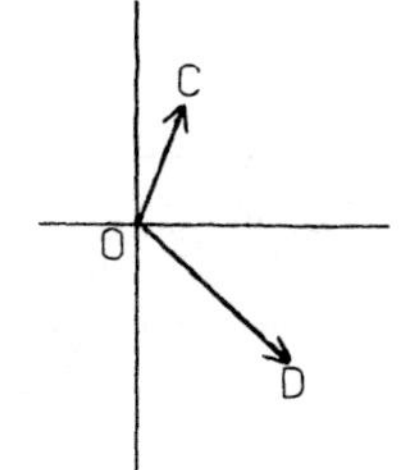

   a) Find the components of their resultant $\overrightarrow{OR}$. Round to tenths.

   b) Find the length of $\overrightarrow{OR}$. Round to tenths.

   c) Find the direction of $\overrightarrow{OR}$. Round to a whole number.

Continued on following page.

SELF-TEST 13 (pages 183-198) - Continued

5. Resultant $\overrightarrow{OR}$ is the sum of vector-addends $\overrightarrow{OF}$ and $\overrightarrow{OG}$. The components of $\overrightarrow{OR}$ and $\overrightarrow{OF}$ are:

$\overrightarrow{OR}$: (-15, 29)
$\overrightarrow{OF}$: (20, 18)

a) Sketch vector-addend $\overrightarrow{OG}$.

b) Find the components of $\overrightarrow{OG}$.

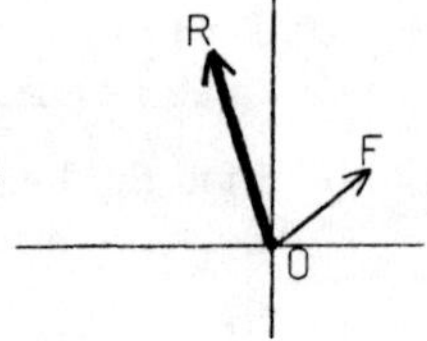

6. Resultant $\overrightarrow{OR}$ is the sum of vector-addends $\overrightarrow{OA}$ and $\overrightarrow{OB}$.

$\overrightarrow{OR}$ is 380 units at 0°.
$\overrightarrow{OA}$ is 250 units at 65°.

a) Sketch vector addend $\overrightarrow{OB}$.

b) Find the components of $\overrightarrow{OB}$. Round to a whole number.

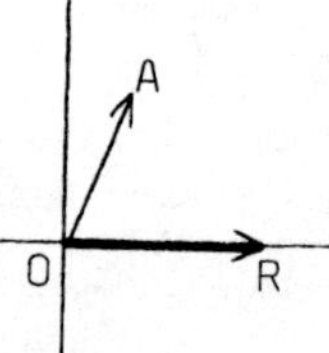

ANSWERS:

1. a) (40, 30)
   b) 50 units
   c) 37°

2. a) (300, -200)
   b) 361 units
   c) -34° or 326°

3. a) (-4.31, 7.40)
   b) 8.56 units
   c) 120°

4. a) (57.6, -9.3)
   b) 58.3 units
   c) -9° or 351°

5. a)

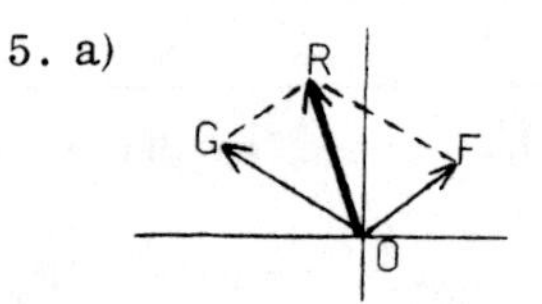

b) (-35, 11)

6. a)

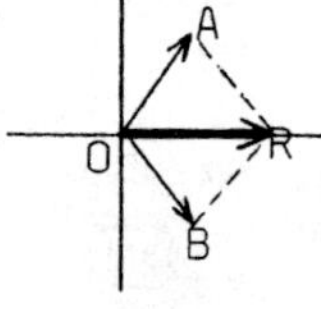

b) (274, -227)

4-15 ADDING THREE OR MORE VECTORS

In this section, we will show how the component method can be used to add three or more vectors.

91. We want to add the three vectors at the right. We can do so in two steps in which we add two vectors at a time. The steps are:

1. Add any two vectors.
2. Then add their resultant to the third vector.

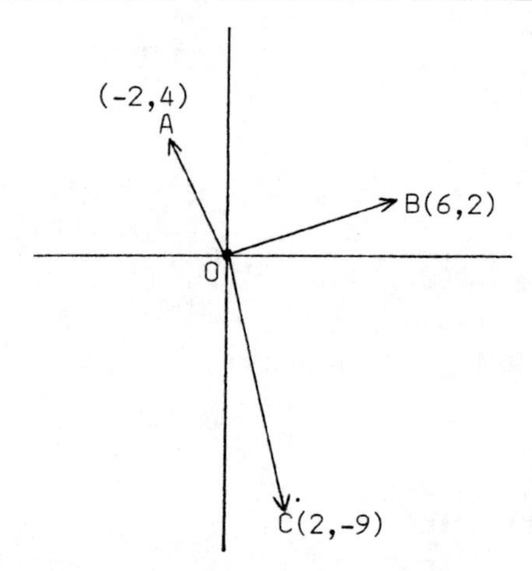

Continued on following page.

91. Continued

The two steps for the addition are described below.

Step 1: Add any two vectors.

We will add $\overrightarrow{OA}$ and $\overrightarrow{OB}$. We have done so at the right. The components of the resultant $\overrightarrow{OT}$ are (4, 6).

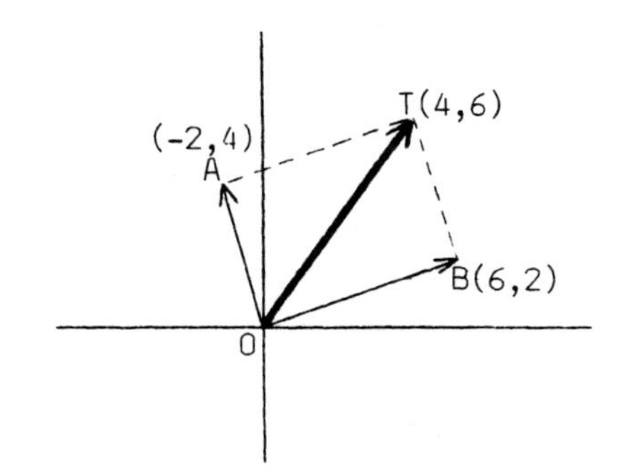

Step 2: Add the resultant $\overrightarrow{OT}$ to $\overrightarrow{OC}$.

We have done so at the right. The components of the resultant $\overrightarrow{OR}$ are (6, -3).

$\overrightarrow{OR}$ is the sum of $\overrightarrow{OA}$, $\overrightarrow{OB}$, and $\overrightarrow{OC}$. It is a single vector that represents the sum of the three vector-addends.

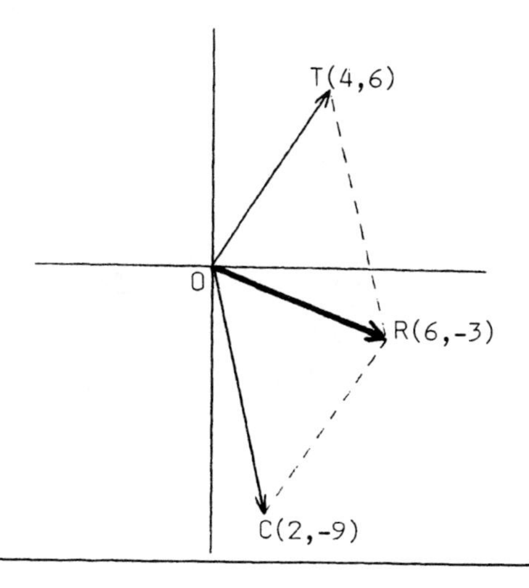

---

92. The three vector-addends ($\overrightarrow{OA}$, $\overrightarrow{OB}$, and $\overrightarrow{OC}$) and the resultant $\overrightarrow{OR}$ from the last frame are shown at the right. The components of all four vectors are given.

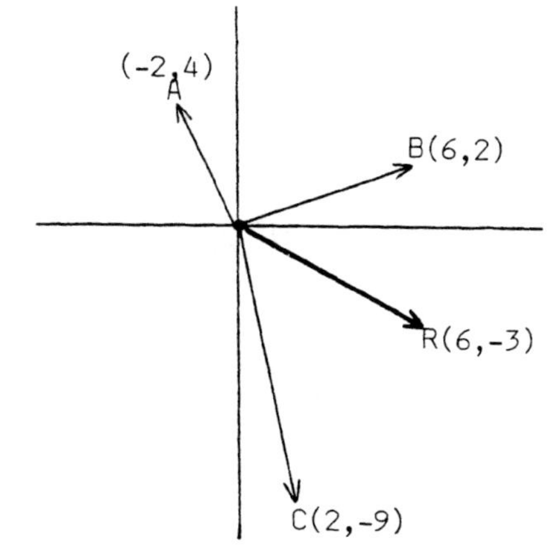

We can find the components of $\overrightarrow{OR}$ by simply adding the three horizontal and vertical components of the vector-addends. We have done so in the table below.

	Horizontal Component	Vertical Component
$\overrightarrow{OA}$	-2	4
$\overrightarrow{OB}$	6	2
$\overrightarrow{OC}$	2	-9
$\overrightarrow{OR}$	6	-3

In the table, we got the same components for $\overrightarrow{OR}$ that are shown on the graph. They are (,).

(6, -3)

93. When adding three or more vectors, we use the component method. Let's use that method to add the three vectors below. Use the table to find the components of the resultant $\overrightarrow{OR}$.

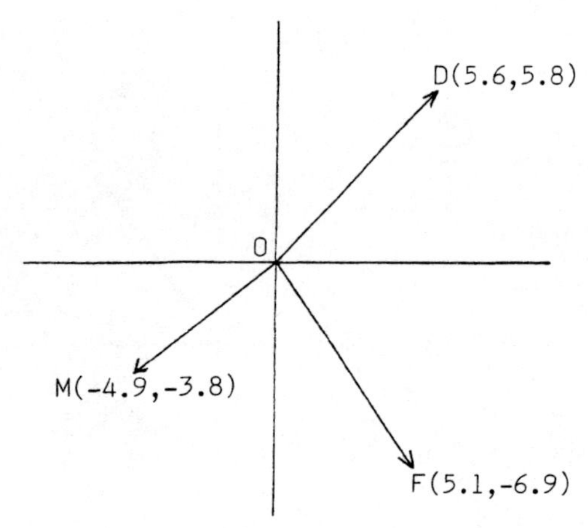

	Horizontal Component	Vertical Component
$\overrightarrow{OD}$		
$\overrightarrow{OF}$		
$\overrightarrow{OM}$		
$\overrightarrow{OR}$		

a) The components of $\overrightarrow{OR}$ are (,).

b) Find the length of $\overrightarrow{OR}$. Round to tenths.

_______ units

c) Find the direction of $\overrightarrow{OR}$. Round to a whole number.

_______°

---

a) (5.8, -4.9) b) 7.6 units c) -40° (or 320°)

94. We want to add the four vectors below. Complete the table to find the components of the resultant $\overrightarrow{OR}$.

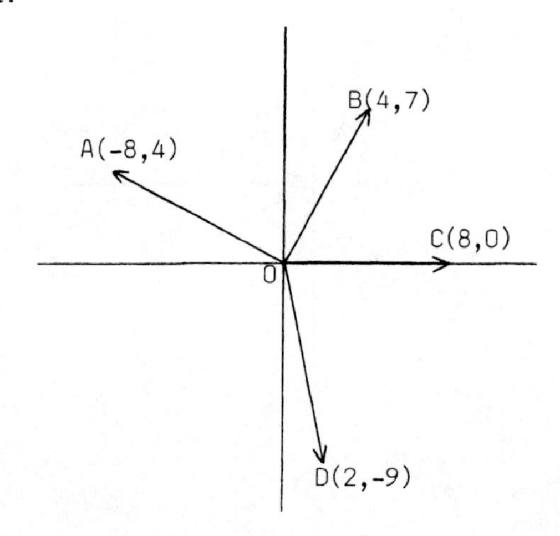

	Horizontal Component	Vertical Component
$\overrightarrow{OA}$		
$\overrightarrow{OB}$		
$\overrightarrow{OC}$		
$\overrightarrow{OD}$		
$\overrightarrow{OR}$		

Continued on following page.

94. Continued

a) The components of $\overrightarrow{OR}$ are (,).

b) Find the length of $\overrightarrow{OR}$. Round to tenths.

_______ units

c) Find the direction of $\overrightarrow{OR}$. Round to tenths.

_______ °

---

a) (6, 2)
b) 6.3 units
c) 18.4°

95. We are given this information about the four vectors at the right:

$\overrightarrow{OF}$ is 8.0 units at 90°.

$\overrightarrow{OG}$ is 5.5 units at 0°.

$\overrightarrow{OH}$ is 7.5 units at -72°.

$\overrightarrow{OT}$ is 9.8 units at -151°.

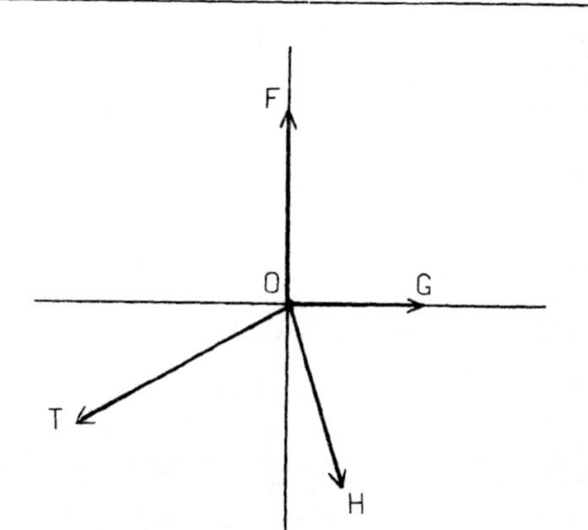

a) Complete this table to find the components of the resultant $\overrightarrow{OR}$. Round to tenths.

	Horizontal Component	Vertical Component
$\overrightarrow{OF}$		
$\overrightarrow{OG}$		
$\overrightarrow{OH}$		
$\overrightarrow{OT}$		
$\overrightarrow{OR}$		

b) Find the length of $\overrightarrow{OR}$. Round to tenths.

_______ units

c) Find the direction of $\overrightarrow{OR}$. Round to a whole number.

_______ °

a)

	Horizontal Component	Vertical Component
$\overrightarrow{OF}$	0	8.0
$\overrightarrow{OG}$	5.5	0
$\overrightarrow{OH}$	2.3	-7.1
$\overrightarrow{OT}$	-8.6	-4.8
$\overrightarrow{OR}$	-0.8	-3.9

b) 4.0 units

c) -102° (or 258°)

4-16 THE STATE OF EQUILIBRIUM

In this section, we will define a state of equilibrium for a system of two or more vectors. Before doing so, we will define vector-opposites and the zero-vector.

---

96. Two numbers are a pair of opposites if their sum is "0".

For example: +6 and -6 are a pair of opposites.

Two vectors are a pair of vector-opposites if both:

1. The sum of their horizontal components is "0".

and 2. The sum of their vertical components is "0".

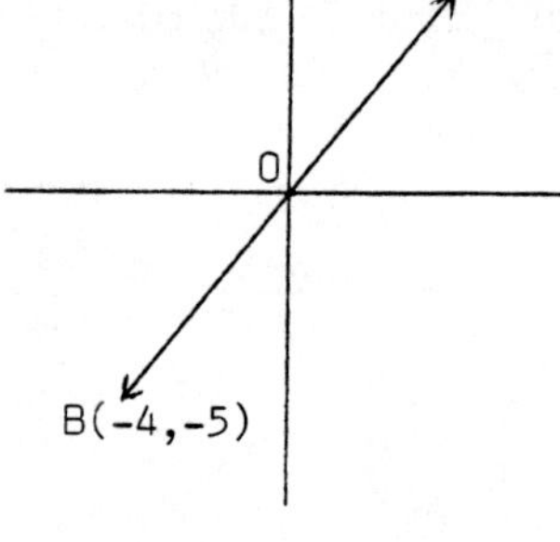

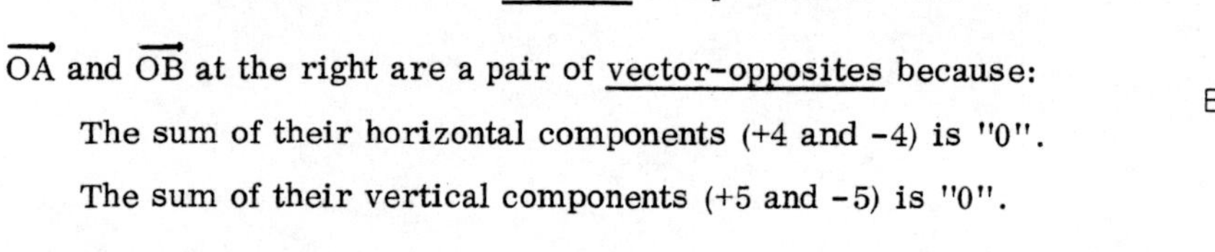

$\overrightarrow{OA}$ and $\overrightarrow{OB}$ at the right are a pair of vector-opposites because:

The sum of their horizontal components (+4 and -4) is "0".

The sum of their vertical components (+5 and -5) is "0".

When a pair of vector-opposites are graphed:

1. They have the same length.
2. They are in opposite directions.

---

97. By examining the graph at the right, find the vector-opposites for each of these:

a) $\overrightarrow{OC}$ ________

b) $\overrightarrow{OD}$ ________

c) $\overrightarrow{OE}$ ________

d) $\overrightarrow{OF}$ ________

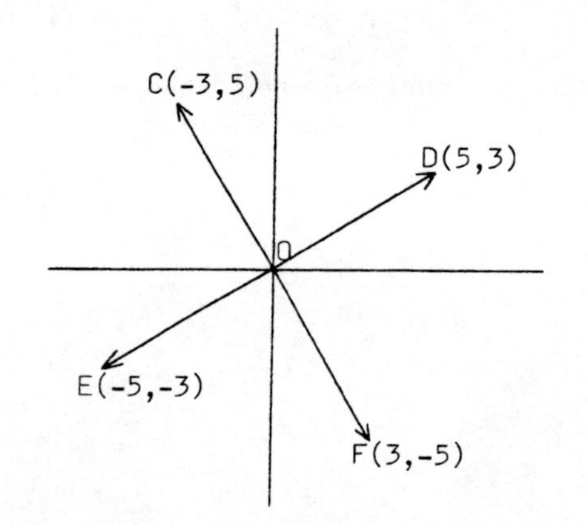

98. $\overrightarrow{OA}$ and $\overrightarrow{OB}$ are a pair of vector-opposites. We used the table below to find the components of their resultant $\overrightarrow{OR}$.

	Horizontal Component	Vertical Component
$\overrightarrow{OA}$	4	2
$\overrightarrow{OB}$	-4	-2
$\overrightarrow{OR}$	0	0

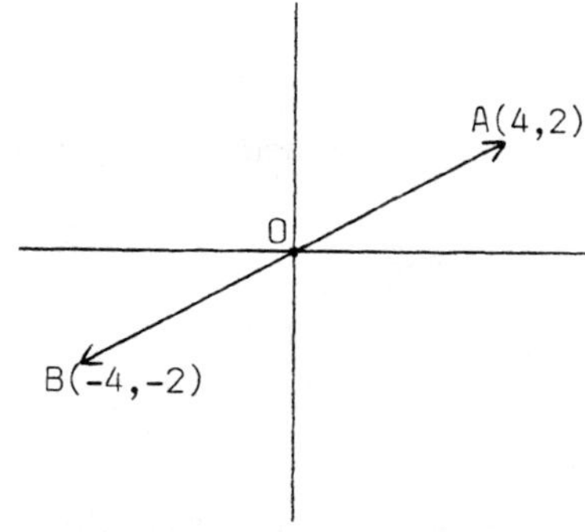

The components of the resultant are (0, 0). Though the resultant has no length or direction, it is a vector. It is called the zero-vector.

a) The components of the zero-vector are (,).

b) The length of the zero-vector is ______ units.

c) Does the zero-vector have a direction? ________

a) $\overrightarrow{OF}$
b) $\overrightarrow{OE}$
c) $\overrightarrow{OD}$
d) $\overrightarrow{OC}$

---

99. The resultant of $\overrightarrow{OA}$ and $\overrightarrow{OB}$ at the right is the zero-vector.

When the resultant of two vectors is the zero-vector, we say that the two vectors are in the "state of equilibrium".

Therefore: $\overrightarrow{OA}$ and $\overrightarrow{OB}$ are in the state of ________________________.

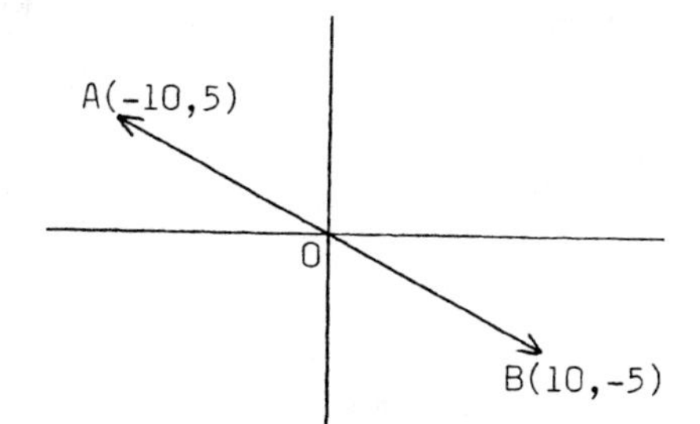

a) (0, 0)
b) 0 units
c) no

---

100. We added the components of the four vectors below in the table at the right. The resultant is $\overrightarrow{OR}$.

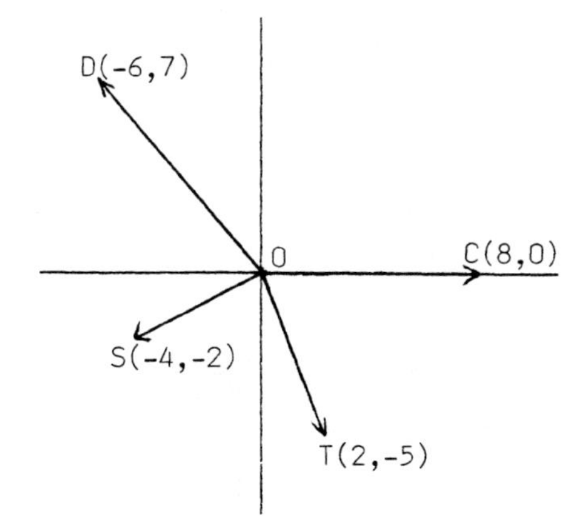

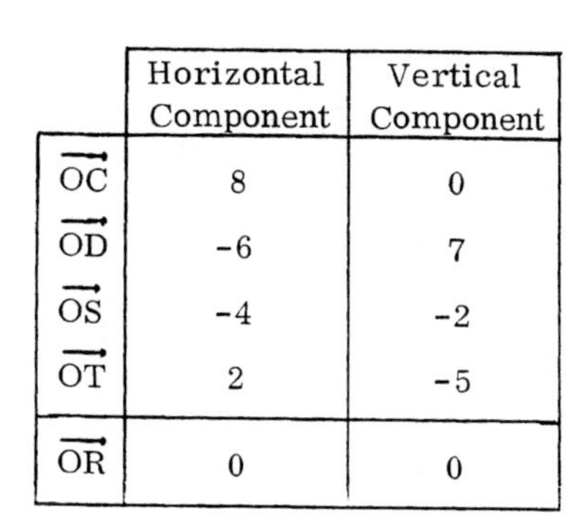

	Horizontal Component	Vertical Component
$\overrightarrow{OC}$	8	0
$\overrightarrow{OD}$	-6	7
$\overrightarrow{OS}$	-4	-2
$\overrightarrow{OT}$	2	-5
$\overrightarrow{OR}$	0	0

Since the resultant of the four vectors is the zero-vector, these four vectors are in a state of ________________________.

equilibrium

---

equilibrium

4-17 EQUILIBRANTS AND THE STATE OF EQUILIBRIUM

An "equilibrant" is a vector added to a system of vectors to produce a state of equilibrium. We will discuss equilibrants in this section.

---

101. As you can see from the table at the left below, $\overrightarrow{OA}$ and $\overrightarrow{OB}$ are not in a state of equilibrium because their resultant $\overrightarrow{OR}$ is not a zero-vector.

	Horizontal Component	Vertical Component
$\overrightarrow{OA}$	-6	3
$\overrightarrow{OB}$	4	2
$\overrightarrow{OR}$	-2	5

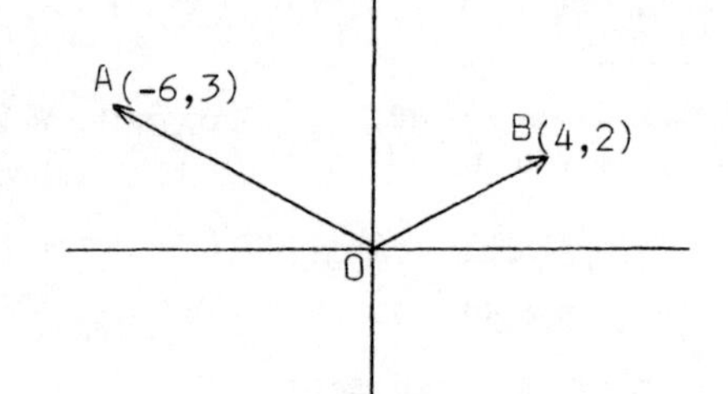

However, we could add a third vector $\overrightarrow{OE}$ to the system to produce a state of equilibrium. The third vector, called an "equilibrant", would have to be the vector-opposite of the resultant $\overrightarrow{OR}$.

That is: The components of the equilibrant $\overrightarrow{OE}$ must be (2, -5).

The equilibrant lies in what quadrant? ______

Answer: Quadrant 4

102. To produce a state of equilibrium, an equilibrant $\overrightarrow{OE}$ must be the vector-opposite of the resultant $\overrightarrow{OR}$. Therefore:

a) If the components of $\overrightarrow{OR}$ are (4, -6), the components of $\overrightarrow{OE}$ must be (,).

b) If the components of $\overrightarrow{OR}$ are (-1, -9), the components of $\overrightarrow{OE}$ must be (,).

Answers: a) (-4, 6) b) (1, 9)

103. In the table below, we found the components of the resultant $\overrightarrow{OR}$ of $\overrightarrow{OC}$ and $\overrightarrow{OD}$.

	Horizontal Component	Vertical Component
$\overrightarrow{OC}$	5	2
$\overrightarrow{OD}$	3	-7
$\overrightarrow{OR}$	8	-5

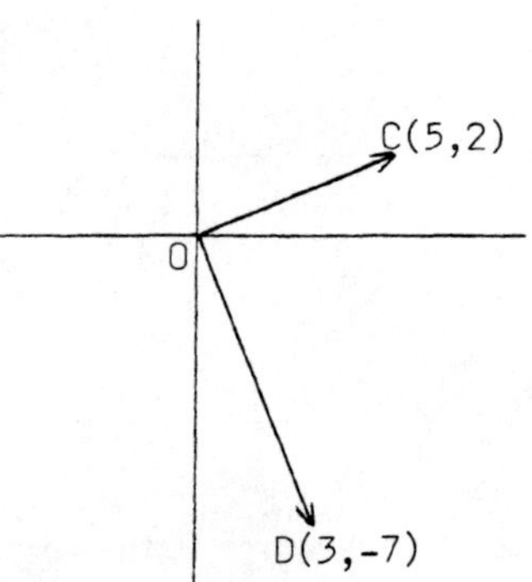

We want to add an equilibrant $\overrightarrow{OE}$ to the system to produce a state of equilibrium.

a) The components of $\overrightarrow{OE}$ must be (,).

b) $\overrightarrow{OE}$ would lie in what quadrant? ______

104. In the table below, we found the components of the resultant $\overrightarrow{OR}$ of $\overrightarrow{OF}$, $\overrightarrow{OM}$, and $\overrightarrow{OS}$.

	Horizontal Component	Vertical Component
$\overrightarrow{OF}$	50	20
$\overrightarrow{OM}$	-20	30
$\overrightarrow{OS}$	35	-15
$\overrightarrow{OR}$	65	35
$\overrightarrow{OE}$		

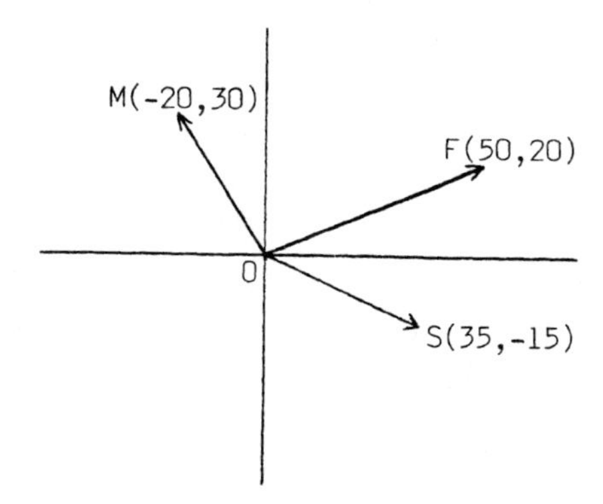

We want to add an equilibrant $\overrightarrow{OE}$ to the system to produce a state of equilibrium.

a) The components of $\overrightarrow{OE}$ must be (,).

b) $\overrightarrow{OE}$ would lie in what quadrant? ________

a) (-8, 5)

b) Quadrant 2

---

105. There are three vectors in the system at the right.

$\overrightarrow{OA}$ is 4.77 units at 90°.

$\overrightarrow{OB}$ is 7.91 units at 0°.

$\overrightarrow{OC}$ is 6.48 units at -90°.

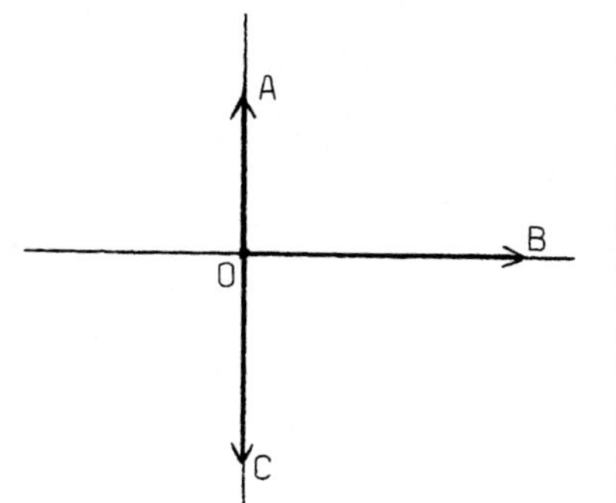

a) Complete the table below to find the components of the resultant $\overrightarrow{OR}$ and the equilibrant $\overrightarrow{OE}$.

	Horizontal Component	Vertical Component
$\overrightarrow{OA}$		
$\overrightarrow{OB}$		
$\overrightarrow{OC}$		
$\overrightarrow{OR}$		
$\overrightarrow{OE}$		

b) Find the length of $\overrightarrow{OE}$. Round to hundredths.

________ units

c) Find the direction of $\overrightarrow{OE}$. Round to a whole number.

________°

a) (-65, -35)

b) Quadrant 3

a)

	Horizontal Component	Vertical Component
$\overrightarrow{OA}$	0	4.77
$\overrightarrow{OB}$	7.91	0
$\overrightarrow{OC}$	0	-6.48
$\overrightarrow{OR}$	7.91	-1.71
$\overrightarrow{OE}$	-7.91	1.71

b) 8.09 units

c) 168°

4-18 APPLIED PROBLEMS

In this section, we will show some applied problems involving vectors.

106. To move a heavy object horizontally, a 240 kg force is applied at a 28° angle. The vector diagram below shows how that force can be resolved into a horizontal component F_h and a vertical component F_v.

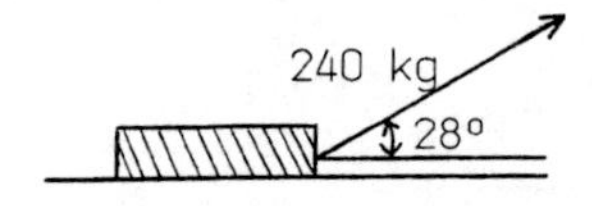

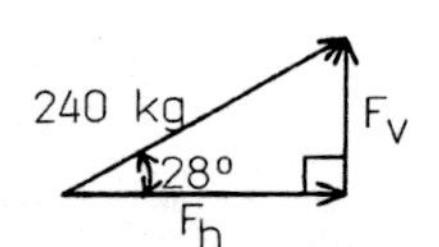

a) Find the horizontal component F_h. Round to a whole number.

F_h = ________ kg

b) Find the vertical component F_v. Round to a whole number.

F_v = ________ kg

107. While traveling due east at 100 mph, a small airplane encounters a 40 mph wind blowing towards the northeast. Find its actual speed with respect to the ground and its angular deviation from east.

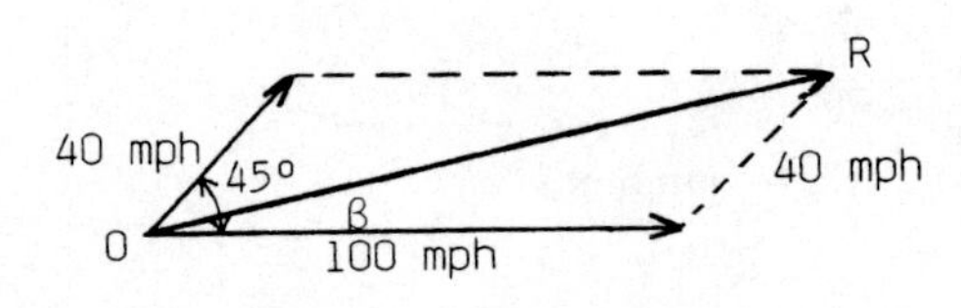

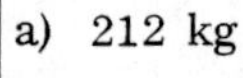

a) 212 kg

b) 113 kg

107. Continued

The resultant $\overrightarrow{OR}$ is the actual speed and angle β is the angular deviation.

a) The actual speed is ________ mph. Round to a whole number.

b) The angular deviation is ________°. Round to a whole number.

---

108. As shown at the left below, a 750 kg load is supported at the midpoint of a cable. We want to find the two equal tension forces in the cable.

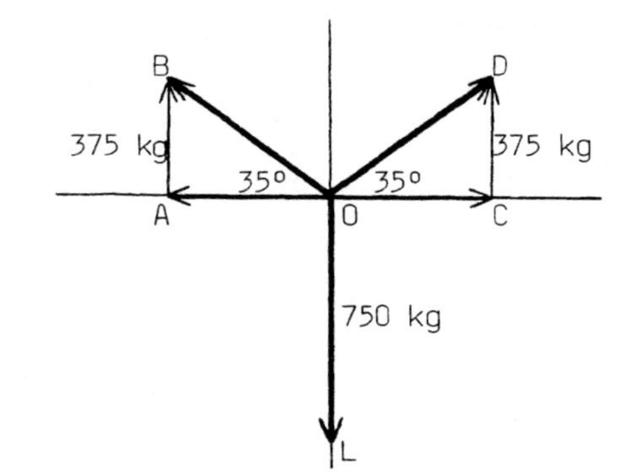

A vector diagram of the three forces is shown at the right above. Because the three forces are in equilibrium, the sum of the vertical components $\overrightarrow{AB}$ and $\overrightarrow{CD}$ must equal 750 kg. And since $\overrightarrow{AB} = \overrightarrow{CD}$, both equal one-half of 750 kg or 375 kg.

Using triangle ODC:

a) Set up the equation needed to find $\overrightarrow{OD}$.

b) Rounded to a whole number, the tension force in each half of the cable is ________ kg.

a) 131 mph

b) $\beta = 12°$

---

109. An a-c voltage V_T is applied to an electrical circuit consisting of a resistor and an inductor connected in series. The voltage V_R across the resistor is 15.7 volts. The voltage V_L across the inductor is 25.3 volts. Find the applied voltage V_T and its phase angle θ.

The vector diagram at the right shows the relationship between V_R, V_L, and V_T. V_T is the resultant of V_R and V_L. θ is the phase angle.

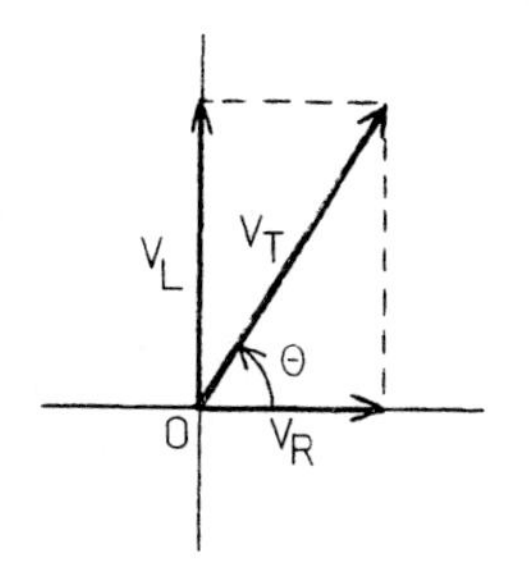

Continued on following page.

a) $\sin 35° = \frac{375}{\overrightarrow{OD}}$

b) 654 kg

109. Continued

a) Find V_T. Round to tenths.

V_T = ________ volts

b) Find θ. Round to a whole number.

θ = ________ °

Answers to 109: a) 29.8 volts b) $\theta = 58°$

110. An a-c electrical circuit consists of a resistor, an inductor, and a capacitor connected in parallel. The current I_R in the resistor is 6.24 milliamperes. The current I_L in the inductor is 9.78 milliamperes. The current I_C in the capacitor is 5.95 milliamperes. Find the total current I_T and its phase angle θ.

The vector diagram at the right shows the relationship between I_R, I_L, and I_C. The total current I_T (not shown) is the resultant of I_R, I_L, and I_C.

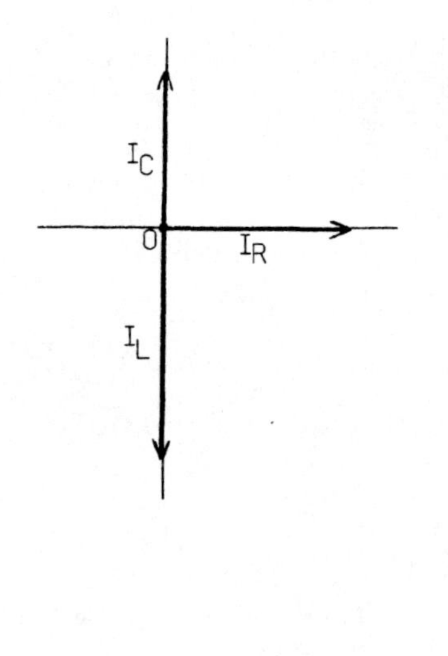

As shown in the table below, the components of the resultant $\vec{I}_T$ are (6.24, -3.83).

	Horizontal Component	Vertical Component
$\vec{I}_R$	6.24	0
$\vec{I}_L$	0	-9.78
$\vec{I}_C$	0	5.95
$\vec{I}_T$	6.24	-3.83

a) Sketch the resultant on the diagram.

b) Find the total current I_T. Round to hundredths.

I_T = ______ milliamperes

c) Find phase angle θ, the angle between I_R and I_T. Round to a whole number.

---

a)

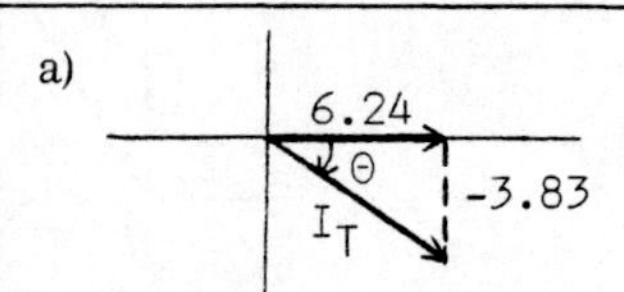

b) I_T = 7.32 milliamperes

c) $\theta = -32°$

SELF-TEST 14 (pages 198-209)

1. To put two vectors $\overrightarrow{ON}$: (48, 14) and $\overrightarrow{OP}$: (31, -27) in the state of equilibrium, a third vector $\overrightarrow{OT}$ is added to the system. Find the coordinates of $\overrightarrow{OT}$. __________

On the axes at the right, sketch these four vectors:

$\overrightarrow{OA}$: (30, 20)
$\overrightarrow{OB}$: (40, -10)
$\overrightarrow{OC}$: (-10, -30)
$\overrightarrow{OD}$: (-20, 10)

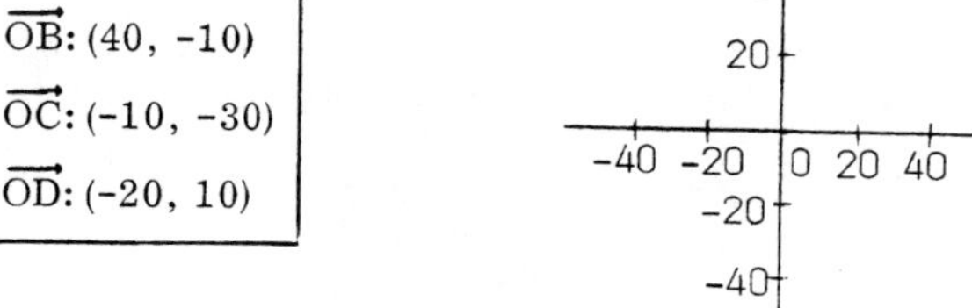

2. Find the components of their resultant $\overrightarrow{OR}$. __________

3. Find the components of their equilibrant $\overrightarrow{OE}$. __________

On the axes at the right, sketch these three vectors:

$\overrightarrow{OF}$: 370 units at 0°
$\overrightarrow{OG}$: 250 units at 90°
$\overrightarrow{OH}$: 540 units at -90°

Reporting all answers as whole numbers, find the following:

4. The components of their resultant $\overrightarrow{OR}$. __________
5. The length of $\overrightarrow{OR}$. __________
6. The direction of $\overrightarrow{OR}$. __________
7. The components of their equilibrant $\overrightarrow{OE}$. __________
8. The length of $\overrightarrow{OE}$. __________
9. The direction of $\overrightarrow{OE}$. __________

As shown at the right, the cable on a crane holds a 250 kg load. We want to find the compression force in the crane beam. A vector diagram of the forces is shown at the far right. The compression force to be found is $\overrightarrow{OF}$.

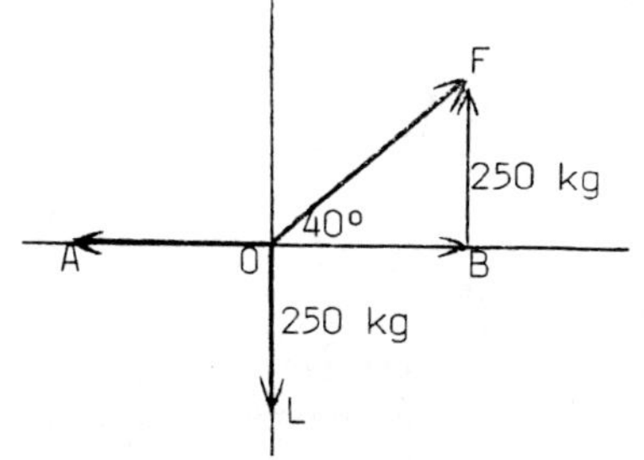

Because the three forces are in equilibrium, vertical components $\overrightarrow{OL}$ and $\overrightarrow{BF}$ are equal in length. Therefore, $\overrightarrow{BF}$ = 250 kg. Using right triangle OBF:

10. Set up the equation needed to find the length of $\overrightarrow{OF}$. __________

11. Rounded to a whole number, the compression force is __________ kg.

ANSWERS:

1. (-79, 13)
2. (40, -10)
3. (-40, 10)
4. (370, -290)
5. 470 units
6. -38° or 322°
7. (-370, 290)
8. 470 units
9. 142° or -218°
10. $\sin 40° = \frac{250}{OF}$
11. 389 kg

SUPPLEMENTARY PROBLEMS - CHAPTER 4

Assignment 11

1. Which of the following are vector quantities?

   a) A 90 km/hr northward velocity
   b) A 115v voltage at 0° phase
   c) A 2,500 lb force upward
   d) A 23°C average daily temperature
   e) A 75 cm^3 volume of water
   f) A 15 lb force at 40°

2. A car travels 19 miles east and 13 miles south. a) Find its diagonal distance from its starting point. b) Find the direction of that distance. Round each answer to a whole number.

3. A 380 kg force and a 230 kg force act on an object at right angles to each other. a) Find the resultant force. b) Find the angle the resultant makes with the 380 kg force. Round each answer to a whole number.

4. Using the parallelogram method, sketch the resultant of each vector pair.

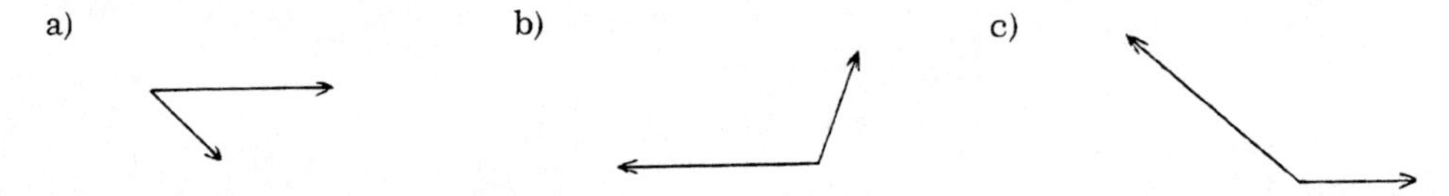

5. Two vectors of 530 units and 470 units act at an angle of 24°. a) Find the length of their resultant. b) Find the angle between the resultant and the 530 unit vector. Round each answer to a whole number.

6. Two vectors of 28.7 units and 41.9 units act at an angle of 155°. a) Find the length of their resultant. Round to tenths. b) Find the angle between the resultant and the 41.9 unit vector. Round to a whole number.

7. Two vectors on the coordinate system are: $\overrightarrow{OG}$ is 22 units at 70° and $\overrightarrow{OH}$ is 34 units at 120°. a) Find the length of their resultant $\overrightarrow{OR}$. b) Find the direction of $\overrightarrow{OR}$ as a standard-position angle. Round each answer to a whole number.

8. Two vectors on the coordinate system are: $\overrightarrow{OA}$ is 3.75 units at 48° and $\overrightarrow{OB}$ is 5.25 units at -67°. a) Find the length of their resultant $\overrightarrow{OR}$. Round to hundredths. b) Find the direction of $\overrightarrow{OR}$ as a standard-position angle. Round to a whole number.

Assignment 12

1. A vector has a horizontal component of -200 units and a vertical component of 350 units. Write the coordinates of the tip of the vector.

2. Vector $\overrightarrow{OA}$ has a length of 785 units and a direction of 128°. a) Find its horizontal component. b) Find its vertical component. Round each answer to a whole number.

3. Vector $\overrightarrow{OP}$ has a length of 8.47 units and a direction of 337°. a) Find its horizontal component. b) Find its vertical component. Round each answer to hundredths.

4. Vector $\overrightarrow{OF}$ has a length of 51.9 units and a direction of -161°. a) Find its horizontal component. b) Find its vertical component. Round each answer to tenths.

Rounding to a whole number, find each standard-position angle:

5. The positive third-quadrant angle whose tangent is 1.8132 .
6. The positive second-quadrant angle whose tangent is -4.6817 .
7. The positive fourth-quadrant angle whose tangent is -0.2915 .
8. The negative third-quadrant angle whose tangent is 0.9346 .
9. The negative fourth-quadrant angle whose tangent is -1.5418 .

10. Vector $\overrightarrow{OT}$ has a horizontal component of 558 units and a vertical component of -392 units. a) Find its length. b) Find its direction. Round each answer to a whole number.

11. Vector $\overrightarrow{OG}$ has a horizontal component of -2.36 and a vertical component of 3.19 . a) Find its length. Round to hundredths. b) Find its direction. Round to a whole number.

12. Vector $\overrightarrow{OM}$ has a horizontal component of -60.5 and a vertical component of -18.9 units. a) Find its length. Round to tenths. b) Find its direction. Round to a whole number.

13. The coordinates of vector $\overrightarrow{OC}$ are (-425, 0). Find its: a) Horizontal component. b) Vertical component. c) Length. d) Direction.

14. Vector $\overrightarrow{OH}$ has a length of 7.14 units and a direction of -90°. Find its: a) Horizontal component. b) Vertical component. c) Length.

Assignment 13

1. The components of vectors $\overrightarrow{OC}$ and $\overrightarrow{OD}$ are (18.0, -23.0) and (-37.0, 42.0), respectively. a) Find the components of their resultant $\overrightarrow{OR}$. b) Find the length of $\overrightarrow{OR}$. Round to tenths. c) Find the direction of $\overrightarrow{OR}$. Round to a whole number.

2. The components of vectors $\overrightarrow{OA}$ and $\overrightarrow{OB}$ are (296, 228) and (174, -415), respectively. Rounding answers to whole numbers, find the following: a) The components of their resultant $\overrightarrow{OR}$. b) The length of $\overrightarrow{OR}$. c) The direction of $\overrightarrow{OR}$.

3. Vector $\overrightarrow{OF}$ is 34.8 units at 50° and vector $\overrightarrow{OG}$ is 61.7 units at 180°. Find the following: a) The components of their resultant $\overrightarrow{OR}$. Round to tenths. b) The length of $\overrightarrow{OR}$. Round to tenths. c) The direction of $\overrightarrow{OR}$. Round to a whole number.

4. Vector $\overrightarrow{OM}$ is 380 units at 138° and vector $\overrightarrow{OP}$ is 520 units at -127°. Rounding answers to whole numbers, find the following: a) The components of their resultant $\overrightarrow{OR}$. b) The length of $\overrightarrow{OR}$. c) The direction of $\overrightarrow{OR}$.

5. Vector $\overrightarrow{OT}$ is 2.96 units at 0° and vector $\overrightarrow{OV}$ is 4.35 units at -90°. Find the following: a) The components of their resultant $\overrightarrow{OR}$. Round to hundredths. b) The length of $\overrightarrow{OR}$. Round to hundredths. c) The direction of $\overrightarrow{OR}$. Round to a whole number.

In the diagrams below, resultant $\overrightarrow{OR}$ is the sum of vectors $\overrightarrow{OA}$ and $\overrightarrow{OB}$. Sketch vector-addend $\overrightarrow{OB}$ on each diagram.

6.

R

O A

7.

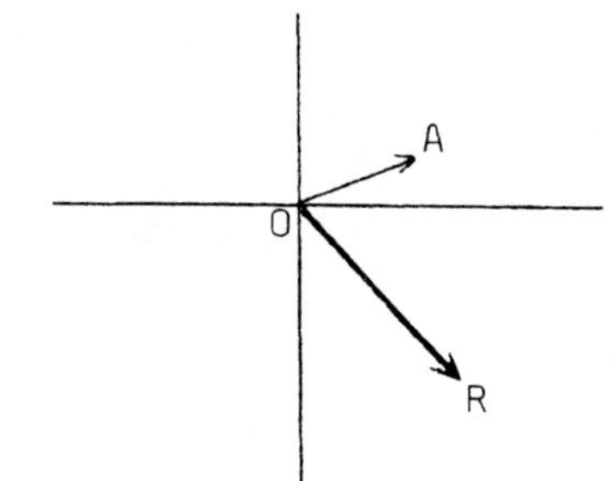

8. Resultant $\overrightarrow{OR}$ is the sum of vectors $\overrightarrow{OG}$ and $\overrightarrow{OH}$. The components of $\overrightarrow{OR}$ are (-25, 15) and the components of $\overrightarrow{OG}$ are (-50, -35). Find the components of vector-addend $\overrightarrow{OH}$.

9. Resultant $\overrightarrow{OR}$ is the sum of vectors $\overrightarrow{OC}$ and $\overrightarrow{OF}$. The components of $\overrightarrow{OR}$ are (-620, 0) and the components of $\overrightarrow{OC}$ are (0, 480). Find the components of vector-addend $\overrightarrow{OF}$.

Assignment 14

1. Find the components of the resultant $\overrightarrow{OR}$ for this system of three vectors:

   $\overrightarrow{OA}$: (5, -4) $\overrightarrow{OB}$: (-3, 2) $\overrightarrow{OC}$: (2, 3)

2. Find the components of the resultant $\overrightarrow{OR}$ for this system of four vectors.

   $\overrightarrow{OD}$: (-700, -450) $\overrightarrow{OF}$: (-375, 825) $\overrightarrow{OG}$: (650, 0) $\overrightarrow{OH}$: (0, -175)

3. The lengths and directions of four vectors are shown below.

   $\overrightarrow{OA}$ is 280 units at 0° $\overrightarrow{OC}$ is 300 units at 150°

   $\overrightarrow{OB}$ is 120 units at -90° $\overrightarrow{OD}$ is 260 units at -120°

   Rounding each answer to a whole number, find: a) The components of resultant $\overrightarrow{OR}$. b) The length of $\overrightarrow{OR}$. c) The direction of $\overrightarrow{OR}$.

4. Vectors $\overrightarrow{OF}$ and $\overrightarrow{OG}$ are a pair of vector-opposites. The components of $\overrightarrow{OF}$ are (70, -50). Find the components of $\overrightarrow{OG}$.

5. Three vectors are in the state of equilibrium. What are the components of their resultant $\overrightarrow{OR}$?

6. Here is a system of two vectors: $\overrightarrow{OA}$: (40, -50) and $\overrightarrow{OB}$: (-10, 30). A third vector $\overrightarrow{OE}$ is added to produce the state of equilibrium. Find the components of $\overrightarrow{OE}$.

7. Here is a system of three vectors: $\overrightarrow{OM}$: (-350, 210), $\overrightarrow{ON}$: (470, 190), and $\overrightarrow{OP}$: (160, -340). Find the components of their equilibrant $\overrightarrow{OE}$.

8. The lengths and directions of three vectors are shown below.

   $\overrightarrow{OQ}$ is 210 units at 0°. $\overrightarrow{OT}$ is 280 units at -90°. $\overrightarrow{OV}$ is 350 units at 180°.

   Rounding each answer to a whole number, find: a) The components of their equilibrant $\overrightarrow{OE}$. b) The length of $\overrightarrow{OE}$. c) The direction of $\overrightarrow{OE}$.

9. Two forces of 315 lb and 184 lb, with a 53° angle between them, are applied to an object. Find their resultant force. Round to a whole number.

10. The total current I_T in an a-c parallel circuit is the vector sum of these two currents:

    I_R is 18 ma at 0° and I_L is 29 ma at -90°. Rounding each answer to a whole number, find:

    a) The total current I_T. b) The angle between I_R and I_T.

11. An airplane traveling due west at 500 mph encounters a 40 mph wind blowing towards the southeast. Rounding each answer to a whole number, find: a) Its actual speed with respect to the ground. b) Its angular deviation from west.

12. Each end of a cable is attached to a horizontal beam. A 600 kg load is suspended from the midpoint of the cable. The angle between each half of the cable and the beam is 28°. Rounding to a whole number, find the tension force in each half of the cable.

13. The applied voltage V_T in an a-c series circuit is the vector sum of these three voltages:

    V_R is 360 mv at 0°, V_L is 490 mv at 90°, and V_C is 280 mv at -90°. Rounding to a whole number, find: a) The applied voltage V_T. b) The angle between V_R and V_T.

Chapter 5 CIRCLE CONCEPTS

In this chapter, we will discuss the following "circle" concepts: central angles, arcs, chords, sectors, segments, tangents, half-tangents, rotational velocity (including angular velocity), and linear velocity. Subdivisions of a degree into decimal numbers and minutes and seconds are also discussed.

5-1 CIRCUMFERENCE

In this section, we will discuss the formulas for the circumference of a circle.

---

1. The distance around a circle is called its "circumference (C)". By breaking the circle below at A and bending its circumference into a straight line, we can see that its circumference is slightly more than 3 times as long as its diameter.

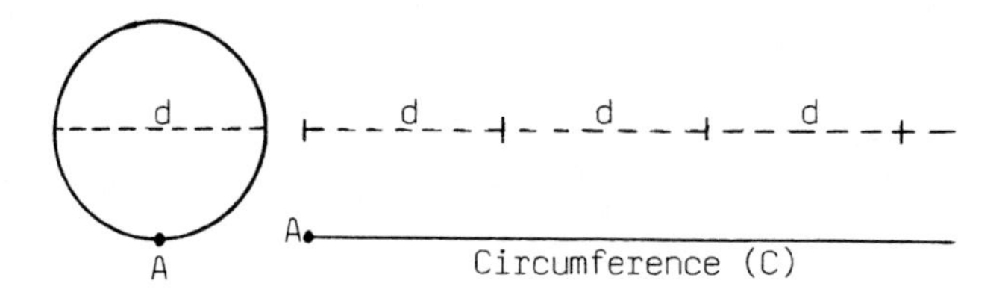

To find the ratio of the circumference of a circle to its diameter, we divide C by d. For any circle, the ratio of C to d is slightly more than 3. The exact value is the unending decimal number 3.1415927... The Greek letter "π" (pronounced "pie") is used for that number. Therefore:

$$\frac{C}{d} = \pi \qquad \text{or} \qquad \frac{C}{d} = 3.1415927...$$

To find the value of π on a calculator, we press [π] or [INV] [π]. Do so.

---

2. Rearranging $\frac{C}{d} = \pi$ to solve for C, we get the following "circumference" formula.

$$\boxed{C = \pi d}$$

Using a calculator and $C = \pi d$, find the circumference of the circles with the following diameters. Round to tenths.

a) If d = 4.28 ft, C = __________ ft

b) If d = 11.7 cm, C = __________ cm

3. Since $d = 2r$, we can also get a circumference formula in terms of the radius. That is:

$$C = \pi d$$

$$C = \pi(2r)$$

$$\boxed{C = 2\pi r}$$

Using a calculator and the bottom formula, complete these. Round to tenths.

a) If $r = 3.64$ in, $C =$ ________ in

b) If $r = 5.97$ m, $C =$ ________ m

Answers: a) 13.4 ft b) 36.8 cm

---

4. The two basic formulas for the circumference of a circle are:

$$C = \pi d \qquad C = 2\pi r$$

By rearranging them, we can solve for "d" and "r". We get:

$$d = \frac{C}{\pi} \qquad r = \frac{C}{2\pi}$$

Using the bottom formulas, we can find either the diameter or radius of a circle if its circumference is known. Complete these. Round to hundredths.

a) If $C = 14.9$ yd, $d =$ ________ yd

b) If $C = 23.7$ cm, $r =$ ________ cm

Answers: a) 22.9 in b) 37.5 m

Answers: a) 4.74 yd b) 3.77 cm

5-2 CENTRAL ANGLES AND ARCS

In this section, we will review some basic facts about central angles and arcs of circles.

---

5. The four angles at the right are central angles because they are formed by radii of the circle. They cut off arcs $\overset{\frown}{AB}$, $\overset{\frown}{BC}$, $\overset{\frown}{CD}$, and $\overset{\frown}{AD}$. We know these facts:

1. The sum of the central angles of a circle is 360°.

2. The sum of the lengths of the arcs of a circle is the circumference of the circle.

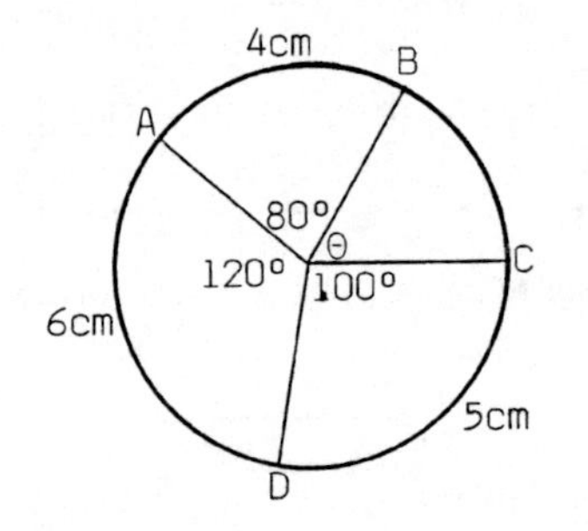

Using the above facts, complete these:

a) Angle θ contains ________°.

b) If the circumference is 18 cm, $\overset{\frown}{BC} =$ ________ cm.

6. The four angles at the right are central angles. We know these facts:

   1. Equal central angles in a circle have equal arcs.

   2. Equal arcs in a circle have equal central angles.

Using the above facts, complete these:

a) Angle θ contains ________ °.

b) $\widehat{EF}$ = ________ cm

c) The circumference of the circle is ________ cm.

a) 60°, from: $360° - (80° + 120° + 100°)$

b) 3 cm, from: 18 cm − (4 cm + 6 cm + 5 cm)

---

7. The circle at the right is divided into six equal central angles and arcs. Therefore:

   1. Each central angle is $\frac{1}{6}$ of 360°.

   2. Each arc is $\frac{1}{6}$ of the circumference.

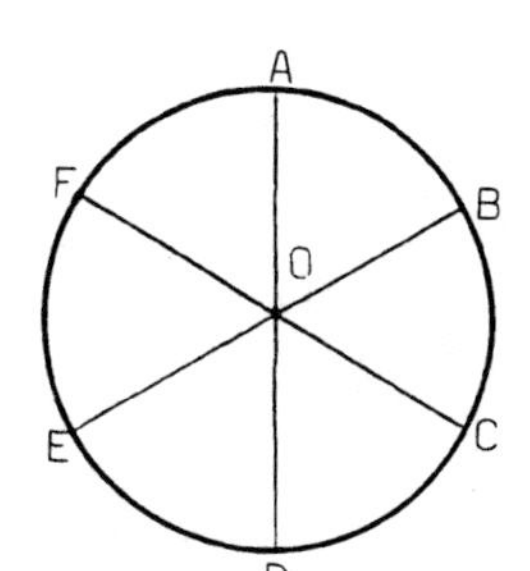

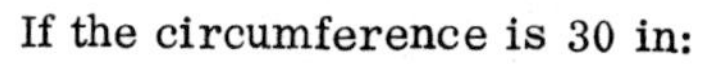

If the circumference is 30 in:

a) How long is $\widehat{AB}$? ________ in

b) How long is $\widehat{ABC}$? ________ in

c) Angle AOB contains ________ °.

a) 60°

b) 10 cm

c) 30 cm

---

8. The semi-circle at the right is divided into five equal central angles and arcs. We know these facts:

   1. The sum of the central angles of a semi-circle is $\frac{1}{2}$ of 360° or 180°.

   2. The arc-length of a semi-circle is $\frac{1}{2}$ of the circumference.

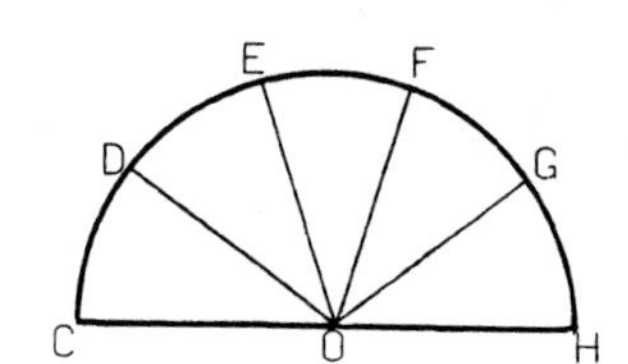

If the circumference of the circle is 100 cm:

a) The arc-length of the semi-circle is ________ cm.

b) The length of $\widehat{EF}$ is ________ cm.

c) Angle DOE contains ________ °.

a) 5 in

b) 10 in

c) 60°

---

a) 50 cm

b) 10 cm ($\frac{1}{5}$ of 50)

c) 36° (from $\frac{1}{5}$ of 180°)

9. The semi-circle at the right is divided into 10 equal arcs.

If the diameter is 11.0 m:

a) The circumference of the circle (to the nearest tenth) is ________ m.

b) The length of the semi-circle is ________ m.

c) The length of $\widehat{AB}$ is ________ m.

---

10. The semi-circular arch at the right is divided into 7 equal parts. OA is the radius of the inner circle; OB is the radius of the outer circle. In the problems below, round each answer to tenths.

If OA = 6.0 ft:

a) How long is the inner semi-circle? ________ ft

b) How long is $\widehat{EF}$? ________ ft

If OB = 9.0 ft:

c) How long is the outer semi-circle? ________ ft

d) How long is $\widehat{CD}$? ________ ft

Answers to 9: a) 34.6 m b) 17.3 m c) 1.73 m

---

11. Any arc is a fractional part of the circumference of a circle. If the circle is divided into equal arcs, it is easy to determine the "fractional part of the circumference" for each arc. For example:

If a circle is divided into 8 equal arcs, each arc is $\frac{1}{8}$ of the circumference or $\frac{1}{8}$(C) .

In terms of C, how long is each arc if the circle is divided into:

a) 4 equal arcs? ________ b) 12 equal arcs? ________

Answers to 10: a) 18.8 ft b) 2.7 ft c) 28.3 ft d) 4.0 ft

---

12. To determine the "fractional part of the circumference" for an arc, we can compare the size of its central angle to 360°. For example:

If the central angle of an arc is 90°, the arc is $\frac{90}{360}$(C) or $\frac{1}{4}$(C) .

If the central angle of an arc is 36°, the arc is $\frac{36}{360}$(C) or ________ .

Answers to 11: a) $\frac{1}{4}$(C) b) $\frac{1}{12}$(C)

	$\frac{1}{10}(C)$

13. When comparing the size of a central angle to 360°, we only reduce to lowest terms in the most obvious cases because a calculator allows us to handle larger numbers easily. For example:

If the central angle of an arc is 84°, the arc is $\frac{84}{360}(C)$.

If the central angle of an arc is 153°, the arc is ________.

Answer: $\frac{153}{360}(C)$

14. The circumference of a circle is 8.68 cm.

a) Find the length of an arc whose central angle is 57°. Round to hundredths.

$$\frac{57}{360}(8.68\text{ cm}) = \frac{57(8.68\text{ cm})}{360} = \underline{\qquad\qquad}\text{ cm}$$

b) Find the length of an arc whose central angle is 121°. Round to hundredths.

$$\frac{121}{360}(8.68\text{ cm}) = \frac{121(8.68\text{ cm})}{360} = \underline{\qquad\qquad}\text{ cm}$$

Answers: a) 1.37 cm b) 2.92 cm

15. The radius of this circle is 1.25 m. Angle AOB is 110°.

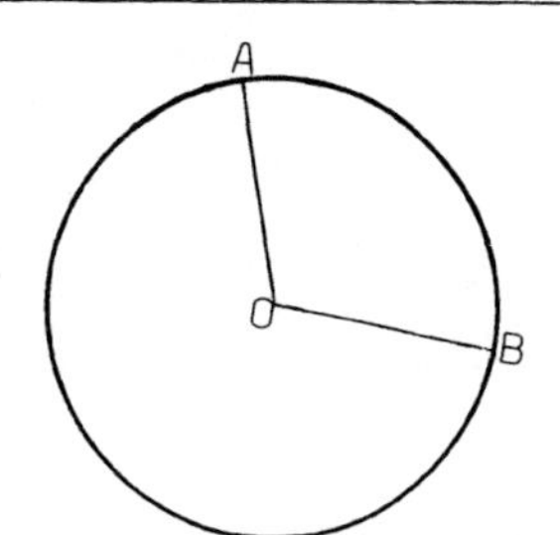

a) Find the circumference. Round to hundredths.

C = ________ m

b) Find the length of $\widehat{AB}$. Round to hundredths.

$\widehat{AB}$ = ________ m

Answers: a) 7.85 m b) 2.40 m

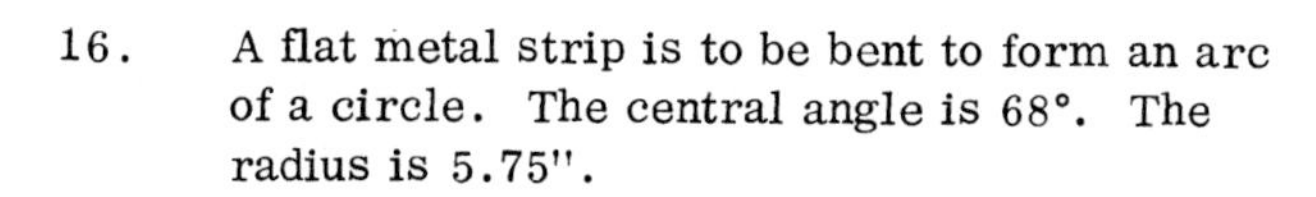

16. A flat metal strip is to be bent to form an arc of a circle. The central angle is 68°. The radius is 5.75".

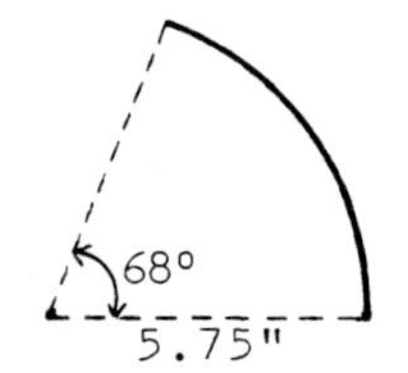

Rounding to hundredths, find the length of the strip. ________

Answer: 6.82"

5-3 CHORDS

In this section, we will review some facts about chords of circles.

---

17. A <u>chord</u> (pronounced "cord") is a straight line that joins any two points on a circle. CD and EF are chords of the circle below.

A <u>diameter</u> is a chord that passes through the center of a circle. EF is a diameter of the circle below.

a) Is OR a chord of the circle? ________

b) If EO is 5 cm, EF is ________ cm.

c) If EF is 16 cm, OF is ________ cm.

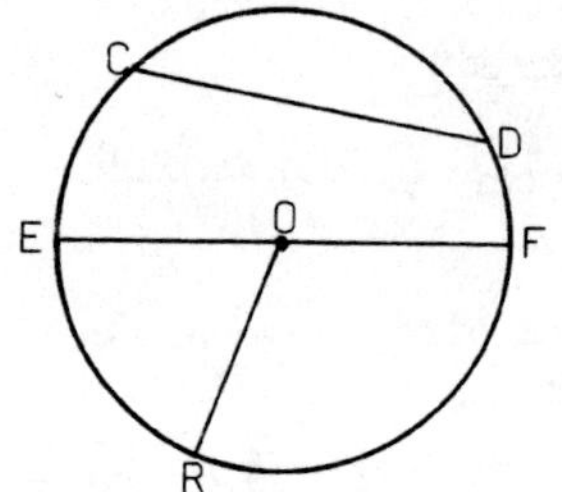

---

18. In the circle at the right, the two radii OA and OB are drawn to the ends of chord AB. Since the two radii are equal:

1. OAB is an <u>isosceles</u> triangle.
2. Angles #1 and #2 are equal.

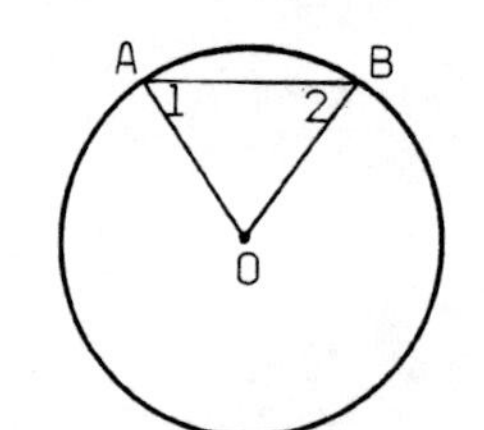

If central angle AOB contains 70°:

a) Angle #1 contains ________°.

b) Angle #2 contains ________°.

a) No. It does not touch two points <u>on</u> the circle.

b) 10 cm

c) 8 cm

---

19. In any given circle, <u>equal arcs have equal chords</u>. All five arcs are equal in the circle at the right.

a) If chord AB is 4 cm, chord AE is ______ cm.

b) Angle COD contains ________°.

c) Angles #1 and #2 each contain ________°.

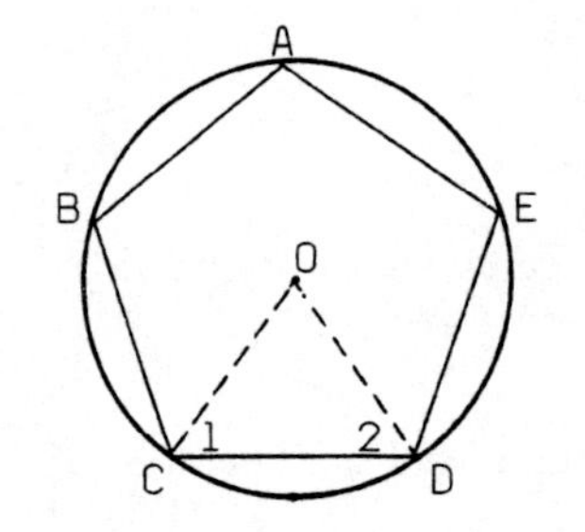

a) 55°

b) 55°

---

20. The circle on the right is divided into eight equal arcs. Three chords (CI, DE, and GF) are drawn.

a) Chord DE is equal to chord ________.

b) Angle COI contains ________°.

c) Angles #1 and #2 each contain ________°.

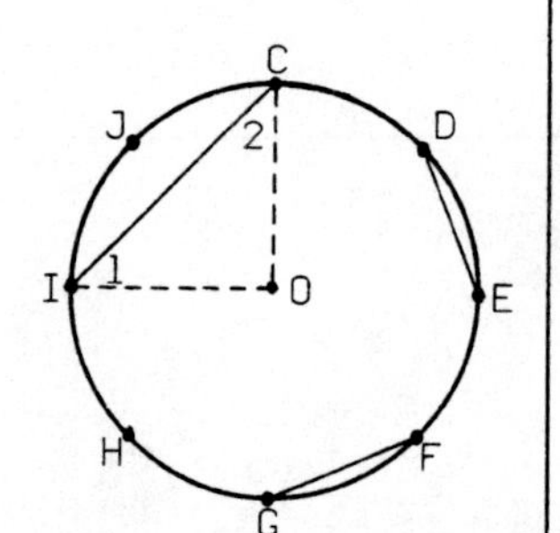

a) 4 cm

b) 72°

c) 54°

21. If a radius (or diameter) of a circle is drawn perpendicular to a chord, it bisects (cuts into two equal parts) the chord.

In this circle, radius OR is drawn perpendicular to chord AB. Therefore:

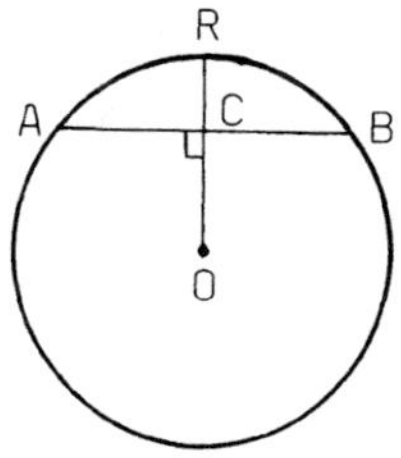

a) Both angle OCA and angle ________ are right angles.

b) If AB is 24 cm, both AC and BC are ________ cm.

c) If OR is 15 cm and OC is 9 cm, CR is ________ cm.

a) GF
b) 90°
c) 45°

---

22. In this circle, radius OR is perpendicular to chord GH. OG and OH are also radii.

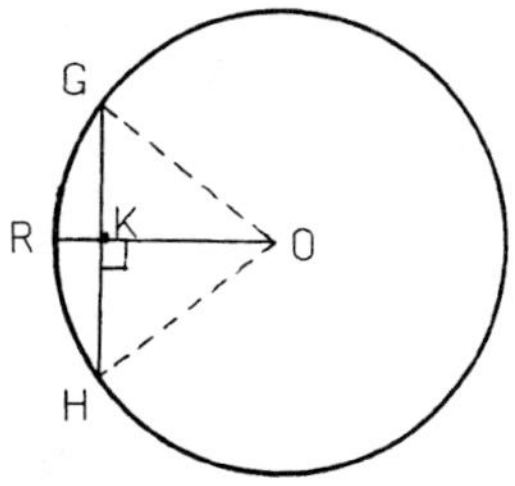

a) If GH is 12 ft, how long are both GK and HK? ________ ft

b) If OR is 10 ft, both OG and OH are ________ ft.

c) If OR is 10 ft and OK is 8 ft, KR is ________ ft.

d) Name the isosceles triangle in this figure. ________

e) OG is the hypotenuse of right triangle ________.

f) OK and KH are the legs of right triangle ________.

a) OCB
b) 12 cm
c) 6 cm, since: CR = OR - OC

---

23. If a radius is drawn perpendicular to a chord, it bisects:

1. the chord.
2. the arc of the chord.
3. the central angle of the chord.

In the circle at the right, radius OA is perpendicular to chord BC. Angle BOC contains 68°.

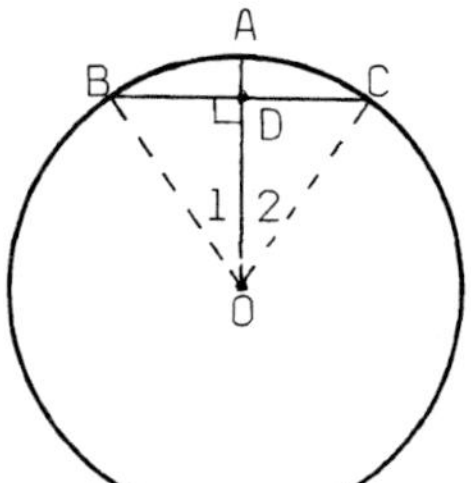

a) Angles #1 and #2 each contain ______°.

b) Angle OCD contains ______°.

c) The hypotenuse of right triangle OBD is ________.

d) The legs of right triangle OCD are ________ and ________.

e) If CD is 7 cm, BC is ________ cm.

a) 6 ft
b) 10 ft
c) 2 ft
d) OGH
e) OGK
f) OHK

---

a) 34° b) 56° c) OB d) OD and CD e) 14 cm

5-4 PROBLEMS INVOLVING CHORDS

In this section, we will discuss some problems that require finding the length or depth of a chord of a circle.

---

24. In the circle at the right, radius OR is perpendicular to chord AB. OC is the distance of the chord from the center of the circle. CR is the distance from the center of the chord to the circle. It is called the "depth" of the chord. We want to find this "depth".

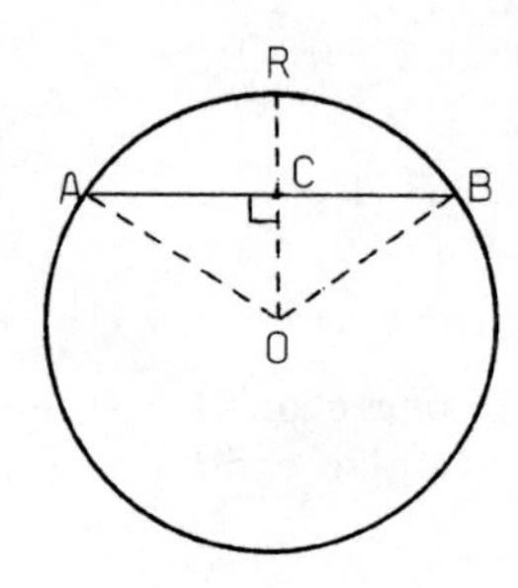

We are given these facts:

1. The radius of the circle is 6.00 in.
2. The length of chord AB is 10.00 in.

We can find the "depth" CR by subtracting OC from OR. That is:

$$CR = OR - OC$$

We know that OR = 6.00 in. To find OC, we can use right triangle AOC. In that triangle, the hypotenuse OA = 6.00 in and AC = 5.00 in since it is half of AB. Using the Pythagorean Theorem, we get:

$$(OC)^2 = (OA)^2 - (AC)^2$$
$$OC = \sqrt{(6.00)^2 - (5.00)^2}$$
$$OC = \sqrt{36 - 25}$$
$$OC = \sqrt{11} = 3.32 \text{ in}$$

Using the values for OR and OC, we get:

$$CR = OR - OC$$
$$= 6.00 \text{ in} - 3.32 \text{ in}$$
$$= \_\_\_\_\_\_ \text{ in}$$

2.68 in

25. In the circle at the right, radius OP is perpendicular to chord MQ. The radius of the circle is 5.00 cm. Central angle MOQ contains 104°. We want to find the "depth" PR.

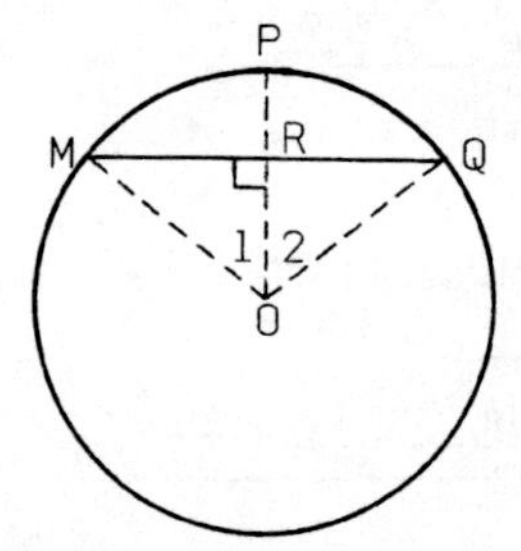

We can use this formula to find PR:

$$PR = OP - OR$$

Continued on following page.

25. Continued

We know that OP = 5.00 cm. To find OR, we can use right triangle OMR. In that triangle, the hypotenuse OM = 5.00 cm and angle #1 = 52°. Using the cosine ratio, we get:

$$\cos \#1 = \frac{OR}{OM}$$

$$\cos 52° = \frac{OR}{5.00}$$

$$OR = 5.00(\cos 52°) = 3.08 \text{ cm}$$

Using the values for OP and OR, we get:

$$PR = OP - OR$$

$$= 5.00 \text{ cm} - 3.08 \text{ cm}$$

$$= \_\_\_\_\_\_\_\_ \text{ cm}$$

1.92 cm

26. Figure 1 shows the end-view of a steel shaft on which a "flat" is to be cut. As you can see from the dimensions:

1. The diameter of the circle is 4.00".
2. We want the length of the "flat" to be 3.20".

We want to find "d", the depth of the cut.

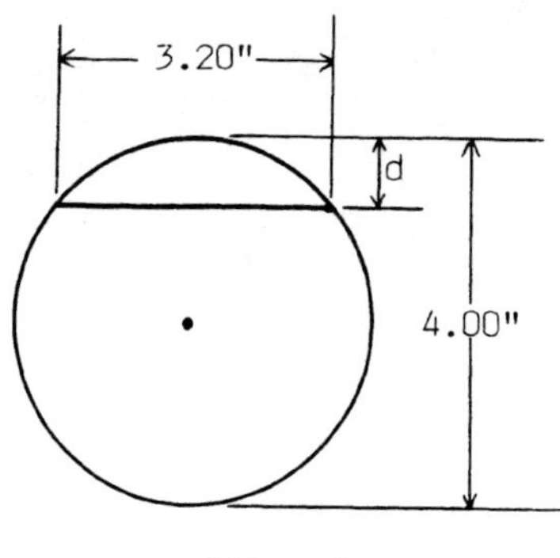

Figure 1

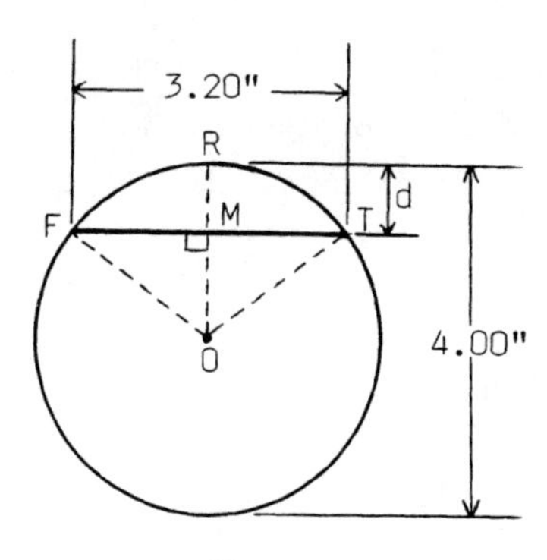

Figure 2

To solve the problem, we have drawn radii OF, OR, and OT in Figure 2. OR is perpendicular to chord FT. We must find MR which equals "d". We can use right triangle OMT.

a) Since the diameter is 4.00", radius OT = ________.

b) Since FT = 3.20", MT = ________.

c) Using the Pythagorean Theorem, find OM. ________

d) Find MR or "d". ________

a) 2.00"

b) 1.60"

c) 1.20"

d) 0.80"

27. We want to find the length of chord AB in the circle at the right. The radius of the circle is 15.0 cm. Central angle AOB is 110°.

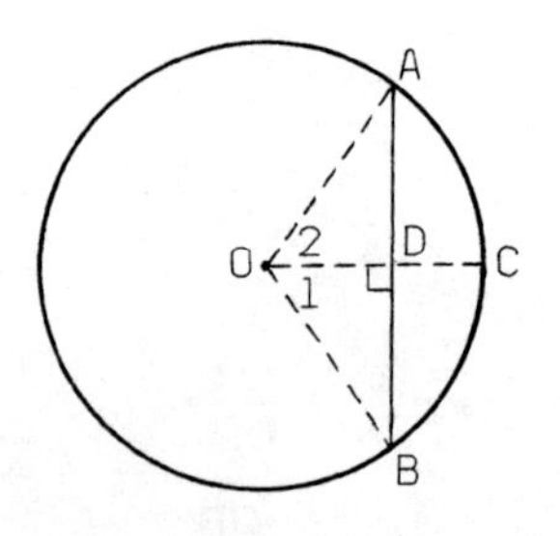

Since AD and BD are half of chord AB, we can find the length of AB if we find the length of either AD or BD. We will find AD. To do so, we will use right triangle AOD. In that triangle:

1. OA is 15.0 cm since it is a radius.
2. Angle #2 contains 55° since it is half of central angle AOB.

We can use the sine ratio to find AD.

$$\sin \#2 = \frac{AD}{OA}$$

$$\sin 55° = \frac{AD}{15.0}$$

$$AD = 15.0 (\sin 55°) = 12.3 \text{ cm}$$

Therefore, AB = ________ cm

24.6 cm

28. The figure from the last frame is shown at the right without radius OC drawn perpendicular to chord AB. The radius of the circle is 15.0 cm. Central angle AOB is 110°. Since AOB is an isosceles triangle, angles #1 and #2 each contain 35°.

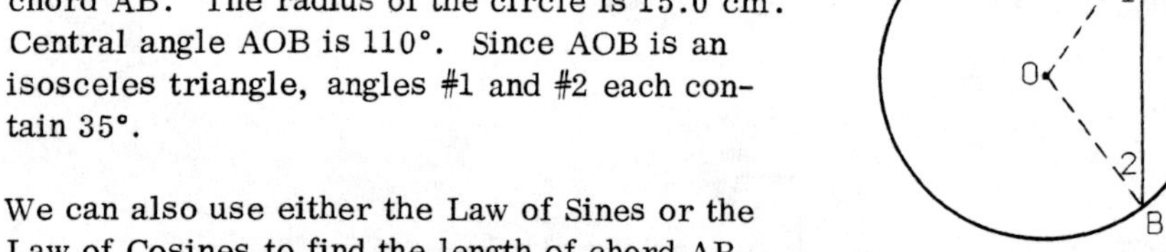

We can also use either the Law of Sines or the Law of Cosines to find the length of chord AB. Both methods are shown below.

Law of Sines

$$\frac{AB}{\sin AOB} = \frac{OA}{\sin \#2}$$

$$\frac{AB}{\sin 110°} = \frac{15.0}{\sin 35°}$$

$$AB = \frac{15.0 (\sin 110°)}{\sin 35°} = 24.6 \text{ cm}$$

Law of Cosines

$$(AB)^2 = (OA)^2 + (OB)^2 - 2(OA)(OB)(\cos AOB)$$

$$AB = \sqrt{(15.0)^2 + (15.0)^2 - 2(15.0)(15.0)(\cos 110°)}$$

$$AB = 24.6 \text{ cm}$$

Are the two answers above identical to the answer in the last frame?

____

Yes

29. Figure 1 below shows a circular metal part with five equally-spaced holes on it. The centers of these five holes lie on a circle (shown in dashed lines) called the "hole-circle". The radius of the hole-circle is 5.00". We want to find the center-to-center distance between two successive holes.

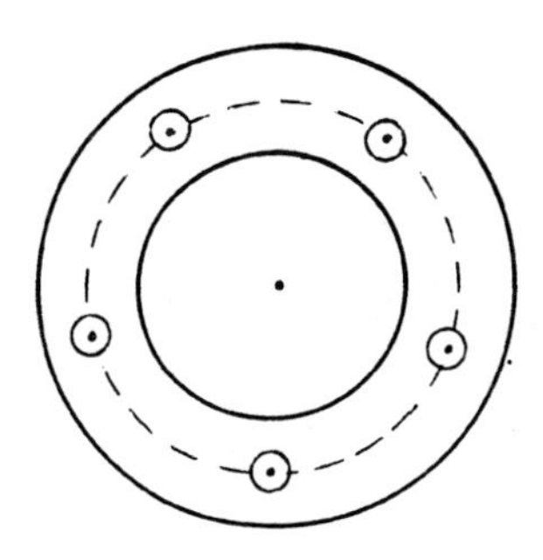

Figure 1 Figure 2

Some extra lines have been drawn in Figure 2. Note that finding the center-to-center distance between two successive holes is the same as finding the length of chord CD.

To find CD, we can find either CF or DF and then double it. Let's find CF. We will use right triangle COF.

a) How long is OC? __________

b) How large is central angle COD? __________

c) How large is angle #1? __________

d) Find the length of CF. Round to hundredths.

CF = __________

e) Find the length of CD, the distance between the center of two successive holes. CD = __________

---

a) 5.00"

b) 72° ($\frac{1}{5}$ of 360°)

c) 36°

d) 2.94", from: $\sin 36° = \frac{CF}{5.00}$

e) 5.88"

---

30. The metal part at the right has ten equally-spaced holes whose centers lie on the "hole-circle". The radius of the hole-circle is 8.75 cm. We want to find the distance between the centers of two alternate holes. That is, we want to find the length of AB.

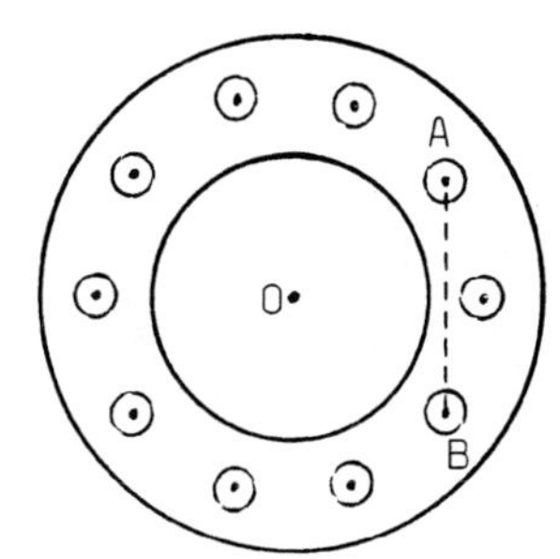

Continued on following page.

30. Continued

Let's use the Law of Sines and the Law of Cosines to find AB. To do so, complete a triangle by drawing radii OA and OB.

a) Use the Law of Sines to find AB. Round to tenths.

AB = ________

b) Use the Law of Cosines to find AB. Round to tenths.

AB = ________

a) 10.3 cm, from:

$$\frac{AB}{\sin 72°} = \frac{8.75}{\sin 54°}$$

b) 10.3 cm, from:

$$AB = \sqrt{(8.75)^2 + (8.75)^2 - 2(8.75)(8.75)(\cos 72°)}$$

SELF-TEST 15 (pp. 213-224)

1. A piece of wire is to be bent to form an arc of a circle. The central angle is 105° and the radius is 14.8 cm. Find the length of the wire. Round to tenths.

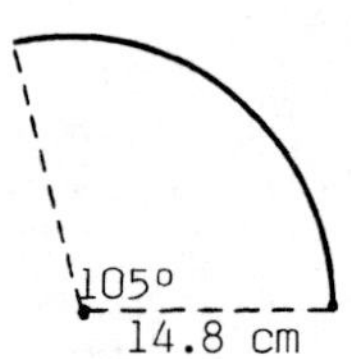

Length = ________

2. The radius of the circle below is 5.18". The length of chord AB is 7.14". Find depth "d". Round to hundredths.

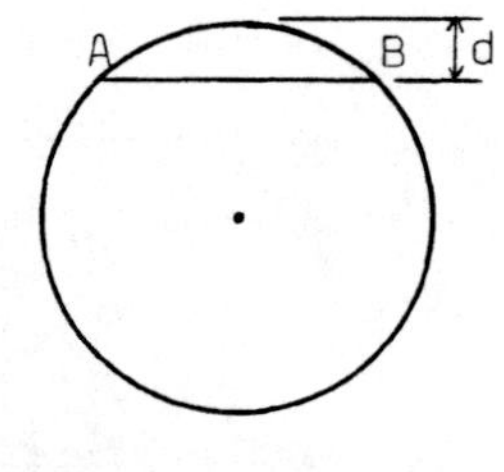

d = ________

Nine equally-spaced holes are placed on the circumference of a "hole-circle" whose radius is 19.2 cm.

3. Find "s", the center-to-center distance between two successive holes. Round to tenths.

4. Find "w", the center-to-center distance between two alternate holes. Round to tenths.

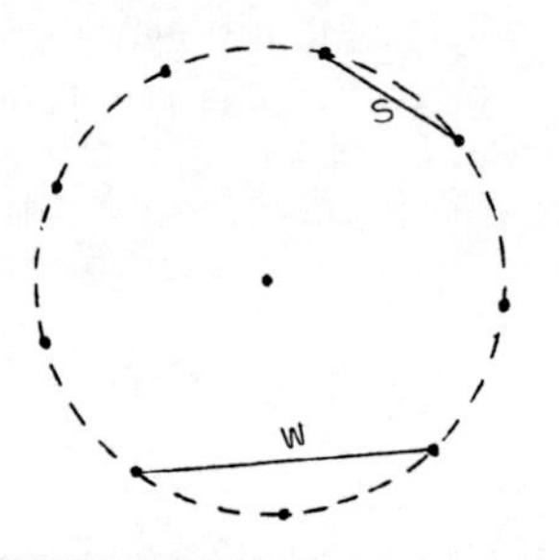

ANSWERS: 1. 27.1 cm   2. d = 1.43"   3. s = 13.1 cm   4. w = 24.7 cm

5-5 SECTORS AND SEGMENTS

In this section, we will define a "sector" and a "segment" of a circle and discuss methods for finding the area of sectors and segments.

31. A sector of a circle is the pie-shaped part that lies between the sides of a central angle. For example, the shaded part in the circle at the right is sector OAB.

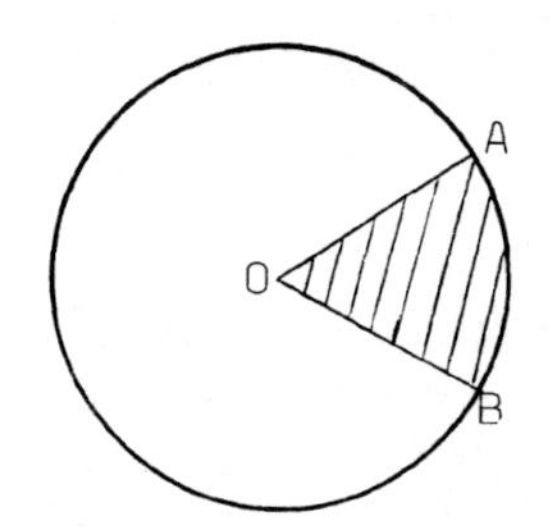

Since central angle AOB is 60°, the area of sector OAB is $\frac{1}{6}$ of the area of the circle.

If the area of the circle is 30 cm^2, the area of sector OAB is ______ cm^2.

---

32. The area of a sector is a fractional part of the total area of a circle. To find the fractional part, we compare the size of the central angle with 360°. For example, if the total area of a circle is 50.0 in^2:

5 cm^2

a) The area of a sector whose central angle is 100° is:

$\frac{100}{360}(50.0) =$ ________ in^2. Round to tenths.

b) The area of a sector whose central angle is 29° is:

$\frac{29}{360}(50.0) =$ ________ in^2. Round to hundredths.

---

33. The formula for the area of a circle in terms of its radius is:

a) 13.9 in^2

b) 4.03 in^2

$$A = \pi r^2$$

The radius of the circle at the right is 8.50 cm. Therefore:

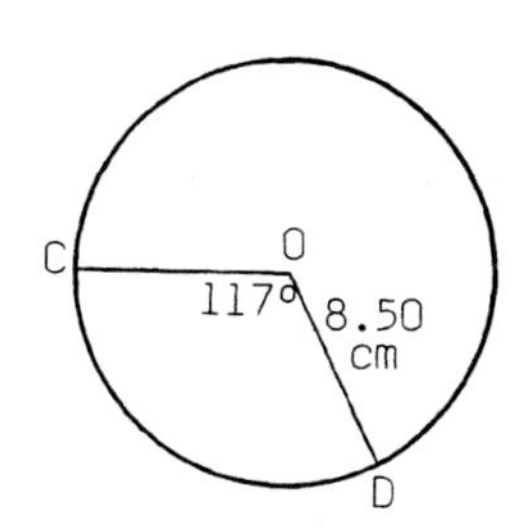

a) The total area is ________ cm^2. Round to a whole number.

b) The area of sector OCD is ________ cm^2. Round to tenths.

---

a) 227 cm^2

b) 73.8 cm^2

34. A segment of a circle is the part that lies between a chord and its arc. For example, the shaded part in the circle at the right is segment AB.

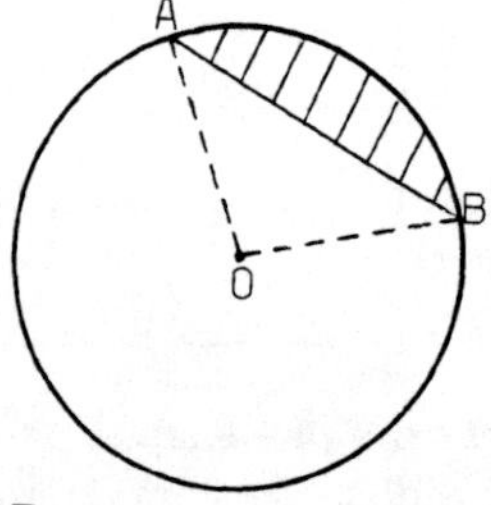

The area of segment AB can be found by subtracting the area of triangle OAB from the area of sector OAB. That is:

Segment AB = Sector OAB - Triangle OAB

Use the above formula for this one:

If the area of sector OAB is 36.0 in^2 and the area of triangle OAB is 24.0 in^2, the area of segment AB is ________ in^2.

---

35. In the circle at the right, the radius is 20.0 in. Central angle AOB contains 110°. Therefore, angles #1 and #2 each contain 55°.

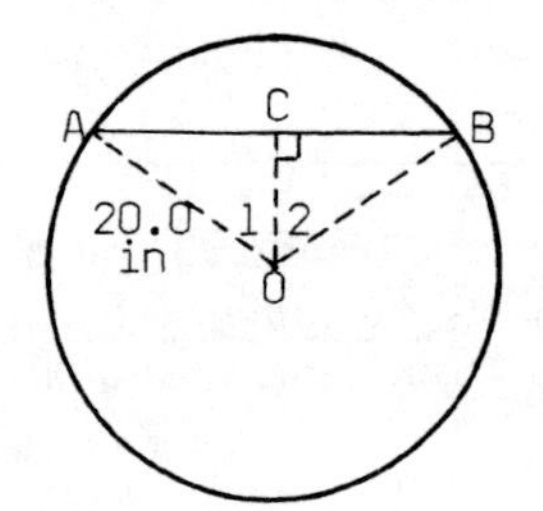

12.0 in^2

To find the area of segment AB, we must find the area of triangle OAB. The formula for the area of any triangle is:

$$\boxed{A = \frac{1}{2}bh}$$ where "h" is the altitude drawn to base "b".

In triangle OAB, OC is an altitude drawn to base AB. Therefore, the area of triangle OAB is:

$$A = \frac{1}{2}(AB)(OC)$$

Therefore, to find the area of the triangle we must find AB and OC.

Finding AB. Using right triangle OAC, we can find AC and then double it to get AB.

$$\sin \#1 = \frac{AC}{OA}$$

$$\sin 55° = \frac{AC}{20.0}$$

$$AC = 20.0 (\sin 55°)$$

$$AC = 16.4 \text{ in}$$

$$AB = 2(AC) = 2(16.4) = 32.8 \text{ in}$$

Continued on following page.

35. Continued

<u>Finding OC</u>. Using right triangle OAC, we can find OC.

$$\cos \#1 = \frac{OC}{OA}$$

$$\cos 55° = \frac{OC}{20.0}$$

$$OC = 20.0\ (\cos 55°)$$

$$OC = 11.5 \text{ in}$$

<u>Finding the area of triangle OAB</u>.

$$A = \frac{1}{2}(AB)(OC)$$

$$= \frac{1}{2}(32.8)(11.5)$$

$= \_\_\_\_\_\_\_\_$ in^2. Round to a whole number.

---

36. In the circle at the right, the radius is 5.00 cm. Central angle DOH contains 88°. We want to find the area of triangle ODH.

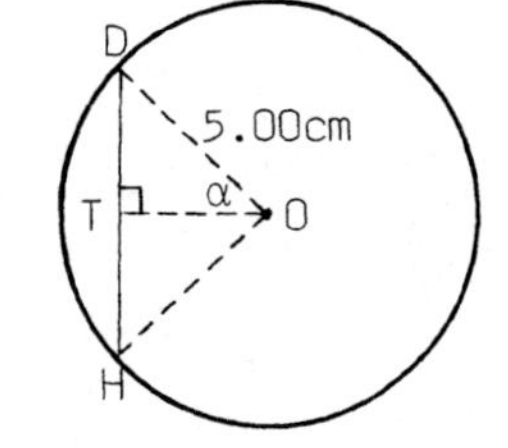

a) How large is angle α? ________

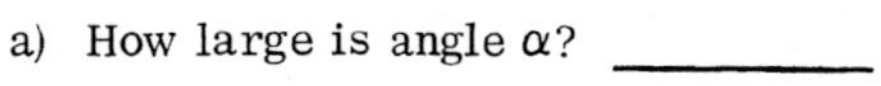

b) Using the cosine ratio, find OT. Round to hundredths.

OT = ________ cm

c) Using the sine ratio, find DT. Round to hundredths.

DT = ________ cm

d) Find DH. ________ cm

e) The area of triangle ODH is ________ cm^2. Round to tenths.

$A = 189\ in^2$

a) 44°

b) 3.60 cm

c) 3.47 cm

d) 6.94 cm

e) 12.5 cm^2

37. The radius of this circle is 12.0 in. Central angle COD contains 114°.

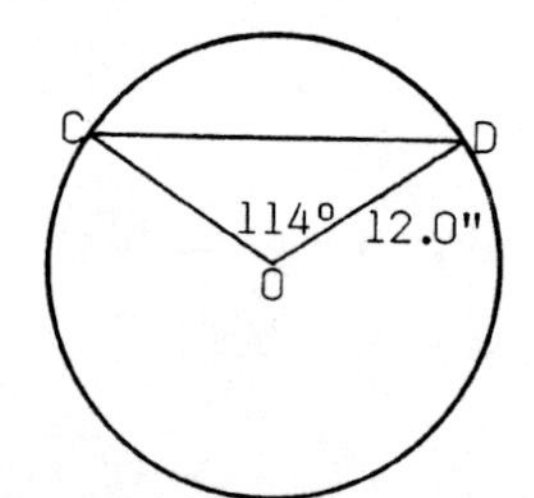

a) Find the area of the circle. Round to a whole number.

____

b) Find the area of sector OCD. Round to a whole number.

____

c) Find the area of triangle COD. Round to tenths.

____

d) Find the area of segment CD.

____

a) 452 in^2 b) 143 in^2 c) 65.8 in^2 d) 77.2 in^2

5-6 TANGENTS AND HALF-TANGENTS

In this section, we will define a "tangent" and a "half-tangent" of a circle. We will also discuss various angle relationships of tangents and half-tangents.

38. A tangent to a circle is a line that touches the circle at only one point. At the right:

Line AB is a tangent. Point D is called the "point of tangency".

Line FH is not a tangent because it touches the circle at two points, M and T.

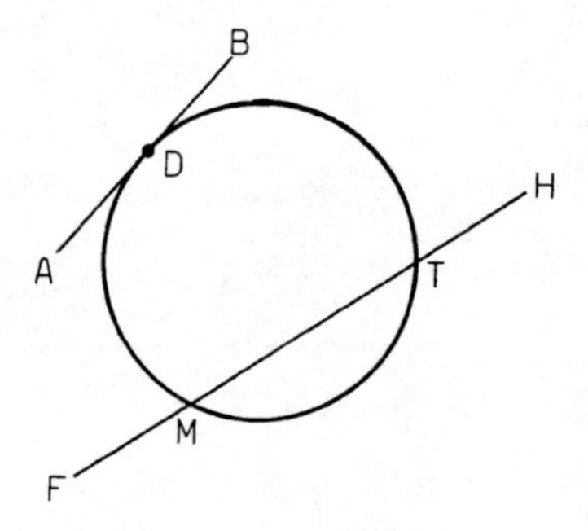

If a radius is drawn to the point of tangency, it is perpendicular to the tangent.

In the figure at the right:

CD is a tangent.
E is the point of tangency.
OE is a radius.

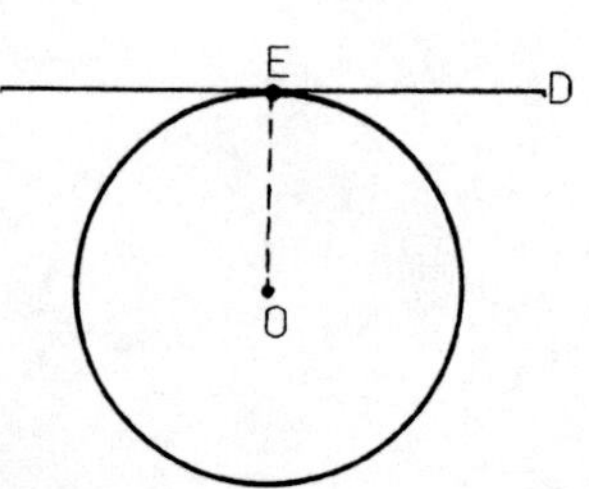

Name the two right angles. ________ and ________

39. In this figure, BD is a tangent. Notice that the line BD is drawn on both sides of the point of tangency.

FR is called a "<u>half-tangent</u>" because it is drawn on only one side of the point of tangency.

How large is angle OFR? ________

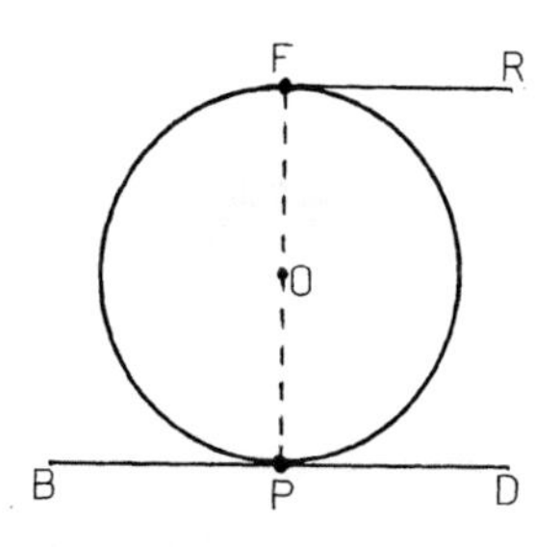

Answer: CEO and OED

---

40. Most applied problems involve half-tangents rather than tangents. For example, two pulleys and a V-belt are shown in the figure below.

AB and CD are the straight-line parts of the V-belt.

AB is a half-tangent:

1. to the large pulley at B.
2. to the small pulley at A.

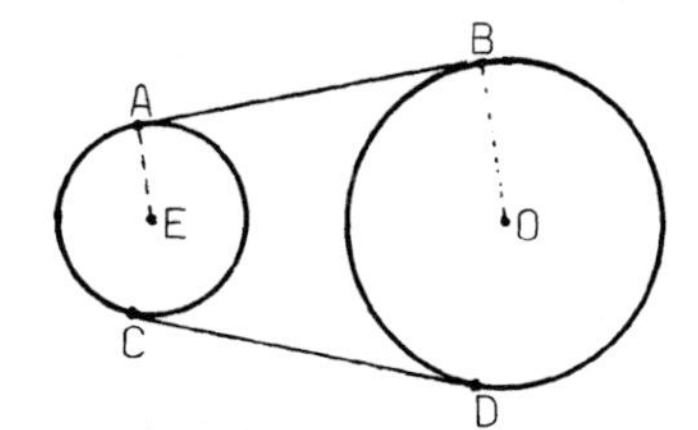

EA and OB are radii. How large are angles EAB and OBA? ________

Answer: 90° (a right angle)

---

41. In the figure at the right:

BD is a tangent to the circle.
M is the point of tangency.
OM is a radius.
MR is a chord.

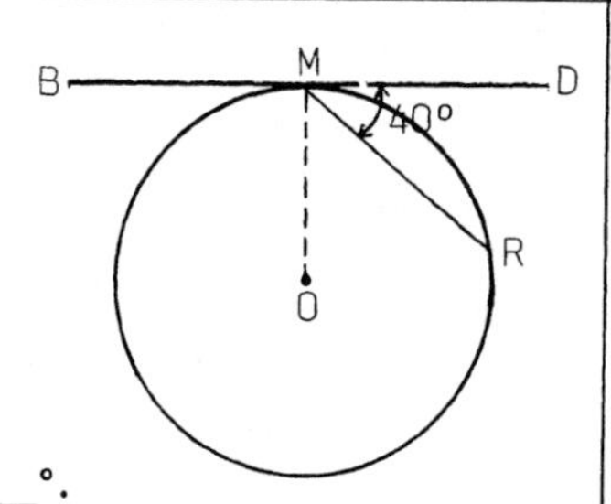

a) Angle OMD contains ________°.

b) Since angle DMR = 40°, angle OMR = ________°.

Answer: 90°

---

42. In the figure at the right, angle ACD is an angle formed by tangent AB and chord CD. We want to show this fact:

> The acute angle formed by a tangent and a chord is one-half as large as the central angle of the chord.

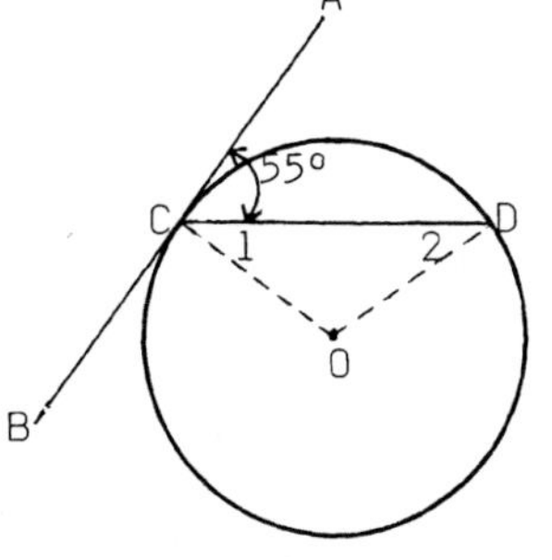

That is, we want to show that:

angle ACD is one-half as large as central angle COD.

Since OC and OD are radii:

1. Triangle OCD is isosceles.
2. Angles #1 and #2 are equal.

Continued on following page.

Answers:
a) 90°
b) 50°, from: 90° - 40°

42. Continued

Using the fact that angle ACD contains 55°, complete these:

a) Angle OCA contains ________°.

b) Angle #1 contains ________°.

c) Angle #2 contains ________°.

d) Angle COD contains ________°.

e) Is angle ACD one-half as large as angle COD? ________

---

a) 90°
b) 35°
c) 35°
d) 110°
e) Yes

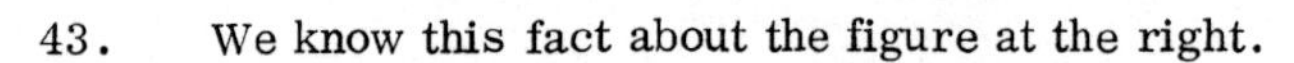

43. We know this fact about the figure at the right.

Angle SFT is one-half of angle FOT.

a) If angle SFT contains 43°, how large is angle FOT? ________°

b) If angle FOT contains 88°, how large is angle SFT? ________°

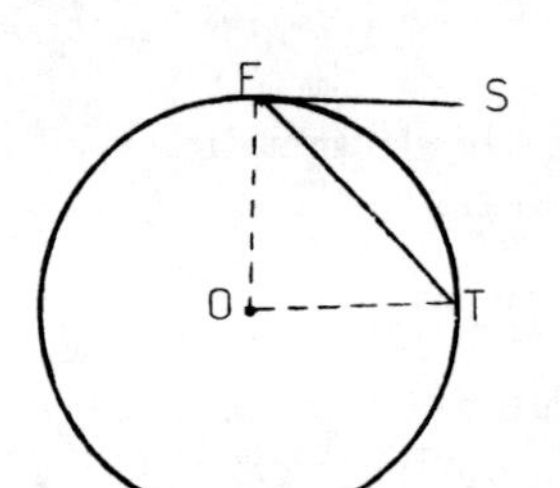

---

a) 86°
b) 44°

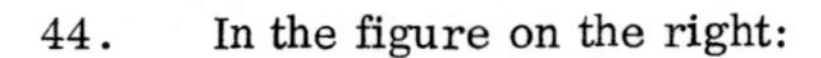

44. In the figure on the right:

CD is a half-tangent.
Angle COE is 78°.
Angle CDE is a right angle.

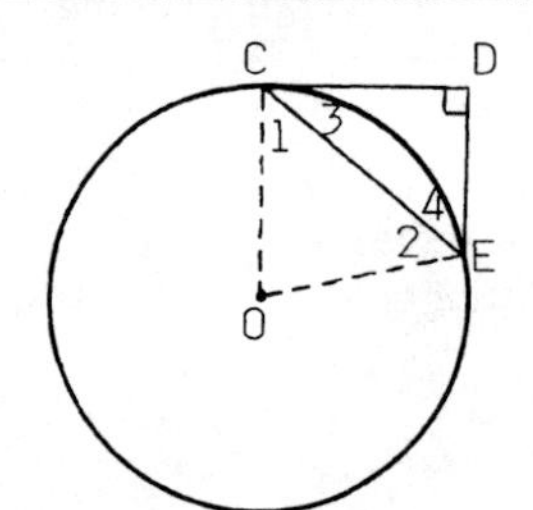

Find the size of each of these angles.

a) Angle #1 = ________°

b) Angle #2 = ________°

c) Angle #3 = ________°.

d) Angle #4 = ________°.

---

a) 51°
b) 51°
c) 39°
d) 51°

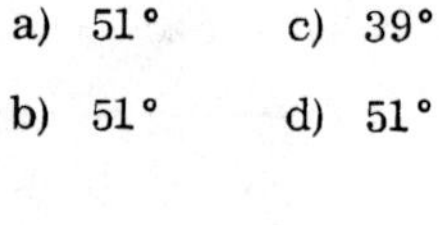

45. In the figure at the right:

AB and AC are half-tangents to the circle from the same external point.

AO connects the external point and the center of the circle.

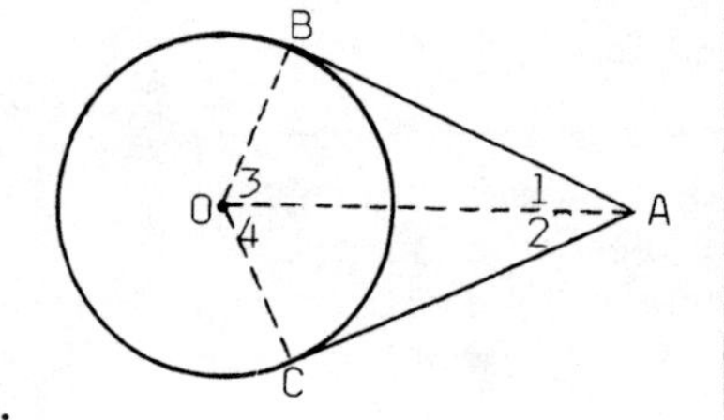

a) OB and OC are ________ of the circle.

b) Angles OBA and OCA each contain ________°.

c) Triangles ABO and ACO are both ________ triangles.

d) The hypotenuse of both right triangles is ________.

---

a) radii
b) 90°
c) right
d) AO

46. FP and FT are half-tangents to the circle from the same external point. When a line (FO) is drawn from the external point to the center of the circle:

1. It bisects the external angle.
2. It bisects the central angle.

That is:

1. Angle #1 equals angle #2.
2. Angle #3 equals angle #4.

If angle #1 contains 30°, find the size of these angles.

a) Angle #2 = ________°
b) Angle #3 = ________°
c) Angle #4 = ________°
d) Angle PFT = ________°
e) Angle POT = ________°
f) Angle FPO = ________°

---

a) 30° b) 60° c) 60° d) 60° e) 120° f) 90°

47.

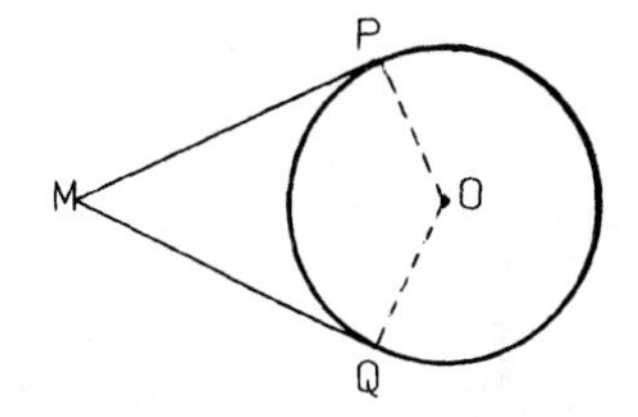

Figure 1

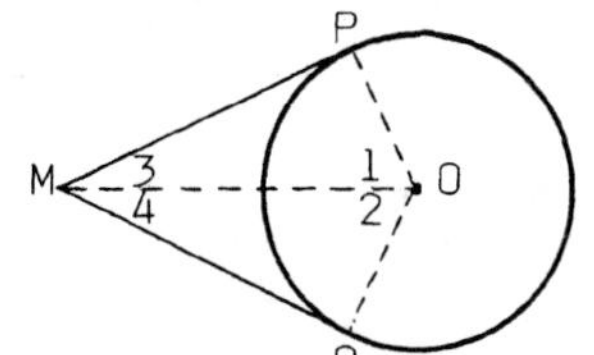
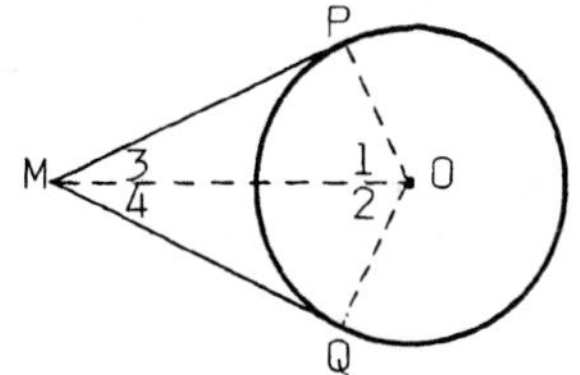

Figure 2

MP and MQ are half-tangents. We want to show that the sum of external angle M and central angle POQ in Figure 1 is 180°. To do so, we drew OM in Figure 2. Both angle #1 and angle #2 contain 65°.

Since angles #1 and #2 each contain 65°,
angles #3 and #4 each contain 25°.

Therefore: a) External angle M contains ________°.

b) Central angle POQ contains ________°.

c) The sum of external angle M and central angle POQ is ________°.

---

a) 50° b) 130° c) 180°

48. AC and AD are half-tangents. In the last frame, we saw that:

Angle A + Angle COD = 180°

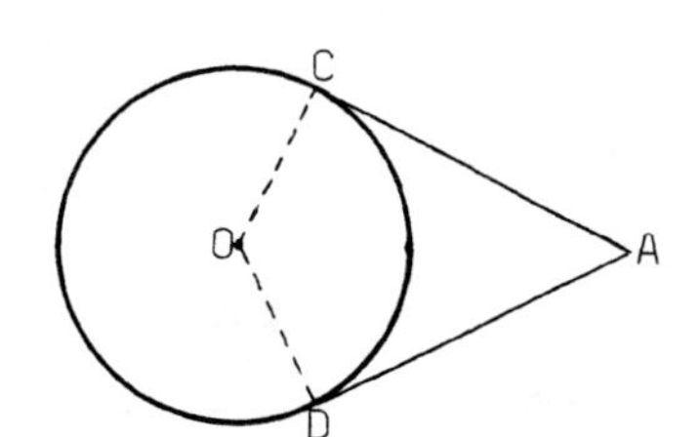

Using that fact, complete these:

a) If external angle A = 55°, central angle COD = ________°.

b) If central angle COD = 122°, external angle A = ________°.

---

a) 125° b) 58°

5-7 PROBLEMS INVOLVING HALF-TANGENTS

In this section, we will solve problems involving half-tangents.

49. In the figure on the right:

    BD is a half-tangent.
    Central angle BOF = 68°.
    Chord BF = 14.0".
    Angle BDF is a right angle.

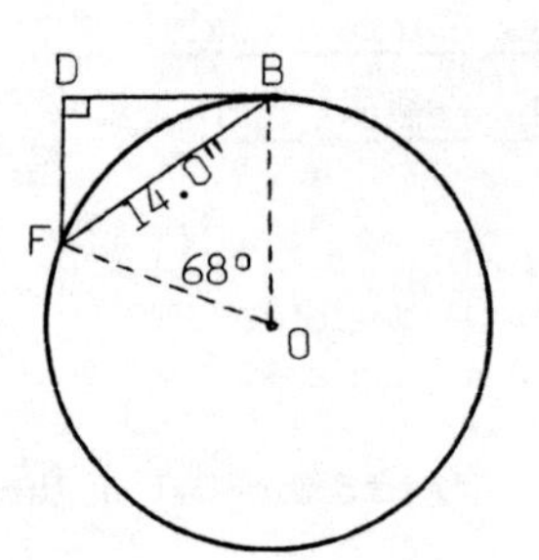

We want to find the length of half-tangent BD.

a) Angle DBF contains _______°.

b) Using right triangle BDF and a trig ratio, find BD.

BD = _________

50. In the figure on the right:

    The radius of the circle is 20.0 cm.
    FT is a half-tangent.
    Angle AFT contains 60°.

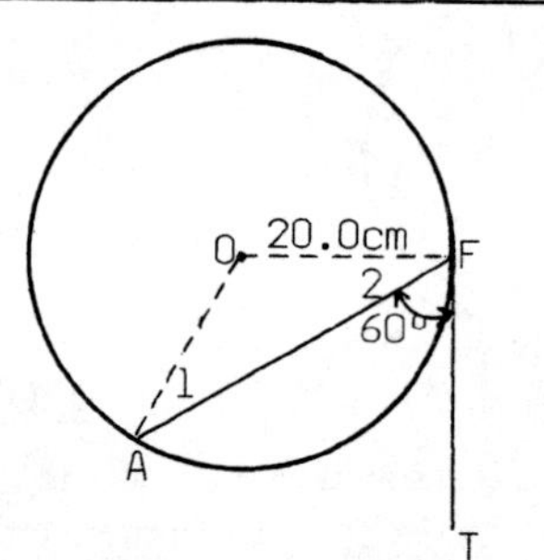

We want to find the length of chord AF.

a) Central angle AOF contains _______°.

b) Both angle #1 and #2 contain _______°.

c) Using the Law of Sines with oblique triangle AOF, find chord AF. Round to tenths.

AF = _________

Answers to 49:

a) 34° (half of 68°)

b) 11.6", from:

$$\cos 34° = \frac{BD}{14.0}$$

51. In the figure at the right:

    MQ is a half-tangent.
    Chord PQ is 4.69".
    Angle MQP is 36°.

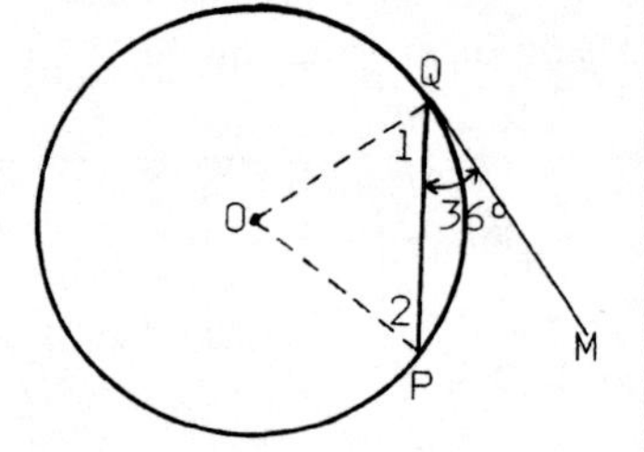

We want to find the radius of the circle.

a) Central angle POQ contains _______°.

b) Both angle #1 and angle #2 contain _______°.

c) Using the Law of Sines with oblique triangle POQ, find radius OP. Round to hundredths.

OP = _________

Answers to 50:

a) 120°

b) 30°

c) 34.6 cm, from:

$$\frac{AF}{\sin 120°} = \frac{20.0}{\sin 30°}$$

52. If two half-tangents are drawn to a circle from the same external point, they are equal. For example, in the figure at the right:

AB = AC

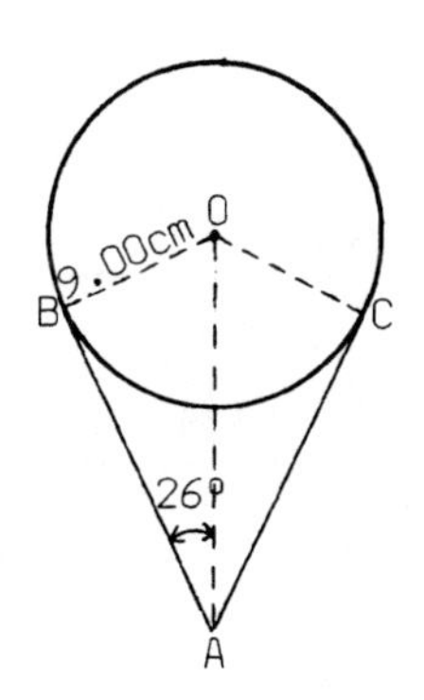

We are given these facts:

The radius is 9.00 cm.
Angle OAB is 26°.

Using these facts, complete these. Round to tenths.

a) How long is AB? AB = ________

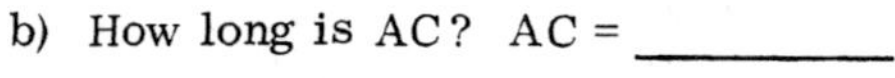

b) How long is AC? AC = ________

c) How long is OA? OA = ________

a) 72°

b) 54°

c) 3.99", from:

$$\frac{OP}{\sin 54°} = \frac{4.69}{\sin 72°}$$

---

53. In the figure at the right:

Radius OP = 10.0".
Half-tangent DP = 18.5".

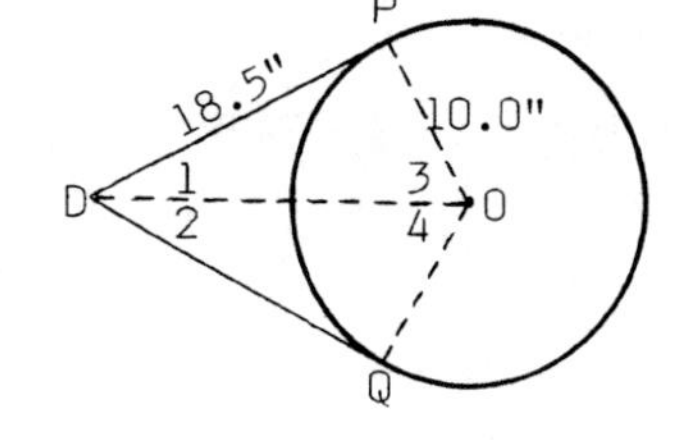

Find the following angles. Round each to a whole number.

a) Angle #1 = ________

b) Angle #2 = ________

c) Angle #3 = ________

d) Angle #4 = ________

e) Find OD. Round to tenths.

OD = ________

a) 18.5 cm, from:

$$\tan 26° = \frac{9.00}{AB}$$

b) 18.5 cm (same as AB)

c) 20.5 cm, from:

$$\sin 26° = \frac{9.00}{OA}$$

---

a) 28°, from:

$$\tan \#1 = \frac{10.0}{18.5}$$

b) 28°

c) 62°

d) 62°

e) 21.0", from:
$(OD)^2 =$
$(10.0)^2 + (18.5)^2$

54. In this figure, BF and BH are half-tangents. Angle B is 58°. Radius OF is 5.00 cm. We want to find the length of the arc $\overset{\frown}{FGH}$.

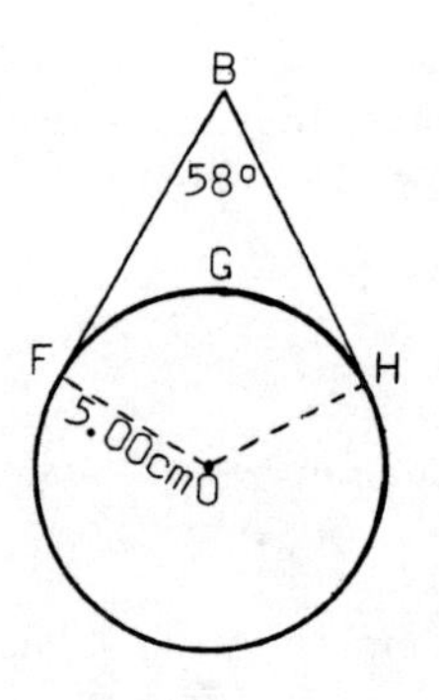

a) Central angle FOH = ________°.

b) Find the circumference. Round to tenths.

C = __________

c) Find $\overset{\frown}{FGH}$. Round to hundredths.

$\overset{\frown}{FGH}$ = __________

---

55. BCDE is a steel wire. It is bent so that CD is a circular arc.

BC and DE are each 2.75". The radius of the circular arc is 1.27". External angle P is 45°. We want to find the total length of the steel wire.

a) Central angle COD is ________°.

b) Rounded to hundredths, the circumference of the circle is ________.

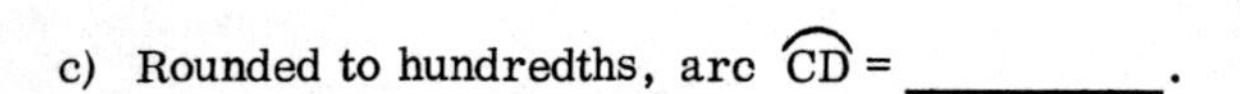

c) Rounded to hundredths, arc $\overset{\frown}{CD}$ = __________.

d) The total length of the steel wire is __________.

Answers to 54:

a) 122°

b) 31.4 cm

c) 10.6 cm, from:

$\frac{122}{360}(31.4)$

---

56. ADMB is a metal strap.

$\overset{\frown}{DM}$ is a circular arc with a 2.10 cm radius.

Angle DTM contains 94°.

AD is 4.00 cm; BM is 4.00 cm.

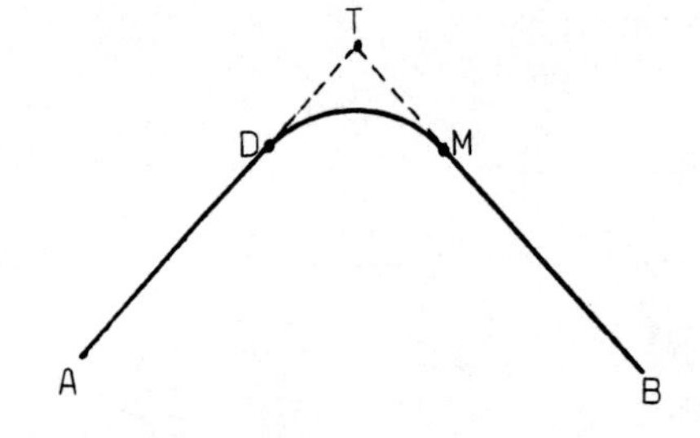

Answers to 55:

a) 135°

b) 7.98"

c) 2.99", from:

$\frac{135}{360}(7.98)$

d) 8.49"

56. Continued

We want to find the total length of the metal strap.

a) Find the length of $\overset{\frown}{DM}$. Round to hundredths.

$\overset{\frown}{DM}$ = ________

b) The total length of the metal strap is ________.

a) 3.15 cm b) 11.15 cm

SELF-TEST 16 (pages 225-235)

In the figure, the radius is 27.5 cm and central angle MON is 102°. Rounding each answer to a whole number, find the area of:

1. Sector OMN. ________
2. Triangle MON. ________
3. Segment MN. ________

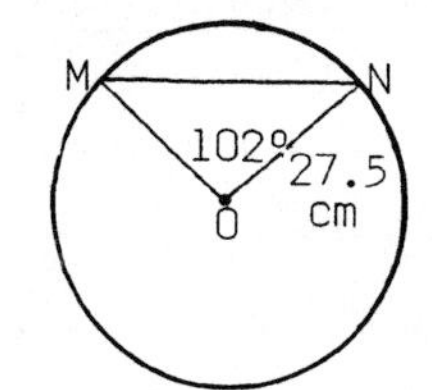

4. In the figure, BC is a half-tangent. The radius is 15.7" and angle CBD is 58°. Rounding to tenths, find the length of chord BD.

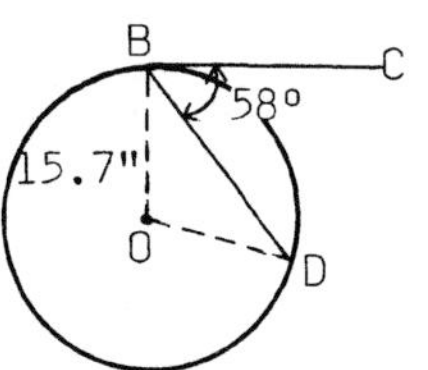

BD = ________

5. In the figure, PG and PF are half-tangents. The radius is 125 mm and angle P = 40°. Rounding to a whole number, find the length of half-tangent PF.

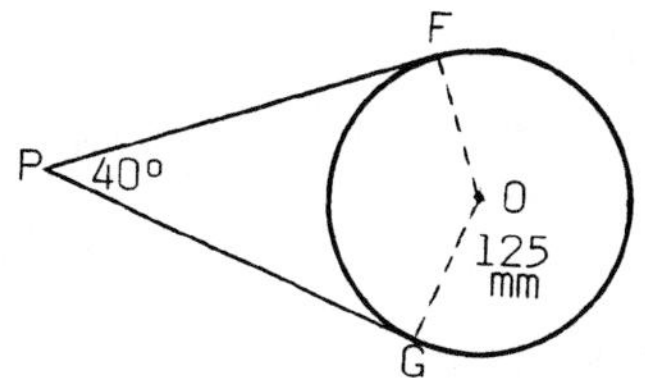

PF = ________

In the figure, PQRT is a metal strap. $\overset{\frown}{QR}$ is a circular arc with a 19.4 cm radius. Angle QGR is 70°. PQ and RT are each 30.0 cm. Rounding to tenths, find:

6. The length of $\overset{\frown}{QR}$. ________
7. The total length of the metal strap. ________

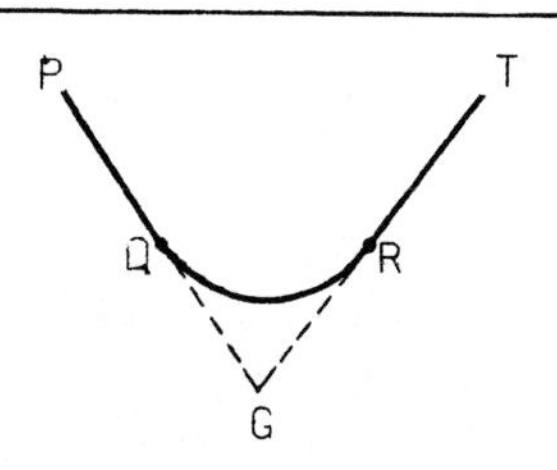

ANSWERS:

1. 673 cm^2
2. 370 cm^2
3. 303 cm^2
4. 26.6"
5. 343 mm
6. 37.2 cm
7. 97.3 cm

5-8 REVOLUTIONS AND ROTATIONAL VELOCITY

The velocity of a rotating object can be measured in terms of the number of revolutions per unit time. We will discuss that measure of rotational velocity in this section.

57. When an object (such as a wheel, gear, drive shaft, or motor armature) turns with a circular motion, it is said to be rotating. Rotations can be measured in terms of "revolutions".

The figure on the right shows what is meant by one revolution for vector OP. One revolution contains 360°.

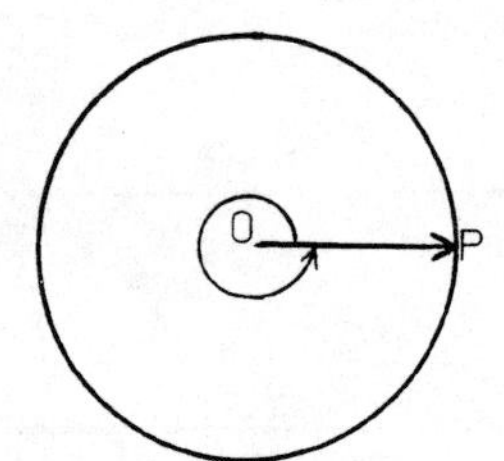

How many degrees are there in:

a) 2 revolutions? ________

b) 10 revolutions? ________

---

58. "Rotational velocity" is the velocity of a rotating object. It is the number of revolutions in a unit of time. The unit of time is either "1 minute" or "1 second". Therefore:

$$\text{Rotational velocity} = \frac{\text{Number of Revolutions}}{\text{1 Unit of Time}}$$

For a rotational velocity of "50 revolutions in 1 minute", we can write:

$$\frac{\text{50 revolutions}}{\text{1 minute}}$$

or

50 revolutions per minute

or

50 revolutions/minute

Though the number "1" is not actually shown:

"50 revolutions per minute" means "50 revolutions per 1 minute".

"50 revolutions/minute" means "50 revolutions/1 minute".

The following abbreviations are used when reporting rotational velocities.

"revs" for "revolutions"
"min" for "minute"
"sec" for "second"

Therefore: 25 revs/min means: 25 revolutions per minute

10 revs/sec means: ________________________

Answers to 57: a) 720° b) 3,600°

Answer to 58: 10 revolutions per second

59. Some even shorter abbreviations are shown below:

"rpm" means: revolutions per minute or revs/min

"rps" means: revolutions per second or revs/sec

Therefore: a) 450 rpm means 450 ______________________

b) 60 rps means 60 ______________________

---

60. The following formula can be used to compute rotational velocity:

$$\text{Rotational Velocity} = \frac{\text{Total Revolutions}}{\text{Total Time}}$$

Let's use the formula to find the rotational velocity of an object that rotates 500 times in 10 minutes.

$$\text{Rotational Velocity} = \frac{500}{10} = \frac{50}{1} = 50 \text{ rpm}$$

Note: 50 rpm means $\frac{50 \text{ revs}}{1 \text{ min}}$

Find the rotational velocity of an object that rotates 40 times in 5 seconds.

Rotational Velocity = ______________

Answers to 59: a) revolutions per minute or revs/min b) revolutions per second or revs/sec

---

61. Since 1 minute = 60 seconds, we use the following unity fractions to convert from "rpm" to "rps" and vice versa.

$$\frac{1 \text{ minute}}{60 \text{ seconds}} = 1 \qquad \frac{60 \text{ seconds}}{1 \text{ minute}} = 1$$

A conversion from "rpm" to "rps" is shown below. Notice how we began by converting "300 rpm" to fraction form.

$$300 \text{ rpm} = \frac{300 \text{ revs}}{\cancel{1 \text{ min}}}\left(\frac{\cancel{1 \text{ min}}}{60 \text{ sec}}\right) = \frac{300 \text{ revs}}{60 \text{ sec}} = \frac{5 \text{ revs}}{1 \text{ sec}} = 5 \text{ rps}$$

Using the same steps, complete this conversion.

$$720 \text{ rpm} = \frac{720 \text{ revs}}{1 \text{ min}}\left(\frac{1 \text{ min}}{60 \text{ sec}}\right) = \text{______________}$$

Answer to 60: 8 rps or $\frac{8 \text{ revs}}{1 \text{ sec}}$

---

62. A conversion from "rps" to "rpm" is shown below. Notice again that we began by converting "10 rps" to fraction form.

$$10 \text{ rps} = \frac{10 \text{ revs}}{\cancel{1 \text{ sec}}}\left(\frac{60 \cancel{\text{ sec}}}{1 \text{ min}}\right) = \frac{600 \text{ revs}}{1 \text{ min}} = 600 \text{ rpm}$$

Using the same steps, complete this conversion.

$$3 \text{ rps} = \frac{3 \text{ revs}}{1 \text{ sec}}\left(\frac{60 \text{ sec}}{1 \text{ min}}\right) = \text{______________}$$

Answer to 61: $\frac{720 \text{ revs}}{60 \text{ sec}} = \frac{12 \text{ revs}}{1 \text{ sec}} = 12 \text{ rps}$

Answer to 62: $\frac{180 \text{ revs}}{1 \text{ min}} = 180 \text{ rpm}$

63. Some completed conversions are shown below:

300 rpm = 5 rps   10 rps = 600 rpm
720 rpm = 12 rps   3 rps = 180 rpm

Since 1 minute is much longer than 1 second, there are more revolutions in 1 minute than in 1 second. Therefore:

a) If 200 rps is converted to "rpm", will the number for "rpm" be larger or smaller than 200? ________

b) If 130 rpm is converted to "rps", will the number for "rps" be larger or smaller than 130? ________

---

64. Complete: a) 47 rps = ________ rpm

b) 2,340 rpm = ________ rps

Answers to 63: a) larger b) smaller

---

65. Complete: a) 977 rpm = ________ rps Round to tenths.

b) 3.47 rps = ________ rpm Round to a whole number.

Answers to 64: a) 2,820 rpm b) 39 rps

Answers to 65: a) 16.3 rps b) 208 rpm

5-9 REVOLUTIONS AND LINEAR VELOCITY

The velocity of a rotating object can be measured by the distance traveled per unit time by a point on the circumference of the object. That measure of velocity is called "linear velocity". We will discuss linear velocity in this section.

---

66. Point P lies on the circumference of a rotating object whose center is point O. In one revolution, point P travels around the circle and ends up where it started. Therefore, the distance traveled by P in one revolution equals the circumference of the circle.

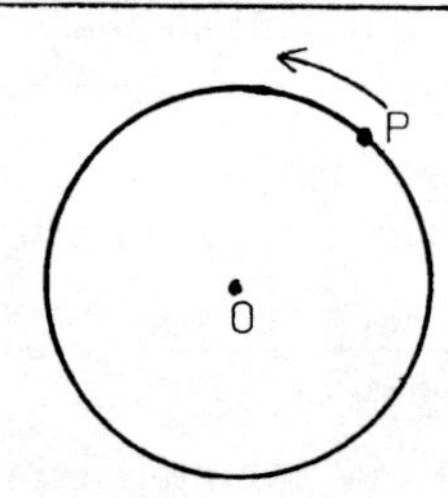

If the circumference is 5.0 cm, how far does point P travel in:

a) 1 revolution? ______ b) 3 revolutions? ______ c) 10 revolutions? ______

67. The distance traveled by a point on the circumference of a rotating object can be computed with this formula:

s = (revs)(C)

where:

s = distance traveled
revs = number of revolutions
C = length of circumference

A rotating wheel has a circumference of 6.28 ft. How far does a point on its circumference travel in:

a) 5 revolutions? ________ ft Round to tenths.

b) 81.7 revolutions? ________ ft Round to a whole number.

a) 5.0 cm
b) 15.0 cm
c) 50.0 cm

---

68. If the diameter of a circle is 12.0 cm:

a) Find the circumference. Round to tenths.

C = ________

b) Find the distance traveled by a point on the circumference in 2 revolutions.

s = ________

a) 31.4 ft
b) 513 ft

---

69. If the radius of a circle is 0.70 m:

a) Find the circumference. Round to hundredths.

C = ________

b) Find the distance traveled by a point on the circumference in 21.5 revolutions.

s = ________

a) C = 37.7 cm
b) s = 75.4 cm

---

70. "Linear velocity" is the distance traveled by a point on the circumference in 1 unit of time. Some possible units are:

feet per minute (or ft/min)
meters per minute (or m/min)

inches per second (or in/sec)
centimeters per second (or cm/sec)

Though the number "1" is not actually shown:

35 ft/min means: 35 ft/1 min

48 cm/sec means: 48 cm/___ sec

a) C = 4.40 m
b) s = 94.6 m

71. The linear velocity of a point can be computed with the following formula:

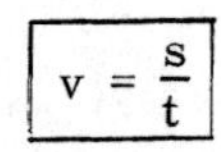

$v = \frac{s}{t}$ where: v = linear velocity
s = distance traveled
t = amount of time

That is, we divide the "total distance traveled" by the "total amount of time". For example:

If a point travels 200 feet in 4 minutes, its linear velocity is:

$$v = \frac{200}{4} \text{ or } \frac{50}{1} \text{ or } 50 \text{ ft/min}$$

If a point travels 456 centimeters in 12 seconds, its linear velocity is:

v = ____________

Answer: 48 cm/1 sec

---

72. The diameter of the circle at the right is 4.75". The circle revolves 20 times in 7.83 seconds. We want to find the linear velocity of point P and the rotational velocity.

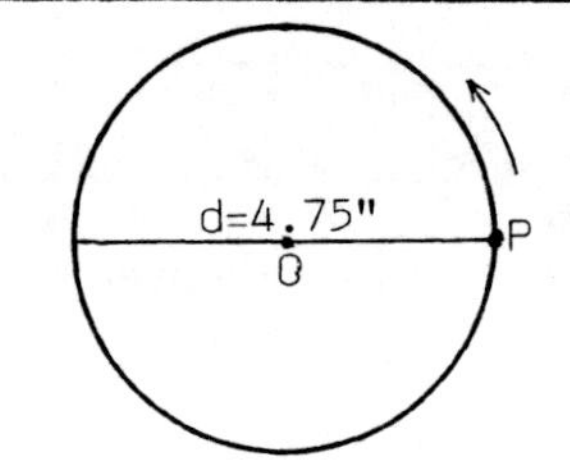

a) Find the circumference. Round to tenths.

C = ________

b) Find the total distance traveled.

s = ________

c) Find the linear velocity. Round to tenths.

v = ________

d) Find the rotational velocity in "rps". Round to hundredths.

Rotational velocity = ________

Answer: $\frac{456}{12}$ or $\frac{38}{1}$ or 38 cm/sec

---

73. If a circular object rotates at a speed of 100 rpm and its circumference is 2.00 ft:

1. We can find the distance traveled by a point on its circumference by using the formula s = (rpm)(C). That is:

$$s = (100 \text{ rpm})(2.00 \text{ ft}) = 200 \text{ ft}$$

2. And since 100 rpm means "100 revolutions in 1 minute", the linear velocity of the point has the same numerical value as "s". That is:

$$v = \frac{200 \text{ ft}}{1 \text{ min}} = 200 \text{ ft/min}$$

Answers:
a) 14.9"
b) 298"
c) 38.1 in/sec, from: $\frac{298}{7.83}$
d) 2.55 rps, from: $\frac{20}{7.83}$

Continued on following page.

73. Continued

Therefore, we can compute the linear velocity of a point by multiplying the rotational velocity (in "rpm" or "rps") by the circumference. That is:

$$\boxed{v = (\text{rpm})(C)} \quad \text{or} \quad \boxed{v = (\text{rps})(C)}$$

Using the formula, we can find the linear velocity of a point on circles with the following rotational velocities and circumferences.

a) 200 rpm with a circumference of 7.5 cm.

$v = (\text{rpm})(C) = (200 \text{ rpm})(7.5 \text{ cm}) =$ _________ cm/min

b) 4 rps with a circumference of 0.82 m.

$v = (\text{rps})(C) = (4 \text{ rps})(0.82 \text{ m}) =$ _________ m/sec

---

74. A wheel is rotating at a velocity of 175 rpm. Its radius is 0.30 m. We want to find the linear velocity of a point on the circumference.

a) Find the circumference. Round to hundredths.

C = _________

b) Find the linear velocity.

v = _________

Answers (73): a) 1,500 cm/min b) 3.28 m/sec

---

75. When computing linear velocity, we can get a very large numerical value for "inches per minute" or "centimeters per minute". To make the numerical value smaller, we can convert "inches" to "feet" and "centimeters" to "meters". Since 12 in = 1 ft and 100 cm = 1 m , we can set up the following unity fractions for the conversions.

$$\frac{1 \text{ ft}}{12 \text{ in}} = 1 \qquad \frac{1 \text{ m}}{100 \text{ cm}} = 1$$

Two examples of conversions are shown below.

$$6{,}000 \text{ in/min} = \frac{6000 \cancel{\text{in}}}{1 \text{ min}}\left(\frac{1 \text{ ft}}{12 \cancel{\text{in}}}\right) = \frac{6000}{12}\left(\frac{\text{ft}}{\text{min}}\right) = 500 \text{ ft/min}$$

$$47{,}800 \text{ cm/min} = \frac{47{,}800 \cancel{\text{cm}}}{1 \text{ min}}\left(\frac{1 \text{ m}}{100 \cancel{\text{cm}}}\right) = \frac{47{,}800}{100}\left(\frac{\text{m}}{\text{min}}\right) = 478 \text{ m/min}$$

Following the examples, complete these conversions.

a) 3,600 in/min = _________ ft/min

b) 57,300 cm/min = _________ m/min

Answers (74): a) 1.88 m b) 329 m/min

76. A grinding wheel with a 8.0" diameter has a turning velocity of 740 rpm. We want to find the linear velocity of a point on its rim.

a) Find "v" in "inches per minute". Round to a whole number.

v = ________ in/min

b) Convert "v" to "feet per minute".

v = ________ ft/min

a) 300 ft/min

b) 573 m/min

---

77. A wheel with a 36 cm radius has a rotational velocity of 225 rpm. We want to find the linear velocity of a point on its circumference.

a) Find "v" in "centimeters per minute". Round to a whole number.

v = ________ cm/min

b) Convert "v" to "meters per minute". Round to a whole number.

v = ________ m/min

a) 18,598 in/min, from: $(8.0)(\pi)(740)$

b) 1,550 ft/min

---

a) 50,894 cm/min b) 509 m/min

5-10 ANGULAR VELOCITY

When rotational velocity is measured in "radians" instead of revolutions, it is called "angular velocity". We will discuss "angular velocity" in this section.

---

78. Instead of measuring rotational velocity in "revolutions per unit time", we can measure it in "radians per unit time". When radians are used, it is called "angular velocity" because radians are a measure of the size of the angle rotated. That is:

$$\text{Angular velocity} = \frac{\text{Number of Radians Rotated}}{\text{1 Unit of Time}}$$

Angular velocity can be computed with the following formula.

$$\omega = \frac{\theta}{t}$$

where: ω = angular velocity
θ = angle rotated (in radians)
t = amount of time

Note: "ω" is the Greek letter "omega" pronounced "oh-may-ga".

Continued on following page.

78. Continued

Using the formula, we can compute the angular velocity for these.

If an object rotates 18 radians in 3 seconds:

$$\omega = \frac{\theta}{t} = \frac{18}{3} \text{ or } \frac{6}{1} \text{ or } 6 \text{ rad/sec}$$

If an object rotates 375 radians in 5 minutes:

$$\omega = \frac{\theta}{t} = \underline{\qquad\qquad}$$

Answer: $\frac{375}{5}$ or $\frac{75}{1}$ or 75 rad/min

79. Angle θ at the right contains 1 radian because it is <u>a central angle whose arc-length equals the radius of the circle</u>.

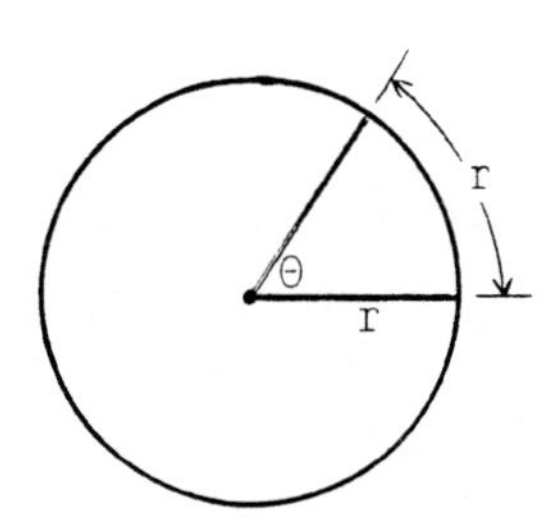

We can use the following formula to compute the number of radians in a central angle:

$$\boxed{\theta = \frac{s}{r}}$$

where: θ = angle size (in radians)
s = arc-length
r = radius

Using the formula, we can compute the number of radians in these.

A central angle whose arc-length is 12 in. if the radius is 4 in.

$$\theta = \frac{12}{4} \text{ or } 3 \text{ radians}$$

A central angle whose arc-length is 51 cm if the radius is 6 cm.

$$\theta = \frac{51}{6} \text{ or } \underline{\qquad\qquad} \text{ radians}$$

Answer: 8.5 radians

80. The arc-length of a complete circle is the circumference of the circle. In terms of the radius "r", $C = 2\pi r$. Therefore, there are 2π radians in a complete circle since:

$$\theta = \frac{C}{r} = \frac{2\pi \not{r}}{\not{r}} = 2\pi \text{ radians}$$

Convert 2π radians to an ordinary number rounded to hundredths.

2π radians = ________ radians

Answer: 6.28 radians

81. Based on the fact that 1 revolution = 2π radians, we can set up the following unity fractions.

$$\frac{1 \text{ rev}}{2\pi \text{ rad}} = 1 \qquad \frac{2\pi \text{ rad}}{1 \text{ rev}} = 1$$

We can use the unity fraction on the left to convert from radians to revolutions.

$$35.6 \text{ rad} = 35.6 \cancel{\text{rad}}\left(\frac{1 \text{ rev}}{2\pi \cancel{\text{rad}}}\right) = \left(\frac{35.6}{2\pi}\right) \text{rev} = 5.67 \text{ rev}$$

We can use the unity fraction on the right to convert from revolutions to radians.

$$18 \text{ rev} = 18 \cancel{\text{rev}}\left(\frac{2\pi \text{ rad}}{1 \cancel{\text{rev}}}\right) = 18(2\pi) \text{ rad} = 113 \text{ rad}$$

Using the steps above, complete each conversion. Round to tenths.

a) 426 rad = __________ rev

b) 6.9 rev = __________ rad

---

a) 67.8 rev

b) 43.4 rad

82. We can use the same unity fractions to convert from one measure of rotational velocity to another <u>when the time unit is the same</u>. For example:

$$16 \text{ rad/sec} = \frac{16 \cancel{\text{rad}}}{1 \text{ sec}}\left(\frac{1 \text{ rev}}{2\pi \cancel{\text{rad}}}\right) = \left(\frac{16}{2\pi}\right)\left(\frac{\text{rev}}{\text{sec}}\right) = 2.55 \text{ rps}$$

$$7.5 \text{ rpm} = \frac{7.5 \cancel{\text{rev}}}{1 \text{ min}}\left(\frac{2\pi \text{ rad}}{1 \cancel{\text{rev}}}\right) = (7.5)(2\pi)\left(\frac{\text{rad}}{\text{min}}\right) = 47.1 \text{ rad/min}$$

Using the steps above, complete each conversion. Round to tenths.

a) 225 rad/min = __________ rpm

b) 13.5 rps = __________ rad/sec

---

a) 35.8 rpm

b) 84.8 rad/sec

83. Since 1 minute = 60 seconds, we can set up the following unity fractions:

$$\frac{1 \text{ min}}{60 \text{ sec}} = 1 \qquad \frac{60 \text{ sec}}{1 \text{ min}} = 1$$

We used the fraction on the left to convert from "rad/min" to "rad/sec" below.

$$180 \text{ rad/min} = \frac{180 \text{ rad}}{\cancel{1 \text{ min}}}\left(\frac{\cancel{1 \text{ min}}}{60 \text{ sec}}\right) = \left(\frac{180}{60}\right)\left(\frac{\text{rad}}{\text{sec}}\right) = 3 \text{ rad/sec}$$

We used the fraction on the right to convert from "rad/sec" to "rad/min" below.

$$1.5 \text{ rad/sec} = \frac{1.5 \text{ rad}}{1 \cancel{\text{sec}}}\left(\frac{60 \cancel{\text{sec}}}{1 \text{ min}}\right) = (1.5)(60)\left(\frac{\text{rad}}{\text{min}}\right) = 90 \text{ rad/min}$$

Using the steps above, complete each conversion.

a) 5,326 rad/min = ________ rad/sec Round to tenths.

b) 6.4 rad/sec = ________ rad/min

---

a) 88.8 rad/sec

b) 384 rad/min

84. To convert 25 rpm to "rad/sec", we can use two steps.

1. Converting "rpm" to "rad/min".

$$25 \text{ rpm} = \frac{25 \cancel{\text{rev}}}{1 \text{ min}}\left(\frac{2\pi \text{ rad}}{1 \cancel{\text{rev}}}\right) = 25(2\pi)\left(\frac{\text{rad}}{\text{min}}\right) \quad 157 \text{ rad/min}$$

2. Converting "rad/min" to "rad/sec".

$$157 \text{ rad/min} = \frac{157 \text{ rad}}{1 \cancel{\text{min}}}\left(\frac{1 \cancel{\text{min}}}{60 \text{ sec}}\right) = \left(\frac{157}{60}\right)\left(\frac{\text{rad}}{\text{sec}}\right) = \underline{2.62 \text{ rad/sec}}$$

The same conversion can be made in one step by using two unity fractions at the same time.

$$25 \text{ rpm} = \frac{25 \cancel{\text{rev}}}{1 \cancel{\text{min}}}\left(\frac{2\pi \text{ rad}}{1 \cancel{\text{rev}}}\right)\left(\frac{1 \cancel{\text{min}}}{60 \text{ sec}}\right) = \frac{25(2\pi)}{60}\left(\frac{\text{rad}}{\text{sec}}\right) = \underline{2.62 \text{ rad/sec}}$$

Complete the following conversion in one step by using two unity fractions at the same time.

140 rpm = ________ rad/sec Round to tenths.

14.7 rad/sec

85. To convert 1,250 rad/min to "rps", we can use two steps.

1. Converting "rad/min" to "rpm".

$$1{,}250 \text{ rad/min} = \frac{1250 \text{ rad}}{1 \text{ min}}\left(\frac{1 \text{ rev}}{2\pi \text{ rad}}\right) = \frac{1250}{2\pi}\left(\frac{\text{rev}}{\text{min}}\right) = \underline{199 \text{ rpm}}$$

2. Converting "rpm" to "rps".

$$199 \text{ rpm} = \frac{199 \text{ rev}}{1 \text{ min}}\left(\frac{1 \text{ min}}{60 \text{ sec}}\right) = \frac{199}{60}\left(\frac{\text{rev}}{\text{sec}}\right) = \underline{3.32 \text{ rps}}$$

The same conversion can be made in one step by using two unity fractions at the same time.

$$1250 \text{ rad/min} = \frac{1250 \text{ rad}}{1 \text{ min}}\left(\frac{1 \text{ rev}}{2\pi \text{ rad}}\right)\left(\frac{1 \text{ min}}{60 \text{ sec}}\right) = \frac{1250}{2\pi(60)}\left(\frac{\text{rev}}{\text{sec}}\right) = \underline{3.32 \text{ rps}}$$

Complete the following conversion in one step by using two unity fractions at the same time.

a) 2,749 rad/min = ________ rps Round to hundredths.

Answer: 7.29 rps

86. Complete each conversion in one step.

a) 6.5 rps = ________ rad/min Round to a whole number.

b) 1.4 rad/sec = ________ rpm Round to tenths.

Answer: a) 2,450 rad/min b) 13.4 rpm

5-11 LINEAR VELOCITY AND ANGULAR VELOCITY

In this section, we will review the meaning of linear velocity and show its relationship to angular velocity.

---

87. We used the formula below to compute the number of radians in an angle. There is 1 radian for each radius in the arc-length.

$$\boxed{\theta = \frac{s}{r}}$$

where: θ = angle size (in radians)
s = arc-length
r = radius

Continued on following page.

87. Continued

When a rotation is measured in radians, we know that a point on the circumference travels 1 radius for each radian of rotation. Therefore, we can find the distance traveled by multiplying the angle size by the radius. That is:

$$\boxed{s = \theta r}$$

Find "s", the distance traveled, if:

a) $\theta = 12$ radians and $r = 5$ cm.

$s =$ ________

b) $\theta = 3.75$ radians and $r = 1.70$ ft. Round to hundredths.

$s =$ ________

---

88. The formula for the linear velocity of a point on the circumference of a rotating object was given earlier. It is:

$$\boxed{v = \frac{s}{t}}$$

where: v = linear velocity
s = distance traveled
t = amount of time

For example, if a point travels 100 cm in 4 seconds, the linear velocity is:

$$v = \frac{100 \text{ cm}}{4 \text{ sec}} = 25 \text{ cm/sec}$$

When the angle of rotation is given in radians, we can use $\boxed{s = \theta r}$ to find the "distance traveled". Use that formula for the problem below.

Find the linear velocity of a point on the circumference of a circular object whose radius is 1.5 cm if it rotates through 30 radians in 2.75 seconds.

a) The distance traveled "s" is ________.

b) The amount of time "t" is ________.

c) The linear velocity "v" is ________. Round to tenths.

Answers to 87:
a) 60 cm
b) 6.38 ft

---

89. A circular object whose radius is 2.25 ft rotates through 419 radians in 1.25 minutes. Let's find the linear velocity of a point on its circumference.

a) Find the distance traveled. Round to a whole number.

$s =$ ________

b) Find the linear velocity. Round to a whole number.

$v =$ ________

Answers to 88:
a) 45 cm
b) 2.75 sec
c) 16.4 cm/sec

90. If a circular object whose radius is 35.0 cm has an angular velocity of 12.4 radians per second:

1. We can find the distance traveled by multiplying the radius by the number of radians.

$$s = \theta r = (12.4)(35.0) = 434 \text{ cm}$$

2. And since 12.4 radians per second means 12.4 radians in 1 second, the linear velocity has the same numerical value as the distance traveled.

$$v = \frac{s}{t} = \frac{434 \text{ cm}}{1 \text{ sec}} = 434 \text{ cm/sec}$$

Therefore, we can compute the linear velocity of a point by multiplying the angular velocity by the radius. That is:

$$\boxed{v = \omega r}$$

where: v = linear velocity
ω = angular velocity
r = radius

Let's use the formula to find the linear velocity when the angular velocity is 18.7 rad/min and the radius is 2.52 ft.

$v = \omega r = (18.7)(2.52) =$ ________ ft/min Round to tenths.

a) 943 ft

b) 754 ft/min

---

91. When using the formula $v = \omega r$, check the units that are given. "r" can be given in "meters", "centimeters", "feet", or "inches". "ω" can be given in "radians per minute" or "radians per second". For example:

If "r" is given in "meters" and "w" is given in "radians per minute", "v" is reported in "m/min".

Use the correct units for these.

a) Find v if $r = 6.45$ in and $\omega = 400$ rad/min.

$v =$ ________

b) Find v if $r = 35.0$ cm and $\omega = 27.4$ rad/sec.

$v =$ ________

47.1 ft/min

---

92. In the formula $v = \omega r$, "r" is the radius of the circle. If the diameter of the circle is given, be sure to substitute the radius in the formula.

The diameter of a circle is 1.2 meters. It rotates at an angular velocity of 540 rad/min. Find the linear velocity of a point on the circle.

$v =$ ________

a) 2,580 in/min

b) 959 cm/sec

---

324 m/min, from:

(0.6)(540)

93. The radius of a circular object is 5.75 in. and its angular velocity is 24.7 rad/sec.

a) Find its linear velocity in "in/sec". Round to a whole number.

v = _________

b) Convert the linear velocity to "ft/min".

v = _________

a) 142 in/sec b) 710 ft/min

SELF-TEST 17 (pages 236-249)

1. A shaft rotates 1,200 times in 8 seconds. Find its rotational velocity in revolutions per second.

_________ rps

Do these conversions.

2. 4,260 rpm = _________ rps

3. 240 rps = _________ rpm

4. A 7.0"-diameter grinding wheel rotates at 240 rpm. Find the linear velocity of a point on its circumference in feet per minute. Round to a whole number.

_________ ft/min

Do these conversions.

5. 9,000 cm/min = _________ m/sec

6. 42 in/sec = _________ ft/min

7. Find the number of radians in a central angle whose arc-length is 81 cm if the radius is 18 cm.

_________ radians

A wheel rotates 60 radians in 2.5 seconds.

8. Find its angular velocity in radians per second.

_________ rad/sec

9. Find its rotational velocity in revolutions per second. Round to hundredths.

_________ rps

Do these conversions. Round each answer to a whole number.

10. 25 rad/sec = _________ rpm

11. 1.85 rps = _________ rad/min

A circular object whose radius is 16.8 cm rotates 125 radians in 5.25 seconds.

12. Find its angular velocity in radians per second. Round to tenths.

_________ rad/sec

13. Find the linear velocity of a point on its circumference in centimeters per second. Round to a whole number.

_________ cm/sec

14. Express the linear velocity in meters per minute.

_________ m/min

ANSWERS:

1. 150 rps
2. 71 rps
3. 14,400 rpm
4. 440 ft/min
5. 1.5 m/sec
6. 210 ft/min
7. 4.5 radians
8. 24 rad/sec
9. 3.82 rps
10. 239 rpm
11. 697 rad/min
12. 23.8 rad/sec
13. 400 cm/sec
14. 240 m/min

5-12 DECIMAL SUBDIVISIONS OF A DEGREE

Degrees can be subdivided into "decimal" subdivisions like "tenths", "hundredths", and so on. We will discuss subdivisions of that type in this section.

---

94. When measuring the size of angles, we can report the size precise to tenths, hundredths, thousandths, and so on. Some examples are shown.

$27.5°$ $139.47°$ $-88.4066°$

We can use a calculator to find the sine, cosine and tangent of angles with "decimal" subdivisions.

Use a calculator for these. Round to four decimal places.

a) $\sin 27.5° =$ ________ c) $\tan 11.607° =$ ________

b) $\cos 139.47° =$ ________ d) $\sin(-88.4066°) =$ ________

---

95. When finding the size of an angle with a given sine, cosine, or tangent, we can also report the size precise to tenths, hundredths, thousandths, and so on. Use a calculator for these.

a) If $\sin \theta = 0.5508$, $\theta =$ ________ Round to tenths.

b) If $\cos \theta = 0.2019$, $\theta =$ ________ Round to hundredths.

c) If $\tan \theta = 1.9872$, $\theta =$ ________ Round to thousandths.

Answers to 94:

a) 0.4617
b) -0.7601
c) 0.2054
d) -0.9996

---

96. Use a calculator for these. Round to four decimal places.

a) If $\sin \theta = 0.0926$, $\theta =$ ________

b) If $\cos \theta = 0.1033$, $\theta =$ ________

c) If $\tan \theta = 0.0069$, $\theta =$ ________

Answers to 95:

a) 33.4°
b) 78.35°
c) 63.288°

Answers to 96:

a) 5.3132° b) 84.0708° c) 0.3953°

5-13 MINUTE-SECOND SUBDIVISIONS OF A DEGREE

Degrees can be subdivided into "minutes" and "seconds". We will discuss subdivisions of that type in this section.

---

97. A degree can be subdivided into "minutes". As a subdivision of a degree, a "minute" is a unit of angle measurement, not a unit of time measurement.

By definition: 1 minute is $\frac{1}{60}$ of 1 degree

or

1 degree = 60 minutes

The symbol ' is used as an abbreviation for "minutes". For example:

17' means: 17 minutes

23°45' means: 23 degrees 45 minutes

Using abbreviation symbols, write each of these.

a) 5 minutes = ________

b) 42 degrees 58 minutes = ________

---

98. Since 1° = 60', we can multiply by 60 to convert degrees to minutes. That is:

$$3° = (3 \times 60)' = 180'$$

Convert each of these to minutes.

a) 2° = ______ b) 5° = ______ c) 10° = ______

Answers to 97: a) 5' b) 42°58'

---

99. When the number of minutes is a multiple of 60 (like 60', 120', 180', 240', and so on), we can convert the minutes to degrees by dividing by 60. For example:

$$180' = \left(\frac{180}{60}\right)^{\circ} = 3°$$

Convert each of these to degrees.

a) 60' = ______ b) 120' = ______ c) 240' = ______

Answers to 98: a) 120' b) 300' c) 600'

Answers to 99: a) 1° b) 2° c) 4°

100. When the number of minutes is greater than 60 but not a multiple of 60, we can convert the minutes to an expression containing degrees and minutes. To do so, we separate the number of minutes into the largest possible multiple of 60 plus the minutes left over. For example:

$$93' = 60' + 33' = 1° + 33' = 1°33'$$

$$145' = 120' + 25' = 2° + 25' = 2°25'$$

Convert each of these to degrees and minutes.

a) 70' = ________ c) 137' = ________

b) 118' = ________ d) 212' = ________

---

101. To add angles stated in degrees and minutes, we add the degrees and minutes separately. For example:

Answers: a) 1°10' b) 1°58' c) 2°17' d) 3°32'

$$\begin{array}{r} 23°38' \\ +\ 40°\ 6' \\ \hline 63°44' \end{array}$$

When the sum of the minutes is greater than 60', it is converted to a degree-minute expression and then simplified. For example:

$$\begin{array}{r} 12°47' \\ +\ 25°28' \\ \hline 37°75' \end{array} = 37° + 75' = 37° + 1°15' = 38°15'$$

Following the examples, do these additions.

a) $\begin{array}{r} 81°44' \\ +\ 12°63' \\ \hline \end{array}$ b) $\begin{array}{r} 104°51' \\ +\ 38°47' \\ \hline \end{array}$

---

102. When the sum of the minutes is exactly 60', it is converted to 1°. However, 0 is written in the sum to show the precision of the sum. An example is shown. Complete the other addition.

Answers: a) 94°47' b) 143°38'

$$\begin{array}{r} 6°20' \\ +\ 10°40' \\ \hline 16°60' \end{array} = 17°0' \qquad \begin{array}{r} 44°19' \\ +\ 6°41' \\ \hline \end{array}$$

---

103. To subtract angles stated in degrees and minutes, we subtract the degrees and minutes separately. An example is shown. Complete the other subtraction.

Answer: 51°0'

$$\begin{array}{r} 48°39' \\ -\ 22°16' \\ \hline 26°23' \end{array} \qquad \begin{array}{r} 124°45' \\ -\ 68°27' \\ \hline \end{array}$$

Answer: 56°18'

104. The subtraction at the left below cannot be done as it stands because 31' is larger than 15'. To do the subtraction, we borrow 1° from 19°, convert the 1° to 60', and add the 60' to 15' as we have done at the right below.

$$\begin{array}{r} 19°15' \\ -\ 7°31' \\ \hline \end{array} \qquad \begin{aligned} 19°15' &= 18° + 1°15' \\ &= 18° + 75' \\ &= 18°75' \end{aligned}$$

We substituted 18°75' for 19°15' and completed the subtraction below.

$$\begin{array}{r} 19°15' \\ -\ 7°31' \\ \hline \end{array} \longrightarrow \begin{array}{r} 18°75' \\ -\ 7°31' \\ \hline 11°44' \end{array}$$

Complete each subtraction.

a) $\begin{array}{r} 118°45' \\ -\ 71°52' \\ \hline \end{array}$ b) $\begin{array}{r} 9°\ 7' \\ -\ 1°58' \\ \hline \end{array}$

Answers: a) 46°53' b) 7° 9'

---

105. To subtract 112°32' from 180°, we borrow 1° or 60' from 180° to get 179°60'. An example is shown. Complete the other subtraction.

$$\begin{array}{r} 180° \\ -\ 112°32' \\ \hline \end{array} \longrightarrow \begin{array}{r} 179°60' \\ -\ 112°32' \\ \hline 67°28' \end{array} \qquad \begin{array}{r} 180° \\ -\ 75°40' \\ \hline \end{array}$$

Answer: 104°20'

---

106. We want to find the third angle of a triangle if the other two angles are 31°45' and 72°51'.

a) Find the sum of the two angles. b) Find the third angle.

________ ________

Answers: a) 104°36' b) 75°24'

---

107. We get a more precise measure of angles by subdividing minutes into "seconds".

By definition: **1 second is $\frac{1}{60}$ of 1 minute**

or

1 minute = 60 seconds

The symbol " is used as an abbreviation for "seconds". For example:

5" means: 5 seconds

Using abbreviations, we can simplify these.

45 degrees 17 minutes 10 seconds = 45°17'10"

12 degrees 38 minutes 55 seconds = ________

108. To convert from minutes to seconds, we multiply by 60. That is:

$3' = (3 \times 60)'' = 180''$ a) $1' =$ ________ b) $2' =$ ________

Answer: 12°38'55"

---

109. When the number of seconds is a multiple of 60, we can convert the seconds to minutes by dividing by 60. For example:

$$60'' = \left(\frac{60}{60}\right)' = 1' \qquad 120'' = \left(\frac{120}{60}\right)' = 2'$$

When the number of seconds is larger than 60 but not a multiple of 60, we can convert the seconds to an expression containing minutes and seconds. To do so, we separate the number of seconds into the largest possible multiple of 60 plus the seconds left over. For example:

$$85'' = 60'' + 25'' = 1' + 25'' = 1'25''$$

$$129'' = 120'' + 9'' = 2' + 9'' = 2'9''$$

Convert each of these to minutes and seconds.

a) $97'' =$ ________ b) $123'' =$ ________

Answer: a) 60" b) 120"

---

110. To add angles containing seconds, we add the degrees, minutes, and seconds separately. An example is shown. Do the other addition.

$$\begin{array}{r} 12°13'41'' \\ +\ 37°44'17'' \\ \hline 49°57'58'' \end{array} \qquad \begin{array}{r} 7°15'\ 9'' \\ +\ 1°12'30'' \\ \hline \end{array}$$

Answer: a) 1'37" b) 2'3"

---

111. When the sum of the minutes or seconds or both is 60 or more, we change the form of the sum. Three examples are discussed.

Answer: 8°27'39"

When the sum of the minutes is 60 or more, we convert to a degree-minute expression and simplify.

$$\begin{array}{r} 13°47'13'' \\ +\ 24°18'26'' \\ \hline 37°65'39'' \end{array} = 38°5'39'' \quad \text{(Since } 65' = 1°5'\text{)}$$

When the sum of the seconds is 60 or more, we convert to a minute-second expression and simplify.

$$\begin{array}{r} 8°21'36'' \\ +\ 10°15'56'' \\ \hline 18°36'92'' \end{array} = 18°37'32'' \quad \text{(Since } 92'' = 1'32''\text{)}$$

When the sum of both the minutes and seconds is 60 or more, we make a double conversion <u>beginning with the seconds</u>.

$$\begin{array}{r} 41°49'59'' \\ +\ 86°35'16'' \\ \hline 127°84'75'' \end{array} \begin{array}{l} = 127°85'15'' \quad \text{(Since } 75'' = 1'15''\text{)} \\ = 128°25'15'' \quad \text{(Since } 85' = 1°25'\text{)} \end{array}$$

Complete each addition.

a) $\begin{array}{r} 7°45'27'' \\ +\ 12°16'55'' \\ \hline \end{array}$ b) $\begin{array}{r} 81°59'42'' \\ +\ 37°38'25'' \\ \hline \end{array}$

112. When we get exactly 60' in a sum precise to seconds, it is converted to 1°. However, 0' is written in the sum so that the seconds are not mistaken for minutes. That is:

27°60'15" is written 28°0'15"

When we get exactly 60" in a sum, it is converted to 1'. However, we write 0" to show the precision of the sum. That is:

12°46'60" is written 12°47'0"

Write each of these in the proper form.

a) 88°60'47" = ____________

b) 120°12'60" = ____________

c) 14°59'98" = ____________

a) 20°2'22"

b) 119°38'7"

---

113. To subtract angles containing seconds, we subtract the degrees, minutes, and seconds separately. Some examples of "borrowing" are discussed below.

We have to borrow 1°(or 60') because 37' is larger than 22'.

```
 17°22'48"  ──→   16°82'48"
-  9°37'31"      -  9°37'31"
                    7°45'17"
```

We have to borrow 1' (or 60") because 47" is larger than 19".

```
 84°25'19"  ──→   84°24'79"
-21°16'47"       -21°16'47"
                  63° 8'32"
```

We have to borrow both 1' and 1°. Notice that we get only 71' since 1' of the 12' was borrowed.

```
 18°12'35"  ──→   17°71'95"
-10°48'51"       -10°48'51"
                   7°23'44"
```

Complete each subtraction.

```
a)  24°17'49"     b)  9°54'28"     c)  98°12'41"
   -  8°42'15"       - 6°27'41"       - 75°37'58"
```

a) 89°0'47"

b) 120°13'0"

c) 15°0'38"

---

114. The third angle in a triangle is found by subtracting the sum of the other two angles from 180°. If the sum contains seconds, we get a subtraction like the one below. Note that 180° was converted to 179°59'60". Complete the other subtraction.

```
   180°          179°59'60"          180°
 - 115°18'35"  - 115°18'35"      -  89°47'50"
                  64°41'25"
```

a) 15°35'34"

b) 3°26'47"

c) 22°34'43"

115. We want to find the third angle in a triangle when the other two angles are 75°12'42" and 50°55'29".

a) Find the sum of the two angles. b) Find the third angle.

________ ________

90°12'10"

a) 126° 8'11" b) 53°51'49"

5-14 TRIG RATIOS OF MINUTE-SECOND SUBDIVISIONS

To find the trig ratios of minute-second subdivisions, we must convert to decimal form first. We will discuss the procedure in this section.

116. There are 60 minutes in 1°. Therefore, to convert from minutes to a decimal part of 1°, we divide by 60. Two examples are shown. For the second conversion, we rounded to hundredths.

$$30' = \left(\frac{30}{60}\right)^{\circ} = .5^{\circ}$$

$$47' = \left(\frac{47}{60}\right)^{\circ} = .78^{\circ}$$

Convert each of these to a decimal part of 1°. Round to hundredths when necessary.

a) 12' = ________ b) 1' = ________ c) 59' = ________

117. To convert degree-minute expressions to decimal form, we convert the minutes to a decimal part of 1°. For example:

37°15' = 37° + 15' = 37° + .25° = 37.25°

89°49' = 89° + 49' = 89° + .82° = 89.82°

Convert each degree-minute expression to decimal form. Round to hundredths when necessary.

a) 17°24' = ________ b) 105°38' = ________

a) .2°

b) .02°

c) .98°

118. To find sin 47°11', we convert 47°11' to decimal form first. The calculator steps are shown. Notice that we pressed [=] before pressing [sin].

Enter	Press	Display
11	[÷]	11
60	[+]	0.1833333
47	[=] [sin]	0.7335322

Rounding to four decimal places, sin 47°11' = ________

a) 17.4°

b) 105.63°

119. Following the steps in the last frame, complete these. Round to four decimal places.

a) sin 85°24' = ________   c) tan 105°6' = ________

b) cos 17°53' = ________   d) sin 140°17' = ________

Answer: 0.7335

---

120. To convert a decimal part of 1° to minutes, we multiply by 60 and then round to the nearest whole-number minute when necessary. For example:

$$.4° = (.4)(60') = 24'$$

$$.69° = (.69)(60') = 41.4' \text{ or } 41'$$

Convert each of these to minutes.

a) .7° = ________   b) .59402° = ________

Answers: a) 0.9968 b) 0.9517 c) -3.7062 d) 0.6390

---

121. To convert 25.87° to degree-minute form, we convert .87° to minutes. We get:

$$25.87° = 25° + .87° = 25° + 52' = 25°52'$$

Convert each of these to degree-minute form.

a) 5.8° = ________   b) 102.4629° = ________

Answers: a) 42' b) 36'

---

122. To find the degree-minute form of the angle whose sine is 0.4825, we find the decimal form first and then convert to degree-minute form. The calculator steps are shown.

Note: We subtracted the whole-number part "28" from 28.848809 and then multiplied the decimal part ".848809" by 60. The "28" should be recorded before doing the subtraction.

Enter	Press	Display
.4825	[INV] [sin] [-]	28.848809
28	[=] [x]	.848809
60	[=]	50.928521

Therefore, if sin θ = 0.4825 , θ = ________

Answers: a) 5°48' b) 102°28'

---

123. Following the steps in the last frame, report θ in degree-minute form. Round minutes to a whole number.

a) If sin θ = 0.0726 , θ = ________

b) If cos θ = 0.6098 , θ = ________

c) If tan θ = 1.2066 , θ = ________

d) If sin θ = 0.9456 , θ = ________

Answer: 28°51'

124. There are 60 seconds in 1'. Therefore, to convert from seconds to a decimal part of 1', we divide by 60. Two examples are shown. For the second conversion, we rounded to hundredths.

$$45'' = \left(\frac{45}{60}\right)' = .75'$$

$$8'' = \left(\frac{8}{60}\right)' = .13'$$

To convert minute-second expressions to a minute expression, we convert the seconds to a decimal part of 1'. For example:

12'45" = 12' + 45" = 12' + .75' = 12.75'

44'8" = 44' + 8" = 44' + .13' = 44.13'

Convert each minute-second expression to a minute expression. Round to hundredths when necessary.

a) 85'48" = ________ b) 3'53" = ________

Answers: a) 4°10' b) 52°25' c) 50°21' d) 71°1'

---

125. To convert 15°29'17" to decimal form, we divide by 60 to convert 17" to a decimal part of 1', add this result to 29', and then divide the minutes by 60 to get a decimal part of 1°. The calculator steps are shown below.

Enter	Press	Display
17	÷	17
60	+	0.2833333
29	= ÷	29.283333
60	+	0.4880556
15	=	15.488056

Therefore, rounding to hundredths, 15°29'17" = 15.49°.

Using the same steps, complete these conversions. Round to hundredths.

a) 3°47'15" = ________ b) 88°4'59" = ________

Answers: a) 85.8' b) 3.88'

Answers: a) 3.79° b) 88.08°

126. To find sin 65°14'43", we convert 65°14'43" to decimal form first. The calculator steps are shown below. They are the same as those shown in the last frame with [sin] pressed at the end.

Enter	Press	Display
43	[÷]	43
60	[+]	0.7166667
14	[=] [÷]	14.716667
60	[+]	0.2452778
65	[=] [sin]	0.9081087

Rounding to four decimal places, sin 65°14'43" = 0.9081 .

Using the same steps, complete these. Round to four decimal places.

a) sin 5°15'25" = ________ c) tan 145°12'13" = ________

b) cos 91°47'3" = ________ d) sin 100°57'1" = ________

---

a) 0.0916

b) -0.0311

c) -0.6949

d) 0.9818

127. To convert 47.538° to degree-minute-second form, we use these steps:

1. Multiply .538 by 60 to get the number of minutes. Rounding to a whole number, we get:

   47.538° = 47°32.28'

2. Multiply .28 by 60 to get the number of seconds. Rounding to a whole number, we get:

   47.538° = 47°32'17"

Using the same steps, convert each of these to degree-minute-second form.

a) 5.911° = ________ b) 84.6075° = ________

---

a) 5°54'40"

b) 84°36'27"

128. To find the degree-minute-second form of the angle whose sine is 0.5194, we find the decimal form first and then convert to degree-minute-second form. The calculator steps are shown.

1. We subtracted the whole-number part "31" from 31.292013° and then multiplied the decimal part ".2920134" by 60. The 31° should be recorded before doing the subtraction.

2. We subtracted the whole-number part "17" from 17.520801' and then multiplied the decimal part ".520801" by 60. The 17' should be recorded before doing the subtraction.

Continued on following page.

128. Continued

Enter	Press	Display
.5194	[INV] [sin] [-]	31.292013
31	[=] [x]	0.2920134
60	[=] [-]	17.520801
17	[=] [x]	0.520801
60	[=]	31.24806

Rounding the seconds to the nearest whole number, we get:

If $\sin\theta = 0.5194$, $\theta =$ ____________

Answer: 31°17'31"

129. Following the steps in the last frame, report θ in degree-minute-second form. Round seconds to a whole number.

a) If $\sin\theta = 0.8012$, $\theta =$ ____________

b) If $\cos\theta = 0.2766$, $\theta =$ ____________

c) If $\tan\theta = 0.0908$, $\theta =$ ____________

d) If $\cos\theta = 0.8451$, $\theta =$ ____________

Answers:

a) 53°14'41"

b) 73°56'33"

c) 5°11'18"

d) 32°19'2"

SELF-TEST 18 (pages 250-261)

Find each acute angle in decimal form. Round to four decimal places.

1. If $\sin \theta = 0.8924$, $\theta =$ ________

2. If $\cos \theta = 0.7029$, $\theta =$ ________

3. If $\tan \theta = 2.9154$, $\theta =$ ________

Do these additions.

4. $78°18' + 31°42'$

5. $17°29'55'' + 66°53'36''$

6. $63°18'47'' + 75°41'53''$

Do these subtractions.

7. $102°15' - 46°37'$

8. $51°23'12'' - 37°55'48''$

9. $87°47'32'' - 19°46'35''$

10. Find the third angle in a triangle when the other two angles are 71°26'45" and 59°38'54". ________

11. Convert 29°47'35" to decimal form. Round to four decimal places. ________

12. Convert 83.7194° to degree-minute-second form. ________

Round each numerical value to four decimal places.

13. $\sin 8°13'53'' =$ ________

14. $\tan 74°45'39'' =$ ________

Find each acute angle in degree-minute-second form.

15. If $\cos \theta = 0.29517$, $\theta =$ ________

16. If $\sin \theta = 0.42839$, $\theta =$ ________

ANSWERS:

1. 63.1764°
2. 45.3399°
3. 71.0677°
4. 110°0'
5. 84°23'31"
6. 139°0'40"
7. 55°38'
8. 13°27'24"
9. 68°0'57"
10. 48°54'21"
11. 29.7931°
12. 83°43'10"
13. 0.1432
14. 3.6707
15. 72°49'56"
16. 25°21'56"

SUPPLEMENTARY PROBLEMS - CHAPTER 5

Assignment 15

1. A circle whose diameter is 35.7 cm is divided into nine equal central angles.
   a) Find the size of a central angle.
   b) Find the length of an arc. Round to tenths.
2. The radius of a circle is 3.38". Find the length of the arc of a 135° central angle. Round to hundredths.
3. A flat metal strip is bent to form a circular arc whose radius is 17.2 cm and whose central angle is 78°. Find the length of the strip. Round to tenths.
4. A chord whose length is 8.72" lies in a circle whose radius is 5.63". Find the following:
   a) The perpendicular distance from the center of the circle to the chord. Round to hundredths.
   b) The "depth" of the chord. Round to hundredths.
   c) The central angle of the chord. Round to a whole number.
5. In a circle of radius 140 cm a chord has a central angle of 100°. Rounding each to a whole number, find the following:
   a) The perpendicular distance from the center of the circle to the chord.
   b) The "depth" of the chord. c) The length of the chord.
6. Eight equally-spaced holes are on the circumference of a circle whose radius is 34.8". Find the center-to-center distance between two successive holes. Round to tenths.
7. Five equally-spaced holes lie on the circumference of a circle whose radius is 43.7 cm. Find the center-to-center distance between two alternate holes. Round to tenths.

Assignment 16

1. In a circle of radius 38.5 cm, find the area of the sector whose central angle is 65°. Round to a whole number.
2. In a circle of radius 21.7", radii OA and OB form a 112° central angle. Their chord is AB. Rounding each answer to a whole number, find the area of: a) The sector. b) Triangle AOB. c) Segment AB.
3. Chord GH lies in a circle of radius 5.38 cm whose center is at point O. At point G on the circle, half-tangent PG forms a 48° angle with chord GH. Sketch the diagram. Find: a) Central angle GOH. b) The length of chord GH. Round to hundredths.
4. In a circle, the length of chord AB is 8.14". At point A on the circle, half-tangent FA forms a 60° angle with chord AB. Sketch the diagram. Find the radius of the circle. Round to hundredths.
5. The center of a circle is at point O. Its radius is 145 mm. From point P outside the circle, half-tangents PS and PT are drawn. External angle SPT is 36°. Sketch the diagram. Rounding each answer to a whole number, find: a) Central angle SOT. b) Half-tangent PS. c) Half-tangent PT. d) Distance OP.
6. The center of a circle is at point O. From point P outside the circle, half-tangents PF and PH are drawn. PF is 36.7" and OP is 41.2". Sketch the diagram. Find: a) Radius OF. Round to tenths. b) Angle FPH. Round to a whole number. c) Central angle FOH. Round to a whole number. d) The length of arc FH. Round to tenths.
7. A straight piece of wire AB is bent into a circular arc with a 43.5 cm radius. From point P outside the arc, half-tangents PA and PB form a 57° angle. Sketch the diagram. Find the length of arc AB. Round to tenths.

Assignment 17

1. A gear rotates 240 times in 16 seconds. Find its rotational velocity in revolutions per minute.

2. The rotor of a jet engine turns 36,300 times in 2.5 minutes. Find its rotational velocity in revolutions per second.

3. Do these conversions involving rotational velocity.

   a) 5.4 rps = ________ rpm   b) 2,520 rpm = ________ rps

4. A grinding wheel whose diameter is 10.5" has a rotational speed of 210 rpm. Find the linear velocity of a point on its circumference in feet per minute.

5. A phonograph record whose diameter is 30.0 cm rotates at $33\frac{1}{3}$ rpm. Find the linear velocity of a point on its circumference in centimeters per second. Round to tenths.

6. Do these conversions involving linear velocity.

   a) 15.6 m/min = _____ cm/sec   b) 21.4 in/sec = _____ ft/min

7. A wheel rotates 70 radians in 1.75 seconds. a) Find its angular velocity in radians per second. b) Convert the angular velocity to revolutions per second. Round to hundredths.

8. A rotating circular object has a radius of 7.20" and an angular velocity of 12.8 radians per second. Find the linear velocity of a point on its circumference in inches per second. Round to tenths.

9. A shaft whose radius is 11.6 cm rotates 160 radians in 3.2 seconds. a) Find its angular velocity in radians per second. b) Find the linear velocity of a point on its circumference in centimeters per second. c) Find its rotational velocity in revolutions per minute. Round to a whole number.

10. Do these conversions involving angular velocity.

    a) Round to tenths.   b) Round to a whole number.

    5,170 rad/min = _____ rps   1,500 rpm = _____ rad/sec

Assignment 18

1. Find each first-quadrant angle in decimal form. Round to three decimal places.

   a) If cos θ = 0.6152, θ = ? b) If sin θ = 0.8907, θ = ? c) If tan θ = 3.8617 , θ = ?

2. Do these additions.

   a) 92°37' + 34°56'   b) 41°48' + 28°12'   c) 15°51'22" + 68°36'55"   d) 87°45'39" + 44°14'26"

3. Do these subtractions.

   a) 50°22' - 37°28'   b) 90° - 17°29'   c) 75°32'16" - 48°51'16"   d) 180° - 117°10'40"

4. Find the third angle in a triangle when the other two angles are 27°55'35" and 80°19'45".

5. Convert each angle to decimal form. Round to three decimal places.
   a) 51°37'19"   b) 138°13'47"   c) 2°54'22"

6. Convert each angle to degree-minute-second form. Round seconds to a whole number.
   a) 15.1372°   b) 80.9751°   c) 127.6815°

7. Find each numerical value. Round to five decimal places.
   a) tan 71°18'   b) sin 43°28'37"   c) cos 15°52'9"

8. Find each first-quadrant angle in degree-minute-second form. Round seconds to a whole number.
   a) If cos θ = 0.71038, θ = ? b) If tan θ = 4.57269, θ = ? c) If sin θ = 0.16625, θ = ?

Chapter 6 IDENTITIES, INVERSE NOTATION, AND EQUATIONS

In this chapter, we will define the other three trigonometric ratios - the cosecant, secant, and cotangent of an angle. Then we will discuss eight basic identities and show how they can be used to prove other identities. We will also discuss inverse trigonometric notation and the solution of trigonometric equations.

6-1 COSECANT, SECANT, AND COTANGENT OF AN ANGLE

In this section, we will define the other three trigonometric ratios - the cosecant, secant, and cotangent of an angle. We will show the calculator steps for finding the numerical values of those three ratios.

---

1. By definition, two quantities are a pair of reciprocals if their product is +1. That is:

"a" and "$\frac{1}{a}$" are a pair of reciprocals, since $(a)\left(\frac{1}{a}\right) = +1$

"$\frac{b}{c}$" and "$\frac{c}{b}$" are a pair of reciprocals, since $\left(\frac{b}{c}\right)\left(\frac{c}{b}\right) = +1$

Write the reciprocal of each of these:

a) $\frac{3}{5}$ ____ b) $\frac{p}{t}$ ____ c) $\frac{1}{9}$ ____ d) 7 ____

---

a) $\frac{5}{3}$ b) $\frac{t}{p}$ c) 9 d) $\frac{1}{7}$

2. There are six trigonometric ratios for any angle. Using the triangle at the right, we defined below the six ratios for angle θ. The three familiar ratios are sine, cosine, and tangent. The three additional ratios are cosecant, secant, and cotangent.

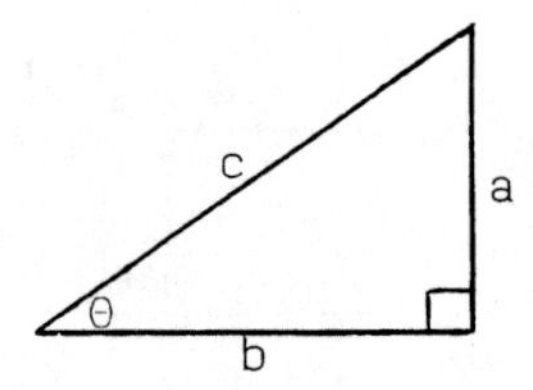

$\text{sine } \theta = \frac{\text{side opposite}}{\text{hypotenuse}} = \frac{a}{c}$	$\text{cosecant } \theta = \frac{\text{hypotenuse}}{\text{side opposite}} = \frac{c}{a}$
$\text{cosine } \theta = \frac{\text{side adjacent}}{\text{hypotenuse}} = \frac{b}{c}$	$\text{secant } \theta = \frac{\text{hypotenuse}}{\text{side adjacent}} = \frac{c}{b}$
$\text{tangent } \theta = \frac{\text{side opposite}}{\text{side adjacent}} = \frac{a}{b}$	$\text{cotangent } \theta = \frac{\text{side adjacent}}{\text{side opposite}} = \frac{b}{a}$

Continued on following page

2. Continued

From the definitions, you can see that "sine θ" and "cosecant θ" are a pair of reciprocals since:

$$\text{sine } \theta = \frac{a}{c}, \quad \text{cosecant } \theta = \frac{c}{a}, \quad \text{and} \quad \left(\frac{a}{c}\right)\left(\frac{c}{a}\right) = +1$$

Using either "secant θ" or "cotangent θ", complete these:

a) The reciprocal of cosine θ is ________________.

b) The reciprocal of tangent θ is ________________.

---

3. The pairs of reciprocals are shown again below. Memorize them.

sine θ	and	cosecant θ
cosine θ	and	secant θ
tangent θ	and	cotangent θ

Write the name of the reciprocal of each ratio.

a) cosine θ ____________ d) secant θ ____________

b) cosecant θ ____________ e) sine θ ____________

c) tangent θ ____________ f) cotangent θ ____________

Answers to 2:
a) secant θ
b) cotangent θ

---

4. Using the terms "side opposite", "side adjacent" and "hypotenuse", define each ratio.

a) cosecant θ =

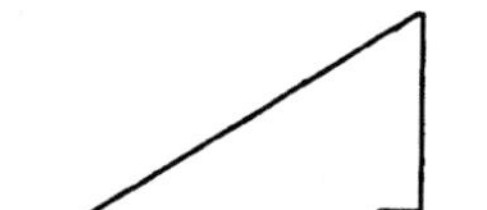

b) secant θ =

c) cotangent θ =

Answers to 3:
a) secant θ
b) sine θ
c) cotangent θ
d) cosine θ
e) cosecant θ
f) tangent θ

Answers to 4:

a) $\dfrac{\text{hypotenuse}}{\text{side opposite}}$

b) $\dfrac{\text{hypotenuse}}{\text{side adjacent}}$

c) $\dfrac{\text{side adjacent}}{\text{side opposite}}$

5. Using the sides of this right triangle, define:

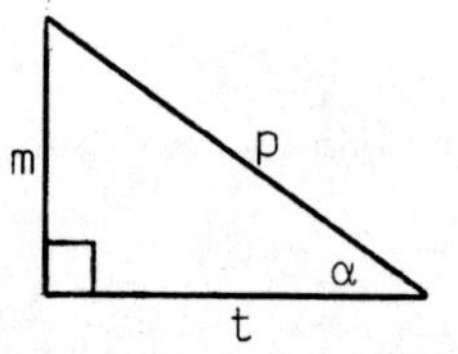

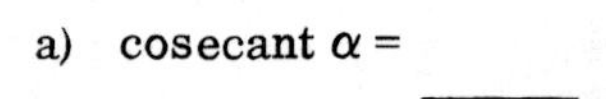

a) cosecant α = ____

b) secant α = ____

c) cotangent α = ____

Answers to frame 5: a) $\frac{p}{m}$ b) $\frac{p}{t}$ c) $\frac{t}{m}$

6. The abbreviations for cosecant θ, secant θ, and cotangent θ are shown below.

	Abbreviation
cosecant θ	csc θ
secant θ	sec θ
cotangent θ	cot θ

In this right triangle:

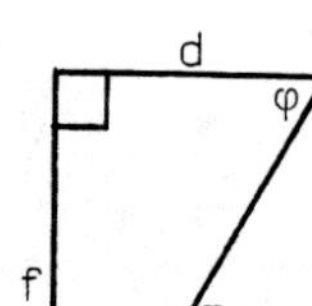
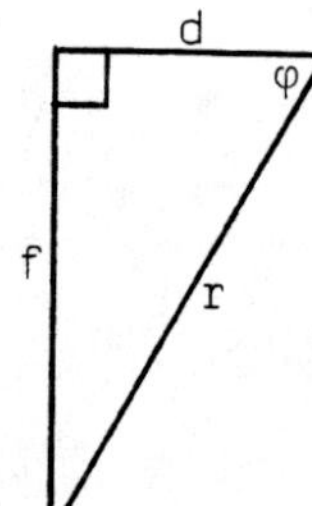

a) csc φ = ____

b) sec φ = ____ c) cot φ = ____

Answers to frame 6: a) $\frac{r}{f}$ b) $\frac{r}{d}$ c) $\frac{d}{f}$

7. In this right triangle:

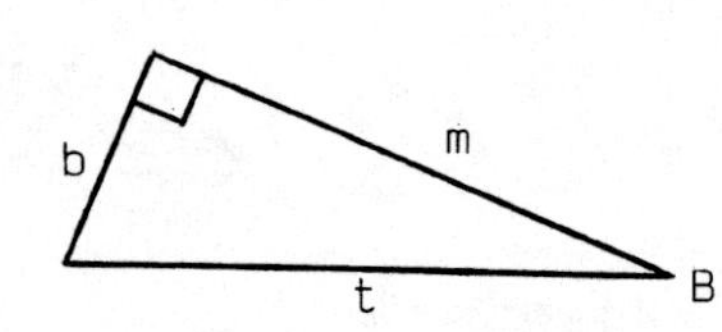

a) cot B = ____

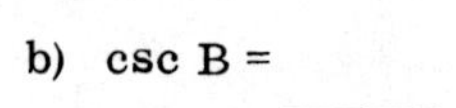

b) csc B = ____

c) sec B = ____

Answers to frame 7: a) $\frac{m}{b}$ b) $\frac{t}{b}$ c) $\frac{t}{m}$

8. Using the unit circle at the right, we defined $\sin\theta$, $\cos\theta$ and $\tan\theta$ below.

$$\sin\theta = \frac{y}{1} \text{ or } y$$

$$\cos\theta = \frac{x}{1} \text{ or } x$$

$$\tan\theta = \frac{y}{x}$$

Using the same circle, complete these definitions.

a) $\csc\theta =$ ______ b) $\sec\theta =$ ______ c) $\cot\theta =$ ______

---

9. To find the numerical values of cosecants of angles on a calculator, we use the reciprocal relationship with sines of angles. For example:

$\csc 50° =$ the reciprocal of $\sin 50°$

$\csc 287° =$ the reciprocal of $\sin 287°$

$\csc(-125°) =$ the reciprocal of $\sin(-125°)$

The calculator steps for the three cosecants above are shown below. Notice that we find the sine for the angle and then press the reciprocal key [1/x].

Enter	Press	Display
50	[sin] [1/x]	1.3054073
287	[sin] [1/x]	-1.0456918
-125	[sin] [1/x]	-1.2207746

Rounding to four decimal places, we get:

$\csc 50° = 1.3054$ $\csc 287° = -1.0457$ $\csc(-125°) =$ ______

a) $\frac{1}{y}$ b) $\frac{1}{x}$ c) $\frac{x}{y}$

---

10. To find the numerical values of secants of angles on a calculator, we use the reciprocal relationship with cosines of angles. For example:

$\sec 67° =$ the reciprocal of $\cos 67°$

$\sec 314° =$ the reciprocal of $\cos 314°$

$\sec(-110°) =$ the reciprocal of $\cos(-110°)$

The calculator steps for the secants above are shown below. Notice that we find the cosine of the angle and then press the reciprocal key [1/x].

Enter	Press	Display
67	[cos] [1/x]	2.5593047
314	[cos] [1/x]	1.4395565
-110	[cos] [1/x]	-2.9238044

Continued on following page.

-1.2208

10. Continued

Rounding to four decimal places, we get:

sec 67° = 2.5593 sec 314° = 1.4396 sec(-110°) = ________

---

11. To find the numerical values of cotangents of angles on a calculator, we use the reciprocal relationship with tangents of angles. For example:

	-2.9238

cot 15° = the reciprocal of tan 15°

cot 339° = the reciprocal of tan 339°

cot(-89°) = the reciprocal of tan(-89°)

The calculator steps for the cotangents above are shown below. Notice that we find the tangent of the angle and then press the reciprocal key [1/x].

Enter	Press	Display
15	[tan] [1/x]	3.7320508
339	[tan] [1/x]	-2.6050891
-89	[tan] [1/x]	-0.0174551

Rounding to four decimal places, we get:

cot 15° = 3.7321 cot 339° = -2.6051 cot(-89°) = ________

---

12. Use a calculator for these. Round to four decimal places.

a) sec 77° = ________ b) csc 230° = ________ c) cot(-112°) = ________

Answer (frame 11): -0.0175

---

13. Use a calculator for these. Round to four decimal places.

a) cot 413° = ______ b) sec(-229°) = ______ c) csc 1,059° = ______

Answers (frame 12): a) 4.4454 b) -1.3054 c) 0.4040

Answers (frame 13): a) 0.7536 b) -1.5243 c) -2.7904

6-2 THE RATIO IDENTITIES

A trigonometric "identity" is a statement of equality that is true for all angles for which the trig ratios in the identity are defined. In this section, we will discuss the two basic "ratio" identities.

---

14. The first ratio identify is shown below. It says that "the tangent of any angle is equal to the ratio of its sine to its cosine".

$$\tan\theta = \frac{\sin\theta}{\cos\theta} \quad \text{or} \quad \frac{\sin\theta}{\cos\theta} = \tan\theta$$

Continued on following page.

14. Continued

We can use the unit circle at the right to prove the identity. The proof applies to angles in all quadrants. From the unit circle, we get:

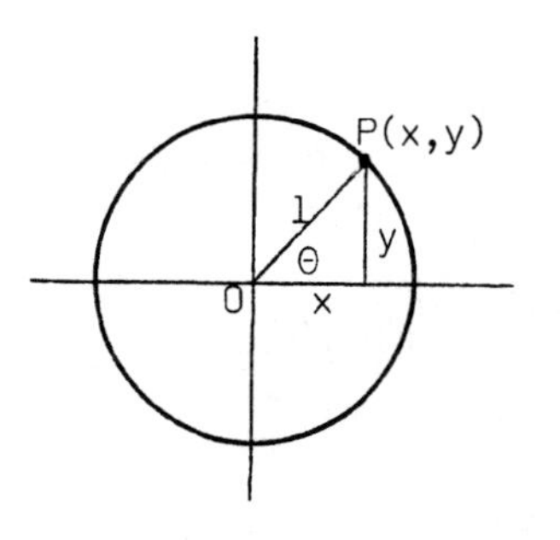

$$\sin\theta = \frac{y}{1} \text{ or } y$$

$$\cos\theta = \frac{x}{1} \text{ or } x$$

$$\tan\theta = \frac{y}{x}$$

Therefore: $\tan\theta = \dfrac{\sin\theta}{\cos\theta}$

---

15. We can use a calculator to show that the first ratio identity is true when $\theta = 47°$, $\theta = 212°$, and $\theta = -68°$. For those angles, the identity becomes:

$$\tan 47° = \frac{\sin 47°}{\cos 47°} \qquad \tan 212° = \frac{\sin 212°}{\cos 212°} \qquad \tan(-68°) = \frac{\sin(-68°)}{\cos(-68°)}$$

Using a calculator, we get these values for the tangents.

$$\tan 47° = 1.0723687$$

$$\tan 212° = 0.6248693$$

$$\tan(-68°) = -2.4750868$$

Use a calculator for these:

a) Does $\dfrac{\sin 47°}{\cos 47°} = 1.0723687$? ________

b) Does $\dfrac{\sin 212°}{\cos 212°} = 0.6248693$? ________

c) Does $\dfrac{\sin(-68°)}{\cos(-68°)} = -2.4750868$? ________

---

a) Yes

b) Yes

c) Yes

16. The second ratio identity is shown below. It says that "the cotangent of any angle is equal to the ratio of its cosine to its sine".

$$\boxed{\cot\theta = \frac{\cos\theta}{\sin\theta} \quad \text{or} \quad \frac{\cos\theta}{\sin\theta} = \cot\theta}$$

Continued on following page.

16. Continued

We can use the unit circle at the right to prove the identity. The proof applies to angles in all quadrants. From the unit circle, we get:

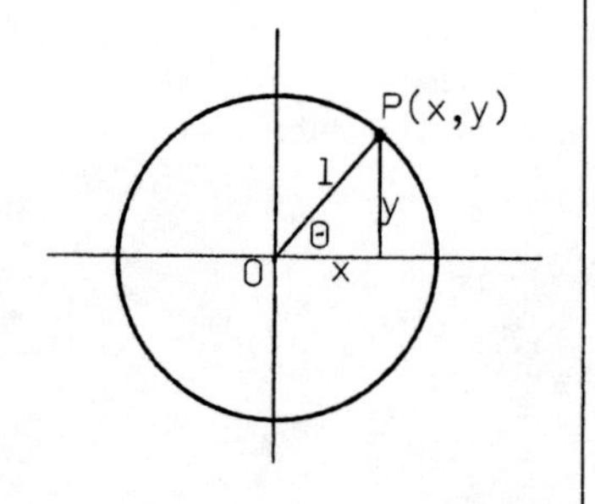

$$\sin \theta = \frac{y}{1} \text{ or } y$$

$$\cos \theta = \frac{x}{1} \text{ or } x$$

$$\cot \theta = \frac{x}{y}$$

Therefore: $\cot \theta = \dfrac{\cos \theta}{\sin \theta}$

---

17. We can use a calculator to show that the second ratio identity is true when $\theta = 17°$, $\theta = 302°$, and $\theta = -98°$. For those angles, the identity becomes:

$$\cot 17° = \frac{\cos 17°}{\sin 17°} \qquad \cot 302° = \frac{\cos 302°}{\sin 302°} \qquad \cot(-98°) = \frac{\cos(-98°)}{\sin(-98°)}$$

Using a calculator, we get these values for the cotangents.

$$\cot 17° = 3.2708526$$

$$\cot 302° = -0.6248694$$

$$\cot(-98°) = 0.1405408$$

Use a calculator for these:

a) Does $\dfrac{\cos 17°}{\sin 17°} = 3.2708526$? ________

b) Does $\dfrac{\cos 302°}{\sin 302°} = -0.6248694$? ________

c) Does $\dfrac{\cos(-98°)}{\sin(-98°)} = 0.1405408$? ________

---

18. Identities can be rearranged like formulas. For example, we solved for "sin θ" at the left below. Notice that we multiplied both sides by "cos θ". Solve for "cos θ" in the other identity.

a) Yes

b) Yes

c) Yes

$$\tan \theta = \frac{\sin \theta}{\cos \theta} \qquad\qquad \cot \theta = \frac{\cos \theta}{\sin \theta}$$

$$\cos \theta(\tan \theta) = \cancel{\cos \theta}\left(\frac{\sin \theta}{\cancel{\cos \theta}}\right)$$

$$\cos \theta(\tan \theta) = \sin \theta$$

or

$$\sin \theta = \cos \theta \tan \theta$$

$\cos \theta = \sin \theta \cot \theta$

19. In the last frame, we got these two solutions:

$$\sin \theta = \cos \theta \tan \theta$$

$$\cos \theta = \sin \theta \cot \theta$$

Though there are no parentheses in the expressions at the right, they are multiplications. That is:

$\cos \theta \tan \theta$ means: $\cos \theta$ <u>times</u> $\tan \theta$

$\sin \theta \cot \theta$ means: $\sin \theta$ <u>times</u> $\cot \theta$

We solved for "cos θ" at the left below. Solve for "sin θ" in the other identity.

$$\tan \theta = \frac{\sin \theta}{\cos \theta} \qquad\qquad \cot \theta = \frac{\cos \theta}{\sin \theta}$$

$$\cos \theta \tan \theta = \sin \theta$$

$$\cos \theta = \frac{\sin \theta}{\tan \theta}$$

Answer to Frame 19: $\sin \theta = \dfrac{\cos \theta}{\cot \theta}$

20. Though $\sin \theta$ and $\cos \theta$ are defined for all values of θ, $\tan \theta$ is not defined for values of θ like those below.

$\tan 90°$ $\qquad$ $\tan 270°$ $\qquad$ $\tan 450°$ $\qquad$ $\tan(-90°)$

Therefore, the identity below is not defined for values of θ where $\tan \theta$ is undefined.

$$\tan \theta = \frac{\sin \theta}{\cos \theta}$$

Four trigonometric ratios ($\tan \theta$, $\csc \theta$, $\sec \theta$, and $\cot \theta$) are not defined for certain values of θ. Therefore, identities containing them are not defined for those values of θ. That fact will not be emphasized in this text.

6-3 THE PYTHAGOREAN IDENTITIES

There are three identities based on the Pythagorean Theorem. We will discuss the three "Pythagorean" identities in this section.

---

21. We will review some symbols and facts about "squaring" before introducing the Pythagorean identities.

<u>Squaring a fraction is equivalent to squaring both its numerator and its denominator</u>. An example is shown. Notice that the expression on the far right does not contain parentheses.

$$\left(\frac{x}{y}\right)^2 = \left(\frac{x}{y}\right)\left(\frac{x}{y}\right) = \frac{x \cdot x}{y \cdot y} = \frac{x^2}{y^2}$$

Continued on following page.

21. Continued

Following the example, write an equivalent expression without parentheses.

$\left(\frac{y}{x}\right)^2 = \frac{y^2}{x^2}$ a) $\left(\frac{b}{c}\right)^2 =$ ______ b) $\left(\frac{p}{q}\right)^2 =$ ______

---

22. Following the example, write an equivalent expression with parentheses.

$\frac{x^2}{y^2} = \left(\frac{x}{y}\right)^2$ a) $\frac{c^2}{d^2} =$ ______ b) $\frac{m^2}{t^2} =$ ______

Answers to 21: a) $\frac{b^2}{c^2}$ b) $\frac{p^2}{q^2}$

---

23. We squared "sin θ" below. Notice that the expression on the far right does not contain parentheses.

$$(\sin\theta)^2 = (\sin\theta)(\sin\theta) = \sin^2\theta$$

Notice in the expression on the far right that the exponent is written between "sin" and "θ", not after "θ". That is:

We write $\sin^2\theta$, not $\sin\theta^2$.

Following the example above, write an equivalent expression without parentheses.

a) $(\cos\theta)^2 =$ ______ b) $(\tan\theta)^2 =$ ______

Answers to 22: a) $\left(\frac{c}{d}\right)^2$ b) $\left(\frac{m}{t}\right)^2$

---

24. Following the example, write an equivalent expression with parentheses.

$\sin^2\theta = (\sin\theta)^2$ a) $\cos^2\theta =$ ______ b) $\cot^2\theta =$ ______

Answers to 23: a) $\cos^2\theta$ b) $\tan^2\theta$

---

25. The first Pythagorean identity is shown below. It says that "the sum of the squares of the sine and cosine of any angle is +1".

$$\boxed{\sin^2\theta + \cos^2\theta = 1}$$

We can use the unit circle at the right to prove the identity. The proof applies to angles in all quadrants. Using the Pythagorean Theorem with the right triangle, we get:

$$y^2 + x^2 = 1$$

But $\sin\theta = y$ and $\cos\theta = x$. Therefore, we can substitute $\sin^2\theta$ for y^2 and $\cos^2\theta$ for x^2. We get the identity:

$$\sin^2\theta + \cos^2\theta = 1$$

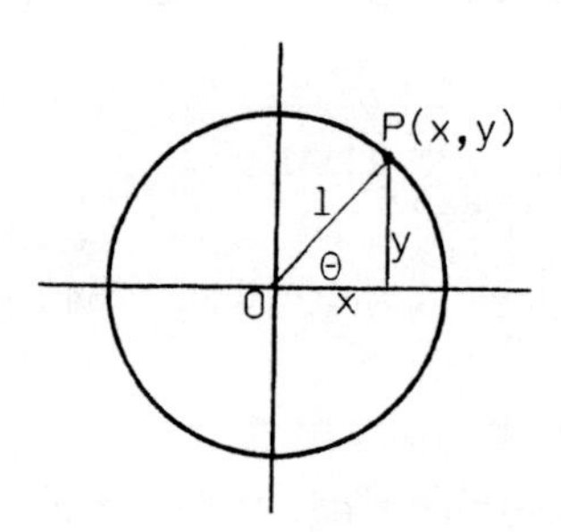

Answers to 24: a) $(\cos\theta)^2$ b) $(\cot\theta)^2$

26. When $\theta = 53°$, the first Pythagorean identity becomes:

$$\sin^2 53° + \cos^2 53° = 1$$

The calculator steps needed to show that the identity is true for $\theta = 53°$ are shown below.

Enter	Press	Display
53	[sin] [x^2] [+]	0.6378187
53	[cos] [x^2] [=]	1

Using the same steps, show that the identity is true when $\theta = 309°$ and $\theta = -24°$.

---

27. The second Pythagorean identity is shown below.

$$\boxed{\tan^2\theta + 1 = \sec^2\theta}$$

To prove the identity, we start with the following equation obtained by applying the Pythagorean Theorem to the triangle in the unit circle in Frame 25.

$$y^2 + x^2 = 1$$

Multiplying both sides by $\frac{1}{x^2}$, we get:

$$\frac{1}{x^2}(y^2 + x^2) = \frac{1}{x^2}(1)$$

$$\frac{1}{x^2}(y^2) + \frac{1}{x^2}(x^2) = \frac{1}{x^2}$$

$$\frac{y^2}{x^2} + \frac{x^2}{x^2} = \frac{1}{x^2}$$

$$\left(\frac{y}{x}\right)^2 + 1 = \left(\frac{1}{x}\right)^2$$

But since $\tan\theta = \frac{y}{x}$ and $\sec\theta = \frac{1}{x}$, we can substitute and get the identity:

$$(\tan\theta)^2 + 1 = (\sec\theta)^2$$

or

$$\tan^2\theta + 1 = \sec^2\theta$$

28. When $\theta = 31°$, the second Pythagorean identity becomes:

$$\tan^2 31° + 1 = \sec^2 31°$$

To show that the identity is true when $\theta = 31°$, we must evaluate the left side and the right side separately. The steps for each are shown below. The value for the left side (1.3610335) should be recorded before evaluating the right side.

Enter	Press	Display
31	[tan] [x^2] [+]	0.3610335
1	[=]	1.3610335

Enter	Press	Display
31	[cos] [1/x] [x^2]	1.3610335

Using the same steps show that the identity is true when $\theta = 242°$ and $\theta = -105°$.

---

29. The third Pythagorean identity is shown below.

$$\boxed{1 + \cot^2\theta = \csc^2\theta}$$

To prove the identity, we again start with the Pythagorean Theorem based on the unit circle (see Frame 25).

$$y^2 + x^2 = 1$$

Multiplying both sides by $\frac{1}{y^2}$, we get:

$$\frac{1}{y^2}(y^2 + x^2) = \frac{1}{y^2}(1)$$

$$\frac{1}{y^2}(y^2) + \frac{1}{y^2}(x^2) = \frac{1}{y^2}$$

$$\frac{y^2}{y^2} + \frac{x^2}{y^2} = \frac{1}{y^2}$$

$$1 + \left(\frac{x}{y}\right)^2 = \left(\frac{1}{y}\right)^2$$

But since $\cot\theta = \frac{x}{y}$ and $\csc\theta = \frac{1}{y}$, we can substitute and get the identity:

$$1 + (\cot\theta)^2 = (\csc\theta)^2$$

or

$$1 + \cot^2\theta = \csc^2\theta$$

30. When $\theta = 74°$, the third Pythagorean identity becomes:

$$1 + \cot^2 74° = \csc^2 74^2$$

To show that the identity is true when $\theta = 74°$, we must evaluate the left side and the right side separately. The steps for each are shown below. The value for the left side (1.0822229) should be recorded before evaluating the right side.

Enter	Press	Display
1	[+]	1
74	[tan] [1/x] [x^2] [=]	1.0822229

Enter	Press	Display
74	[sin] [1/x] [x^2]	1.0822229

Using the same steps, show that the identity is true when $\theta = 327°$ and $\theta = -49°$.

---

31. We can solve for "$\sin^2\theta$" and "$\sin\theta$" in the identity below.

$$\sin^2\theta + \cos^2\theta = 1$$

To solve for "$\sin^2\theta$", we add the opposite of "$\cos^2\theta$" to both sides and get:

$$\sin^2\theta = 1 - \cos^2\theta$$

To solve for "$\sin\theta$", we now take the positive square root of both sides and get:

$$\sin\theta = \sqrt{1 - \cos^2\theta}$$

Similarly, solve the same identity for "$\cos^2\theta$" and "$\cos\theta$".

a) $\cos^2\theta =$ ______________ b) $\cos\theta =$ ______________

---

32. We solved the identity below for "$\tan^2\theta$" and "$\tan\theta$". The positive square root is shown. Similarly, solve for "$\cot^2\theta$" and "$\cot\theta$" in the other identity.

$\tan^2\theta + 1 = \sec^2\theta$	$1 + \cot^2\theta = \csc^2\theta$
$\tan^2\theta = \sec^2\theta - 1$	$\cot^2\theta =$ ______________
$\tan\theta = \sqrt{\sec^2\theta - 1}$	$\cot\theta =$ ______________

Answers to 31:

a) $\cos^2\theta = 1 - \sin^2\theta$

b) $\cos\theta = \sqrt{1 - \sin^2\theta}$

Answers to 32:

$\cot^2\theta = \csc^2\theta - 1$

$\cot\theta = \sqrt{\csc^2\theta - 1}$

6-4 THE RECIPROCAL IDENTITIES

There are three identities based on the three reciprocal relationships among the trigonometric ratios. We will discuss the three "reciprocal" identities in this section.

---

33. By definition, the "sine" and "cosecant" of an angle are a pair of reciprocals. Therefore, we can write the following identity:

$$\boxed{\sin\theta\csc\theta = 1}$$

We solved for "csc θ" below. Solve for "sin θ" in the same identity.

$$\sin\theta\csc\theta = 1 \qquad\qquad \sin\theta\csc\theta = 1$$

$$\left(\frac{1}{\sin\theta}\right)(\sin\theta\csc\theta) = 1\left(\frac{1}{\sin\theta}\right)$$

$$(1) \qquad \csc\theta = \frac{1}{\sin\theta}$$

$$\csc\theta = \frac{1}{\sin\theta}$$

Answer: $\sin\theta = \dfrac{1}{\csc\theta}$

34. By definition, the "cosine" and "secant" of an angle are a pair of reciprocals. Therefore, we can write the following identity:

$$\boxed{\cos\theta\sec\theta = 1}$$

a) Solve for "cos θ".

$\cos\theta\sec\theta = 1$

b) Solve for "sec θ".

$\cos\theta\sec\theta = 1$

Answers:

a) $\cos\theta = \dfrac{1}{\sec\theta}$

b) $\sec\theta = \dfrac{1}{\cos\theta}$

35. By definition, the "tangent" and "cotangent" of an angle are a pair of reciprocals. Therefore, we can write the following identity:

$$\boxed{\tan\theta\cot\theta = 1}$$

a) Solve for "$\tan\theta$".

$\tan\theta\cot\theta = 1$

b) Solve for "$\cot\theta$".

$\tan\theta\cot\theta = 1$

---

a) $\tan\theta = \dfrac{1}{\cot\theta}$

b) $\cot\theta = \dfrac{1}{\tan\theta}$

36. The three "reciprocal" identities are summarized below:

$$\sin\theta\csc\theta = 1$$
$$\cos\theta\sec\theta = 1$$
$$\tan\theta\cot\theta = 1$$

$\sin\theta = \dfrac{1}{\csc\theta}$	and	$\csc\theta = \dfrac{1}{\sin\theta}$
$\cos\theta = \dfrac{1}{\sec\theta}$	and	$\sec\theta = \dfrac{1}{\cos\theta}$
$\tan\theta = \dfrac{1}{\cot\theta}$	and	$\cot\theta = \dfrac{1}{\tan\theta}$

6-5 THE EIGHT BASIC IDENTITIES

In this section, we will summarize the eight basic identities and give some practice with them.

---

37. The eight basic identities are listed below.

Reciprocal Identities	Ratio Identities	Pythagorean Identities
$\sin\theta\csc\theta = 1$	$\tan\theta = \dfrac{\sin\theta}{\cos\theta}$	$\sin^2\theta + \cos^2\theta = 1$
$\cos\theta\sec\theta = 1$		$\tan^2\theta + 1 = \sec^2\theta$
$\tan\theta\cot\theta = 1$	$\cot\theta = \dfrac{\cos\theta}{\sin\theta}$	$1 + \cot^2\theta = \csc^2\theta$

Continued on following page.

37. Continued

Using the reciprocal identities, complete these:

a) The reciprocal of $\sin\theta$ = ________.

b) The reciprocal of $\cot\theta$ = ________.

c) $\frac{1}{\sec\theta}$ = ________ d) $\frac{1}{\tan\theta}$ = ________

---

38. Using the ratio identities, complete these:

a) $\tan\theta = \frac{\square}{\square}$ b) $\cot\theta = \frac{\square}{\square}$

Answers to 37: a) $\csc\theta$ b) $\tan\theta$ c) $\cos\theta$ d) $\cot\theta$

---

39. Using the Pythagorean identities, complete these:

a) $\sin^2\theta$ + ________ = 1

b) ________ + 1 = $\sec^2\theta$ c) $1 + \cot^2\theta$ = ________

Answers to 38: a) $\frac{\sin\theta}{\cos\theta}$ b) $\frac{\cos\theta}{\sin\theta}$

---

40. Using all types of identities, complete these:

a) $\frac{\sin\theta}{\cos\theta}$ = ________ d) $\frac{1}{\csc\theta}$ = ________

b) $\frac{1}{\tan\theta}$ = ________ e) $\frac{\cos\theta}{\sin\theta}$ = ________

c) $\cot^2\theta$ = ________ f) $\sec^2\theta$ = ________

Answers to 39: a) $\cos^2\theta$ b) $\tan^2\theta$ c) $\csc^2\theta$

---

41. One expression for "$\tan\theta$" is given below. Write another expression for "$\tan\theta$".

$\tan\theta = \frac{1}{\cot\theta}$ $\tan\theta$ = ________

Answers to 40: a) $\tan\theta$ b) $\cot\theta$ c) $\csc^2\theta - 1$ d) $\sin\theta$ e) $\cot\theta$ f) $\tan^2\theta + 1$

---

42. One expression for "$\cot\theta$" is given below. Write another expression for "$\cot\theta$".

$\cot\theta = \frac{\cos\theta}{\sin\theta}$ $\cot\theta$ = ________

Answer to 41: $\tan\theta = \frac{\sin\theta}{\cos\theta}$

Answer to 42: $\cot\theta = \frac{1}{\tan\theta}$

43. By rearranging the basic identities, we can write four different expressions for "cos θ":

$$\cos\theta = \frac{1}{\sec\theta} \quad , \text{ from: } \cos\theta\sec\theta = 1$$

$$\cos\theta = \sin\theta\cot\theta \quad , \text{ from: } \cot\theta = \frac{\cos\theta}{\sin\theta}$$

$$\cos\theta = \frac{\sin\theta}{\tan\theta} \quad , \text{ from: } \tan\theta = \frac{\sin\theta}{\cos\theta}$$

$$\cos\theta = \sqrt{1-\sin^2\theta} \quad , \text{ from: } \sin^2\theta + \cos^2\theta = 1$$

By rearranging the basic identities, write four different expressions for "sin θ".

$\sin\theta =$ ______ $\sin\theta =$ ______

$\sin\theta =$ ______ $\sin\theta =$ ______

Answers to frame 43:

$\sin\theta = \frac{1}{\csc\theta}$

$\sin\theta = \cos\theta\tan\theta$

$\sin\theta = \frac{\cos\theta}{\cot\theta}$

$\sin\theta = \sqrt{1-\cos^2\theta}$

44. Using the Pythagorean identities, write expressions for "sec θ" and "cot θ".

a) $\sec\theta =$ ______ b) $\cot\theta =$ ______

Answers to frame 44:

a) $\sec\theta = \sqrt{\tan^2\theta + 1}$ b) $\cot\theta = \sqrt{\csc^2\theta - 1}$

SELF-TEST 19 (pages 264-280)

Using sides "h", "t", and "w" of this right triangle, define:

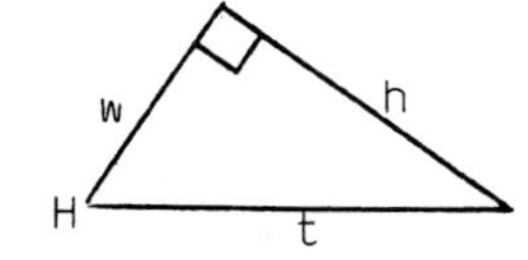

1. sec H = ______
2. cot H = ______
3. csc H = ______

Rounding to four decimal places, find the numerical value of:

4. cot 14° = ______
5. csc 296° = ______
6. sec(-158°) = ______

Using the ratio identities, complete these:

7. $\frac{\sin\theta}{\cos\theta} =$ ______

8. $\sin\theta\cot\theta =$ ______

Using the Pythagorean identities, complete these:

9. ______ $+ \cos^2\theta = 1$

10. $\csc^2\theta - 1 =$ ______

Continued on following page.

SELF-TEST 19 (pages 264-280) - Continued

Using the reciprocal identities, complete these:

11. $\csc \theta =$ ________

12. $(\sec \theta)($________$) = 1$

13. Solve this identity for "tan θ".

$$\tan^2\theta + 1 = \sec^2\theta$$

$\tan \theta =$ ________

ANSWERS:

1. $\frac{t}{w}$	4. 4.0108	7. $\tan \theta$	11. $\frac{1}{\sin \theta}$
2. $\frac{w}{h}$	5. -1.1126	8. $\cos \theta$	12. $\cos \theta$
3. $\frac{t}{h}$	6. -1.0785	9. $\sin^2\theta$	13. $\tan \theta = \sqrt{\sec^2\theta - 1}$
		10. $\cot^2\theta$	

6-6 PROVING IDENTITIES

"Proving an identity" means showing that both sides of a statement are equal. To do so, we use the basic identities to make substitutions and then simplify. We will discuss the process in this section.

45. To prove that the boxed statement below is an identity, we showed that the right side also equals "sin θ" by substituting $\frac{\sin \theta}{\cos \theta}$ for $\tan \theta$ and simplifying. By substituting $\frac{\cos \theta}{\sin \theta}$ for $\cot \theta$ and simplifying, prove the other identity.

$$\boxed{\sin \theta = \cos \theta \tan \theta}$$

$$= \cancel{\cos \theta}\left(\frac{\sin \theta}{\cancel{\cos \theta}}\right)$$

$$= \sin \theta$$

$$\boxed{\cos \theta = \sin \theta \cot \theta}$$

Answer to 45:

$$\cos \theta = \sin \theta \cot \theta$$

$$= \cancel{\sin \theta}\left(\frac{\cos \theta}{\cancel{\sin \theta}}\right)$$

$$= \cos \theta$$

46. To prove the identity below, we substituted $\frac{\cos \theta}{\sin \theta}$ for $\cot \theta$ and simplified the right side. By substituting $\frac{\sin \theta}{\cos \theta}$ for $\tan \theta$ and simplifying, prove the other identity.

$$\boxed{\sin \theta = \frac{\cos \theta}{\cot \theta}}$$

$$= \frac{\cos \theta}{\frac{\cos \theta}{\sin \theta}}$$

$$= \cancel{\cos \theta}\left(\frac{\sin \theta}{\cancel{\cos \theta}}\right)$$

$$= \sin \theta$$

$$\boxed{\cos \theta = \frac{\sin \theta}{\tan \theta}}$$

47. Sometimes we must make a double substitution to prove an identity. An example is shown. By making the suggested substitutions, prove the other identity.

To prove the identity below, we substituted:

$\frac{\sin\theta}{\cos\theta}$ for $\tan\theta$

$\frac{1}{\cos\theta}$ for $\sec\theta$

$$\boxed{\sin\theta = \frac{\tan\theta}{\sec\theta}}$$

$$= \frac{\frac{\sin\theta}{\cos\theta}}{\frac{1}{\cos\theta}}$$

$$= \left(\frac{\sin\theta}{\cancel{\cos\theta}}\right)\left(\frac{\cancel{\cos\theta}}{1}\right)$$

$$= \sin\theta$$

To prove the identity below, substitute:

$\frac{1}{\sin\theta}$ for $\csc\theta$

$\frac{\sin\theta}{\cos\theta}$ for $\tan\theta$

$$\boxed{1 = \cos\theta\csc\theta\tan\theta}$$

Answer:

$$\cos\theta = \frac{\sin\theta}{\tan\theta}$$

$$= \frac{\sin\theta}{\frac{\sin\theta}{\cos\theta}}$$

$$= \cancel{\sin\theta}\left(\frac{\cos\theta}{\cancel{\sin\theta}}\right)$$

$$= \cos\theta$$

---

48. To prove the identity below, we substituted $(1 + \cot^2\theta)$ for $\csc^2\theta$ from a Pythagorean identity. By making a similar substitution for $\sec^2\theta$ from a Pythagorean identity, prove the other identity.

$$\boxed{1 = \csc^2\theta - \cot^2\theta}$$

$$= (1 + \cot^2\theta) - \cot^2\theta$$

$$= 1 + \cot^2\theta - \cot^2\theta$$

$$= 1$$

$$\boxed{1 = \sec^2\theta - \tan^2\theta}$$

Answer:

$$1 = \cos\theta\csc\theta\tan\theta$$

$$= \cancel{\cos\theta}\left(\frac{1}{\cancel{\sin\theta}}\right)\frac{\cancel{\sin\theta}}{\cancel{\cos\theta}}$$

$$= 1$$

---

49. To prove the identity below, we factored by the distributive principle before substituting $\sin^2\theta$ for $(1 - \cos^2\theta)$. We also substituted for $\csc\theta$ in a later step. Using the same type of steps, prove the other identity.

$$\boxed{\sin\theta = \csc\theta - \csc\theta\cos^2\theta}$$

$$= \csc\theta(1 - \cos^2\theta)$$

$$= \csc\theta(\sin^2\theta)$$

$$= \left(\frac{1}{\sin\theta}\right)(\sin^2\theta)$$

$$= \frac{\sin^2\theta}{\sin\theta}$$

$$= \sin\theta$$

$$\boxed{\cos\theta = \sec\theta - \sec\theta\sin^2\theta}$$

Answer:

$$1 = \sec^2\theta - \tan^2\theta$$

$$= (\tan^2\theta + 1) - \tan^2\theta$$

$$= 1 + \tan^2\theta - \tan^2\theta$$

$$= 1$$

50. The proof of the identity below includes a subtraction of fractions. Following the same type of steps, prove the other identity.

$$\boxed{\sin\theta = \csc\theta - \cos\theta\cot\theta}$$

$$\boxed{\cos\theta = \sec\theta - \sin\theta\tan\theta}$$

$$= \frac{1}{\sin\theta} - \cos\theta\left(\frac{\cos\theta}{\sin\theta}\right)$$

$$= \frac{1}{\sin\theta} - \frac{\cos^2\theta}{\sin\theta}$$

$$= \frac{1 - \cos^2\theta}{\sin\theta}$$

$$= \frac{\sin^2\theta}{\sin\theta}$$

$$= \sin\theta$$

$$\cos\theta = \sec\theta - \sec\theta\sin^2\theta$$

$$= \sec\theta(1 - \sin^2\theta)$$

$$= \sec\theta(\cos^2\theta)$$

$$= \left(\frac{1}{\cos\theta}\right)(\cos^2\theta)$$

$$= \frac{\cos^2\theta}{\cos\theta}$$

$$= \cos\theta$$

---

51. We proved the same identity in two different ways below. At the left, we substituted and rearranged only on the right side. At the right, we substituted and rearranged on both sides.

$$\boxed{\sin\theta\tan\theta = \sec\theta - \cos\theta}$$

$$= \frac{1}{\cos\theta} - \cos\theta$$

$$= \frac{1}{\cos\theta} - \frac{\cos^2\theta}{\cos\theta}$$

$$= \frac{1 - \cos^2\theta}{\cos\theta}$$

$$= \frac{\sin^2\theta}{\cos\theta}$$

$$= \sin\theta\left(\frac{\sin\theta}{\cos\theta}\right)$$

$$= \sin\theta\tan\theta$$

$$\boxed{\sin\theta\tan\theta = \sec\theta - \cos\theta}$$

$$\sin\theta\left(\frac{\sin\theta}{\cos\theta}\right) = \frac{1}{\cos\theta} - \cos\theta$$

$$\frac{\sin^2\theta}{\cos\theta} = \frac{1}{\cos\theta} - \frac{\cos^2\theta}{\cos\theta}$$

$$= \frac{1 - \cos^2\theta}{\cos\theta}$$

$$= \frac{\sin^2\theta}{\cos\theta}$$

Using either method, prove this identity:

$$\boxed{\cos\theta\cot\theta = \csc\theta - \sin\theta}$$

$$\cos\theta = \sec\theta - \sin\theta\tan\theta$$

$$= \frac{1}{\cos\theta} - \sin\theta\left(\frac{\sin\theta}{\cos\theta}\right)$$

$$= \frac{1}{\cos\theta} - \frac{\sin^2\theta}{\cos\theta}$$

$$= \frac{1 - \sin^2\theta}{\cos\theta}$$

$$= \frac{\cos^2\theta}{\cos\theta}$$

$$= \cos\theta$$

Answer to Frame 51:

$$\cos\theta\cot\theta = \csc\theta - \sin\theta$$
$$= \frac{1}{\sin\theta} - \sin\theta$$
$$= \frac{1}{\sin\theta} - \frac{\sin^2\theta}{\sin\theta}$$
$$= \frac{1-\sin^2\theta}{\sin\theta}$$
$$= \frac{\cos^2\theta}{\sin\theta}$$
$$= \cos\theta\left(\frac{\cos\theta}{\sin\theta}\right)$$
$$= \cos\theta\cot\theta$$

or

$$\cos\theta\cot\theta = \csc\theta - \sin\theta$$
$$\cos\theta\left(\frac{\cos\theta}{\sin\theta}\right) = \frac{1}{\sin\theta} - \sin\theta$$
$$\frac{\cos^2\theta}{\sin\theta} = \frac{1}{\sin\theta} - \frac{\sin^2\theta}{\sin\theta}$$
$$= \frac{1-\sin^2\theta}{\sin\theta}$$
$$= \frac{\cos^2\theta}{\sin\theta}$$

---

52. The need to prove an identity does not occur often. Identities are mainly used in solving trig equations, in simplifying trig formulas, and in making trig substitutions in calculus. In such uses, either the necessary identity is given or it can be found in a "table" of identities. A partial table of identities is given on page 284. It includes the identities most frequently used in calculus.

6-7 INVERSE TRIGONOMETRIC NOTATION

Expressions like arcsin 0.5026 or $\cos^{-1}$ 0.9311 are examples of inverse trigonometric notation. We will discuss notation of that type in this section.

---

53. The equation sin 50° = 0.7660 can be stated in words in two ways:

"The sine of 50° is 0.7660."
"50° is the angle whose sine is 0.7660."

The abbreviation "arcsin" is used for the phrase "the angle whose sine is". Therefore:

50° is [the angle whose sine is] 0.7660.

↓

50° is [arcsin] 0.7660

Substituting "=" for "is", we get the following equation:

50° = arcsin 0.7660

Write each statement as an equation in arcsin notation.

a) 25° is the angle whose sine is 0.4226. ______________

b) 71° is the angle whose sine is 0.9455. ______________

---

a) 25° = arcsin 0.4226

b) 71° = arcsin 0.9455

TRIGONOMETRIC IDENTITIES

Reciprocal Identities

$\sin\theta \csc\theta = 1$

$\cos\theta \sec\theta = 1$

$\tan\theta \cot\theta = 1$

Ratio Identities

$$\tan\theta = \frac{\sin\theta}{\cos\theta}$$

$$\cot\theta = \frac{\cos\theta}{\sin\theta}$$

Pythagorean Identities

$\sin^2\theta + \cos^2\theta = 1$

$\tan^2\theta + 1 = \sec^2\theta$

$1 + \cot^2\theta = \csc^2\theta$

Complementary-Angle Identities

$\sin\theta = \cos(90° - \theta)$

$\cos\theta = \sin(90° - \theta)$

$\tan\theta = \cot(90° - \theta)$

Supplementary-Angle Identities

$\sin\theta = +\sin(180° - \theta)$

$\cos\theta = -\cos(180° - \theta)$

$\tan\theta = -\tan(180° - \theta)$

Negative-Angle Identities

$\sin(-\theta) = -\sin\theta$

$\cos(-\theta) = +\cos\theta$

$\tan(-\theta) = -\tan\theta$

Double-Angle Identities

$\sin 2\theta = 2\sin\theta\cos\theta$

$\cos 2\theta = \cos^2\theta - \sin^2\theta$

$\cos 2\theta = 2\cos^2\theta - 1$

$\cos 2\theta = 1 - 2\sin^2\theta$

$$\tan 2\theta = \frac{2\tan\theta}{1 - \tan^2\theta}$$

Half-Angle Identities

$$\sin\frac{\theta}{2} = \pm\sqrt{\frac{1 - \cos\theta}{2}}$$

$$\cos\frac{\theta}{2} = \pm\sqrt{\frac{1 + \cos\theta}{2}}$$

$$\tan\frac{\theta}{2} = \frac{\sin\theta}{1 + \cos\theta}$$

Power Identities

$$\sin^2\theta = \frac{1 - \cos 2\theta}{2}$$

$$\cos^2\theta = \frac{1 + \cos 2\theta}{2}$$

$$\sin^3\theta = \frac{3\sin\theta - \sin 3\theta}{4}$$

Angle Sum-Difference Identities

$\sin(\alpha + \beta) = \sin\alpha\cos\beta + \cos\alpha\sin\beta$

$\sin(\alpha - \beta) = \sin\alpha\cos\beta - \cos\alpha\sin\beta$

$\cos(\alpha + \beta) = \cos\alpha\cos\beta - \sin\alpha\sin\beta$

$\cos(\alpha - \beta) = \cos\alpha\cos\beta + \sin\alpha\sin\beta$

Function Sum-Difference Identities

$$\sin\alpha + \sin\beta = 2\sin\left(\frac{\alpha + \beta}{2}\right)\cos\left(\frac{\alpha - \beta}{2}\right)$$

$$\sin\alpha - \sin\beta = 2\cos\left(\frac{\alpha + \beta}{2}\right)\sin\left(\frac{\alpha - \beta}{2}\right)$$

$$\cos\alpha + \cos\beta = 2\cos\left(\frac{\alpha + \beta}{2}\right)\cos\left(\frac{\alpha - \beta}{2}\right)$$

$$\cos\alpha - \cos\beta = -2\sin\left(\frac{\alpha + \beta}{2}\right)\sin\left(\frac{\alpha - \beta}{2}\right)$$

54. The two equations below are equivalent. The one on the right is in arcsin notation.

$\sin 30° = 0.5000$ $30° = \arcsin 0.5000$

Write each equation in arcsin notation.

a) $\sin 17° = 0.2924$ ______

b) $\sin 82° = 0.9903$ ______

c) $\sin \theta = b$ ______

Answers: a) $17° = \arcsin 0.2924$ b) $82° = \arcsin 0.9903$ c) $\theta = \arcsin b$

---

55. The two equations below are equivalent. The one on the right is in ordinary notation.

$90° = \arcsin 1.000$ $\sin 90° = 1.0000$

Write each equation in ordinary notation.

a) $7° = \arcsin 0.1219$ ______

b) $66° = \arcsin 0.9135$ ______

c) $\alpha = \arcsin t$ ______

Answers: a) $\sin 7° = 0.1219$ b) $\sin 66° = 0.9135$ c) $\sin \alpha = t$

---

56. To find θ below, we converted to ordinary notation first. We also rounded to the nearest whole-number degree.

Find acute angle θ if: $\theta = \arcsin 0.8191$

$\sin \theta = 0.8191$

$\theta = 55°$

Use the same method for this one. Round to the nearest whole-number degree.

Find acute angle B if: $B = \arcsin 0.1026$

a) $\sin B =$ ______

b) $B =$ ______

Answers: a) $\sin B = 0.1026$ b) $B = 6°$

---

57. To find M below, we converted to ordinary notation first. We also rounded to four decimal places.

Find M if: $47° = \arcsin M$

$\sin 47° = M$

$M = 0.7314$

Use the same method for this one. Round to four decimal places.

Find T if: $81° = \arcsin T$

a) $\sin 81° =$ ______

b) $T =$ ______

58. Complete these. Round either to the nearest whole-number degree or to four decimal places.

a) Find acute angle A if: A = arcsin 0.5099

A = ________

b) Find R if: 69° = arcsin R

R = ________

Answers: a) sin 81° = T b) T = 0.9877

---

59. The symbol $\sin^{-1}$ is sometimes used for the symbol arcsin. That is:

$\boxed{\sin^{-1}}$ means $\boxed{\text{arcsin}}$

Therefore, $\sin^{-1}$ also means "the angle whose sine is".

Note: In $\sin^{-1}$, the -1 is not an exponent. The entire expression means "the angle whose sine is".

Since $\sin^{-1}$ means arcsin, the following two equations are equivalent.

$$55° = \arcsin 0.8192$$

$$55° = \sin^{-1} 0.8192$$

Write each of these in $\sin^{-1}$ notation.

a) 13° = arcsin 0.2250 ________________

b) 48° = arcsin 0.7431 ________________

Answers: a) A = 31° b) R = 0.9336

---

60. Complete these. Round either to the nearest whole-number degree or to four decimal places.

a) Find acute angle F if: F = $\sin^{-1}$ 0.2112. F = ________

b) Find "d" if: 75° = $\sin^{-1}$d. d = ________

Answers: a) 13° = $\sin^{-1}$ 0.2250 b) 48° = $\sin^{-1}$ 0.7431

---

61. The symbols $\boxed{\text{arccos}}$ and $\boxed{\cos^{-1}}$ mean "the angle whose cosine is". Therefore, the three equations below are equivalent.

$$\cos 45° = 0.7071$$
$$45° = \arccos 0.7071$$
$$45° = \cos^{-1} 0.7071$$

Write the equation below in two equivalent forms.

cos 78° = 0.2079 ________________

____

Answers: a) F = 12° b) d = 0.9659

Answers: 78° = arccos 0.2079

78° = $\cos^{-1}$ 0.2079

62. Complete these. Round either to the nearest whole-number degree or to four decimal places.

a) Find acute angle θ if: $\theta = \arccos 0.6129$. θ = ________

b) Find P if: $12° = \arccos P$. P = ________

c) Find acute angle A if: $A = \cos^{-1} 0.0877$. A = ________

d) Find "y" if: $80° = \cos^{-1} y$. y = ________

---

63. The symbols [arctan] and [$\tan^{-1}$] mean "the angle whose tangent is". Therefore the three equations below are equivalent.

$$\tan 65° = 2.1445$$
$$65° = \arctan 2.1445$$
$$65° = \tan^{-1} 2.1445$$

Write the equation below in two equivalent forms.

$\tan 15° = 0.2679$ ________________

____

Answers to 62:
a) $\theta = 52°$
b) $P = 0.9781$
c) $A = 85°$
d) $y = 0.1736$

---

64. Complete these. Round either to the nearest whole-number degree or to four decimal places.

a) Find acute angle α if: $\alpha = \arctan 3.7318$. α = ________

b) Find R if: $40° = \arctan R$. R = ________

c) Find acute angle C if: $C = \tan^{-1} 0.9066$ C = ________

d) Find "b" if: $87° = \tan^{-1} b$ b = ________

Answers to 63:
$15° = \arctan 0.2679$
$15° = \tan^{-1} 0.2679$

Answers to 64: a) $\alpha = 75°$ b) $R = 0.8391$ c) $C = 42°$ d) $b = 19.0811$

6-8 FORMULAS CONTAINING INVERSE TRIGONOMETRIC NOTATION

A few formulas contain inverse trigonometric notation like arccos 0.5199 or $\tan^{-1} 1.2644$. We will do evaluations with formulas of that type in this section.

---

65. The formula below contains inverse trigonometric notation. We want to find θ when R = 18.7 and Z = 39.5 . The calculator steps are shown.

$$\theta = \cos^{-1}\left(\frac{R}{Z}\right)$$

$$\theta = \cos^{-1}\left(\frac{18.7}{39.5}\right)$$

Enter	Press	Display
18.7	[÷]	18.7
39.5	[=] [INV] [cos]	61.743622

Rounding to the nearest whole-number degree, θ = ________.

66. Using the steps from the last frame, complete these. Round to the nearest whole-number degree.

a) Find θ when $X = 2,595$ and $R = 1,180$.

$$\theta = \tan^{-1}\left(\frac{X}{R}\right)$$

θ = ________

b) Find α when $X = 247$ and $Z = 289$.

$$\alpha = \arcsin\left(\frac{X}{Z}\right)$$

α = ________

Answer: 62°

---

67. To find X in the formula below when $\theta = 44°$ and $Z = 625$, we begin by substituting as we have done on the right.

$$\theta = \sin^{-1}\left(\frac{X}{Z}\right) \qquad 44° = \sin^{-1}\left(\frac{X}{625}\right)$$

Knowing that $\sin 44° = \frac{X}{625}$ and $\sin 44° = 0.6946584$, we can set up the following equation:

$$\frac{X}{625} = 0.6946584$$

Rounding to a whole number, X = ________

Answer: a) 66° b) 59°

---

68. To find Z in the formula below when $\alpha = 60°$ and $R = 5,140$, we begin by substituting as we have done on the right.

$$\alpha = \arctan\left(\frac{R}{Z}\right) \qquad 60° = \arctan\left(\frac{5,140}{Z}\right)$$

Knowing that $\tan 60° = \frac{5,140}{Z}$ and $\tan 60° = 1.7320508$, we can set up the following equation.

$$\frac{5,140}{Z} = 1.7320508$$

Rounding to a whole number, Z = ________

Answer: 434

---

69. Using the steps from the last two frames, complete these. Round to a whole number.

a) Find R when $\theta = 15°$ and $Z = 584$.

$$\theta = \arccos\left(\frac{R}{Z}\right)$$

R = ________

b) Find Z when $\alpha = 54°$ and $X = 215$.

$$\alpha = \sin^{-1}\left(\frac{X}{Z}\right)$$

Z = ________

Answer: 2,968

---

a) R = 564 b) Z = 266

SELF-TEST 20 (pages 280-289)

Prove the following identities.

1. $\sec\theta \sin\theta = \tan\theta$

2. $\sin\theta \cos\theta = \tan\theta - \tan\theta \sin^2\theta$

3. Write this equation in arcsin notation.

   $\sin\theta = w$

   ______

4. Write this equation in ordinary notation.

   $52° = \tan^{-1} 1.2799$

   ______

5. Write this equation in ordinary notation.

   $B = \arccos h$

   ______

Find an acute angle satisfying each equation. Round to a whole number.

6. $A = \cos^{-1} 0.8718$

   A = ______

7. $\theta = \arcsin 0.4525$

   θ = ______

Find the numerical value of each letter. Round to four decimal places.

8. $85° = \arctan p$

   p = ______

9. $52° = \sin^{-1} w$

   w = ______

10. Find θ when $X = 475$ and $Z = 805$. Round to a whole number.

    $$\theta = \sin^{-1}\left(\frac{X}{Z}\right)$$

    θ = ______

11. Find R when $\theta = 63°$ and $X = 44.9$. Round to tenths.

    $$\theta = \arctan\left(\frac{X}{R}\right)$$

    R = ______

ANSWERS:

1. $\sec\theta \sin\theta = \tan\theta$

   $\left(\frac{1}{\cos\theta}\right)\sin\theta =$

   $\frac{\sin\theta}{\cos\theta} =$

   $\tan\theta =$

2. $\sin\theta\cos\theta = \tan\theta - \tan\theta\sin^2\theta$

   $= \tan\theta(1 - \sin^2\theta)$

   $= \left(\frac{\sin\theta}{\cos\theta}\right)(\cos^2\theta)$

   $= \sin\theta\cos\theta$

3. $\theta = \arcsin w$

4. $\tan 52° = 1.2799$

5. $\cos B = h$

6. $A = 29°$

7. $\theta = 27°$

8. $p = 11.4301$

9. $w = 0.7880$

10. $\theta = 36°$

11. $R = 22.9$

6-9 FINDING ANGLES WITH KNOWN TANGENTS

In this section, we will review the procedure for finding angles with known tangents. Angles from 0° to 360° are emphasized.

70. A 50° angle and a 230° angle are shown in standard position below.

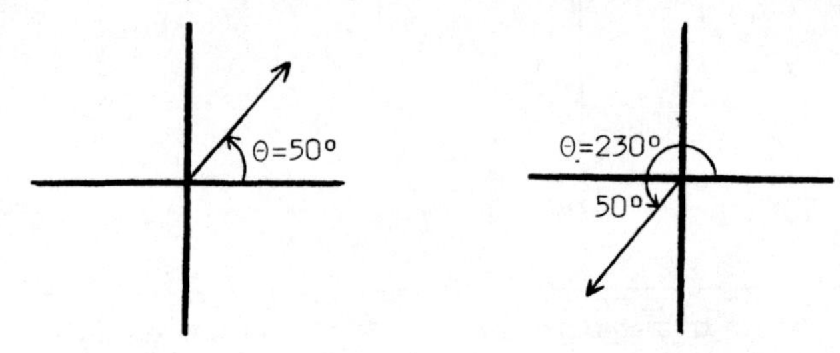

The tangents of angles are positive in both the first and third quadrants. Since the reference angle for both angles above is 50°, both have the same tangent. That is:

$$\tan 50° = \tan 230° = 1.1917536$$

In all three equations below, θ is the angle whose tangent is 1.1917536 .

$$\tan \theta = 1.1917536$$

$$\theta = \arctan 1.1917536$$

$$\theta = \tan^{-1} 1.1917536$$

To find θ on a calculator, we enter 1.1917536 and press [INV] [tan] . We get:

Enter	Press	Display
1.1917536	[INV] [tan]	50

The calculator gives us the first-quadrant angle. Since 50° is the reference angle, we can add 50° to 180° to find the third-quadrant angle. We get:

$$50° + 180° = 230°$$

Find the first-quadrant and third-quadrant angles that satisfy each equation below. Round to the nearest whole-number degree.

a) $\tan \theta = 0.5967$ $\theta =$ ________ and ________

b) $\theta = \arctan 0.1027$ $\theta =$ ________ and ________

c) $\theta = \tan^{-1} 12.507$ $\theta =$ ________ and ________

---

a) 31° and 211°

b) 6° and 186°

c) 85° and 265°

71. A 130° angle and a 310° angle are shown in standard position below.

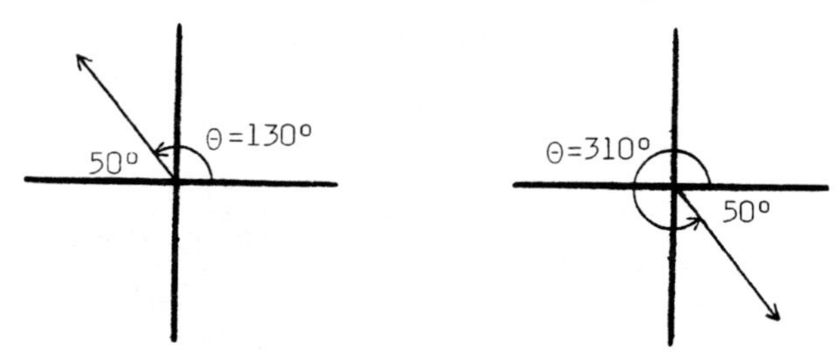

The tangents of angles are negative in both the second and fourth quadrants. Since the reference angle for both angles above is 50°, both have the same tangent. That is:

$$\tan 130° = \tan 310° = -1.1917536$$

In all three equations below, θ is the angle whose tangent is -1.1917536 .

$$\tan \theta = -1.1917536$$

$$\theta = \arctan(-1.1917536)$$

$$\theta = \tan^{-1}(-1.1917536)$$

To find θ on a calculator, we enter -1.1917536 and press [INV] [tan] . We get:

Enter	Press	Display
1.1917536	[+/-] [INV] [tan]	-50

The calculator gives us -50°, a negative fourth-quadrant angle. Since 50° is the reference angle, we can subtract 50° from 180° and 360° to get the positive second-quadrant and fourth-quadrant angles. That is:

$$180° - 50° = 130°$$

$$360° - 50° = 310°$$

Find the positive second-quadrant and fourth-quadrant angles that satisfy each equation below. Round to the nearest whole-number degree.

a) $\tan \theta = -1.5609$ $\theta =$ ________ and ________

b) $\theta = \arctan(-0.2086)$ $\theta =$ ________ and ________

c) $\theta = \tan^{-1}(-8.4731)$ $\theta =$ ________ and ________

---

a) 123° and 303°

b) 168° and 348°

c) 97° and 277°

---

72. Find the two angles between 0° and 360° that satisfy each equation below. Round to the nearest whole-number degree.

a) $\theta = \tan^{-1} 1.3055$ $\theta =$ ________ and ________

b) $\theta = \arctan(-0.9174)$ $\theta =$ ________ and ________

---

a) 53° and 233°

b) 137° and 317°

6-10 FINDING ANGLES WITH KNOWN SINES

In this section, we will discuss the procedure for finding angles with known sines. Angles from 0° to 360° are emphasized.

73. A 60° angle and a 120° angle are shown in standard position below.

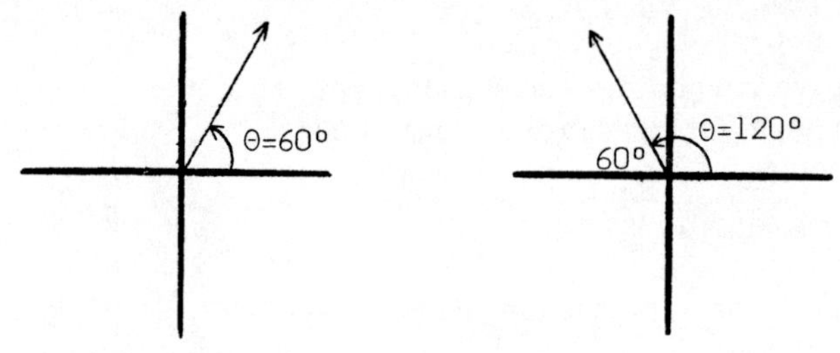

The sines of angles are positive in both the first and second quadrants. Since the reference angle for both angles above is 60°, both have the same sine. That is:

$$\sin 60° = \sin 120° = 0.8660254$$

In all three equations below, θ is the angle whose sine is 0.8660254 .

$$\sin \theta = 0.8660254$$

$$\theta = \arcsin 0.8660254$$

$$\theta = \sin^{-1} 0.8660254$$

To find θ on a calculator, we enter 0.8660254 and press [INV] [sin] . We get:

Enter	Press	Display
0.8660254	[INV] [sin]	60

The calculator gives us the first-quadrant angle. Since 60° is the reference angle, we can subtract 60° from 180° to get the second-quadrant angle. We get:

$$180° - 60° = 120°$$

Find the first-quadrant and second-quadrant angles that satisfy each equation below. Round to the nearest whole-number degree.

a) $\sin \theta = 0.4827$ θ = ________ and ________

b) $\theta = \arcsin 0.0988$ θ = ________ and ________

c) $\theta = \sin^{-1} 0.9541$ θ = ________ and ________

a) 29° and 151°

b) 6° and 174°

c) 73° and 107°

74. A 240° angle and a 300° angle are shown in standard position below.

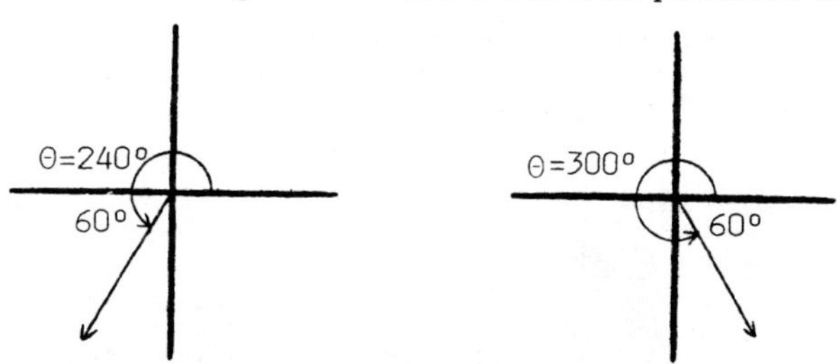

The sines of angles are negative in both the third and fourth quadrants. Since the reference angle for both angles above is 60°, both have the same sine. That is:

$$\sin 240° = \sin 300° = -0.8660254$$

In all three equations below, θ is the angle whose sine is -0.8660254 .

$$\sin \theta = -0.8660254$$

$$\theta = \arcsin(-0.8660254)$$

$$\theta = \sin^{-1}(-0.8660254)$$

To find θ on a calculator, we enter -0.8660254 and press [INV] [sin] . We get:

Enter	Press	Display
0.8660254	[+/-] [INV] [sin]	-60

The calculator gives us -60°, a negative fourth-quadrant angle. However, since 60° is the reference angle, we can add 60° to 180° and subtract 60° from 360° to get the positive third-quadrant and fourth-quadrant angles. That is:

$$180° + 60° = 240°$$

$$360° - 60° = 300°$$

Find the positive third-quadrant and fourth-quadrant angles that satisfy each equation below. Round to the nearest whole-number degree.

a) $\sin \theta = -0.5821$ $\theta =$ ________ and ________

b) $\theta = \arcsin(-0.1233)$ $\theta =$ ________ and ________

c) $\theta = \sin^{-1}(-0.9877)$ $\theta =$ ________ and ________

---

a) 216° and 324°

b) 187° and 353°

c) 261° and 279°

75. Find the two angles between 0° and 360° that satisfy each equation below. Round to the nearest whole-number degree.

a) $\theta = \arcsin 0.7043$ $\theta =$ ________ and ________

b) $\theta = \sin^{-1}(-0.2195)$ $\theta =$ ________ and ________

---

a) 45° and 135°

b) 193° and 347°

6-11 FINDING ANGLES WITH KNOWN COSINES

In this section, we will discuss the procedure for finding angles with known cosines. Angles from 0° to 360° are emphasized.

76. A 60° angle and a 300° angle are shown in standard position below.

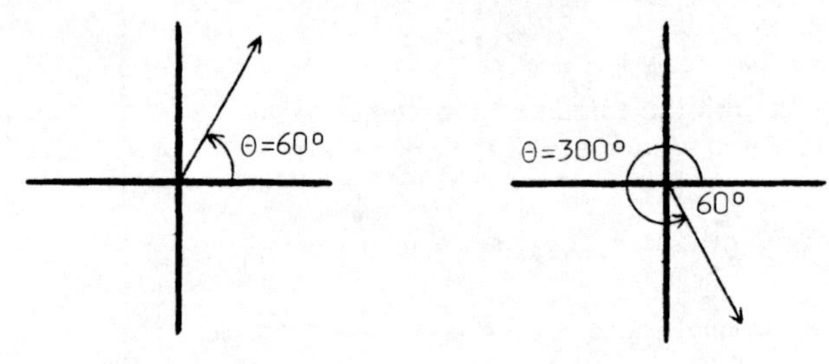

The cosines of angles are positive in both the first and fourth quadrants. Since the reference angle for both angles above is 60°, both have the same cosine. That is:

$$\cos 60° = \cos 300° = 0.5$$

In all three equations below, θ is the angle whose cosine is 0.5 .

$$\cos \theta = 0.5$$

$$\theta = \arccos 0.5$$

$$\theta = \cos^{-1} 0.5$$

To find θ on a calculator, we enter 0.5 and press [INV] [cos] . We get:

Enter	Press	Display
0.5	[INV] [cos]	60

The calculator gives us the first-quadrant angle. Since 60° is the reference angle, we can subtract 60° from 360° to get the fourth-quadrant angle. We get:

$$360° - 60° = 300°$$

Find the first-quadrant and fourth-quadrant angles that satisfy each equation. Round to the nearest whole-number degree.

a) $\cos \theta = 0.4297$ $\theta =$ ________ and ________

b) $\theta = \arccos 0.9561$ $\theta =$ ________ and ________

c) $\theta = \cos^{-1} 0.0925$ $\theta =$ ________ and ________

a) 65° and 295°

b) 17° and 343°

c) 85° and 275°

77. A 120° angle and a 240° angle are shown in standard position below.

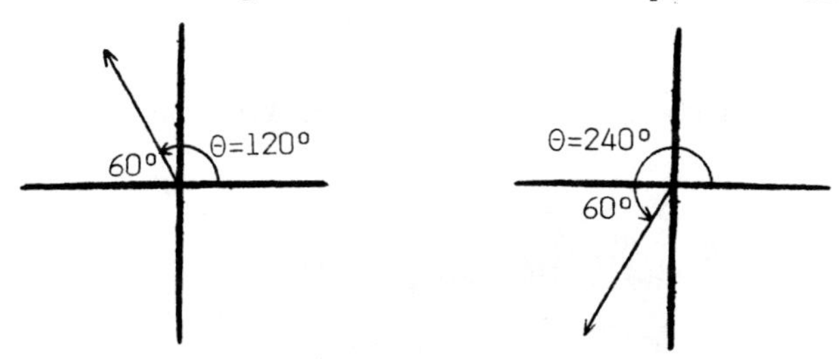

The cosines of angles are negative in both the second and third quadrants. Since the reference angle for both angles above is 60°, both have the same cosine. That is:

$$\cos 120° = \cos 240° = -0.5$$

In all three equations below, θ is the angle whose cosine is -0.5 .

$$\cos \theta = -0.5$$

$$\theta = \arccos(-0.5)$$

$$\theta = \cos^{-1}(-0.5)$$

To find θ on a calculator, we enter -0.5 and press [INV] [cos]. We get:

Enter	Press	Display
0.5	[+/-] [INV] [cos]	120

The calculator gives us 120°, the positive second-quadrant angle. We can find the third-quadrant angle in two steps:

1. Subtract 120° from 180° to find the reference angle.

$$180° - 120° = 60°$$

2. Add the reference angle to 180°.

$$180° + 60° = 240°$$

Find the second-quadrant and third-quadrant angles that satisfy each equation. Round to the nearest whole-number degree.

a) $\cos \theta = -0.5527$ θ = ________ and ________

b) $\theta = \arccos(-0.9251)$ θ = ________ and ________

c) $\theta = \cos^{-1}(-0.0273)$ θ = ________ and ________

---

a) 124° and 236°

b) 158° and 202°

c) 92° and 268°

---

78. Find the two angles between 0° and 360° that satisfy each equation below. Round to the nearest whole-number degree.

a) $\theta = \cos^{-1} 0.7164$ θ = ________ and ________

b) $\theta = \arccos(-0.2048)$ θ = ________ and ________

---

a) 44° and 316°

b) 102° and 258°

6-12 TRIGONOMETRIC EQUATIONS

In this section, we will solve some simple trigonometric equations. The solutions are limited to angles between 0° and 360°.

---

79. The equation below is a trigonometric equation of the simplest type.

$$\tan \theta = 1.1917536$$

Solving the equation means finding the angle or angles that satisfy it. There are many solutions for the equation because all of the following angles satisfy it.

1. Two positive angles between 0° and 360°.

   They are: 50° and 230°

2. Two positive angles in each revolution beyond 360°.

   They are: 410° and 590°, 770° and 950°, 1130° and 1310°, etc.

   Note: To get the pairs, we added 360°, 720°, and 1080° to 50° and 230°.

3. Two negative angles between 0° and -360°.

   They are: -130° and -310°.

4. Two negative angles in each revolution beyond -360°.

   They are: -490° and -670°, -850° and -1030°, -1210° and -1390°, etc.

   Note: To get the pairs, we added -360°, -720°, and -1080° to -130° and -310°.

As you can see, the number of solutions for the equation is very large.

---

80. Since the number of solutions for a trigonometric equation can be quite large, we will ordinarily ask for only one or two angles of the solution.

Round each answer to the nearest whole-number degree.

a) Find the second-quadrant angle between 0° and 360° that satisfies:

   $\tan \theta = -0.8466$ $\theta =$ ________

b) Find the fourth-quadrant angle between 0° and 360° that satisfies:

   $\cos \theta = 0.2951$ $\theta =$ ________

c) Find the two angles between 0° and 360° that satisfy:

   $\sin \theta = -0.0466$ $\theta =$ ________ and ________

---

a) 140°

b) 287°

c) 183° and 357°

81. Round each answer to the nearest whole-number degree.

a) Find the third-quadrant angle between 0° and 360° that satisfies:

$\cos\theta = -0.7052$ $\theta =$ ________

b) Find the fourth-quadrant angle between 0° and 360° that satisfies:

$\sin\theta = -0.4132$ $\theta =$ ________

c) Find the two angles between 0° and 360° that satisfy:

$\tan\theta = 11.466$ $\theta =$ ________ and ________

---

82. Find the angle (or angles) between 0° and 360° that satisfies each equation.

a) $\sin\theta = +1$ $\theta =$ ________

b) $\cos\theta = 0$ $\theta =$ ________

c) $\tan\theta = 0$ $\theta =$ ________

Answers to 81: a) 225° b) 336° c) 85° and 265°

---

83. To solve the equation below, we isolated "sin θ" first. Only the acute-angle (0° to 90°) solution is given. Find the acute-angle solution for the other equation. Round to the nearest whole-number degree.

$2\sin\theta = 1$

$\sin\theta = \frac{1}{2}$ or 0.5

$\theta = 30°$

$4\cos\theta = 3$

Answers to 82: a) 90° b) 90° and 270° c) 180°

---

84. To solve the equation below, we began by replacing each side with its opposite. Find an acute-angle solution to the other equation. Round to the nearest whole-number degree.

$-6\sin\theta = -5$

$6\sin\theta = 5$

$\sin\theta = \frac{5}{6}$

$\theta = 56°$

$-8\tan\theta = -1$

Answer to 83: $\theta = 41°$, from: $\cos\theta = 0.75$

---

85. To solve the equation below, we began by replacing each side with its opposite. Solve the other equation. Give two solutions between 0° and 360°. Round to the nearest whole-number degree.

$-\cos\theta = 0.4187$

$\cos\theta = -0.4187$

$\theta = 115°$ and $245°$

$-\tan\theta = 1.8362$

Answer to 84: $\theta = 7°$, from: $\tan\theta = \frac{1}{8}$

86. To isolate "sin θ" below, we added -4 to both sides. Solve the other equation. Give a solution between 0° and 360°.

$$4 + \sin\theta = 5$$
$$(-4) + 4 + \sin\theta = 5 + (-4)$$
$$\sin\theta = 1$$
$$\theta = 90°$$

$$3 + \cos\theta = 2$$

Answer: $\theta = 119°$ and $299°$, from: $\tan\theta = -1.8362$

---

87. To isolate "cos θ" below, we began by isolating "3 cos θ". Solve the other equation. Give two solutions between 0° and 360°. Round to the nearest whole-number degree.

$$3\cos\theta + 2 = 0$$
$$3\cos\theta + 2 + (-2) = 0 + (-2)$$
$$3\cos\theta = -2$$
$$\cos\theta = -\frac{2}{3}$$
$$\theta = 132° \text{ and } 228°$$

$$5\tan\theta - 7 = 0$$

Answer: $\theta = 180°$, from: $\cos\theta = -1$

---

88. Following the example, solve the other equation. Give two solutions between 0° and 360°. Round to the nearest whole-number degree.

$$4 - 2\cos\alpha = 5$$
$$-2\cos\alpha = 1$$
$$2\cos\alpha = -1$$
$$\cos\alpha = -\frac{1}{2} \text{ or } -0.5$$
$$\alpha = 120° \text{ and } 240°$$

$$10 - 5\tan\varphi = 8$$

Answer: $\theta = 54°$ and $234°$, from: $\tan\theta = \frac{7}{5}$

---

89. To solve the equation below, we began by getting both "tan θ" terms on the same side.

$$7 + 3\tan\theta = 5\tan\theta$$
$$7 + 3\tan\theta + (-3\tan\theta) = 5\tan\theta + (-3\tan\theta)$$
$$7 = 2\tan\theta$$
$$\tan\theta = \frac{7}{2} \text{ or } 3.5$$
$$\theta = 74° \text{ and } 254°$$

Answer: $\varphi = 22°$ and $202°$, from: $\tan\varphi = \frac{2}{5}$ or 0.4

Continued on following page.

89. Continued

Solve the following equation. Give two solutions between 0° and 360°. Round to the nearest whole-number degree.

$$5 - 2\cos\alpha = 3\cos\alpha + 9$$

Answer: $\alpha = 143°$ and $217°$, from: $\cos\alpha = -0.8$

90. To solve the equation below, we began by substituting "1 cos α" for "cos α". Find two solutions between 0° and 360° for the other equation. Round to the nearest whole-number degree.

$$2 + \cos\alpha = 4\cos\alpha$$
$$2 + 1\cos\alpha = 4\cos\alpha$$
$$2 = 3\cos\alpha$$
$$\cos\alpha = \frac{2}{3}$$
$$\alpha = 48° \text{ and } 312°$$

$$5 + 9\sin\theta = \sin\theta$$

Answer: $\theta = 219°$ and $321°$, from: $\sin\theta = -\frac{5}{8}$

91. To solve the equation below, we began by substituting "-1 sin θ" for "-sin θ". Find two solutions between 0° and 360° for the other equation. Round to the nearest whole-number degree.

$$4\sin\theta = 3 - \sin\theta$$
$$4\sin\theta = 3 - 1\sin\theta$$
$$5\sin\theta = 3$$
$$\sin\theta = \frac{3}{5}$$
$$\theta = 37° \text{ and } 143°$$

$$11 - \tan\theta = 5\tan\theta$$

Answer: $\theta = 61°$ and $241°$, from: $\tan\theta = \frac{11}{6}$

92. The equation below contains two trigonometric ratios. To solve it, we must get an equation that contains <u>only one</u> trigonometric ratio. To do so, we used an identity to substitute for "tan θ". Find two solutions between 0° and 360° for the other equation.

$$\cos\theta\tan\theta = 0.2019$$

$$\cancel{\cos\theta}\left(\frac{\sin\theta}{\cancel{\cos\theta}}\right) = 0.2019$$

$$\sin\theta = 0.2019$$

$$\theta = 12^\circ \text{ and } 168^\circ$$

$$4\sin\theta\cot\theta = 3$$

---

93. To get an equation with only one trigonometric ratio below, we substituted "tan θ" for $\frac{\sin\theta}{\cos\theta}$. Find two solutions between 0° and 360° for the other equation.

Answer to 92: $\theta = 41^\circ$ and 319°, from $\cos\theta = 0.75$

$$1.5\cos\theta = \sin\theta$$

$$1.5 = \frac{\sin\theta}{\cos\theta}$$

$$1.5 = \tan\theta$$

$$\theta = 56^\circ \text{ and } 236^\circ$$

$$4\sin A = 5\cos A$$

Answer to 93: $A = 51^\circ$ and 231°, from: $\tan\theta = 1.25$

SELF-TEST 21 (pages 290-301)

Find the angles between 0° and 360° that satisfy each equation below. Round each answer to the nearest whole-number degree.

1. $\theta = \arcsin 0.5279$ ______	2. $\theta = \cos^{-1} 0.1095$ ______	3. $\theta = \arctan(-0.8712)$ ______	4. $\theta = \sin^{-1}(-0.2768)$ ______

5. $3 \cos \theta = -2$ ______	6. $3 - \sin \theta = 4$ ______	7. $5 - 2 \tan \theta = 8$ ______
8. $\sin \alpha = 3 \sin \alpha - 1$ ______	9. $6 \cos \theta = 4 - \cos \theta$ ______	10. $1.8 - \tan \phi = 2 \tan \phi$ ______
11. $\sin A = 2.4 \cos A$ ______	12. $2 \cos \alpha = 7 \sin \alpha$ ______	13. $8 \cos \theta \tan \theta = 5$ ______

ANSWERS:

1. $\theta = 32°$ and $148°$
2. $\theta = 84°$ and $276°$
3. $\theta = 139°$ and $319°$
4. $\theta = 196°$ and $344°$
5. $\theta = 132°$ and $228°$
6. $\theta = 270°$
7. $\theta = 124°$ and $304°$
8. $\alpha = 30°$ and $150°$
9. $\theta = 55°$ and $305°$
10. $\phi = 31°$ and $211°$
11. $A = 67°$ and $247°$
12. $\alpha = 16°$ and $196°$
13. $\theta = 39°$ and $141°$

SUPPLEMENTARY PROBLEMS - CHAPTER 6

Assignment 19

State whether "sec θ", "csc θ", or "cot θ" is the reciprocal of:

1. $\tan \theta$ 2. $\sin \theta$ 3. $\cos \theta$

Using "opp", "adj", and "hyp", define:

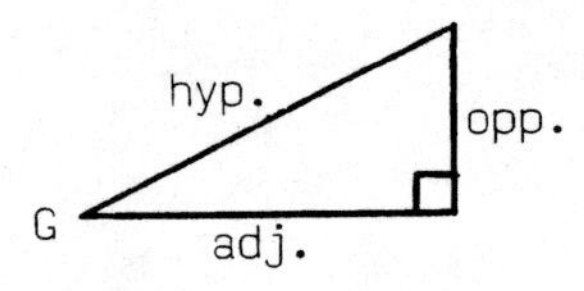

4. csc G 5. sec G 6. cot G

Using sides "p", "r", and "s" in right-triangle PRS, define:

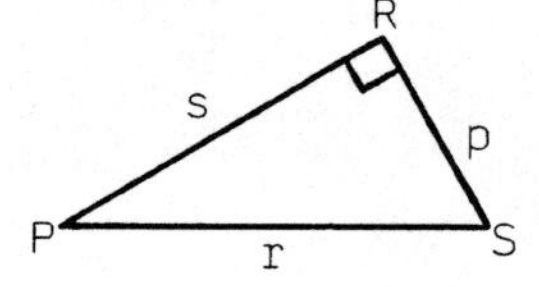

7. sec P 9. csc P 11. cot S

8. cot P 10. csc S 12. sec S

Rounding to four decimal places, find the value of:

13. sec 48° 15. cot 75° 17. sec(-60°)

14. csc 21° 16. csc 213° 18. cot 339°

Using the ratio identities, complete the following.

19. $\frac{\cos \theta}{\sin \theta} = \_\_\_\_$ 20. $\frac{\sin \theta}{\tan \theta} = \_\_\_\_$ 21. $\frac{\sin \theta}{\cos \theta} = \_\_\_\_$ 22. $\frac{\cos \theta}{\cot \theta} = \_\_\_\_$

Using the Pythagorean identities, complete the following.

23. $1 - \sin^2\theta = \_\_\_\_$ 24. $\sec^2\theta - 1 = \_\_\_\_$ 25. $\csc^2\theta - \cot^2\theta = \_\_\_\_$

Using the reciprocal identities, complete the following.

26. $(\sec \theta)(\_\_\_) = 1$ 27. $(\tan \theta)(\_\_\_) = 1$ 28. $(\cos \theta)(\_\_\_) = 1$ 29. $(\csc \theta)(\_\_\_) = 1$

30. Solve for "$\sin \theta$". $\frac{\cos \theta}{\sin \theta} = \cot \theta$

31. Solve for "$\csc \theta$". $\sin \theta = \frac{1}{\csc \theta}$

32. Solve for "$\cos \theta$". $\sin^2\theta + \cos^2\theta = 1$

33. Solve for "$\sec \theta$". $\sec^2\theta - \tan^2\theta = 1$

Assignment 20

Prove the following identities.

1. $\cos \theta \csc \theta = \cot \theta$
2. $\sin \theta \sec \theta \cot \theta = 1$
3. $\tan \theta \csc \theta = \sec \theta$
4. $\sin \theta = \csc \theta - \csc \theta \cos^2\theta$
5. $\sec \theta = \cos \theta + \tan \theta \sin \theta$
6. $\cot \theta \cos \theta = \csc\theta - \sin\theta$

Write each equation in arcsin, arccos, or arctan notation.

7. $\sin 30° = 0.5$ 8. $\cos \theta = r$ 9. $\tan A = 2.7384$

Write each equation in ordinary notation.

10. $52° = \arccos 0.6157$ 11. $\theta = \tan^{-1} 1.2517$ 12. $\alpha = \sin^{-1} w$

Write each equation in $\sin^{-1}$, $\cos^{-1}$, or $\tan^{-1}$ notation.

13. $\tan 60° = 1.732$ 14. $\sin 12° = d$ 15. $\cos \theta = N$

Rounding to a whole number, find an acute angle satisfying each equation.

16. $\theta = \cos^{-1} 0.5336$
17. $\alpha = \arctan 3.1918$
18. $A = \sin^{-1} 0.9917$

Rounding to four decimal places, find the numerical value of each letter.

19. $39° = \tan^{-1}T$
20. $86° = \arcsin h$
21. $45° = \arccos x$

Using the formula below and rounding each answer to a whole number:

$$\boxed{\theta = \arccos\left(\frac{R}{Z}\right)}$$

22. Find θ when $R = 395$ and $Z = 473$.
23. Find Z when $\theta = 75°$ and $R = 2{,}250$.

Using the formula below and rounding each answer to a whole number:

$$\boxed{\alpha = \sin^{-1}\left(\frac{X}{Z}\right)}$$

24. Find X when $\alpha = 55°$ and $Z = 27{,}355$.
25. Find Z when $\alpha = 14°$ and $X = 138$.

Assignment 21

Find the angles between 0° and 360° that satisfy each equation. Round each answer to the nearest whole-number degree.

1. $\theta = \arcsin 0.5279$
2. $\theta = \tan^{-1} 5.7360$
3. $\theta = \cos^{-1} 0.1095$
4. $\theta = \arctan(-0.8712)$
5. $\theta = \sin^{-1}(-0.2648)$
6. $\theta = \arccos(-0.7044)$

7. $3 \cos \theta = 1$
8. $-\sin \theta = 0.92$
9. $-2 \tan \theta = 5$
10. $-2 \cos \theta = 1$
11. $9 \sin \theta = 4$
12. $-7 \cos \theta = -5$
13. $4 \tan \theta = -15$
14. $-5 \sin \theta = -5$
15. $6 + \tan \theta = 8$
16. $3 - \sin \theta = 4$
17. $4 \cos \theta + 1 = 0$
18. $1 - 2 \tan \theta = 7$
19. $5 \sin \theta + 3 = 0$
20. $2 - 9 \cos \theta = 6$
21. $3 - \tan \theta = 4$
22. $2 \cos \theta + 5 = 3$
23. $1 + \sin \theta = 3 \sin \theta$
24. $1 - 6 \cos \theta = \cos \theta$
25. $2 \tan \theta - 1 = 3 \tan \theta + 3$
26. $2 \cos \theta + 3 = 1 - \cos \theta$
27. $8 \cos \theta \tan \theta = 1$
28. $5 \sin \theta \cot \theta = 4$
29. $\sin \theta = 5 \cos \theta$
30. $3 \cos \theta = 2 \sin \theta$

Chapter 7 COMPLEX NUMBERS

In this chapter, we will define complex numbers and show how they can be represented by vectors on the complex plane. Methods for adding, subtracting, multiplying, and dividing complex numbers are shown. Since complex numbers can be expressed in either rectangular form or polar form, the four basic operations are shown in both forms. Some formula evaluations requiring combined operations with complex numbers are included.

7-1 REAL AND IMAGINARY NUMBERS

In this section, we will define "real" and "imaginary" numbers.

---

1. Any positive or negative whole number, decimal, or fraction is a "real" number. That is:

   37 , 5.3 , 0.068 , and $\frac{4}{7}$ are real numbers.

   -37 , -5.3 , -0.068 , and $-\frac{4}{7}$ are also ________ numbers.

---

2. Before defining "imaginary" numbers, we must review some basic ideas about squares and square roots. | real

   1. When a number is squared, the "square" is always a positive number. That is:

      The "square" of both +5 and -5 is +25.

   2. Any positive number has two square roots, a principal (or positive) square root and a negative square root. The following symbols are used:

      $\sqrt{}$ or $+\sqrt{}$ for the principal square root

      $-\sqrt{}$ for the negative square root.

      Therefore: $\sqrt{9}$ or $+\sqrt{9} = +3$

      $-\sqrt{9} = -3$

   a) What number can be squared to get -36? ________

   b) The principal square root of 64 is ________.

   c) The negative square root of 81 is ________.

3. If we take either square root of a positive number, we get a real number. For example:

$\sqrt{49} = +7$, and +7 is a real number.

$-\sqrt{49} = -7$, and -7 is a real number.

However, we cannot take the square root of a negative number and get a real number. For example:

$\sqrt{-9}$ does not equal +3, since $(+3)^2 = +9$.

$\sqrt{-9}$ does not equal -3, since $(-3)^2 = +9$.

a) What two real numbers do we get if we take the two square roots of 100? ________ and ________

b) Do we get a real number if we take the square root of -4? ________

Answers: a) There is none. b) +8 c) -9

---

4. The square roots of positive numbers are real numbers. For example:

$\sqrt{16}$ and $-\sqrt{39}$ are real numbers.

The square roots of negative numbers are not real numbers. To contrast them with "real" numbers, mathematicians call them "imaginary" numbers. For example:

$\sqrt{-16}$ and $\sqrt{-39}$ are imaginary numbers.

Note: The term "imaginary" was probably a bad choice. It suggests that the numbers do not exist or that they are useless. However, they do exist and they are useful.

Write either "real" or "imaginary" in each blank.

a) $\sqrt{27}$ and $-\sqrt{85}$ are ________________ numbers.

b) $\sqrt{-8}$ and $\sqrt{-56}$ are ________________ numbers.

Answers: a) +10 and -10 b) No

---

5. Using the principles for radicals, we can write any imaginary number as a real number times $\sqrt{-1}$. For example:

$$\sqrt{-9} = \sqrt{(9)(-1)} = \sqrt{9} \cdot \sqrt{-1} = 3\sqrt{-1}$$
$$\sqrt{-25} = \sqrt{(25)(-1)} = \sqrt{25} \cdot \sqrt{-1} = 5\sqrt{-1}$$

Following the examples, write each of these as a real number times $\sqrt{-1}$.

a) $\sqrt{-4}$ = ____________ b) $\sqrt{-64}$ = ____________

Answers: a) real b) imaginary

---

6. To write $\sqrt{-47}$ as a real number times $\sqrt{-1}$, we used a calculator. We rounded to hundredths.

$$\sqrt{-47} = \sqrt{(47)(-1)} = \sqrt{47} \cdot \sqrt{-1} = 6.86\sqrt{-1}$$

Use a calculator to write each of these as a real number times $\sqrt{-1}$.

a) Round to hundredths. $\sqrt{-83}$ = ____________

b) Round to tenths. $\sqrt{-407}$ = ____________

Answers: a) $2\sqrt{-1}$, from: $\sqrt{4} \cdot \sqrt{-1}$ b) $8\sqrt{-1}$, from: $\sqrt{64} \cdot \sqrt{-1}$

7. Instead of writing $\sqrt{-1}$, the letter "j" can be used.

Instead of $6\sqrt{-1}$, we can write 6j .

Write each of these with "j" instead of $\sqrt{-1}$.

a) $2\sqrt{-1}$ = ______ b) $25\sqrt{-1}$ = ______ c) $1.76\sqrt{-1}$ = ______

Answers: a) $9.11\sqrt{-1}$, from: $\sqrt{83} \cdot \sqrt{-1}$ b) $20.2\sqrt{-1}$, from: $\sqrt{407} \cdot \sqrt{-1}$

---

8. Remember that "j" stands for $\sqrt{-1}$. Therefore:

10j means $10\sqrt{-1}$ 4.9j means ______

Answers: a) 2j b) 25j c) 1.76j

---

9. Three forms of the same imaginary number are shown below.

$\sqrt{-4}$ $2\sqrt{-1}$ 2j

<u>The non-radical or "j" form is always used because it is more efficient in mathematical operations</u>.

We can add, subtract, multiply, divide, and square imaginary numbers. When doing so, the letter "j" is handled just like any other letter. For example:

$3j + 4j = 7j$ $(5j)^2 = (5j)(5j) = 25j^2$

$9j - 5j = 4j$ $\frac{4j}{8j} = \left(\frac{4}{8}\right)\left(\frac{j}{j}\right) = \frac{1}{2}(1) = \frac{1}{2}$

$8(3j) = 24j$

Complete these:

a) $3j - 8j$ = ______ c) $(9j)^2$ = ______

b) $(-4)(9j)$ = ______ d) $\frac{32j}{8j}$ = ______

Answer: $4.9\sqrt{-1}$

---

10. Two points should be mentioned about the use of the letter "j" for imaginary numbers.

1. In pure mathematics, the letter "i" is used instead of "j" for $\sqrt{-1}$. That is:

"3i" is used instead of "3j" for $3\sqrt{-1}$.

However, the letter "j" will be used in this text because the letter "j" is used in practical applications.

2. Sometimes the letter "j" is written in front of the real number instead of after it. That is:

5j is written j5 , with the "j" first.

We will avoid the j5 type of notation in this text <u>and always write the real number first</u>.

Answers: a) -5j b) -36j c) $81j^2$ d) 4

7-2 COMPLEX NUMBERS

In this section, we will define "complex" numbers and show how they can be represented graphically by a vector on the complex plane.

---

11. A "complex" number is an addition of a real number and an imaginary number. Each number below, for example, is a complex number.

$$4 + 5j \qquad 8 + 3j$$

As you can see, a "complex" number is one number with two parts: a real-number part and an imaginary-number part.

In $4 + 5j$: the real-number part is 4.
the imaginary-number part is $5j$.

In $8 + 3j$: a) the real-number part is ______.

b) the imaginary-number part is ______.

---

12. In a complex number, the coefficient of "j" can be negative. Two examples are shown.

Answers to frame 11: a) 8 b) 3j

$$2 + (-7j) \qquad 9 + (-6j)$$

When the coefficient of "j" is negative, the complex number is usually written as a subtraction. For example:

Instead of $2 + (-7j)$, we write: $2 - 7j$

Instead of $9 + (-6j)$, we write: ______________

---

13. In a complex number, the real-number part can be negative. Two examples are shown:

Answer to frame 12: $9 - 6j$

$$-5 + 3j \qquad -2 + (-9j)$$

Ordinarily, we would write $-2 + (-9j)$ as a subtraction. That is:

Instead of $-2 + (-9j)$, we write: ______________

---

14. Any complex number can be represented graphically by a vector on the complex plane. A diagram of the complex plane is shown at the right. Notice these points:

Answer to frame 13: $-2 - 9j$

1. The horizontal-axis is called the R-axis (or real-number axis).

2. The vertical-axis is called the j-axis (or imaginary-number axis).

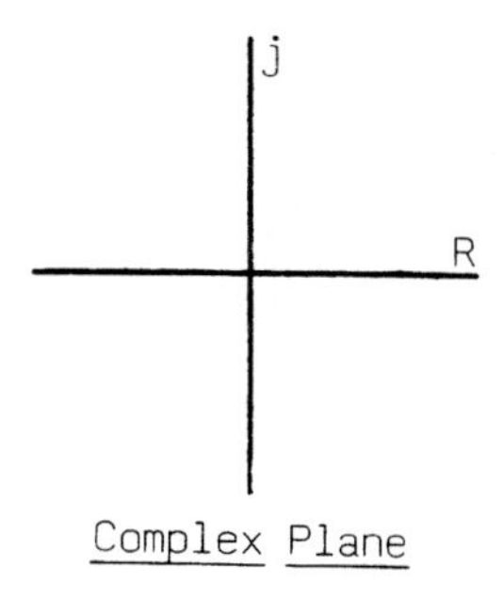

Complex Plane

15. On the complex plane at the right, we have drawn the vector that represents:

$$4 + 3j$$

Notice these points:

1. The vector begins at the origin.
2. The real-number part ("4") is graphed on the horizontal or R-axis. "4" means 4-units to the right.
3. The imaginary-number part ("3j") is graphed on the vertical or j-axis. "3j" means 3-units up.

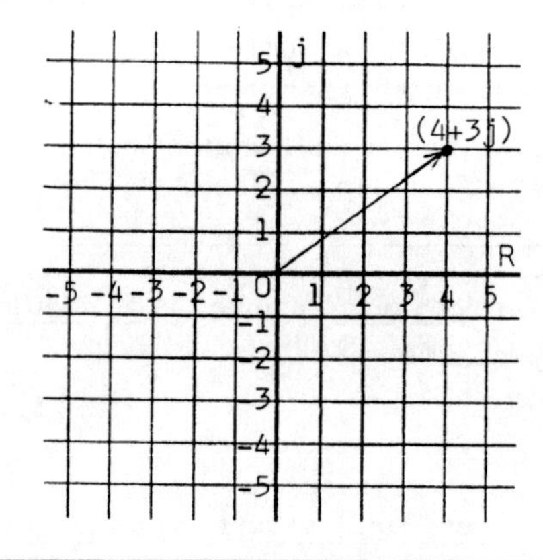

---

16. Four vectors are drawn on the complex plane at the right. Write the complex number which each vector represents.

#1: 

#2: ____________________

#3: ____________________

#4: ____________________

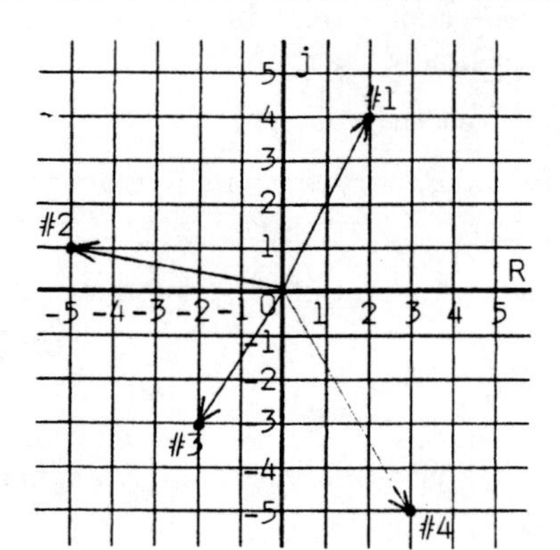

---

17. Four vectors are drawn on the complex plane at the right. Write the complex number which each vector represents.

#1: ____________________

#2: ____________________

#3: ____________________

#4: ____________________

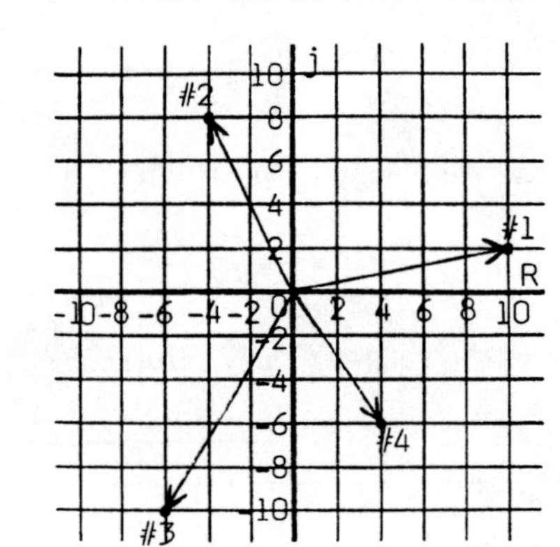

Answers to 16:

#1: 2 + 4j

#2: -5 + 1j (or -5 + j)

#3: -2 - 3j

#4: 3 - 5j

Answers to 17:

#1: 10 + 2j

#2: -4 + 8j

#3: -6 - 10j

#4: 4 - 6j

7-3 COMPLEX NUMBERS IN WHICH THE R-TERM OR j-TERM IS "0"

In this section, we will discuss complex numbers in which either the R-term or j-term is "0". We will show that complex numbers of that type can be written in either "complete" or "incomplete" form.

---

18. Four vectors are drawn on the complex plane at the right. Vector #2 and vector #4 lie on the real (horizontal) axis. Write the complex number which represents each vector.

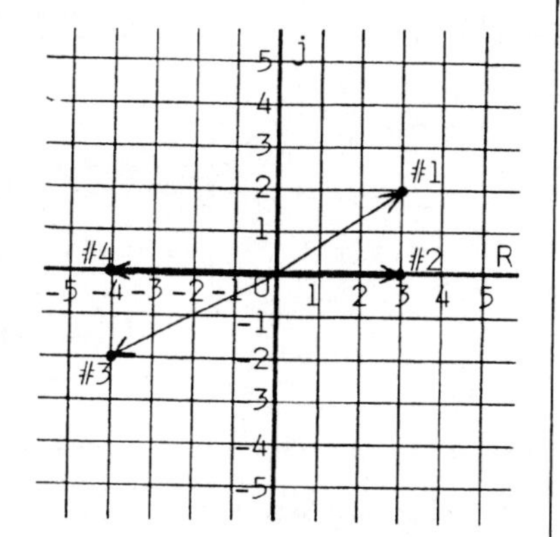

#1: ____________

#2: ____________

#3: ____________

#4: ____________

---

19. In the last frame, we got the complex numbers below for vectors that lie on the horizontal (real) axis. In each one, the coefficient of the "j" term is "0".

$$3 + 0j \qquad\qquad -4 - 0j$$

Since "0j" = 0, we can put either an addition sign ("+") or a subtraction sign ("-") in front of 0j. That is:

$$-4 - 0j = -4 + 0j$$

However, we use the form with the addition sign. That is:

Instead of $-4 - 0j$, we use $-4 + 0j$.

Instead of $7 - 0j$, we use ____________ :

Answers to frame 18:
#1: $3 + 2j$
#2: $3 + 0j$
#3: $-4 - 2j$
#4: $-4 - 0j$

---

20. Four vectors are drawn on the complex plane at the right. Vector #2 and vector #4 lie on the imaginary (vertical) axis. Write the complex number that represents each vector.

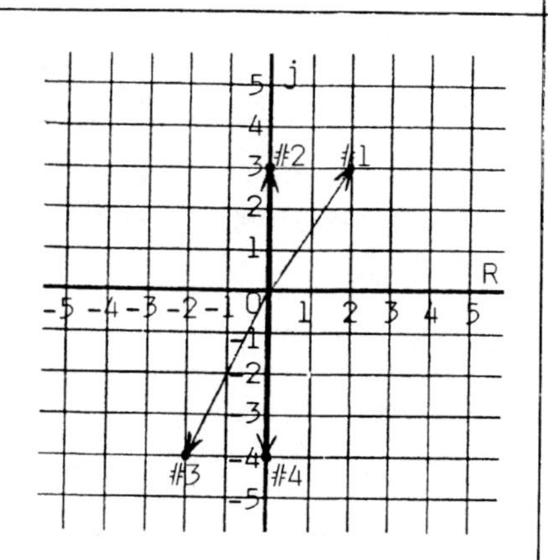

#1: ____________

#2: ____________

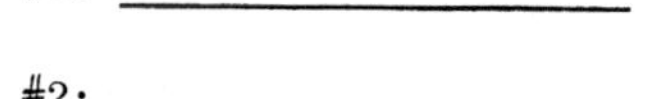

#3: ____________

#4: ____________

Answer to frame 19: $7 + 0j$

21. In the last frame, we got the complex numbers below for vectors that lie on the vertical (imaginary) axis. In each one, the real-number term is "0".

$0 + 3j$ $0 - 4j$

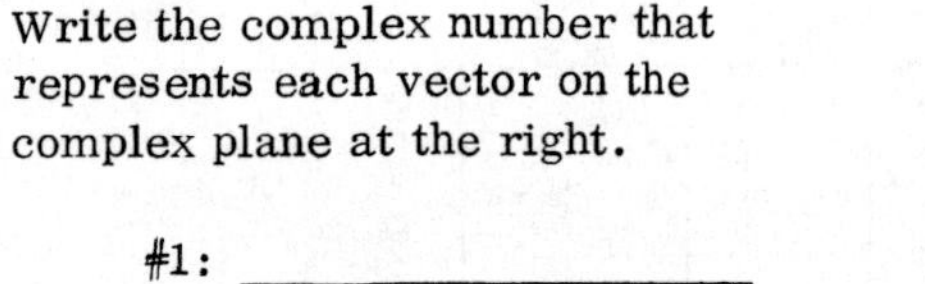

Write the complex number that represents each vector on the complex plane at the right.

#1: ______________

#2: ______________

#3: ______________

#4: ______________

#1: $2 + 3j$

#2: $0 + 3j$

#3: $-2 - 4j$

#4: $0 - 4j$

---

22. The general form of a complex number is $\boxed{a + bj}$.

If the real-number term "a" is "0", we get complex numbers like those below. Their vectors lie on the imaginary (vertical) axis.

$0 + 5j$ $0 - 9j$

If "b", the coefficient of "j", is "0", we get complex numbers like those below. Their vectors lie on the real (horizontal) axis.

$3 + 0j$ $-10 + 0j$

a) Which one of the three complex numbers below has a vector on the horizontal axis? ______________

$0 + 4j$ $8 + 0j$ $0 - 7j$

b) Which one of the three complex numbers below has a vector on the vertical axis? ______________

$-3 + 0j$ $0 - 2j$ $10 + 0j$

#1: $0 + 4j$

#2: $8 + 0j$

#3: $0 - 10j$

#4: $-6 + 0j$

---

23. When either "a" or "b" in a complex number is "0", we can write the complex number without the "0" term. For example:

Instead of $0 + 3j$, we can write "3j".

Instead of $7 + 0j$, we can write "7".

Write each of these without the "0" term.

a) $0 - 5j$ = ______ b) $-1 + 0j$ = ______ c) $0 + j$ = ______

a) $8 + 0j$

b) $0 - 2j$

---

a) $-5j$

b) -1

c) j

24. When either "a" or "b" in a complex number is "0", the "0" term may or may not be written.

A "<u>complete</u>" complex number is one in which <u>both terms are written</u>.

$0 + 7j$ and $8 + 0j$ are "<u>complete</u>" complex numbers.

An "<u>incomplete</u>" complex number is one in which the "0" term is not written.

$7j$ and 8 are "<u>incomplete</u>" complex numbers.

Write each of these "incomplete" complex numbers as a "complete" complex number.

a) $5j$ = ______________ c) -9 = ______________

b) 4 = ______________ d) $-j$ = ______________

Answers: a) $0 + 5j$ b) $4 + 0j$ c) $-9 + 0j$ d) $0 - j$

25. a) Since $27 = 27 + 0j$, the vector for "27" lies on the ______________ (horizontal/vertical) axis of the complex plane.

b) Since $-11j = 0 - 11j$, the vector for "$-11j$" lies on the ______________ (horizontal/vertical) axis of the complex plane.

Answers: a) horizontal b) vertical

26. If the following "incomplete" complex numbers were graphed on the complex plane, which ones would have vectors on the <u>horizontal</u> axis? ____________

a) $12j$ b) 10 c) -1 d) $-j$

Answer: (b) and (c)

7-4 ADDING COMPLEX NUMBERS

In this section, we will discuss the addition of complex numbers. We will show how vectors can be added by adding complex numbers.

---

27. To add two complex numbers, we can add their vectors. For example, we added $(1 + 3j)$ and $(4 + 1j)$ at the right by the parallelogram method. The complex number of the resultant is $(5 + 4j)$. Therefore:

$$(1 + 3j) + (4 + 1j) = 5 + 4j$$

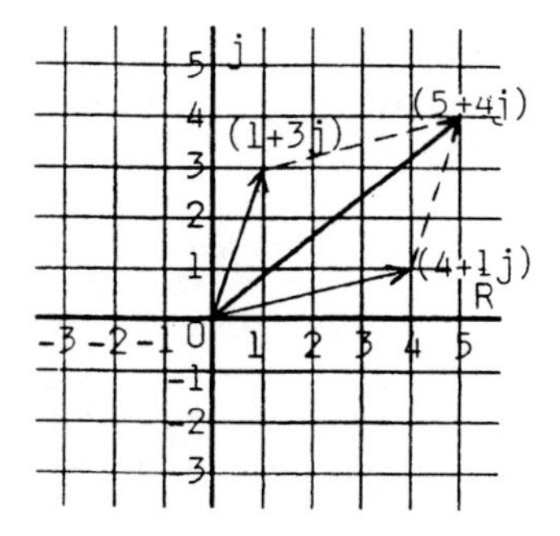

The sum can be obtained directly by adding the real-number parts and the imaginary-number parts of the addends. That is:

$$5 = 1 + 4$$
$$4j = 3j + 1j$$

Continued on following page.

27. Continued

Do these additions by simply adding the real-number and imaginary-number parts.

a) $(5 + 3j) + (6 + 4j) =$ ______________

b) $(9 + 2j) + (-4 + 6j) =$ ______________

---

28. In the addition below, both $(4 - 3j)$ and $(-7 - 2j)$ contain a subtraction.

$$(4 - 3j) + (-7 - 2j) = -3 - 5j$$

To get the sum, we again added the real-number parts and the imaginary-number parts. That is:

$$-3 = 4 + (-7)$$

$$-5j = (-3j) + (-2j)$$

Do these additions.

a) $(8 - 2j) + (-4 + 5j) =$ ______________

b) $(-1 - 6j) + (-10 - 7j) =$ ______________

Answers: a) $11 + 7j$ b) $5 + 8j$

---

29. The coefficient "1" for the j-term is not usually written. However:

$$3 + j \text{ means } 3 + 1j$$

$$-4 - j \text{ means } -4 - 1j$$

Do these additions.

a) $(-5 + j) + (3 + j) =$ ______________

b) $(12 - j) + (-7 - 2j) =$ ______________

Answers: a) $4 + 3j$ b) $-11 - 13j$

---

30. To add three or more complex numbers, we add their real-number parts and their imaginary-number parts. Do these additions.

a) $(3 + 5j) + (7 - 4j) + (-4 - 6j) =$ ______________

b) $(-1 - j) + (1 - j) + (-2 + j) + (4 - j) =$ ______________

Answers: a) $-2 + 2j$ b) $5 - 3j$

---

31. When "a" or "b" in one or both complex numbers is "0", we add them in the usual way. For example:

$$(8 + 5j) + (0 - 3j) = 8 + 2j$$

$$(1 + 0j) + (-3 - 4j) = -2 - 4j$$

$$(0 - 8j) + (-1 + 0j) = \text{______________}$$

Answers: a) $6 - 5j$ b) $2 - 2j$

Answer: $-1 - 8j$

32. When one or both complex numbers is "incomplete", we add them in the usual way. For example:

$$(4 + 3j) + (-2j) = 4 + j$$

$$(-1) + (7 - j) = 6 - j$$

$$(-9j) + (10) = \_\_\_\_\_\_$$

---

33. When both complex numbers contain only real-number terms or imaginary-number terms, the sum is an "incomplete" complex number. For example:

$(3) + (-5) = -2$ (which is $-2 + 0j$)

$(-6j) + (j) = -5j$ (which is $0 - 5j$)

a) The vector for "-2" lies on the ______________ (horizontal/vertical) axis of the complex plane.

b) The vector for "-5j" lies on the ______________ (horizontal/vertical) axis of the complex plane.

Answer to 32: 10 - 9j

---

34. When vectors are represented by complex numbers, we can add the vectors by adding their complex numbers. An example is discussed below.

We added two vectors by the parallelogram method on the complex plane at the right. The complex numbers of the vector-addends are $(3 + 2j)$ and $(1 - 4j)$. The complex number of the resultant is $(4 - 2j)$. The resultant can be obtained by adding the complex numbers of the vector-addends. That is:

$$(3 + 2j) + (1 - 4j) = 4 - 2j$$

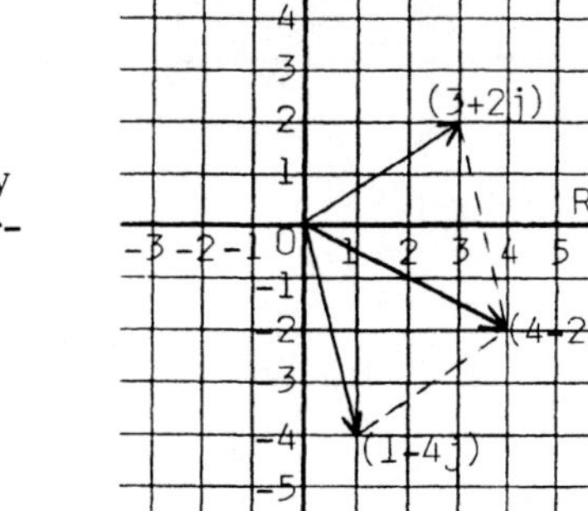

Find each resultant below by adding the complex numbers of the vector-addends.

	Vector 1	Vector 2	Resultant
a)	6.5 + 1.8j	2.3 - 4.9j	______
b)	375 - 15j	-150 + 40j	______

Answers to 33: a) horizontal b) vertical

---

35. a) The complex numbers of two vectors are $(12 + j)$ and $(-9 - 4j)$. Their resultant is ______________.

b) The complex numbers of three vectors are $(1 - j)$, $(3 + j)$, and $(-7 + 2j)$. Their resultant is ______________.

Answers to 34: a) 8.8 - 3.1j b) 225 + 25j

Answers to 35: a) 3 - 3j b) -3 + 2j

SELF-TEST 22 (pages 304-314)

1. In "j" form, $\sqrt{-183}$ = ________. Round to tenths.

Referring to the complex plane at the right, write the complex number which each vector represents.

2. #1: ________ 4. #3: ________

3. #2: ________ 5. #4: ________

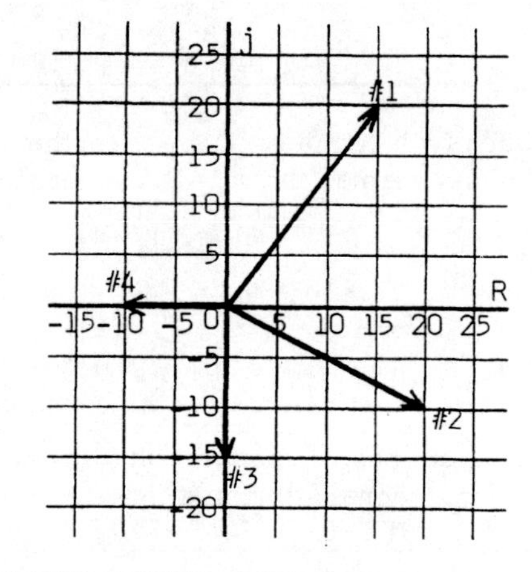

6. When graphed, which of the following complex numbers would be vectors lying on the vertical axis? ________

a) 2 + 0j b) 0 + 2j c) -3j d) 5 e) 1 - j f) 0 - j

Do these additions.

7. (8 - 2j) + (-3 + j) ________

8. (-37.3 + 25.4j) + (19.7 - 12.6j) ________

Do these additions.

9. (-5 + 3j) + (1 - j) + (-2 - 2j) ________

10. (0 - 3j) + (5 + 2j) + (-6 + 0j) ________

11. Find the resultant of two vectors whose complex numbers are (0 - 370j) and (180 + 0j). ________

12. Find the resultant of three vectors whose complex numbers are (50 + 35j), (-60 - 50j), and (25 - 30j). ________

ANSWERS:

1. 13.5j	4. 0 - 15j	7. 5 - j	10. -1 - j
2. 15 + 20j	5. -10 + 0j	8. -17.6 + 12.8j	11. 180 - 370j
3. 20 - 10j	6. b, c, and f	9. -6 + 0j	12. 15 - 45j

7-5 SUBTRACTING COMPLEX NUMBERS

In this section, we will discuss the subtraction of complex numbers. The use of complex numbers for vector subtraction is shown.

---

36. A subtraction of complex numbers is shown below. It is the equivalent of a subtraction of a binomial which is an addition.

$$(7 + 6j) - (3 + 4j)$$

To perform the subtraction, we convert it to an addition by "adding the opposite of (3 + 4j)". To do so, we add the opposites of both 3 and 4j and then simplify. We get:

$$(7 + 6j) + (\text{the opposite of } 3 + 4j)$$

$$(7 + 6j) + (-3) + (-4j) = 4 + 2j$$

Following the example, complete these subtractions.

a) $(6 - 4j) - (2 + 5j) = (6 - 4j) + (\quad) + (\quad) =$ ________

b) $(-8 + 9j) - (1 + 3j) = (-8 + 9j) + (\quad) + (\quad) =$ ________

---

37. To perform the subtraction below, we added the opposite of (-3 + 2j).

$$(4 - j) - (-3 + 2j) = (4 - j) + (3) + (-2j) = 7 - 3j$$

Following the example, complete these subtractions.

a) $(10 + 3j) - (-5 + 8j) = (10 + 3j) + (\quad) + (\quad) =$ ________

b) $(-4 - 2j) - (-3 + j) = (-4 - 2j) + (\quad) + (\quad) =$ ________

Answers to 36:

a) $(-2) + (-5j) = 4 - 9j$

b) $(-1) + (-3j) = -9 + 6j$

---

38. The subtraction below is equivalent to the subtraction of a binomial which is a subtraction.

$$(8 + 5j) - (3 - 4j)$$

To perform the subtraction, we convert it to an addition by "adding the opposite of (3 - 4j)". To do so, we add the opposites of 3 and -4j and then simplify. We get:

$$(8 + 5j) + (\text{the opposite of } 3 - 4j)$$

$$(8 + 5j) + (-3) + (4j) = 5 + 9j$$

Following the example, complete these subtractions.

a) $(7 - 9j) - (5 - 6j) = (7 - 9j) + (\quad) + (\quad) =$ ________

b) $(-1 - 5j) - (6 - 9j) = (-1 - 5j) + (\quad) + (\quad) =$ ________

Answers to 37:

a) $(5) + (-8j) = 15 - 5j$

b) $(3) + (-1j) = -1 - 3j$

Answers to 38:

a) $(-5) + (6j) = 2 - 3j$

b) $(-6) + (9j) = -7 + 4j$

39. To perform the subtraction below, we added the opposite of $(-2 - 2j)$.

$$(6 - j) - (-2 - 2j) = (6 - j) + (2) + (2j) = 8 + j$$

Following the example, complete these subtractions.

a) $(10 - 7j) - (-4 - 5j) = (10 - 7j) + (\quad) + (\quad) =$ ________

b) $(-5 + 3j) - (-1 - 6j) = (-5 + 3j) + (\quad) + (\quad) =$ ________

Answers to 39: a) $(4) + (5j) = 14 - 2j$ b) $(1) + (6j) = -4 + 9j$

40. In the subtractions below, each complex number contains one "0" term. Notice how we converted to additions.

$$(6 + 0j) - (0 - 3j) = (6 + 0j) + (0) + (3j) = 6 + 3j$$

$$(0 - 4j) - (5 + 0j) = (0 - 4j) + (-5) + (0j) = -5 - 4j$$

Following the examples, complete these subtractions.

a) $(-7 + 0j) - (0 - 5j) =$ ________

b) $(0 + 2j) - (-1 + 0j) =$ ________

Answers to 40: a) $-7 + 5j$ b) $1 + 2j$

41. In each subtraction below, the first complex number is "incomplete". Notice how we converted to additions.

$$(-2j) - (4 + 3j) = (-2j) + (-4) + (-3j) = -4 - 5j$$

$$(10) - (-1 - 5j) = (10) + (1) + (5j) = 11 + 5j$$

Following the examples, complete these.

a) $(7j) - (6 - j) =$ ________

b) $(-8) - (-3 + j) =$ ________

Answers to 41: a) $-6 + 8j$ b) $-5 - j$

42. In each subtraction below, the second complex number is "incomplete". Notice how we converted to additions.

$$(7 - 3j) - (4) = (7 - 3j) + (-4) = 3 - 3j$$

$$(-2 + 4j) - (-6j) = (-2 + 4j) + (6j) = -2 + 10j$$

Following the examples, complete these.

a) $(9 + 2j) - (8) =$ ________

b) $(10 - 5j) - (-j) =$ ________

Answers to 42: a) $1 + 2j$ b) $10 - 4j$

43. In the subtraction below, the answer is a complex number with a "0" term. It can be written as an "incomplete" complex number.

$$(6 + 5j) - (2 + 5j) = 4 + 0j \quad \text{(or 4)}$$

Following the example, complete these.

a) $(9 - 3j) - (9) =$ ________

b) $(4) - (4 - 7j) =$ ________

44. When vectors are represented by complex numbers, we can use a subtraction of complex numbers to find the second vector-addend when the resultant and the first vector-addend are given. An example is discussed below.

A vector addition is shown on the complex plane at the right. The vector-addends are $(1 + 2j)$ and $(3 - 4j)$. The resultant is $(4 - 2j)$.

By subtracting one vector-addend from the resultant, we can find the other vector-addend. The two possible subtractions are shown.

$$(4 - 2j) - (1 + 2j)$$
$$= (4 - 2j) + (-1) + (-2j)$$
$$= 3 - 4j$$

$$(4 - 2j) - (3 - 4j) = (4 - 2j) + (-3) + 4j = 1 + 2j$$

Are the two answers, $(3 - 4j)$ and $(1 + 2j)$, the original vector-addends? ________

a) $0 - 3j$ (or $-3j$)

b) $0 + 7j$ (or $7j$)

---

45. In each problem below, the Resultant and Vector #1 are given. Use a subtraction to find Vector #2.

	Resultant	Vector #1	Vector #2
a)	$7 - 3j$	$1 - 5j$	________
b)	$6.4 - 3.9j$	$-1.4 + 2.8j$	________
c)	$340 - 25j$	$-180 - j$	________
d)	$1.79j$	$2.58 + 2.44j$	________

Yes

---

a) $6 + 2j$

b) $7.8 - 6.7j$

c) $520 - 24j$

d) $-2.58 - 0.65j$

46. a) Vector A and vector B are added. Their resultant is (-5 + 4j). If vector A is (3 - 9j), find vector B.

Vector B = ____________

b) The sum of vector C and vector D is (-12 + 0j). If vector D is (-20 + 5j), find vector C.

Vector C = ____________

a) -8 + 13j b) 8 - 5j

7-6 CONVERTING FROM RECTANGULAR FORM TO POLAR FORM

Complex numbers can be expressed in either "rectangular" form or "polar" form. We will define those two forms in this section, and show the method for converting from "rectangular" form to "polar" form.

47. The form we have been using for complex numbers is called "rectangular" form. The "rectangular" form is:

$$\boxed{a + bj}$$ where "a" and "b" are real numbers, and "b" is the coefficient of "j".

The vector for a complex number is shown on the complex plane at the right. Notice these points:

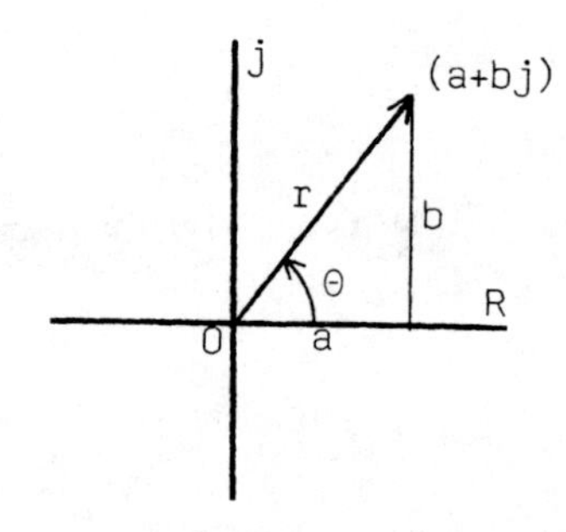

1. The complex number is (a + bj).
2. "r" is the length of its vector.
3. "θ" is the direction of its vector as a standard-position angle.
4. "a" is the horizontal component of its vector.
5. "b" is the vertical component of its vector.

To find the length "r", we use the Pythagorean Theorem.

$$r^2 = a^2 + b^2$$

$$r = \sqrt{a^2 + b^2}$$

To find the direction θ, we use the tangent ratio.

$$\tan \theta = \frac{b}{a}$$

$$\theta = \arctan \frac{b}{a}$$

48. When the length (r) and direction (θ) of the vector of a complex number are known, we can write the complex number in the form $r\underline{/\theta}$. That form is called the "polar" form of a complex number.

The length and direction of the vector of the complex number at the right are:

r = 5 units

θ = 140°

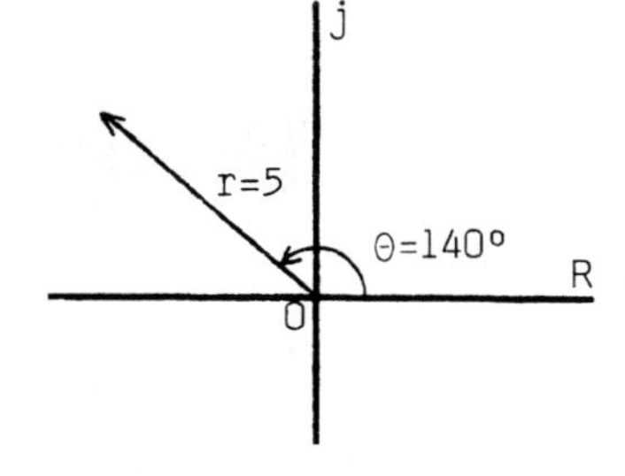

Therefore, the "polar" form of the complex number is $5\underline{/140°}$. Notice these points about $5\underline{/140°}$:

1. The symbol $\underline{/\quad}$ means "angle".
2. The "angle" symbol is extended under "140°".
3. Though 5 and $\underline{/140°}$ are written next to each other, they are not a multiplication.

If the length and direction of a vector representing a complex number are 27.5 units and 228°, write the complex number in polar form.

____

Answer: $27.5\underline{/228°}$

49. A complex number can be written in either of two forms.

1. The "rectangular form" is a + bj .
2. The "polar form" is $r\underline{/\theta}$.

The vector at the right represents a complex number. The complex number can be written in either of two forms:

a) The "rectangular form" is ______________.

b) The "polar form" is ______________.

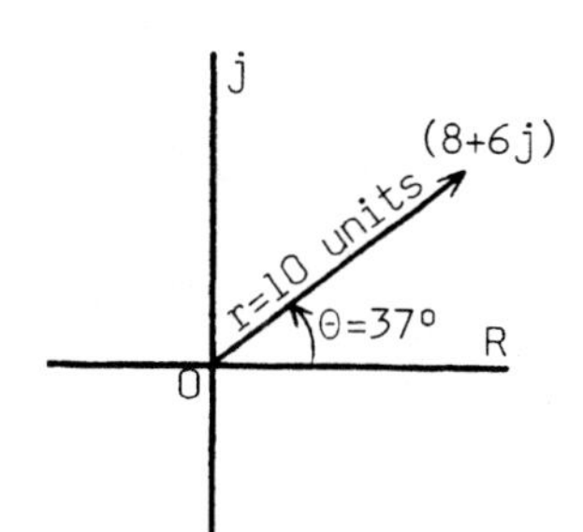

Answers:

a) 8 + 6j

b) $10\underline{/37°}$

50. The second-quadrant vector at the right represents a complex number. The rectangular form of the complex number is $(-5 + 3j)$. We want to find its polar form. We know these facts:

1. The horizontal component ("a") is -5.
2. The vertical component ("b") is 3.

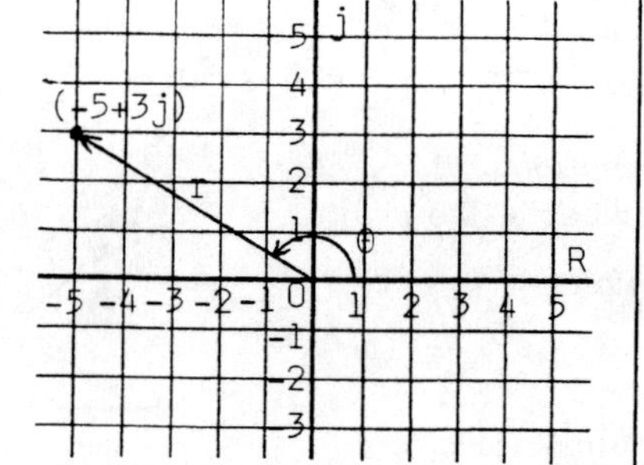

a) Use the Pythagorean Theorem to find the length "r". Round to tenths.

r = ________ units

b) Use the tangent ratio to find θ. Round to a whole number.

θ = ________

c) The polar form of the complex number is ________________.

---

a) 5.8 units

b) 149°

c) $5.8\underline{/149°}$

51. The third-quadrant vector below represents a complex number.

a) Find the length of the vector. Round to hundredths.

r = ________ units

b) Find the direction of the vector. Round to a whole number.

θ = ________

c) The polar form of the complex number is ________________.

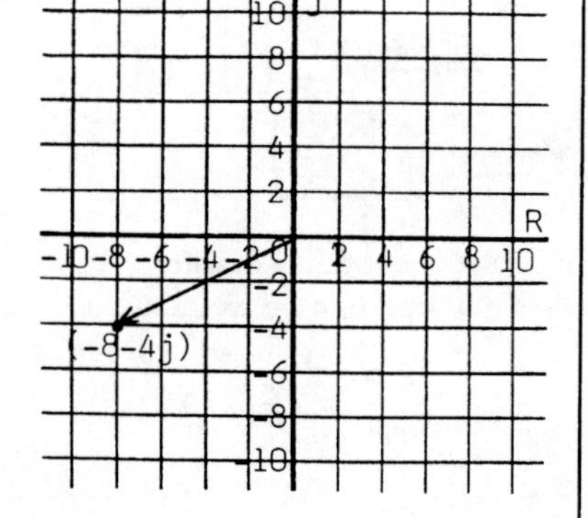

---

a) 8.94 units

b) 207°

c) $8.94\underline{/207°}$

52. The fourth-quadrant vector below represents a complex number.

a) Find "r". Round to tenths.

r = __________ units

b) Find θ. Round to a whole number.

θ = __________

c) The polar form of the complex number is __________________.

j
R
(15-25j)

---

53. Find the polar form of (-2.75 + 4.68j). Round "r" to hundredths and θ to a whole number.

j
R

Polar Form = ______________

a) 29.2 units

b) 301°

c) 29.2/301°

---

54. Find the polar form of (240 - 185j). Round both "r" and θ to a whole number.

j
R

Polar Form = ______________

5.43/120°

---

55. For a third-quadrant or fourth-quadrant complex number, we sometimes use a negative angle for θ. For example:

Instead of 17/300° , we use 17/-60° .

Instead of 45/235° , we use 45/-125° .

Write each polar form with a negative angle.

a) 1.98/315° b) 675/280° c) 47.8/207°

______________ ______________ ______________

303/322°

	a) $1.98\angle{-45^\circ}$
	b) $675\angle{-80^\circ}$
	c) $47.8\angle{-153^\circ}$

56. Each of the four complex numbers on the complex plane below has a "0" term. Therefore, the vector of each is on an axis.

We converted two of the complex numbers from rectangular form to polar form below.

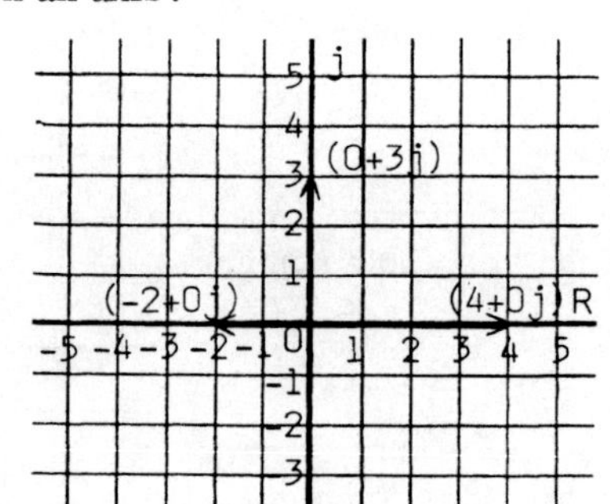

The length of (-2 + 0j) is 2 units.
The direction of (-2 + 0j) is 180°.
Therefore: (-2 + 0j) = $2\angle{180^\circ}$.

The length of (0 + 3j) is 3 units.
The direction of (0 + 3j) is 90°.
Therefore: (0 + 3j) = $3\angle{90^\circ}$.

Following the examples, convert the other two complex numbers to polar form.

a) (4 + 0j) = ___________ b) (0 - 5j) = ___________

---

	a) $4\angle{0^\circ}$
	b) $5\angle{270^\circ}$ or $5\angle{-90^\circ}$

57. Convert each of these to polar form. <u>It is helpful to make a sketch of each vector first</u>.

a) (0 - 10j) = ___________

b) (1.68 + 0j) = ___________

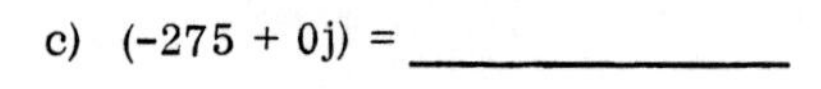

c) (-275 + 0j) = ___________

d) (0 + 9.7j) = ___________

---

a) $10\angle{270^\circ}$ or $10\angle{-90^\circ}$

b) $1.68\angle{0^\circ}$

c) $275\angle{180^\circ}$

d) $9.7\angle{90^\circ}$

58. Any "incomplete" complex number has a vector that lies on an axis. Convert each of these incomplete complex numbers from rectangular form to polar form.

a) (16.5) = ______________

b) (12j) = ______________

c) (-3.52) = ______________

d) (-450j) = ______________

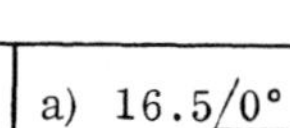

a) $16.5\angle 0^\circ$ b) $12\angle 90^\circ$ c) $3.52\angle 180^\circ$ d) $450\angle 270^\circ$ or $450\angle -90^\circ$

7-7 CONVERTING FROM POLAR FORM TO RECTANGULAR FORM

In this section, we will discuss the method used to convert complex numbers from polar form to rectangular form. The method for adding or subtracting complex numbers in polar form is also discussed.

59. The complex number $20.0\angle 40^\circ$ is graphed at the right. Its vector is in the first quadrant.

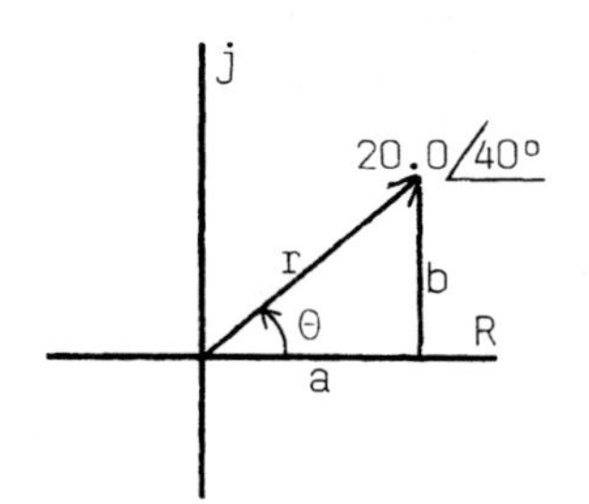

To convert $20.0\angle 40^\circ$ to rectangular form, we must find the components "a" and "b". We have done so below. Each component was rounded to tenths.

To find "a", we use $\cos\theta$.

$$\cos 40^\circ = \frac{a}{20.0}$$

$$a = 20.0(\cos 40^\circ)$$

$$a = 15.3 \text{ units}$$

To find "b", we use $\sin\theta$.

$$\sin 40^\circ = \frac{b}{20.0}$$

$$b = 20.0(\sin 40^\circ)$$

$$b = 12.9 \text{ units}$$

Since "a" = 15.3 and "b" = 12.9, the rectangular form of $20.0\angle 40^\circ$ is ______________.

15.3 + 12.9j

60. The complex number $15.0\underline{/145°}$ is graphed at the right. Its vector is in the second quadrant.

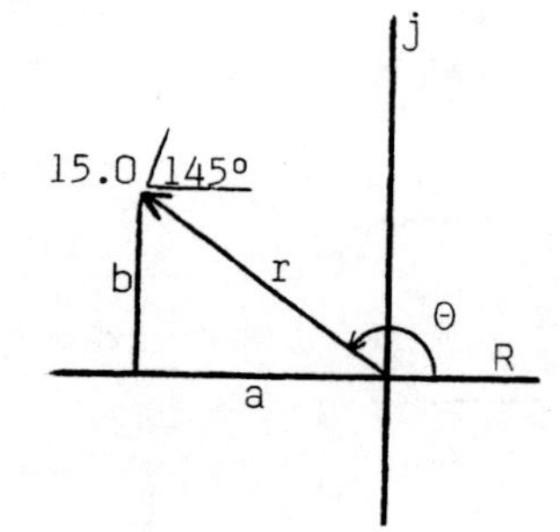

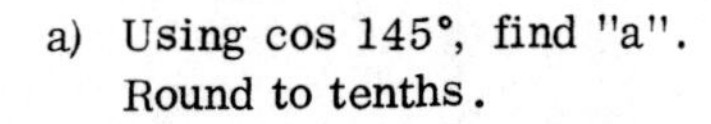

a) Using cos 145°, find "a". Round to tenths.

a = __________ units

b) Using sin 145°, find "b". Round to tenths.

b = __________ units

c) The rectangular form of $15.0\underline{/145°}$ is ____________________.

---

61. The complex number $30.0\underline{/217°}$ is graphed at the right. Its vector is in the third quadrant.

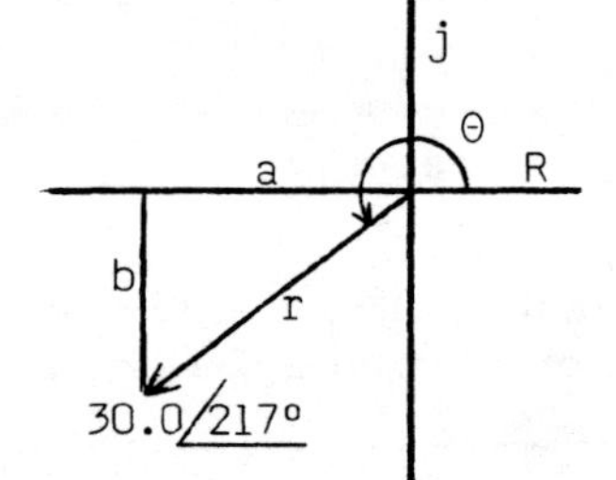

a) Using cos 217°, find "a". Round to tenths.

a = __________ units

b) Using sin 217°, find "b". Round to tenths.

b = __________ units

c) The rectangular form of $30.0\underline{/217°}$ is  ____________________.

Answers to 60:

a) -12.3 units

b) 8.6 units

c) -12.3 + 8.6j

---

62. The complex number $25.0\underline{/318°}$ is graphed at the right. Its vector is in the fourth quadrant.

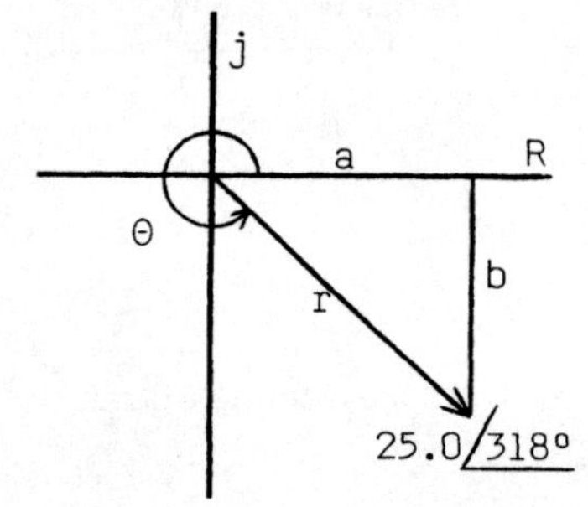

a) Find "a". Round to tenths.

a = __________ units

b) Find "b". Round to tenths.

b = __________ units

c) The rectangular form of $25.0\underline{/318°}$ is ____________________.

Answers to 61:

a) -24.0 units

b) -18.1 units

c) -24.0 - 18.1j

63. The complex number $475\angle{-37^\circ}$ is graphed at the right. Its vector is in the fourth quadrant.

a) Find "a". Round to a whole number.

a = ________ units

b) Find "b". Round to a whole number.

b = ________ units

c) The rectangular form of $475\angle{-37^\circ}$ is ____________.

a) 18.6 units

b) -16.7 units

c) 18.6 - 16.7j

---

64. Find the rectangular form of the complex number $1.68\angle{171^\circ}$. Round "a" and "b" to hundredths. It is helpful to make a sketch.

$1.68\angle{171^\circ}$ = ____________

a) 379 units

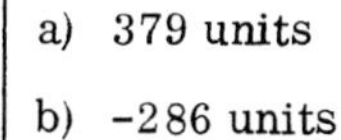

b) -286 units

c) 379 - 286j

---

65. Find the rectangular form of the complex number $9.64\angle{-112^\circ}$. Round "a" and "b" to hundredths.

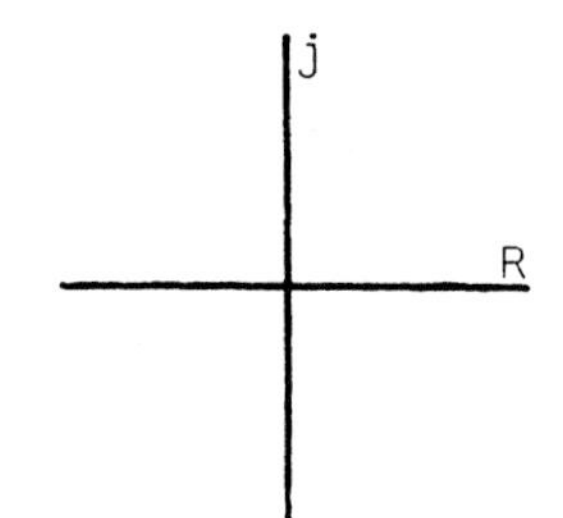

$9.64\angle{-112^\circ}$ = ____________

-1.66 + 0.26j

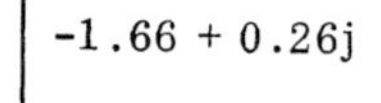

---

66. The vector of each complex number at the right is on an axis. Following the examples, write the rectangular form of the other complex numbers.

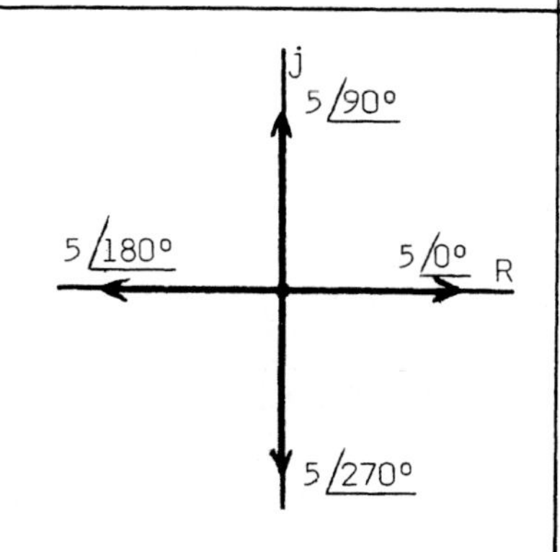

$5\angle{0^\circ}$ = 5 + 0j

$5\angle{90^\circ}$ = 0 + 5j

a) $5\angle{180^\circ}$ = ____________

b) $5\angle{270^\circ}$ = ____________

-3.61 - 8.94j

67. Write the rectangular form of each complex number below. Make a sketch if it helps.

a) $6.5\angle 180^\circ$ = ____________

b) $12\angle 90^\circ$ = ____________

c) $445\angle 0^\circ$ = ____________

d) $2.87\angle 270^\circ$ = ____________

a) $-5 + 0j$

b) $0 - 5j$

---

68. Write the rectangular form of each complex number below.

a) $209\angle 90^\circ$ = ____________

b) $8.66\angle 180^\circ$ = ____________

c) $45.7\angle 0^\circ$ = ____________

d) $81.9\angle 270^\circ$ = ____________

a) $-6.5 + 0j$

b) $0 + 12j$

c) $445 + 0j$

d) $0 - 2.87j$

---

69. When complex numbers are given in polar form, they cannot be added or subtracted directly. The following steps are used to add or subtract them.

1. Convert each complex number to rectangular form.
2. Perform the addition or subtraction in rectangular form.
3. Convert the answer from rectangular form back to polar form.

Using the steps above, let's perform $25.0\angle 77^\circ + 35.0\angle 303^\circ$.

a) Convert $25.0\angle 77^\circ$ to rectangular form: ____________
Round to tenths.

b) Convert $35.0\angle 303^\circ$ to rectangular form: ____________
Round to tenths.

c) Write the resultant in rectangular form: ____________

d) Convert the resultant back to polar form: ____________
Round "r" to tenths and θ to a whole number.

a) $0 + 209j$

b) $-8.66 + 0j$

c) $45.7 + 0j$

d) $0 - 81.9j$

70. Using the same steps, let's perform $300\underline{/0°} - 200\underline{/90°}$.

a) Convert $300\underline{/0°}$ to rectangular form: ________

b) Convert $200\underline{/90°}$ to rectangular form: ________

c) Find the answer in rectangular form: ________

d) Convert the answer to polar form: ________
Round both "r" and θ to a whole number.

a) 5.6 + 24.4j
b) 19.1 - 29.4j
c) 24.7 - 5.0j
d) $25.2\underline{/349°}$

---

a) 300 + 0j b) 0 + 200j c) 300 - 200j d) $361\underline{/326°}$ or $361\underline{/-34°}$

SELF-TEST 23 (pages 315-327)

Do these subtractions. Write each answer as a "complete" complex number.

1. (1 + 2j) - (2 - j) = ________
2. (30 - 20j) - (10 + 40j) = ________
3. (6 + 0j) - (6 + 8j) = ________
4. (3 - j) - (4 - j) = ________

5. Vector F and vector G are added. Their resultant is (30 - 50j). If vector G is (60 + 20j), find vector F.

Vector F = ________

Convert to polar form.

6. Round "r" and θ to whole numbers.

(278 + 194j) ________

7. Round "r" to hundredths and θ to a whole number.

(3.85 - 5.48j) ________

Convert to rectangular form.

8. Round to tenths.

$70.0\underline{/236°}$ ________

9. Round to a whole number.

$483\underline{/112°}$ ________

10. Convert (0 - 420j) to polar form. ________

11. Convert $15.3\underline{/0°}$ to rectangular form. ________

12. Convert (-6.95 + 0j) to polar form. ________

13. Add. Write the sum in polar form. Round "r" and θ to whole numbers.

$240\underline{/60°} + 410\underline{/160°}$ ________

14. Subtract. Write the difference in polar form. Round "r" and θ to whole numbers.

$250\underline{/90°} - 400\underline{/180°}$ ________

ANSWERS:

1. -1 + 3j
2. 20 - 60j
3. 0 - 8j
4. -1 + 0j
5. -30 - 70j
6. $339\underline{/35°}$
7. $6.70\underline{/-55°}$ or $6.70\underline{/305°}$
8. -39.1 - 58.0j
9. -181 + 448j
10. $420\underline{/-90°}$ or $420\underline{/270°}$
11. 15.3 + 0j
12. $6.95\underline{/180°}$
13. $437\underline{/127°}$
14. $472\underline{/32°}$

7-8 MULTIPLYING IN RECTANGULAR FORM

In this section, we will show the method for multiplying complex numbers in rectangular form.

---

71. Two multiplications of complex numbers in rectangular form are shown below.

$$(5 + 2j)(8 + 6j) \qquad (9 - j)(4 - 7j)$$

Since each complex number is a binomial, each multiplication is a multiplication of two binomials. The pattern for a multiplication of two binomials is:

$$(a + b)(c + d) = a(c + d) + b(c + d)$$
$$= ac + ad + bc + bd$$

Note: Both terms in $(c + d)$ are multiplied first by "a" and then by "b".

We used the pattern above for the multiplication below. Notice how we combined 30j and 16j in Step 3.

Step 1: $(5 + 2j)(8 + 6j) = 5(8 + 6j) + 2j(8 + 6j)$

Step 2: $= 40 + 30j + 16j + 12j^2$

Step 3: $= 40 + 46j + 12j^2$

Following the example, complete this multiplication.

$$(4 + j)(3 + 2j) = 4(3 + 2j) + j(3 + 2j)$$

= ______________

= ______________

---

72. Following the example, complete the multiplication below.

Answer to 71: $= 12 + 8j + 3j + 2j^2$; $= 12 + 11j + 2j^2$

$$(5 + 4j)(8 - j) = 5(8 - j) + 4j(8 - j)$$
$$= 40 - 5j + 32j - 4j^2$$
$$= 40 + 27j - 4j^2$$

$$(3 + j)(9 - j) = 3(9 - j) + j(9 - j)$$

= ______________

= ______________

Answer to 72: $= 27 - 3j + 9j - j^2$; $= 27 + 6j - j^2$

73. Following the example, complete the other multiplication.

$$(2 - 4j)(10 + j) = 2(10 + j) - 4j(10 + j)$$
$$= 20 + 2j - 40j - 4j^2$$
$$= 20 - 38j - 4j^2$$

$$(7 - j)(5 - 6j) = 7(5 - 6j) - j(5 - 6j)$$
$$= \underline{\hspace{4cm}}$$
$$= \underline{\hspace{4cm}}$$

Answer: $= 35 - 42j - 5j + 6j^2$

$= 35 - 47j + 6j^2$

74. Complete each multiplication.

a) $(9 + 7j)(1 - j) =$

b) $(1 - 8j)(6 - j) =$

Answers: a) $9 - 2j - 7j^2$

b) $6 - 49j + 8j^2$

75. When two complex numbers in rectangular form are multiplied, the product is a trinomial containing a "j^2" term. For example:

$$(5 + j)(3 + 7j) = 15 + 38j + 7j^2$$
$$(2 - j)(9 - 4j) = 18 - 17j + 4j^2$$

Each product above can be simplified because $\boxed{j^2 = -1}$. The fact that $j^2 = -1$ is demonstrated below.

Since $j = \sqrt{-1}$, $j^2 = \sqrt{-1} \cdot \sqrt{-1} = -1$

Note: The statement above uses the pattern for multiplying two square-root radicals containing the same radicand. That is:

Just as $\sqrt{5} \cdot \sqrt{5} = 5$,

$\sqrt{-1} \cdot \sqrt{-1} = -1$.

Using the fact that $j^2 = -1$, we can simplify each product above to a complex number. One simplification is shown. Complete the other one.

$15 + 38j + 7j^2$	$18 - 17j + 4j^2$
$15 + 38j + 7(-1)$	$18 - 17j + 4(-1)$
$15 + 38j - 7$	____________
$8 + 38j$	____________

Answer: $18 - 17j - 4$

$14 - 17j$

76. The product below contains a negative j^2-term. Notice how we simplified it. Complete the other simplification.

$10 + 13j - 4j^2$	$9 - 7j - 8j^2$
$10 + 13j - 4(-1)$	$9 - 7j - 8(-1)$
$10 + 13j + 4$	______________
$14 + 13j$	______________

Answer: $9 - 7j + 8$
$17 - 7j$

77. Simplify each product to a complex number.

a) $4 + 7j + 5j^2$ = __________

c) $-6 - 3j + 7j^2$ = __________

b) $9 - 6j - j^2$ = __________

d) $-11 + 4j - 9j^2$ = __________

Answers:
a) $-1 + 7j$
b) $10 - 6j$
c) $-13 - 3j$
d) $-2 + 4j$

78. Write each product as a complex number.

a) $(3 + 4j)(7 + j)$ =

b) $(20 - 3j)(40 - j)$ =

Answers:
a) $17 + 31j$
b) $797 - 140j$

79. Write each product as a complex number.

a) $(-1 - 4j)(3 + j)$ =

b) $(10 + 6j)(20 - 4j)$ =

Answers: a) $1 - 13j$ b) $224 + 80j$

7-9 MULTIPLICATIONS INVOLVING "INCOMPLETE" COMPLEX NUMBERS

In this section, we will discuss multiplications in which one or both complex numbers has a "0" term. Multiplications of that type involve "incomplete" complex numbers.

80. To perform the multiplication below, we began by writing $(4 + 0j)$ as the "incomplete" complex number "4".

$$(4 + 0j)(6 - j) = 4(6 - j) = 24 - 4j$$

Following the example, complete this multiplication.

$$(-7 + 0j)(8 + 9j) = \_\_\_\_\_\_\_\_\_\_$$

81. In the multiplication below, we began by writing $(0 + 3j)$ as the "incomplete" complex number "3j". Notice that we simplified the product to get a complex number.

Answer to 80: $-56 - 63j$

$$(0 + 3j)(-5 + j) = 3j(-5 + j)$$
$$= -15j + 3j^2$$
$$= -15j - 3$$
$$= -3 - 15j$$

Following the example, write each product as a complex number.

a) $(0 + 4j)(3 + 2j) =$ b) $(0 + 2j)(7 - j) =$

82. Write each product as a complex number.

a) $(0 - 5j)(1 + 6j) =$ b) $(0 - j)(10 + j) =$

Answers to 81:
a) $-8 + 12j$, from: $12j + 8j^2$
b) $2 + 14j$, from: $14j - 2j^2$

83. In the multiplications below, we began by writing both factors as an "incomplete" complex number. Notice that we wrote each product as a "complete" complex number with a "0" term.

$$(7 + 0j)(0 + 2j) = 7(2j) = 14j = 0 + 14j$$
$$(9 + 0j)(-1 + 0j) = 9(-1) = -9 = -9 + 0j$$
$$(0 + 2j)(0 - 4j) = 2j(-4j) = -8j^2 = 8 = 8 + 0j$$

Answers to 82:
a) $30 - 5j$, from: $-5j - 30j^2$
b) $1 - 10j$, from: $-10j - j^2$

Continued on following page.

83. Continued

Write each product as a "complete" complex number with a "0" term.

a) $(0 - 2j)(9 + 0j) =$

b) $(-1 + 0j)(-1 + 0j) =$

c) $(0 - 3j)(0 - 6j) =$

---

84. Write each product as a "complete" complex number and then convert it to polar form.

	Complex Number		Polar Form
a) $(3 + 0j)(0 + 5j) =$	________	=	________
b) $(-4 + 0j)(6 + 0j) =$	________	=	________
c) $(0 - 7j)(0 + j) =$	________	=	________
d) $(0 - 2j)(4 + 0j) =$	________	=	________

Answers to 83:

a) $0 - 18j$

b) $1 + 0j$

c) $-18 + 0j$

Answers to 84: a) $0 + 15j = 15\underline{/90°}$ b) $-24 + 0j = 24\underline{/180°}$ c) $7 + 0j = 7\underline{/0°}$ d) $0 - 8j = 8\underline{/270°}$ or $8\underline{/-90°}$

7-10 MULTIPLYING CONJUGATES

In this section, we will define the "conjugate" of a complex number and discuss multiplication of conjugates.

---

85. In the multiplication below, both complex numbers contain a "4" and a "3j". One complex number is a sum; the other is a difference. Since $-12j + 12j = 0$, the product simplifies to an "incomplete" complex number.

$$(4 + 3j)(4 - 3j) = 16 - 12j + 12j - 9j^2$$
$$= 16 - 9j^2$$
$$= 16 - 9(-1)$$
$$= 16 + 9 = 25 \quad \text{(or } 25 + 0j)$$

Complex numbers like $(4 + 3j)$ and $(4 - 3j)$ are called a pair of conjugates.

$(4 + 3j)$ is called the conjugate of $(4 - 3j)$

$(4 - 3j)$ is called the conjugate of $(4 + 3j)$

Write the conjugate of each complex number.

a) $7 + j$ ________ b) $10 - 5j$ ________ c) $1.4 - 2.7j$ ________

86. Whenever a pair of conjugates are multiplied, the product simplifies to an "incomplete" complex number because the sum of the j-terms is "0". For example:

$$(8 + j)(8 - j) = 64 - 8j + 8j - j^2$$
$$= 64 - j^2$$
$$= 64 - (-1)$$
$$= 64 + 1 = 65 \quad \text{(or } 65 + 0j)$$

Following the example, complete these:

a) $(5 + 2j)(5 - 2j) =$

b) $(10 - j)(10 + j) =$

Answers (previous frame): a) $7 - j$ b) $10 + 5j$ c) $1.4 + 2.7j$

---

87. The complex numbers in the multiplication below are a pair of conjugates.

$$(-2 + 5j)(-2 - 5j) = 4 + 10j - 10j - 25j^2$$
$$= 4 - 25j^2$$
$$= 4 - 25(-1)$$
$$= 4 + 25 = 29 \quad \text{(or } 29 + 0j)$$

Following the example, complete these:

a) $(-4 + 6j)(-4 - 6j) =$

b) $(-1 - j)(-1 + j) =$

Answers (frame 86): a) 29 (or $29 + 0j$) b) 101 (or $101 + 0j$)

---

88. Use a calculator for these:

a) $(15.5 + 12.8j)(15.5 - 12.8j) =$
Round to a whole number.

b) $(-275 - 140j)(-275 + 140j) =$
Round to hundreds.

Answers (frame 87): a) 52 (or $52 + 0j$) b) 2 (or $2 + 0j$)

Answers (frame 88): a) 404 (or $404 + 0j$) b) 95,200 (or $95,200 + 0j$)

89. Since the product of a pair of conjugates is always a positive number with a "0j" term, its vector lies on the positive side of the R-axis. The standard-position angle of the vector is 0°.

Using the fact above, do this multiplication and convert the product to polar form.

$$(3 + j)(3 - j) =$$

Answer: $10\underline{/0°}$, from: (10 + 0j)

7-11 MULTIPLYING IN POLAR FORM

When complex numbers are given in polar form, they can be multiplied directly. We will discuss the method in this section.

90. Here is a multiplication of complex numbers in rectangular form.

$$(4 + 3j)(6 + 8j) = 0 + 50j$$

The polar forms of each factor and the product are:

$$4 + 3j = 5\underline{/37°}$$
$$6 + 8j = 10\underline{/53°}$$
$$0 + 50j = 50\underline{/90°}$$

By examining the numbers in the polar forms, you can see these facts:

1. The length of the product can be obtained by multiplying the lengths of the factors. That is:

$$(5)(10) = 50$$

2. The direction of the product can be obtained by adding the directions of the factors. That is:

$$37° + 53° = 90°$$

Let's check the same facts with the multiplication below.

$$(4 + 2j)(3 + 5j) = 2 + 26j$$

$$4 + 2j = 4.47\underline{/27°}$$
$$3 + 5j = 5.83\underline{/59°}$$
$$2 + 26j = 26.1\underline{/86°}$$

a) Does $(4.47)(5.83) = 26.1$? ________

b) Does $27° + 59° = 86°$? ________

91. To multiply complex numbers in polar form:

1. We multiply their lengths to get the length of the product.
2. We add their directions to get the direction of the product.

Find each product below. Notice how we wrote parentheses around each polar form to signify a multiplication.

a) $(3\angle 20°)(4\angle 50°)$ = ________ b) $(7\angle 65°)(10\angle 75°)$ = ________

a) Yes

b) Yes

---

92. When multiplying in polar form, we can get an angle greater than 360°. For example:

$$(3\angle 305°)(2\angle 81°) = 6\angle 386°$$

We do not ordinarily use angles greater than 360° when writing a complex number in polar form. Therefore, we subtract 360° from 386° to get an equivalent angle smaller than 360°. That is:

Instead of $6\angle 386°$, we write $\mathbf{6\angle 26°}$.

Complete each multiplication. Use an angle smaller than 360° in the product.

a) $(12\angle 325°)(5\angle 82°)$ = ________

b) $(9\angle 347°)(10\angle 112°)$ = ________

a) $12\angle 70°$

b) $70\angle 140°$

---

93. When the vector of a complex number is in the third or fourth quadrant, a negative angle is sometimes used for its direction. Adding signed angles is the same as adding signed numbers. That is:

$$80° + (-20°) = 60°$$

$$(-90°) + 40° = -50°$$

$$(-25°) + (-31°) = -56°$$

Complete each multiplication.

a) $(10\angle 57°)(3\angle -22°)$ = ________ b) $(5\angle -40°)(2\angle -35°)$ = ________

a) $60\angle 47°$

b) $90\angle 99°$

---

94. Write each product with a negative angle and then write it with a positive angle between 0° and 360°.

a) $(4\angle 40°)(3\angle -67°)$ = ________ = ________

b) $(2\angle -45°)(7\angle -55°)$ = ________ = ________

a) $30\angle 35°$

b) $10\angle -75°$

---

a) $12\angle -27° = 12\angle 333°$

b) $14\angle -100° = 14\angle 260°$

95. Use a calculator for these.

a) $(25.9\underline{/120^\circ})(14.1\underline{/93^\circ}) =$
Round to a whole number.

b) $(1.59\underline{/47^\circ})(8.04\underline{/-98^\circ}) =$
Round to tenths.

---

96. In each multiplication below, the product has a vector that lies on an axis. Write each product in polar form and then convert the product to rectangular form.

		Polar Form		Rectangular Form
a) $(10\underline{/40^\circ})(5\underline{/-40^\circ})$	=	________	=	________
b) $(4\underline{/210^\circ})(2\underline{/-30^\circ})$	=	________	=	________
c) $(3\underline{/220^\circ})(8\underline{/50^\circ})$	=	________	=	________
d) $(6\underline{/200^\circ})(7\underline{/250^\circ})$	=	________	=	________

Answers (frame 95):
a) $365\underline{/213^\circ}$
b) $12.8\underline{/-51^\circ}$ or $12.8\underline{/309^\circ}$

---

97. To multiply three or more complex numbers in rectangular form, we multiply two at a time. An example is shown.

$$\begin{aligned}(4 + j)(3 - 2j)(5 - j) &= (12 - 5j - 2j^2)(5 - j)\\ &= (14 - 5j)(5 - j)\\ &= 70 - 39j + 5j^2\\ &= 65 - 39j\end{aligned}$$

Following the example, complete this multiplication.

$(1 - j)(6 + j)(3 - 2j) =$

Answers (frame 96):
a) $50\underline{/0^\circ} = 50 + 0j$
b) $8\underline{/180^\circ} = -8 + 0j$
c) $24\underline{/270^\circ} = 0 - 24j$
d) $42\underline{/90^\circ} = 0 + 42j$

---

98. To multiply three or more complex numbers in polar form, we multiply their lengths and add their directions. An example is shown.

$$(2\underline{/30^\circ})(5\underline{/10^\circ})(3\underline{/20^\circ}) = 30\underline{/60^\circ}$$

Note: The length of the product is: $(2)(5)(3) = 30$
The direction of the product is: $30^\circ + 10^\circ + 20^\circ = 60^\circ$

Find each product in polar form.

a) $(4\underline{/50^\circ})(10\underline{/-25^\circ})(2\underline{/45^\circ}) =$ ________

b) $(3\underline{/-60^\circ})(2\underline{/40^\circ})(10\underline{/-50^\circ})(4\underline{/20^\circ}) =$ ________

Answer (frame 97): $11 - 29j$

Answers (frame 98): a) $80\underline{/70^\circ}$ b) $240\underline{/-50^\circ}$ or $240\underline{/310^\circ}$

SELF-TEST 24 (pages 328-337)

Do these multiplications. Write each product as a "complete" complex number.

1. $(2 + 3j)(1 + 2j) =$ ________
2. $(8 + 7j)(9 - 8j) =$ ________
3. $(30 - 10j)(20 + 40j) =$ ________
4. $(3 - j)(1 - 3j) =$ ________
5. $(0 + 8j)(5 + 2j) =$ ________
6. $(1 - 4j)(2 + 0j) =$ ________
7. $(20 + 0j)(0 + 15j) =$ ________
8. $(7 + j)(7 - j) =$ ________

9. Find the product of $(5 - 2j)$ and its conjugate. ________

Do these multiplications. Write each product in polar form, with angle θ less than 360°.

10. $(15\underline{/84^\circ})(12\underline{/158^\circ})$ ________
11. $(1.5\underline{/190^\circ})(4\underline{/250^\circ})$ ________
12. Round "r" to hundredths. $(3.79\underline{/68^\circ})(1.63\underline{/-124^\circ})$ ________

13. Multiply: $(5\underline{/50^\circ})(4\underline{/130^\circ})$ Write the product in rectangular form. ________

14. Find the product in polar form. $(20\underline{/240^\circ})(5\underline{/-70^\circ})(7\underline{/50^\circ})$ ________

15. Find the product in rectangular form. $(2 - j)(3 + j)(1 - 2j)$ ________

ANSWERS:

1. $-4 + 7j$
2. $128 - j$
3. $1,000 + 1,000j$
4. $0 - 10j$
5. $-16 + 40j$
6. $2 - 8j$
7. $0 + 300j$
8. $50 + 0j$
9. 29 or $29 + 0j$
10. $180\underline{/242^\circ}$
11. $6\underline{/80^\circ}$
12. $6.18\underline{/-56^\circ}$ or $6.18\underline{/304^\circ}$
13. $-20 + 0j$
14. $700\underline{/220^\circ}$
15. $5 - 15j$

7-12 DIVIDING IN RECTANGULAR FORM

In this section, we will discuss the procedure for dividing complex numbers in rectangular form.

99. When a complex number is divided by itself, the quotient is "1". For example:

$$\frac{3 + 2j}{3 + 2j} = 1 \qquad \frac{10 - j}{10 - j} = 1$$

If we multiply any division by a fraction like those above, the product equals the original division since we simply multiplied by "1". For example:

$$\left(\frac{7 + 3j}{4 - 2j}\right)\left(\frac{4 + 2j}{4 + 2j}\right) = \frac{7 + 3j}{4 - 2j}, \text{ since } \frac{4 + 2j}{4 + 2j} = 1$$

$$\left(\frac{8 - j}{5 + j}\right)\left(\frac{5 - j}{5 - j}\right) = \frac{8 - j}{5 + j}, \text{ since } \frac{5 - j}{5 - j} = \_\_\_\_$$

---

100. To divide complex numbers, we must eliminate the j-term in the denominator. To do so, we multiply both terms of the division by the conjugate of the denominator. Two examples are given.

Answer: 1

To eliminate the j-term in the denominator (5 - j) below, we multiply both terms by its conjugate (5 + j). We get:

$$\frac{3 + 4j}{5 - j} = \left(\frac{3 + 4j}{5 - j}\right)\left(\frac{5 + j}{5 + j}\right) = \frac{(3 + 4j)(5 + j)}{(5 - j)(5 + j)}$$

$$= \frac{15 + 23j + 4j^2}{25 - j^2}$$

$$= \frac{11 + 23j}{26}$$

To eliminate the j-term in the denominator (3 + 2j) below, we multiply both terms by its conjugate (3 - 2j). We get:

$$\left(\frac{6 - j}{3 + 2j}\right) = \left(\frac{6 - j}{3 + 2j}\right)\left(\frac{3 - 2j}{3 - 2j}\right) = \frac{(6 - j)(3 - 2j)}{(3 + 2j)(3 - 2j)}$$

$$= \_\_\_\_\_\_\_\_$$

$$= \_\_\_\_\_\_\_\_$$

Answer:

$$= \frac{18 - 15j + 2j^2}{9 - 4j^2}$$

$$= \frac{16 - 15j}{13}$$

101. After eliminating the j-term in the denominator, we get expressions like those below in which the denominator is a real number.

$$\frac{6 + 7j}{10} \qquad \frac{4 - 5j}{2} \qquad \frac{10 + 6j}{4}$$

Expressions like those above can be converted to a single complex number. For example:

$$\frac{6 + 7j}{10} = \frac{6}{10} + \frac{7j}{10} = 0.6 + 0.7j$$

$$\frac{4 - 5j}{2} = \frac{4}{2} - \frac{5j}{2} = 2 - 2.5j$$

$$\frac{10 + 6j}{4} = \frac{10}{4} + \frac{6j}{4} = \underline{\qquad\qquad}$$

---

102. Use a calculator to complete each conversion to a single complex number. Round to hundredths.

Answer to 101: 2.5 + 1.5j

a) $\frac{11 + 23j}{26} = \frac{11}{26} + \frac{23j}{26} = \underline{\qquad\qquad}$

b) $\frac{27.9 - 35.6j}{18.7} = \frac{27.9}{18.7} - \frac{35.6j}{18.7} = \underline{\qquad\qquad}$

Answer to 102: a) 0.42 + 0.88j b) 1.49 - 1.90j

103. A complete division is shown below. The quotient is a complex number.

$$\frac{5 + 4j}{3 - j} = \left(\frac{5 + 4j}{3 - j}\right)\left(\frac{3 + j}{3 + j}\right) = \frac{15 + 17j + 4j^2}{9 - j^2} = \frac{11 + 17j}{10} = \frac{11}{10} + \frac{17j}{10} = 1.1 + 1.7j$$

Following the example, complete this division.

$$\frac{7 - 6j}{1 + j} =$$

Answer to 103: 0.5 - 6.5j

104. Use a calculator for these. Round to thousandths.

a) $\frac{5 - 7j}{8 + 3j} =$

b) $\frac{-2.3 - 4.7j}{5.1 - 2.8j} =$

Answer to 104: a) 0.260 - 0.973j , from: $\frac{19 - 71j}{73}$ b) 0.042 - 0.898j , from: $\frac{1.43 - 30.41j}{33.85}$

7-13 DIVISIONS INVOLVING "INCOMPLETE" COMPLEX NUMBERS

In this section, we will discuss divisions involving one or two "incomplete" complex numbers.

105. In each division below, the numerator was written as an "incomplete" complex number by dropping the "0" term. Notice that we then multiplied both terms by the conjugate of the denominator.

$$\frac{0 + 2j}{3 - 4j} = \frac{2j}{3 - 4j}\left(\frac{3 + 4j}{3 + 4j}\right) = \frac{6j + 8j^2}{9 - 16j^2} = \frac{-8 + 6j}{25} = -0.32 + 0.24j$$

$$\frac{3 + 0j}{5 + 2j} = \frac{3}{5 + 2j}\left(\frac{5 - 2j}{5 - 2j}\right) = \frac{15 - 6j}{25 - 4j^2} = \frac{15 - 6j}{29} = 0.52 - 0.21j$$

Using the same steps, complete each division.

a) $\frac{6 + 0j}{2 - j} =$

b) $\frac{0 + 4j}{1 + 3j} =$

106. In the division below, the denominator was written as an "incomplete" complex number by dropping the "0j" term. Complete the second division.

$$\frac{7 - 9j}{10 + 0j} = \frac{7 - 9j}{10} = \frac{7}{10} - \frac{9j}{10} = 0.7 - 0.9j$$

$$\frac{8 + 6j}{-4 + 0j} =$$

Answers to 105:

a) $2.4 + 1.2j$, from: $\frac{12 + 6j}{5}$

b) $1.2 + 0.4j$, from: $\frac{12 + 4j}{10}$

107. In each division below, the denominator was written as an "incomplete" complex number by dropping the "0" term. Notice how we then eliminated the j-term in the denominator by multiplying both terms by "j".

Answer to 106: $-2 - 1.5j$

$$\frac{7 + 3j}{0 + 4j} = \frac{7 + 3j}{4j} = \frac{7 + 3j}{4j}\left(\frac{j}{j}\right) = \frac{7j + 3j^2}{4j^2} = \frac{-3 + 7j}{-4} = \frac{-3}{-4} + \frac{7j}{-4} = 0.75 - 1.75j$$

$$\frac{5 - 6j}{0 - 10j} = \frac{5 - 6j}{-10j} = \frac{5 - 6j}{-10j}\left(\frac{j}{j}\right) = \frac{5j - 6j^2}{-10j^2} = \frac{6 + 5j}{10} = \frac{6}{10} + \frac{5j}{10} = 0.6 + 0.5j$$

Continued on following page.

107. Continued

Following the examples, do these divisions.

a) $\frac{10 - 7j}{0 + 2j} =$

b) $\frac{40 + 30j}{0 - 5j} =$

---

108. Following the example, complete the other division.

$$\frac{9 - j}{0 + j} = \frac{9 - j}{j} = \left(\frac{9 - j}{j}\right)\left(\frac{j}{j}\right) = \frac{9j - j^2}{j^2} = \frac{1 + 9j}{-1} = \frac{1}{-1} + \frac{9j}{-1} = -1 - 9j$$

$\frac{-8 + j}{0 - j} =$

Answers to 107: a) $-3.5 - 5j$ b) $-6 + 8j$

---

109. Complete each division.

a) $\frac{45 - 60j}{15 + 0j} =$

b) $\frac{60 + 36j}{0 - 12j} =$

Answer to 108: $\frac{-1 - 8j}{1} = -1 - 8j$

---

110. In each division below, we began by writing both terms as "incomplete" complex numbers.

$$\frac{0 + 10j}{5 + 0j} = \frac{10j}{5} = 2j \quad \text{(or } 0 + 2j\text{)}$$

$$\frac{13 + 0j}{-10 + 0j} = \frac{13}{-10} = -1.3 \quad \text{(or } -1.3 + 0j\text{)}$$

Complete: a) $\frac{0 - 7j}{-2 + 0j} =$

b) $\frac{-15 + 0j}{3 + 0j} =$

Answers to 109: a) $3 - 4j$ b) $-3 + 5j$

Answers to 110: a) $3.5j$ (or $0 + 3.5j$) b) -5 (or $-5 + 0j$)

111. When both terms can be written as "incomplete" complex numbers and the denominator is a j-term, two different methods are used.

1. If the numerator is also a j-term, we can cancel the "j's". For example:

$$\frac{0 + 12j}{0 + 3j} = \frac{12j}{3j} = \frac{12\not{j}}{3\not{j}} = \frac{12}{3} = 4 \qquad \text{(or } 4 + 0j\text{)}$$

$$\frac{0 - 7j}{0 + 10j} = \frac{-7j}{10j} = \frac{-7\not{j}}{10\not{j}} = \frac{-7}{10} = -0.7 \qquad \text{(or } -0.7 + 0j\text{)}$$

2. If the numerator is an R-term, we multiply both terms by "j". For example:

$$\frac{6 + 0j}{0 + 2j} = \frac{6}{2j} = \frac{6}{2j}\left(\frac{j}{j}\right) = \frac{6j}{2j^2} = \frac{6j}{-2} = -3j \qquad \text{(or } 0 - 3j\text{)}$$

$$\frac{10 + 0j}{0 - 4j} = \frac{10}{-4j} = \frac{10}{-4j}\left(\frac{j}{j}\right) = \frac{10j}{-4j^2} = \frac{10j}{4} = 2.5j \qquad \text{(or } 0 + 2.5j\text{)}$$

Following the examples, complete these:

a) $\frac{0 + 48j}{0 - 6j} =$

b) $\frac{-9 + 0j}{0 + 3j} =$

c) $\frac{0 - 14j}{0 - 5j} =$

d) $\frac{-7 + 0j}{0 - 4j} =$

Answers to 111:

a) -8 (or -8 + 0j)

b) 3j (or 0 + 3j)

c) 2.8 (or 2.8 + 0j)

d) -1.75j (or 0 - 1.75j)

112. Write each quotient as an "incomplete" complex number, a "complete" complex number, and in polar form.

	Incomplete Complex Number		Complete Complex Number		Polar Form
a) $\frac{42}{7}$ =	____________	=	____________	=	____________
b) $\frac{24j}{-6}$ =	____________	=	____________	=	____________
c) $\frac{20}{-4j}$ =	____________	=	____________	=	____________
d) $\frac{-80j}{10j}$ =	____________	=	____________	=	____________

113. The first steps of the methods used to divide complex numbers in rectangular form are summarized below.

1. If the denominator is a "complete" complex number with two non-zero terms, we multiply both terms by the conjugate of the denominator. This first step is used with the divisions below.

$$\frac{7+5j}{4-j} \qquad \frac{-10}{6+3j} \qquad \frac{9j}{2-5j}$$

2. If the denominator is an "incomplete" complex number, three different first steps are used.

a) If the denominator contains only an R-term, we divide by the R-term. Some examples are:

$$\frac{10-4j}{2} \qquad \frac{-12}{6} \qquad \frac{25j}{-15}$$

b) If the denominator contains only a j-term, we multiply both terms by "j". Some examples are:

$$\frac{20-6j}{3j} \qquad \frac{8}{7j} \qquad \frac{-9}{-3j}$$

c) If both the numerator and denominator contain only j-terms, we can cancel the "j's". Some examples are:

$$\frac{12j}{2j} \qquad \frac{-8j}{4j} \qquad \frac{-3j}{-15j}$$

a) $6 = 6 + 0j = 6\underline{/0°}$

b) $-4j = 0 - 4j = 4\underline{/270°}$ or $4\underline{/-90°}$

c) $5j = 0 + 5j = 5\underline{/90°}$

d) $-8 = -8 + 0j = 8\underline{/180°}$

---

114. In which of the following divisions would we begin by multiplying both terms by the conjugate of the denominator? ________

a) $\frac{15-10j}{0+5j}$ b) $\frac{20+j}{1-j}$ c) $\frac{0-9j}{3+2j}$ d) $\frac{36+0j}{-6+0j}$

---

115. In which of the following divisions would we begin by multiplying both terms by "j"? ________

a) $\frac{10-5j}{0-2j}$ b) $\frac{0-8j}{0+4j}$ c) $\frac{7+0j}{0+j}$ d) $\frac{0+25j}{-5+0j}$

(b) and (c)

---

116. In which of the following divisions would we begin by cancelling the "j's"? ________

a) $\frac{16+14j}{2+0j}$ b) $\frac{0+8j}{0-4j}$ c) $\frac{25-15j}{0+5j}$ d) $\frac{0-90j}{0+15j}$

(a) and (c)

(b) and (d)

SELF-TEST 25 (pages 338-344)

Divide and write each quotient as a complex number in rectangular form.

1. $\frac{3 - j}{2 + j}$ ________

2. $\frac{3 - 2j}{1 - 3j}$ ________

3. $\frac{-40 + 30j}{20 - 10j}$ ________

4. Round to thousandths.

   $\frac{1.2 - 4.5j}{2.5 + 1.4j}$ ________

The divisions below involve complex numbers in which the R-term or the j-term is "0". Divide and write each quotient as a complex number in rectangular form.

5. $\frac{0 - 2j}{4 + 3j}$ ________

6. $\frac{30 + 0j}{7 - j}$ ________

7. $\frac{7.2 + 2.8j}{0 - 4j}$ ________

8. $\frac{-960 + 0j}{0 + 500j}$ ________

Divide and write each quotient in polar form.

9. $\frac{0 - 15j}{4 + 0j}$ ________

10. $\frac{2400 + 0j}{-12 + 0j}$ ________

11. $\frac{0 - 38j}{0 - 5j}$ ________

ANSWERS:

1. 1 - j
2. 0.9 + 0.7j
3. -2.2 + 0.40j
4. -0.402 - 1.575j

   from: $\frac{-3.3 - 12.93j}{8.21}$
5. -0.24 - 0.32j
6. 4.2 + 0.6j
7. -0.7 + 1.8j
8. 0 + 1.92j
9. $3.75\angle{-90°}$ or $3.75\angle{270°}$
10. $200\angle{180°}$
11. $7.6\angle{0°}$

7-14 DIVIDING IN POLAR FORM

In this section, we will discuss the procedure for dividing complex numbers in polar form.

117. The division of complex numbers in rectangular form at the left below has been converted to polar form at the right.

$$\frac{0 + 50j}{4 + 3j} = 6 + 8j \longrightarrow \frac{50\underline{/90°}}{5\underline{/37°}} = 10\underline{/53°}$$

By examining the numbers in the polar forms, you can see these facts:

1. The length of the quotient can be obtained by dividing the length of the numerator by the length of the denominator. That is:

$$\frac{50}{5} = 10$$

2. The direction of the quotient can be obtained by subtracting the direction of the denominator from the direction of the numerator. That is:

$$90° - 37° = 53°$$

Let's check the same facts with the division below.

$$\frac{9 + 12j}{0 + 3j} = 4 - 3j \longrightarrow \frac{15\underline{/53°}}{3\underline{/90°}} = 5\underline{/-37°}$$

In the polar form: a) Does $\frac{15}{3} = 5$? ______

b) Does $53° - 90° = -37°$? ______

---

118. To divide complex numbers in polar form:

1. We divide the length of the numerator by the length of the denominator to get the length of the quotient.

2. We subtract the direction of the denominator from the direction of the numerator to get the direction of the quotient.

Using the procedure above, find each quotient.

a) $\frac{20\underline{/80°}}{2\underline{/30°}} =$ ______ b) $\frac{42\underline{/60°}}{7\underline{/100°}} =$ ______

Answers to 117: a) Yes b) Yes

---

119. Do these. Round each length to hundredths.

a) $\frac{35.3\underline{/99°}}{14.7\underline{/31°}} =$ ______ b) $\frac{4.66\underline{/57°}}{3.14\underline{/88°}} =$ ______

Answers to 118: a) $10\underline{/50°}$ b) $6\underline{/-40°}$ or $6\underline{/320°}$

120. When the vector of a complex number is in the third or fourth quadrant, a <u>negative</u> angle is sometimes used for its direction. Subtracting signed angles is the same as subtracting signed numbers. That is:

$$80^\circ - (-20^\circ) = 80^\circ + 20^\circ = 100^\circ$$

$$-50^\circ - 30^\circ = -50^\circ + (-30^\circ) = -80^\circ$$

$$-100^\circ - (-40^\circ) = -100^\circ + 40^\circ = -60^\circ$$

Complete each division.

a) $\dfrac{48\angle -20^\circ}{6\angle 40^\circ}$ = ________ b) $\dfrac{24\angle 50^\circ}{12\angle -60^\circ}$ = ________

a) $2.40\angle 68^\circ$

b) $1.48\angle -31^\circ$
or $1.48\angle 329^\circ$

---

121. When dividing, we can get a negative angle beyond -180° for the direction. For example:

$$\frac{24\angle 50^\circ}{3\angle 300^\circ} = 8\angle -250^\circ$$

We do not ordinarily use negative angles beyond -180° to report the direction of a quotient. We convert negative angles of that size to positive angles. Therefore:

$8\angle -250^\circ$ is written $8\angle 110^\circ$

Do these divisions. Convert negative angles beyond -180° to a positive angle.

a) $\dfrac{40\angle 100^\circ}{8\angle 330^\circ}$ = ________ b) $\dfrac{75\angle 30^\circ}{25\angle 350^\circ}$ = ________

a) $8\angle -60^\circ$ or $8\angle 300^\circ$

b) $2\angle 110^\circ$

---

122. Write each quotient in polar form and then convert it to rectangular form.

	Polar Form	Rectangular Form
a) $\dfrac{24\angle 180^\circ}{6\angle 90^\circ}$ =	________ =	________
b) $\dfrac{12\angle 93^\circ}{2\angle 93^\circ}$ =	________ =	________
c) $\dfrac{18\angle 90^\circ}{6\angle -90^\circ}$ =	________ =	________
d) $\dfrac{40\angle 90^\circ}{10\angle 180^\circ}$ =	________ =	________

a) $5\angle 130^\circ$
(not $5\angle -230^\circ$)

b) $3\angle 40^\circ$
(not $3\angle -320^\circ$)

---

a) $4\angle 90^\circ = 0 + 4j$ b) $6\angle 0^\circ = 6 + 0j$ c) $3\angle 180^\circ = -3 + 0j$ d) $4\angle -90^\circ$ or $4\angle 270^\circ = 0 - 4j$

7-15 FORMULA EVALUATIONS INVOLVING COMBINED OPERATIONS

Some evaluations with formulas related to electrical circuits involve combined operations with complex numbers in either rectangular form or polar form. We will discuss formula evaluations of that type in this section.

---

123. The formula for total impedance in an a-c electrical circuit is:

$$Z_t = \frac{Z_1 Z_2}{Z_1 + Z_2}$$

where Z_t , Z_1 , and Z_2 are complex numbers in either rectangular form or polar form.

If we are given Z_1 and Z_2 , we can find Z_t by an evaluation that involves several operations with complex numbers. For example:

If $Z_1 = 0 - 40j$ and $Z_2 = 10 + 20j$, by substitution we get:

$$Z_t = \frac{(0 - 40j)(10 + 20j)}{(0 - 40j) + (10 + 20j)}$$

Step 1: To simplify the numerator, we do a multiplication and get:

$$Z_t = \frac{800 - 400j}{(0 - 40j) + (10 + 20j)}$$

Step 2: To simplify the denominator, we do an addition and get:

$$Z_t = \frac{800 - 400j}{10 - 20j}$$

Step 3: To get Z_t in rectangular form, we divide and get:

$$Z_t = \left(\frac{800 - 400j}{10 - 20j}\right)\left(\frac{10 + 20j}{10 + 20j}\right)$$

$$= \frac{8,000 + 12,000j - 8,000j^2}{100 - 400j^2}$$

$$= \frac{16,000 + 12,000j}{500}$$

= __________

Answer: 32 + 24j

124. The same formula is given at the left below. To find Z_t when $Z_1 = 40\angle 50°$ and $Z_2 = 30\angle 20°$, we substitute as we have done at the right below.

$$Z_t = \frac{Z_1 Z_2}{Z_1 + Z_2} \qquad Z_t = \frac{(40\angle 50°)(30\angle 20°)}{40\angle 50° + 30\angle 20°}$$

Continued on following page.

124. Continued

Step 1: To simplify the numerator, we do a multiplication and get:

$$Z_t = \frac{1200\angle 70°}{40\angle 50° + 30\angle 20°}$$

Step 2: To simplify the denominator, we do an addition. However, to add complex numbers in polar form, we convert to rectangular form, do the addition, and then convert back to polar form. We get:

$$40\angle 50° = 25.7 + 30.6j$$

$$30\angle 20° = 28.2 + 10.3j$$

Therefore: $Z_t = \dfrac{1200\angle 70°}{53.9 + 40.9j} = \dfrac{1200\angle 70°}{67.7\angle 37°}$

Step 3: To get Z_t in polar form, we divide. Do so. Round the length to tenths.

$$Z_t = \frac{1200\angle 70°}{67.7\angle 37°} = \_\_\_\_\_\_\_\_$$

---

125. When the formula in the last frame is rearranged to solve for Z_1, we get the formula below. Notice that the denominator is a subtraction.

$$\boxed{Z_1 = \frac{Z_t Z_2}{Z_2 - Z_t}}$$

If $Z_t = 4 + j$ and $Z_2 = 5 + 2j$, substitute and find Z_1.

$Z_1 =$ ________

Answer to 124: $17.7\angle 33°$

Answer to 125:

$$Z_1 = \frac{(4 + j)(5 + 2j)}{(5 + 2j) - (4 + j)}$$

$$= \left(\frac{18 + 13j}{1 + j}\right)\left(\frac{1 - j}{1 - j}\right)$$

$$= \frac{31 - 5j}{2}$$

$$= 15.5 - 2.5j$$

126. The formula at the right shows the relationship among current, voltage, and two impedances in an a-c electrical circuit.

$$I = \frac{E}{Z_1 + Z_2}$$

Find I when: $E = 8 + 5j$, $Z_1 = 2 + 3j$, $Z_2 = 4 + 4j$. Round to thousandths.

I = ____________

---

127. When the formula in the last frame is rearranged to solve for E, we get the formula below.

$$E = I(Z_1 + Z_2)$$

Find E in polar form when: $I = 15\angle{-20°}$, $Z_1 = 40\angle{0°}$, $Z_2 = 30\angle{90°}$.

E = ____________

Answer to 126:

$$I = \frac{8 + 5j}{6 + 7j}$$

$$= \frac{83 - 26j}{85}$$

$$= 0.976 - 0.306j$$

---

128. The formula below involves voltages in an a-c electrical circuit.

$$E = I_1Z_1 + I_2Z_2$$

Find E when: $I_1 = 3\angle{40°}$, $Z_1 = 2\angle{50°}$, $I_2 = 5\angle{30°}$, $Z_2 = 4\angle{-30°}$.

Note: Though the values are given in polar form, report your answer in rectangular form.

E = ____________

Answer to 127:

$$E = (15\angle{-20°})[(40 + 0j) + (0 + 30j)]$$

$$= (15\angle{-20°})(40 + 30j)$$

$$= (15\angle{-20°})(50\angle{37°})$$

$$= 750\angle{17°}$$

Answer to 128:

$$E = 6\angle{90°} + 20\angle{0°}$$

$$= (0 + 6j) + (20 + 0j)$$

$$= 20 + 6j$$

129. Here is the same formula: $E = I_1Z_1 + I_2Z_2$

Find E when: $I_1 = 10 + 0j$, $Z_1 = 6 + 4j$, $I_2 = -4 + 2j$, $Z_2 = 5 + 0j$.

Note: Though the values given in rectangular form, report your answer in polar form. Round "r" and θ to whole numbers.

E = ______

$E = 40 + 50j = 64\angle 51°$

SELF-TEST 26 (pages 345-351)

Find each quotient in polar form.

1. $\frac{72\angle 113°}{18\angle 48°}$ ______

2. $\frac{1.80\angle 45°}{1.44\angle -95°}$ ______

3. $\frac{14\angle -52°}{25\angle 238°}$ ______

Find each quotient in polar form and in rectangular form.

4. $\frac{64\angle 0°}{32\angle -180°}$ ______ and ______

5. $\frac{39\angle -180°}{26\angle 90°}$ ______ and ______

In the formula at the right, $Z_1 = 3 - 2j$ and $Z_2 = 4 + 3j$.

$$Z_t = \frac{Z_1Z_2}{Z_1 + Z_2}$$

6. Find Z_t in rectangular form.

Z_t = ______

7. Find Z_t in polar form. Round "r" to hundredths and θ to a whole number.

Z_t = ______

Continued on following page.

SELF-TEST 26 (pages 345-351) - Continued

In the formula at the right, $I = 2\underline{/-10°}$, $Z_1 = 4\underline{/0°}$, and $Z_2 = 3\underline{/90°}$.

$$E = I(Z_1 + Z_2)$$

8. Find E in polar form.
   Round θ to a whole number.

E = ____________

9. Find E in rectangular form.
   Round to tenths.

E = ____________

ANSWERS:

1. $4\underline{/65°}$
2. $1.25\underline{/140°}$
3. $0.56\underline{/70°}$
4. $2\underline{/180°}$ and $(-2 + 0j)$
5. $1.5\underline{/90°}$ and $(0 + 1.5j)$
6. $2.54 - 0.22j$
7. $2.55\underline{/-5°}$ or $2.55\underline{/355°}$
8. $10\underline{/27°}$
9. $8.9 + 4.5j$

SUPPLEMENTARY PROBLEMS - CHAPTER 7

Assignment 22

1. Which of the following are "imaginary" numbers?

   a) -9   b) $\sqrt{-9}$ c) $-\sqrt{36}$ d) $-\sqrt{-25}$ e) 4j

Write each of the following in "j" form.

2. $\sqrt{-64}$ 3. $-\sqrt{-4}$ 4. Round to hundredths: $\sqrt{-31.7}$

In Problems 5-10, write the complex number which each vector represents.

5. Vector #1
6. Vector #2
7. Vector #3

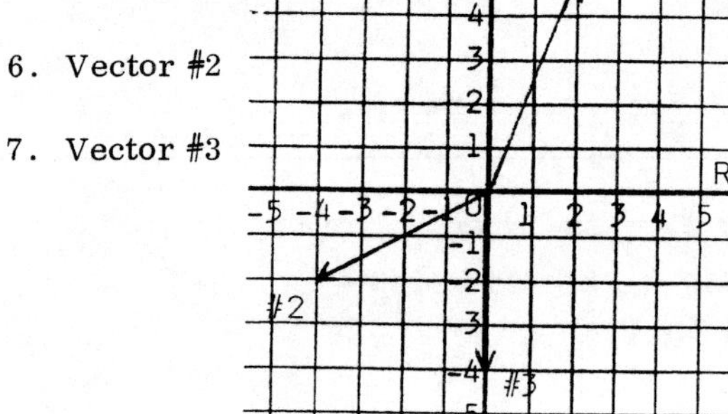

8. Vector #4
9. Vector #5
10. Vector #6

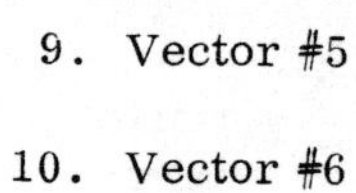

11. Which of the following complex numbers would graph as vectors on the horizontal axis?

    a) 4   b) 1 + j   c) 0 - 2j   d) -2   e) 4j   f) 2 + 0j

12. Which of the following complex numbers would graph as vectors on the vertical axis?

    a) 1 - j   b) 2j   c) -j   d) 72   e) 0 + 25j   f) -5 + 0j

Do these additions. Write each sum as a "complete" complex number.

13. (2 - 2j) + (-5 + 3j)
14. (-28 + 16j) + (72 - 39j)
15. (5.6 - 3.2j) + (-5.6 + 0j)
16. (0 - 410j) + (630 - 150j)
17. (1 + j) + (1 - 2j) + (-2 + 3j)
18. (3 - 5j) + (6 + 6j) + (-10 - 2j)
19. (0 + 16.8j) + (27.2 + 0j) + (-51.7 - 16.8j)
20. (-2j) + (-5) + (6j) + 12

Find the resultant of two vectors whose complex numbers are:

21. (-16 + 11j) and (30 - 23j)   22. (2.7 + 3.5j) and (-6.3 - 1.8j)   23. (0 - 50j) and (30 + 0j)

Find the resultant of three vectors whose complex numbers are:

24. (8 - 5j), (-4 + 3j), and (2 - 2j)   25. (0 - 280j), (510 + 0j), and (0 + 430j)

Assignment 23

Do these subtractions. Write each answer as a "complete" complex number.

1. (5 - 2j) - (3 + 7j)
2. (-18 + 11j) - (12 - 17j)
3. (40 + 60j) - (30 - 10j)
4. (6 + 5j) - (3 + 0j)
5. (32 + 0j) - (0 + 16j)
6. (0 - 200j) - (300 + 0j)

7. Vector A and vector B are added. Their resultant is $(50 - 10j)$. If vector B is $(40 + 30j)$, find vector A.

8. The resultant of vector F and vector G is $(0 + 380j)$. If vector F is $(-460 + 380j)$, find vector G.

Convert to polar form. Round θ to a whole number in each problem.

9. Round "r" to a whole number.

   $572 + 419j$

10. Round "r" to tenths.

    $-17.5 + 36.3j$

11. Round "r" to hundredths.

    $1.78 - 2.95j$

12. Round "r" to hundredths.

    $-25{,}100 - 13{,}400j$

Convert to rectangular form. Round "a" and "b" as directed.

13. Round to hundredths.

    $6.59\angle 73^\circ$

14. Round to a whole number.

    $285\angle -27^\circ$

15. Round to tenths.

    $47.6\angle 235^\circ$

16. Round to hundreds.

    $8{,}120\angle 142^\circ$

Convert each rectangular form to polar form.

17. $0 + 12j$ 18. $-57 + 0j$ 19. $0 - 210j$

Convert each polar form to rectangular form.

20. $6\angle 0^\circ$ 21. $150\angle 90^\circ$ 22. $35\angle 270^\circ$

Add and write each sum in polar form. Round "r" and θ to whole numbers.

23. $300\angle 90^\circ + 500\angle 0^\circ$ 24. $450\angle 60^\circ + 140\angle 270^\circ$

Subtract and write each difference in polar form. Round "r" and θ to whole numbers.

25. $160\angle 0^\circ - 240\angle 90^\circ$ 26. $520\angle 90^\circ - 350\angle 30^\circ$

Assignment 24

Find each product as a "complete" complex number.

1. $(1 + j)(1 + 2j)$
2. $(1 - j)(1 - 3j)$
3. $(5 + j)(3 - j)$
4. $(9 - 4j)(7 - 5j)$
5. $(-10 - 5j)(30 + 16j)$
6. $(16 + 4j)(-8 + 2j)$
7. $(20 - 7j)(20 + 7j)$
8. $(2.4 + j)(1.5 - j)$
9. $(0 - 8j)(-5 + 0j)$
10. $(-20 - 13j)(0 + 10j)$
11. $(0 + 50j)(0 + 50j)$
12. $(4.8 - 5j)(6.5 + 7j)$

Multiply each complex number below by its conjugate. Write each product as a "complete" complex number.

13. $1 + j$ 14. $5 - 3j$ 15. $-7 - 8j$ 16. $120 + 60j$

Find each product in polar form with angle θ between -180° and 360°.

17. $(9\angle 28^\circ)(6\angle 75^\circ)$
18. $(15\angle 140^\circ)(20\angle 120^\circ)$
19. Round "r" to hundredths.

    $(2.62\angle 165^\circ)(3.18\angle 239^\circ)$

20. $(8\angle 80^\circ)(5\angle -50^\circ)$
21. $(12\angle -126^\circ)(25\angle 48^\circ)$
22. Round "r" to tenths.

    $(5.9\angle -34^\circ)(8.7\angle -79^\circ)$

23. $(14\angle 162^\circ)(30\angle 90^\circ)$
24. $(60\angle 120^\circ)(10\angle -150^\circ)$
25. Round "r" to hundreds.

    $(258\angle 170^\circ)(139\angle 260^\circ)$

Find each product in rectangular form as a "complete" complex number.

26. $(3\angle 55^\circ)(2\angle 35^\circ)$
27. $(20\angle 160^\circ)(8\angle 110^\circ)$
28. $(4.5\angle 80^\circ)(1.2\angle -80^\circ)$
29. $(50\angle 40^\circ)(30\angle 140^\circ)$
30. $(12\angle 0^\circ)(8\angle -90^\circ)$
31. $(4.5\angle -60^\circ)(1.8\angle 150^\circ)$

Find each product in polar form.

32. $(4\angle 10^\circ)(3\angle -40^\circ)(6\angle 50^\circ)$
33. $(2\angle 160^\circ)(5\angle 80^\circ)(7\angle 60^\circ)$
34. $(5\angle -90^\circ)(8\angle 135^\circ)(5\angle 90^\circ)$

Find each product in rectangular form as a "complete" complex number.

35. $(2 - j)(1 + j)(3 - j)$
36. $(1 + 2j)(4 + 3j)(1 - 2j)$
37. $(5 + 0j)(8 - 3j)(0 - j)$

Assignment 25

Find each quotient as a "complete" complex number.

1. $\frac{2 + 4j}{1 + j}$
2. $\frac{1 - 3j}{1 - j}$
3. $\frac{-3 - j}{1 + 2j}$
4. $\frac{3 - 11j}{3 - j}$
5. $\frac{-14 + 8j}{2 - 4j}$
6. $\frac{13 - 9j}{4 + 3j}$
7. $\frac{6 + 17j}{3 - 4j}$
8. $\frac{1.1 + 1.3j}{-1 + 2j}$
9. Round to thousandths. $\frac{-1 + j}{3 - 2j}$
10. Round to hundredths. $\frac{6 - 9j}{8 + 7j}$
11. Round to thousandths. $\frac{2.6 + 3.2j}{1.5 - 4.5j}$

Use "incomplete" complex numbers to find each quotient. Write each quotient as a "complete" complex number.

12. $\frac{0 + 10j}{1 - 2j}$
13. $\frac{8 + 0j}{4 + 2j}$
14. $\frac{-3 + 0j}{0 + 4j}$
15. $\frac{12 - 7j}{5 + 0j}$
16. $\frac{-240 + 600j}{0 - 30j}$
17. $\frac{0 - 28j}{30 + 40j}$

Find each quotient in polar form with angle θ between -180° and 360°.

18. $\frac{0 + 8j}{2 + 0j}$
19. $\frac{15 + 0j}{0 + 4j}$
20. $\frac{-240 + 0j}{50 + 0j}$
21. $\frac{0 - 8.4j}{0 - 2.1j}$
22. $\frac{-48 + 0j}{0 + 20j}$
23. $\frac{0 - 850j}{250 + 0j}$

Assignment 26

Find each quotient in polar form with angle θ between -180° and 360°.

1. $\frac{32\angle 115^\circ}{20\angle 47^\circ}$
2. $\frac{8.5\angle 75^\circ}{1.7\angle -25^\circ}$
3. $\frac{174\angle -18^\circ}{500\angle 24^\circ}$
4. $\frac{1{,}230\angle 60^\circ}{1{,}640\angle 300^\circ}$
5. $\frac{48.6\angle -110^\circ}{1.50\angle 260^\circ}$
6. $\frac{620\angle -73^\circ}{250\angle -19^\circ}$
7. Round "r" to hundredths. $\frac{583\angle -28^\circ}{167\angle 17^\circ}$
8. Round "r" to thousandths. $\frac{2.97\angle 53^\circ}{4.36\angle -79^\circ}$

Find each quotient in polar form and in rectangular form.

9. $\dfrac{8\angle -90^\circ}{2\angle 90^\circ}$ 10. $\dfrac{3\angle 120^\circ}{4\angle 30^\circ}$ 11. $\dfrac{6\angle 60^\circ}{2\angle 60^\circ}$ 12. $\dfrac{50\angle -20^\circ}{20\angle 70^\circ}$

13. $I = \dfrac{E}{Z}$ Find I in polar form when $E = 120\angle 0^\circ$ and $Z = 480\angle -22^\circ$.

14. $Z_t = \dfrac{Z_1 Z_2}{Z_1 + Z_2}$ Find Z_t in polar form when $Z_1 = 400\angle 0^\circ$ and $Z_2 = 200\angle 60^\circ$. Round "r" and θ to whole numbers.

15. $E = I(Z_1 + Z_2)$ Find E in polar form when $I = 1.0\angle 45^\circ$, $Z_1 = 40\angle 0^\circ$, and $Z_2 = 20\angle 30^\circ$. Round "r" and θ to whole numbers.

16. $Z_2 = \dfrac{Z_1 Z_t}{Z_1 - Z_t}$ Find Z_2 in rectangular form when $Z_1 = 6 + j$ and $Z_t = 4 - 2j$. Round to hundredths.

17. $E = I_1 Z_1 + I_2 Z_2$ Find E in rectangular form when $I_1 = 1 - j$, $I_2 = 2 + j$, $Z_1 = 4 - 3j$, and $Z_2 = 3 + 2j$.

Chapter 8 SINE WAVES

In this chapter, we will discuss fundamental and non-fundamental sine waves and their harmonics. The following properties of sine waves are defined: basic cycles, amplitude, phase difference and phase shifts. Cosine waves are introduced. Some additions involving sine waves are also shown.

8-1 FUNDAMENTAL SINE WAVES

The graphs of equations of the form $\boxed{y = A \sin \theta}$ where A is positive are sine waves with a 360° cycle (or period) with a basic cycle (or period) from 0° to 360°. Sine waves of that type are called "fundamental" sine waves. We will discuss fundamental sine waves in this section.

---

1. Each equation below is of the form $\boxed{y = A \sin \theta}$ where A is positive.

$$y = 2 \sin \theta \qquad y = 7.5 \sin \theta \qquad y = \frac{1}{2} \sin \theta$$

The equation $y = \sin \theta$ is also of the form $y = A \sin \theta$ since:

$$y = \sin \theta \text{ can be written } y = 1 \sin \theta .$$

That is, though the constant A is not explicitly shown in $y = \sin \theta$, the constant A is really ______.

---

2. We graphed $\boxed{y = \sin \theta}$ in an earlier chapter. To do so, we found pairs of values for y and θ. For example:

If $\theta = 0°$, $y = \sin 0° = 0$

If $\theta = 30°$, $y = \sin 30° = 0.5$

Use a calculator to complete these for the equation above. If necessary, round to two decimal places.

a) If $\theta = 90°$, y = ________ c) If $\theta = 390°$, y = ________

b) If $\theta = 120°$, y = ________ d) If $\theta = -135°$, y = ________

Answer to 1: 1

Answers to 2:

a) 1 c) 0.5

b) 0.87 d) -0.71

3. To graph $\boxed{y = 2 \sin \theta}$, we must find pairs of values for y and θ. For example:

If $\theta = 0°$, $y = 2 \sin 0° = 2(0) = 0$

If $\theta = 30°$, $y = 2 \sin 30° = 2(0.5) = 1$

Use a calculator to complete these for the equation above. If necessary, round to two decimal places.

a) If $\theta = 90°$, $y = 2 \sin 90° =$ ________

b) If $\theta = 270°$, $y = 2 \sin 270° =$ ________

c) If $\theta = -60°$, $y = 2 \sin(-60°) =$ ________

---

4. To graph $\boxed{y = \frac{1}{2} \sin \theta}$, we must find pairs of values for y and θ. For example:

If $\theta = 0°$, $y = \frac{1}{2} \sin 0° = \frac{1}{2}(0) = 0$

If $\theta = 90°$, $y = \frac{1}{2} \sin 90° = \frac{1}{2}(1) = \frac{1}{2}$ or 0.5

Use a calculator to complete these for the equation above. If necessary, round to two decimal places.

a) If $\theta = 30°$, $y = \frac{1}{2} \sin 30° =$ ________

b) If $\theta = 315°$, $y = \frac{1}{2} \sin 315° =$ ________

c) If $\theta = -180°$, $y = \frac{1}{2} \sin(-180°) =$ ________

Answers to 3: a) 2 b) -2 c) -1.73

---

5. Any sine wave can be divided into "cycles" or "periods". A basic cycle (or period) contains a complete positive loop on the left and a complete negative loop on the right as shown.

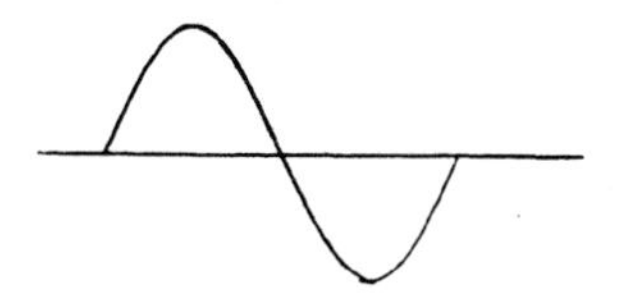

Basic Cycle (or Period) Of a Sine Wave

In the figure below, the graph of $\boxed{y = \sin \theta}$ is shown from -360° to 720°.

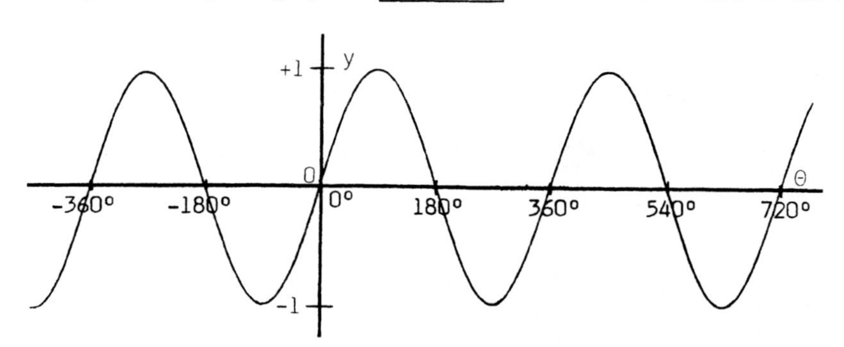

Answers to 4: a) 0.25 b) -0.35 c) 0

Continued on following page.

5. (Continued)

The part of the graph shown contains three basic cycles (or periods). The first cycle begins at -360° and ends at 0°.

a) The second cycle begins at 0° and ends at ________°.

b) The third cycle begins at 360° and ends at ________°.

---

a) 360°

b) 720°

---

6. The basic cycle of $y = \sin \theta$ from 0° to 360° is graphed at the right.

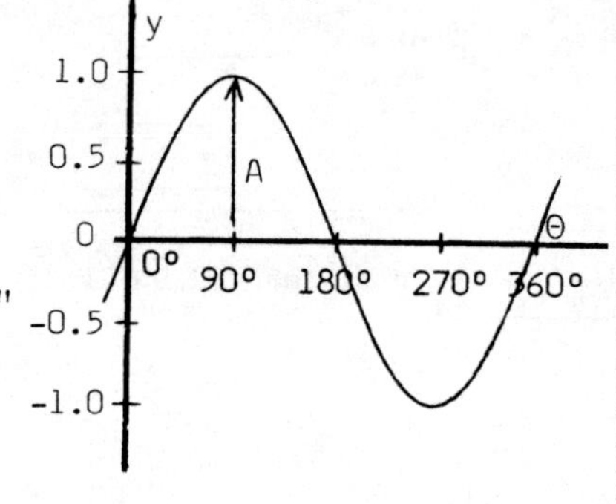

The arrow labeled "A" in the positive loop shows that the "maximum value of y" occurs when $\theta = 90°$. The "maximum value of y" is called the "amplitude" of the sine wave.

For $y = \sin \theta$:

a) The "maximum value of y" is ________.

b) The "amplitude" of the sine wave is ________.

---

a) 1.0 or 1 b) 1.0 or 1

---

7. The basic cycle of $y = 2 \sin \theta$ from 0° to 360° and the basic cycle of $y = \frac{1}{2} \sin \theta$ from 0° to 360° are graphed below. The arrow labeled "A" in the positive loop shows that the "maximum value of y" or "amplitude" of each sine wave occurs when $\theta = 90°$.

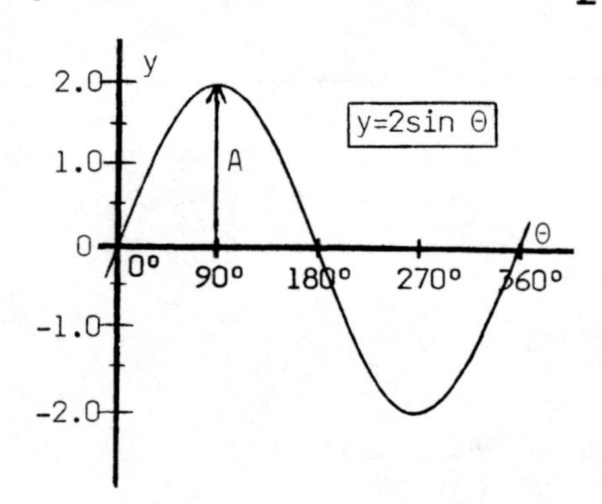

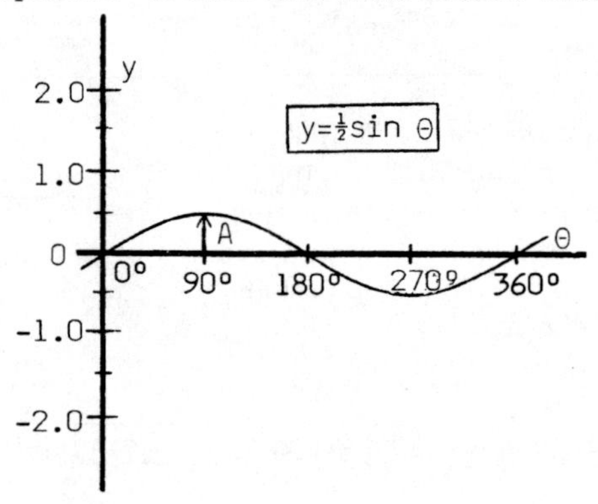

For $y = 2 \sin \theta$: a) The maximum value of y is ________.

b) The amplitude of the sine wave is ________.

For $y = \frac{1}{2} \sin \theta$: c) The maximum value of y is ________.

d) The amplitude of the sine wave is ________.

---

a) 2.0 or 2 c) 0.5 or $\frac{1}{2}$

b) 2.0 or 2 d) 0.5 or $\frac{1}{2}$

8. A sine wave with a 360° cycle (or period) with a basic cycle (or period) from 0° to 360° is called a "fundamental" sine wave. Fundamental sine waves have the equation $y = A \sin \theta$. They also have these properties:

1. They have a positive peak (or maximum value of y) at 90°.
2. Their amplitude equals the constant A in the equation.

For $y = 1 \sin \theta$, the amplitude is 1.

For $y = 2 \sin \theta$, the amplitude is 2.

For $y = \frac{1}{2} \sin \theta$, the amplitude is $\frac{1}{2}$.

If the following equations were graphed, what would the amplitude be for each sine wave?

a) $y = 4 \sin \theta$ ________ c) $y = 300 \sin \theta$ ________

b) $y = 10 \sin \theta$ ________ d) $y = 7.41 \sin \theta$ ________

Answers: a) 4 c) 300 b) 10 d) 7.41

9. Write the equation of the fundamental sine wave:

a) Whose amplitude is 25. ____________________

b) Whose amplitude is 0.69 . ____________________

Answers: a) $y = 25 \sin \theta$ b) $y = 0.69 \sin \theta$

10. Both sine waves at the right are fundamental sine waves. Write the equation of each.

a) Sine wave C.

____

b) Sine wave D.

____

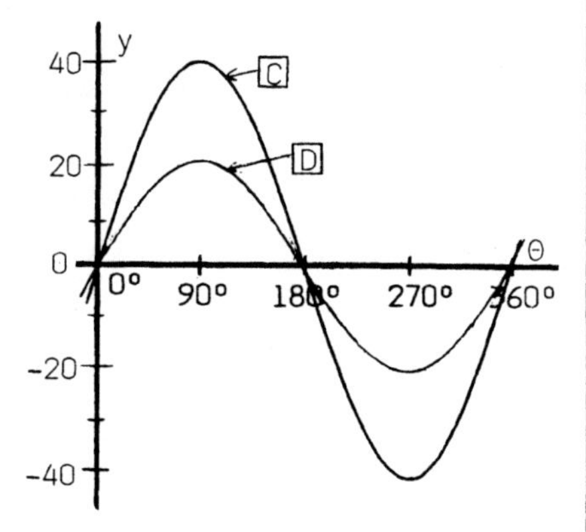

Answers: a) $y = 40 \sin \theta$ b) $y = 20 \sin \theta$

8-2 NON-FUNDAMENTAL SINE WAVES

The graphs of equations of the form $y = A\sin(\theta \pm \varphi)$ where A is positive are sine waves with a 360° cycle (or period) without a basic cycle (or period) from 0° to 360°. Sine waves of that type are called "non-fundamental" sine waves. We will discuss non-fundamental sine waves in this section.

---

11. Each equation below is of the form $y = A\sin(\theta + \varphi)$ or $y = A\sin(\theta - \varphi)$ where A is positive.

$$y = 3\sin(\theta + 90°) \qquad y = 20\sin(\theta - 30°)$$

The equation $y = \sin(\theta + 45°)$ is also of the form $y = A\sin(\theta + \varphi)$ since:

$$y = \sin(\theta + 45°) \text{ can be written } y = 1\sin(\theta + 45°)$$

That is, though the constant A is not explicitly shown in $y = \sin(\theta + 45°)$, the constant A is really ______.

---

12. To find pairs of values for $y = \sin(\theta - 60°)$, we subtract 60° from each value substituted for θ. That is:

If $\theta = 90°$, $y = \sin(90° - 60°) = \sin 30° = 0.5$

If $\theta = -30°$, $y = \sin(-30° - 60°) = \sin(-90°) = -1$

Complete these for the equation above. If necessary, round to two decimal places.

a) If $\theta = 180°$, $y = \sin(180° - 60°) = \sin$ ______ ° = ______

b) If $\theta = -210°$, $y = \sin(-210° - 60°) = \sin$ ______ ° = ______

Answer to 11: 1

---

13. To find pairs of values for $y = 4\sin(\theta + 30°)$, we add 30° to each value substituted for θ. That is:

If $\theta = 120°$, $y = 4\sin(120° + 30°) = 4\sin 150° = 2$

If $\theta = -90°$, $y = 4\sin(-90° + 30°) = 4\sin(-60°) = -3.46$

Complete these for the equation above. If necessary, round to two decimal places.

a) If $\theta = 300°$, $y = 4\sin(300° + 30°) = 4\sin$ ______ ° = ______

b) If $\theta = -150°$, $y = 4\sin(-150° + 30°) = 4\sin$ ______ ° = ______

Answers to 12:

a) $\sin 120° = 0.87$

b) $\sin(-270°) = 1$

Answers to 13:

a) $4\sin 330° = -2$

b) $4\sin(-120°) = -3.46$

14. The graph of $y = \sin(\theta + 60°)$ is given below.

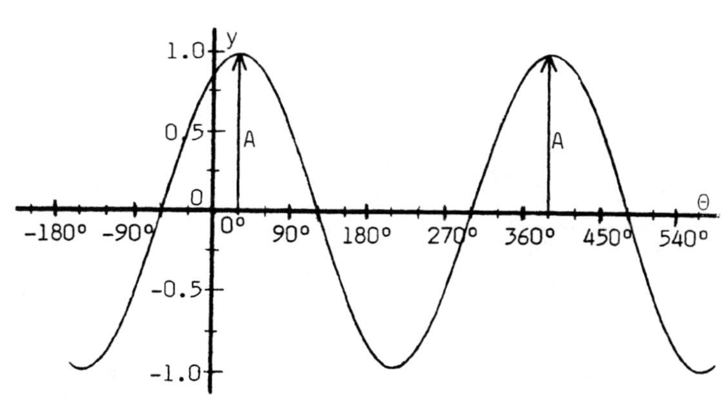

Notice these points about the sine wave.

1. Though it has a 360° cycle (or period), it does not have a basic cycle from 0° to 360°. Therefore, it is not a fundamental sine wave.
2. It has a basic cycle (or period) from -60° to 300°.
3. It has positive peaks at 30° and 390°. See the arrows labeled "A".

The amplitude of the sine wave is ________.

1.0 or 1

---

15. The graph of $y = 2\sin(\theta - 90°)$ is given below.

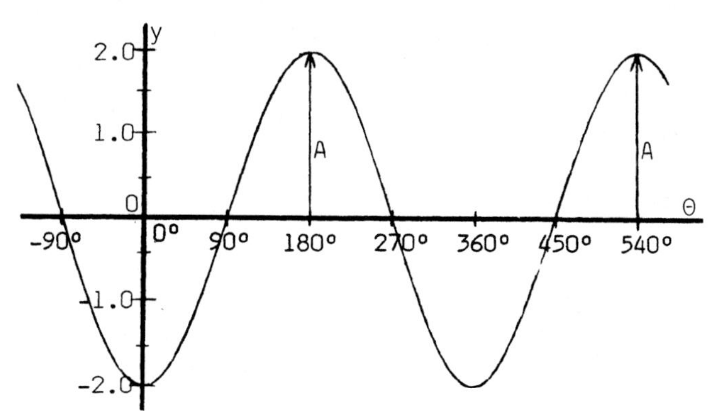

The sine wave is a non-fundamental sine wave because it does not have a basic cycle (or period) from 0° to 360°.

a) The only basic cycle on the graph is from ________° to ________°.

b) There are positive peaks when θ = ________° and θ = ________°.

c) The amplitude of the sine wave is ________.

---

a) 90° to 450°

b) 180° and 540°

c) 2.0 or 2

16. The graph of $y = 40 \sin(\theta + 45°)$ is given below.

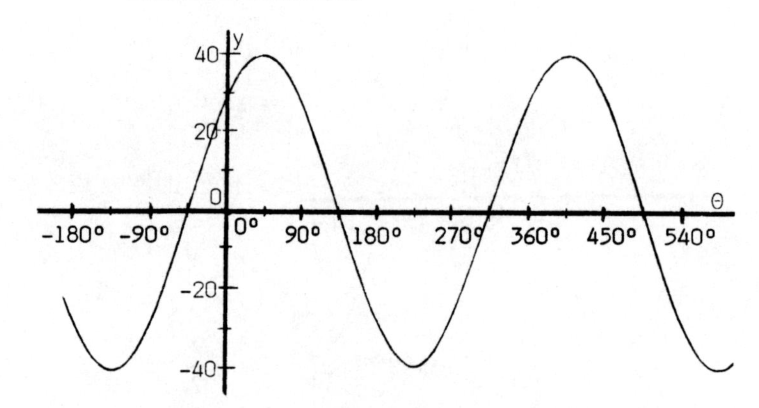

a) Why is the sine wave a non-fundamental sine wave?

____

b) The only basic cycle on the graph is from ________° to ________°.

c) There are positive peaks when θ = ________° and θ = ________°.

d) The amplitude of the sine wave is ________.

Answers (16):

a) It does not have a basic cycle from 0° to 360°.

b) -45° to 315°

c) 45° and 405°

d) 40

---

17. In the last few frames, we saw that the amplitude of a non-fundamental sine wave is also given by the constant A in the equation $y = A \sin(\theta \pm \varphi)$. That is:

The amplitude of $y = \sin(\theta + 60°)$ is 1.

The amplitude of $y = 2 \sin(\theta - 90°)$ is 2.

The amplitude of $y = 40 \sin(\theta + 45°)$ is 40.

If the following equations were graphed, what would the amplitude be for each non-fundamental sine wave?

a) $y = 3 \sin(\theta + 30°)$ ________

b) $y = 500 \sin(\theta - 45°)$ ________

c) $y = 7.5 \sin(\theta + 90°)$ ________

d) $y = 83.1 \sin(\theta - 60°)$ ________

Answers (17):

a) 3　　c) 7.5

b) 500　　d) 83.1

---

18. Which of the following equations have graphs that are non-fundamental sine waves? ____________

a) $y = 60 \sin \theta$

b) $y = \sin(\theta - 50°)$

c) $y = 14.5 \sin(\theta + 15°)$

d) $y = 1.97 \sin \theta$

Answer (18): (b) and (c)

8-3 PHASE DIFFERENCES BETWEEN NON-FUNDAMENTAL AND FUNDAMENTAL SINE WAVES

Two sine waves are either "in-phase" or "out-of-phase". Any non-fundamental sine wave is "out-of-phase" with a fundamental sine wave. The shift needed to put a non-fundamental "in-phase" with a fundamental is called the "phase difference" (or "phase shift" or "phase angle") of the non-fundamental. We will discuss phase differences of that type in this section.

---

19. Two sine waves are graphed in each figure below. Those at the left are "in-phase". Those at the right are "out-of-phase".

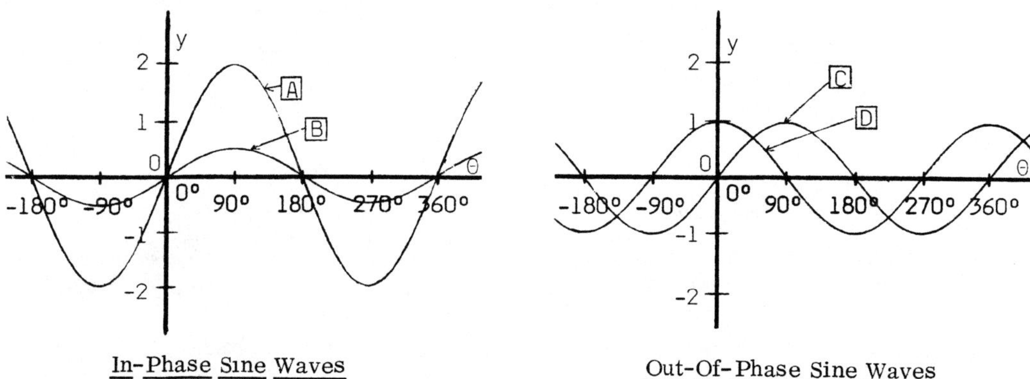

In-Phase Sine Waves — Out-Of-Phase Sine Waves

In-Phase Sine Waves. At the left, both A and B are fundamental sine waves. Each has a cycle from 0° to 360° with a positive peak at 90°. Since A and B cross the horizontal axis at the same points and peak at the same points, we say that they are "in-phase".

Out-Of-Phase Sine Waves. At the right, C is a fundamental sine wave but D is a non-fundamental sine wave because it does not have a basic cycle from 0° to 360° with a positive peak at 90°. Since C and D do not cross the horizontal axis at the same point or peak at the same point, we say that they are "out-of-phase".

a) If sine waves A and C were graphed together, would they be in-phase or out-of-phase?

____

b) If sine waves B and D were graphed together, would they be in-phase or out-of-phase?

____

---

a) In-phase

b) Out-of-phase

20. The two sine waves below are out-of-phase. The solid line is the graph of the non-fundamental sine wave $y = \sin(\theta + 90°)$. The dashed line is the graph of the fundamental sine wave $y = \sin \theta$.

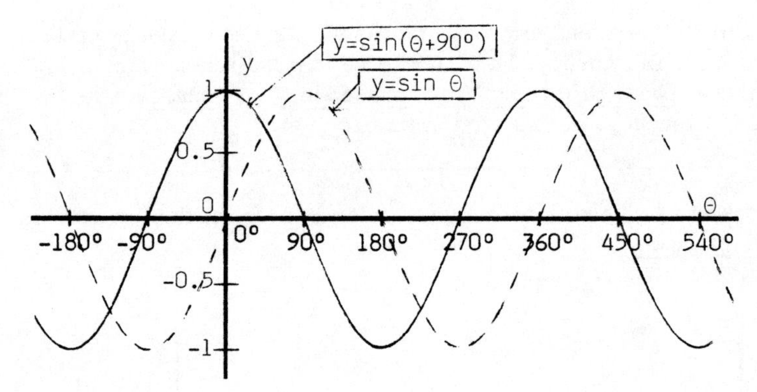

Notice these points:

1. Though the two graphs have the same shape, the positive peaks and horizontal-axis crossings of $y = \sin(\theta + 90°)$ are to the left of those for $y = \sin \theta$.
2. The graph of $y = \sin(\theta + 90°)$ could be put in-phase with the graph of $y = \sin \theta$ by shifting the graph of $y = \sin(\theta + 90°)$ to the right along the horizontal axis.

Since the horizontal axis is calibrated in degrees, horizontal shifts are stated in degrees.

How many degrees to the right must we shift the graph of $y = \sin(\theta + 90°)$ to put it in-phase with the graph of $y = \sin \theta$? ________ ° to the right

---

21. The same two out-of-phase sine waves are shown again below.

90° to the right

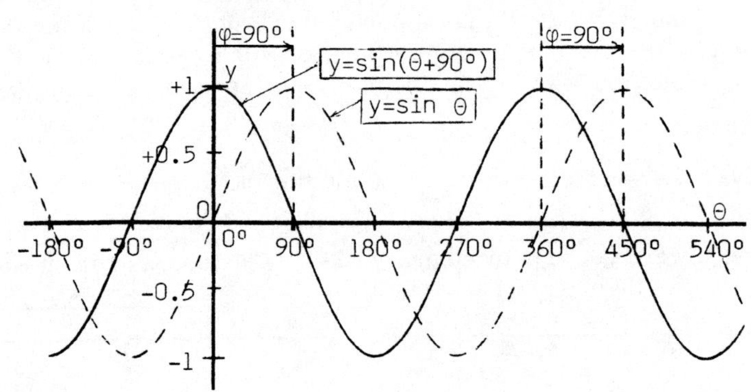

We used adjacent peaks in two places to show the horizontal shift needed to put $y = \sin(\theta + 90°)$ in-phase with $y = \sin \theta$. The two pairs of peaks used were at 0° and 90° and also at 360° and 450°. The arrow labeled φ (pronounced "fee") represents the shift needed.

The amount that the two sine waves are out-of-phase is called their "phase difference". The "phase difference" is the size of the shift needed to put the non-fundamental in-phase with the fundamental. The terms "phase shift" or "phase angle" are sometimes used instead of "phase difference".

On the graph above, the arrow φ represents the phase difference (or phase shift or phase angle). For the sine waves above, the phase difference is ________ ° to the right.

22. The two sine waves below are also out-of-phase. The solid line is the graph of $y = \sin(\theta + 60°)$. The dashed line is the graph of the fundamental $y = \sin \theta$.

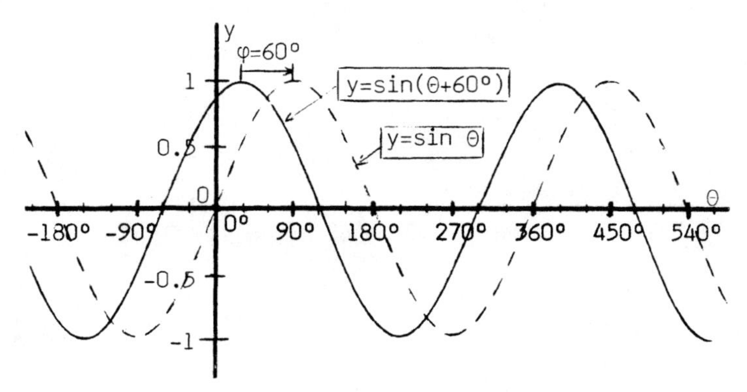

Using the adjacent peaks at 30° and 90°, we showed the shift needed to put $y = \sin(\theta + 60°)$ in-phase with $y = \sin \theta$.

a) How many degrees to the right must $y = \sin(\theta + 60°)$ be shifted to put it in-phase with $y = \sin \theta$? ________

b) Therefore, the phase difference or phase angle is how many degrees? φ = ________ ° to the right

90° to the right

---

23. There are two other technical terms that are sometimes used instead of the term "phase difference". They are: ______________________ and ______________________.

a) 60°

b) φ = 60° to the right

---

24. We can also have phase shifts to the left. As an example, we graphed $y = \sin(\theta - 90°)$ with the graph of $y = \sin \theta$ below.

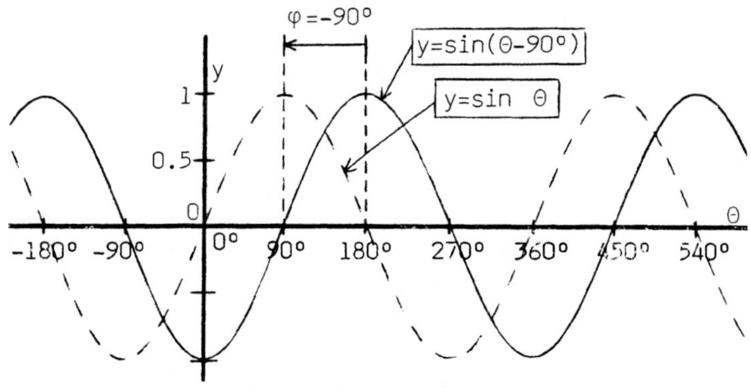

As you can see, the positive peaks and horizontal-axis crossings of $y = \sin(\theta - 90°)$ lie to the right of those of $y = \sin \theta$. Using the peaks at 90° and 180°, we diagrammed the shift needed to put $y = \sin(\theta - 90°)$ in-phase with $y = \sin \theta$.

a) To put the two sine waves in-phase, $y = \sin(\theta - 90°)$ must be shifted ________ ° to the left.

b) Since the shift is to the left, φ , the phase difference or phase angle, is represented by what signed number? ________ °

phase shift and phase angle

25. We can also have phase shifts with sine waves whose amplitude is other than "1". For example, the graph of $y = 2\sin(\theta - 60°)$ is shown below with the graph of the fundamental $y = 2\sin\theta$.

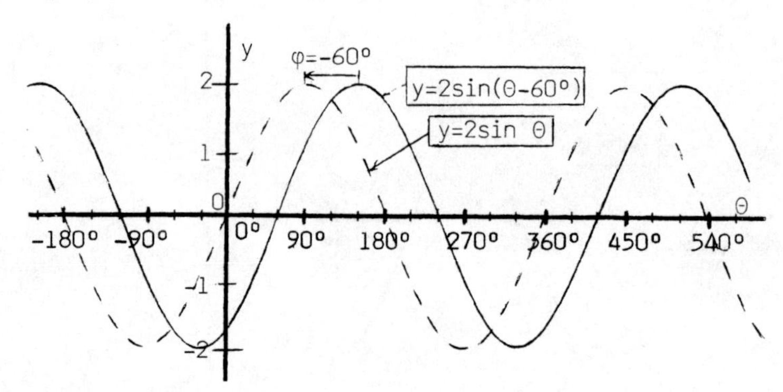

The peaks and horizontal-axis crossings of $y = 2\sin(\theta - 60°)$ lie to the right of those of $y = 2\sin\theta$. Therefore, we must shift $y = 2\sin(\theta - 60°)$ to the left to put it in-phase with $y = 2\sin\theta$. We used the peaks at 90° and 150° to show the shift.

a) To put $y = 2\sin(\theta - 60°)$ in-phase with $y = 2\sin\theta$, we must shift the non-fundamental ________° to the left.

b) Since the shift is to the left, φ, the phase difference or phase angle, is represented by what signed number? ________°

a) 90° to the left

b) -90°

---

26. The graph of $y = \sin(\theta + 135°)$ is given below with the graph of $y = \sin\theta$.

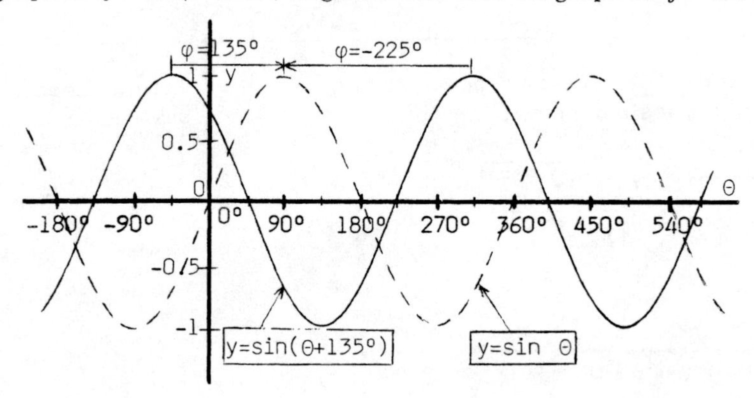

$y = \sin(\theta + 135°)$ can be put in-phase with $y = \sin\theta$ by a horizontal shift in either direction.

1. Using the peaks at -45° and 90°, we get a 135° shift to the right. $\varphi = +135°$

2. Using the peaks at 90° and 315°, we get a -225° shift to the left. $\varphi = -225°$

Though two phase shifts are possible, we always use the one with the smaller number of degrees. Therefore, for the non-fundamental sine wave wave above, we use ________° as φ.

a) 60° to the left

b) -60°

27. On the graph below, the solid sine wave is a non-fundamental and the dashed sine wave is a fundamental. The amplitude of each sine wave is 4.

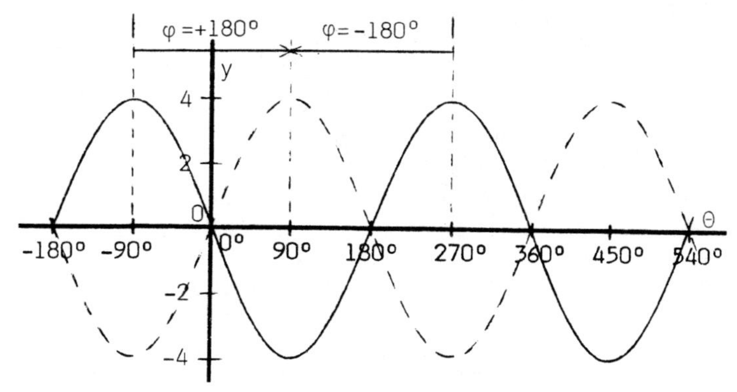

The two possible phase shifts for putting the non-fundamental in-phase with the fundamental are shown. They are:

$\varphi = +180°$ (Using the peaks at -90° and 90°)

$\varphi = -180°$ (Using the peaks at 90° and 270°)

Since each shift involves the same number of degrees, we can use either phase shift. Therefore, the two possible equations for the non-fundamental sine wave are:

$y = 4\sin(\theta + 180°)$ $y = 4\sin(\theta - 180°)$

What is the equation of the fundamental sine wave? ______________

+135°

---

28. The two general equations for non-fundamental sine waves are:

$$y = A\sin(\theta + \varphi) \qquad y = A\sin(\theta - \varphi)$$

In each equation:

1. "A" is the amplitude of the sine wave.
2. "φ" is the phase difference or phase shift or phase angle of the sine wave with respect to a fundamental sine wave.

   A "$+\varphi$" means that the phase difference is positive. Therefore, to put the sine wave in-phase with a fundamental, the phase shift is to the right.

   A "$-\varphi$" means that the phase difference is negative. Therefore, to put the sine wave in-phase with a fundamental, the phase shift is to the left.

Given this equation: $y = \sin(\theta + 120°)$

a) The amplitude of the sine wave is ________.

b) As a signed angle, the phase difference with a fundamental sine wave is ________°.

c) To put the graph of the sine wave in-phase with a fundamental sine wave, the graph must be shifted 120° to the ____________ (right/left).

$y = 4\sin\theta$

29. Given this equation: $y = 1.5 \sin(\theta - 30°)$

a) The amplitude of the sine wave is ________.

b) As a signed angle, the phase angle with a fundamental sine wave is ________°.

c) To put the graph of the sine wave in-phase with a fundamental sine wave, the graph must be shifted 30° to the __________ (right/left).

a) 1

b) +120°

c) right

---

30. a) A non-fundamental sine wave has an amplitude of 10 and a phase angle $\varphi = +75°$ with respect to a fundamental sine wave. Write its equation.

____

b) A non-fundamental sine wave has an amplitude of 0.63 and a phase angle $\varphi = -150°$ with respect to a fundamental sine wave. Write its equation.

____

a) 1.5

b) -30°

c) left

---

a) $y = 10 \sin(\theta + 75°)$ b) $y = 0.63 \sin(\theta - 150°)$

8-4 PHASE DIFFERENCES BETWEEN NON-FUNDAMENTAL SINE WAVES

There can also be phase differences between two non-fundamental sine waves. We will discuss phase differences of that type in this section.

---

31. Two non-fundamental sine waves are graphed in each figure below. The two at the left are in-phase. The two at the right are out-of-phase.

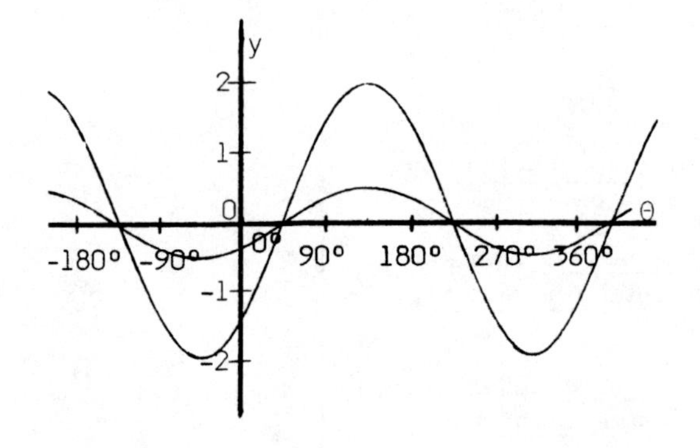

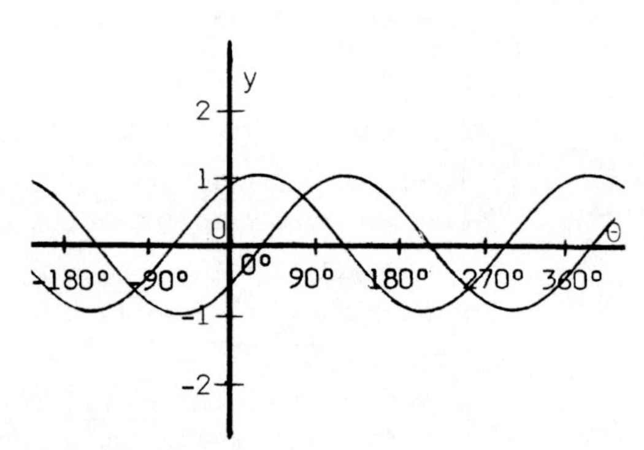

In-Phase Sine Waves. The two sine waves at the left cross the horizontal axis at the same point and peak at the same points.

Out-Of-Phase Sine Waves. The two sine waves at the right do not cross the horizontal axis at the same points or peak at the same points.

32. The following two non-fundamental sine waves are graphed below.

A: $y = \sin(\theta + 30°)$
B: $y = \sin(\theta - 90°)$

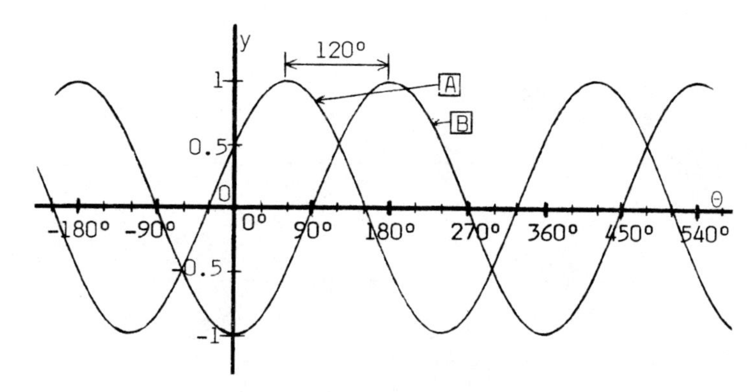

The phase angle φ in each equation represents the shift needed to put the non-fundamental in-phase with a fundamental.

For A, we must shift 30° to the right.
For B, we must shift 90° to the left.

We can also talk about the phase difference between the two non-fundamentals. It is 120° (see the arrow).

It is +120° if we consider the shift needed to put A in-phase with B.
It is -120° if we consider the shift needed to put B in-phase with A.

However, we ordinarily do not use a signed angle to represent the phase difference between two non-fundamentals. We simply say that the phase difference is 120°.

Does the 120° phase difference appear in the equation of either non-fundamental sine wave? ________

No

33. The following two non-fundamental sine waves are graphed below.

C: $y = 2 \sin(\theta + 60°)$
D: $y = 2 \sin(\theta - 30°)$

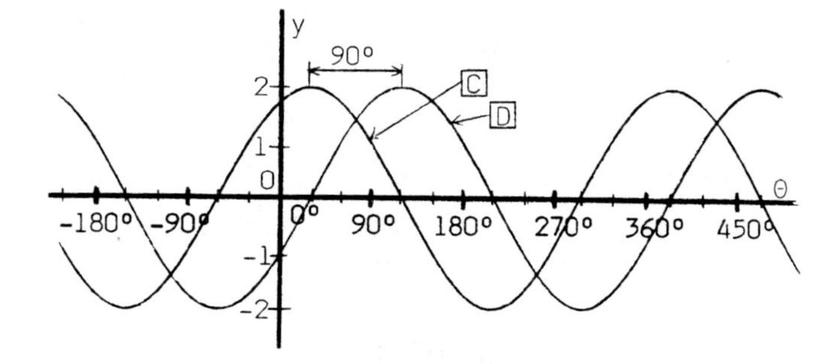

Continued on following page.

33. Continued

The arrow shows that the phase difference between C and D is 90°.

a) To put C in-phase with D, we would have to shift C 90° to the ____________ (right/left).

b) To put D in-phase with C, we would have to shift D 90° to the ____________ (right/left).

---

a) right b) left

SELF-TEST 27 (pp. 356-370)

1. Write the equation of the fundamental sine wave whose amplitude is 170. ____________

2. The graph of $y = \sin\theta$ has a "basic cycle" that begins at $\theta = -360°$ and ends at $\theta =$ ________.

What is the amplitude of the graph for each equation below?

3. $y = 0.35 \sin\theta$ ________

4. $y = 52.8 \sin(\theta - 20°)$ ________

5. $y = \sin(\theta + 150°)$ ________

If $y = 12 \sin\theta$ is graphed:

6. What is the "maximum value of y"? ________

7. For what angle θ between 0° and 360° is y a maximum? $\theta =$ ________

8. To put it in-phase with a fundamental sine wave, the graph of $y = 8 \sin(\theta - 60°)$ must be shifted 60° to the ____________ (right/left).

9. To put it in-phase with the graph of $y = 80 \sin\theta$, the graph of $y = 25 \sin(\theta + 140°)$ must be shifted 140° to the ____________ (right/left).

If $y = 16.2 \sin(\theta - 160°)$ is graphed:

10. What is the "maximum value of y"? ________

11. Its signed phase angle with a fundamental sine wave is ________.

12. A non-fundamental sine wave has an amplitude of 6.47 and a phase angle $\varphi = -90°$ with a fundamental sine wave. Write its equation. ____________

13. A non-fundamental sine wave has a "maximum value of y" = 350 and a phase angle $\varphi = 130°$ with the graph of $y = 200 \sin\theta$. Write its equation. ____________

14. The graph of $y = 50 \sin(\theta + 60°)$ is in-phase with the graph of which equation below? ________

a) $y = 50 \sin\theta$ b) $y = 50 \sin(\theta - 60°)$ c) $y = 25 \sin(\theta + 60°)$ d) $y = 25 \sin(\theta + 30°)$

ANSWERS:

1. $y = 170 \sin\theta$
2. 0°
3. 0.35
4. 52.8
5. 1
6. 12
7. 90°
8. left
9. right
10. 16.2
11. -160°
12. $y = 6.47 \sin(\theta - 90°)$
13. $y = 350 \sin(\theta + 130°)$
14. c

8-5 SINE-WAVE HARMONICS

Sine waves in which there are two or more basic cycles in a 360° period are called "harmonics". The general equation for harmonics of fundamental sine waves is: $\boxed{y = A \sin n\theta}$. The general equation for harmonics shifted horizontally is: $\boxed{y = A \sin(n\theta \pm \phi)}$. In both equations, "n" is a whole number like 2, 3, 4, and so on. We will discuss sine-wave "harmonics" in this section.

---

34. To graph a sine-wave harmonic like $\boxed{y = \sin 2\theta}$, we must find pairs of values for y and θ. To do so, we multiply each value of θ by 2 and then find the sine of the new angle. For example:

$$\text{If } \theta = 30°, \quad y = \sin(2)(30°) = \sin 60° = 0.87$$

$$\text{If } \theta = 90°, \quad y = \sin(2)(90°) = \sin 180° = 0$$

Following the examples, complete these. If necessary, round to two decimal places.

a) If $\theta = 45°$, $y = \sin(2)(45°) = \sin$ ______° = ______

b) If $\theta = 150°$, $y = \sin(2)(150°) = \sin$ ______° = ______

---

a) $\sin 90° = 1$

b) $\sin 300° = -0.87$

35. To graph $\boxed{y = 3 \sin 5\theta}$, we must find pairs of values for y and θ. To do so:

1. We multiply the value substituted for θ by 5.

$$\text{If } \theta = 30°, \quad y = 3 \sin(5)(30°) = 3 \sin 150°$$

2. We then multiply the sine of the new angle by 3.

$$y = 3 \sin 150° = 3(0.5) = 1.5$$

Using the same equation, complete these. If necessary, round to two decimal places.

a) Find y when $\theta = 12°$. y = ______

b) Find y when $\theta = -45°$. y = ______

---

a) $y = 2.60$

b) $y = 2.12$

36. To find pairs of values for $\boxed{y = \sin(3\theta + 60°)}$, we use these steps:

1. Multiply the value substituted for θ by 3.
2. Then add 60° before finding the sine.

If $\theta = 20°$, we get:

$$y = \sin[3(20°) + 60°] = \sin(60° + 60°) = \sin 120° = 0.87$$

Using the same equation, complete these. If necessary, round to two decimal places.

a) Find y when $\theta = 10°$. y = ________

b) Find y when $\theta = -45°$. y = ________

Answers to 36:

a) y = 1, from: $\sin 90°$

b) y = -0.97, from: $\sin(-75°)$

37. To find pairs of values for $\boxed{y = 5\sin(2\theta - 30°)}$, we use these steps:

1. Multiply the value substituted for θ by 2 and then subtract 30°.
2. Find the sine of the new angle and then multiply that sine by 5.

If $\theta = 60°$, we get:

$$y = 5\sin[2(60°) - 30°] = 5\sin(120° - 30°) = 5\sin 90° = 5(1) = 5$$

Using the same equation, complete these. If necessary, round to two decimal places.

a) Find y when $\theta = 90°$. y = ________

b) Find y when $\theta = 220°$. y = ________

Answers to 37:

a) y = 2.5, from: $5\sin 150°$

b) y = 3.83, from: $5\sin 410°$

38. The graph at the left below is a basic cycle because it contains a complete positive loop and a complete negative loop. The sine wave on the right below can be divided into two basic cycles (see the dotted line).

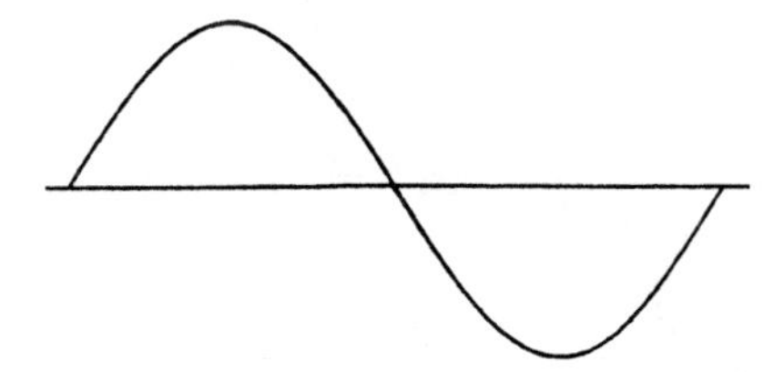

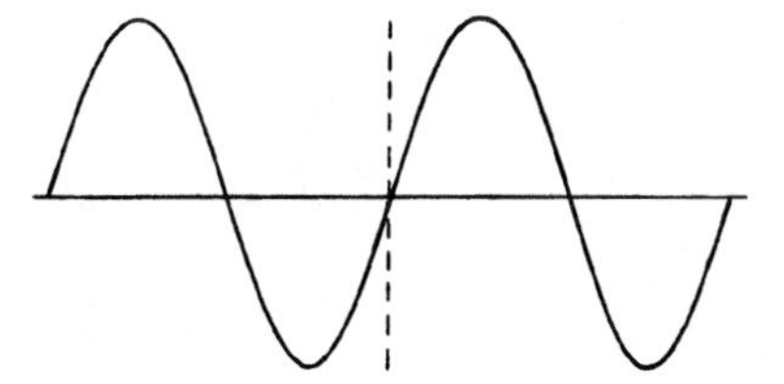

Into how many basic cycles can we divide each of the following sine waves?

a)

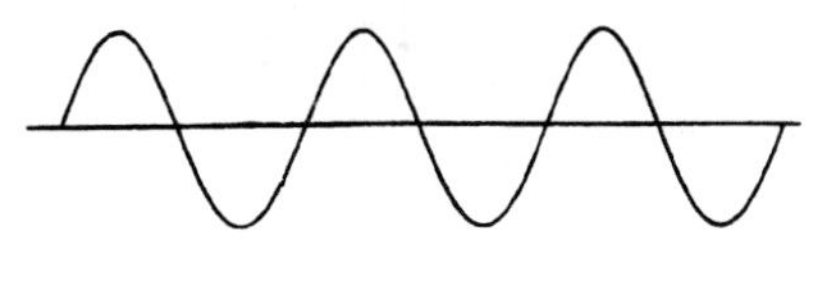

_____ basic cycles

b)

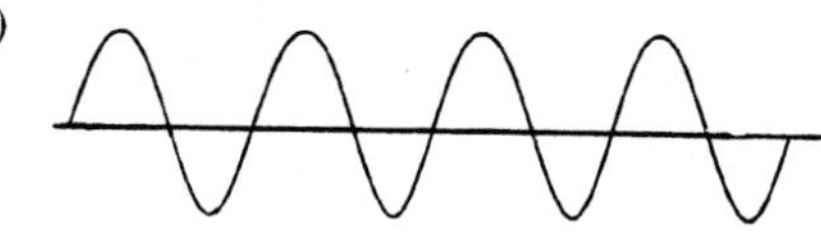

_____ basic cycles

---

39. The graph of the harmonic $y = \sin 2\theta$ from 0° to 360° is given below with the graph of $y = \sin \theta$.

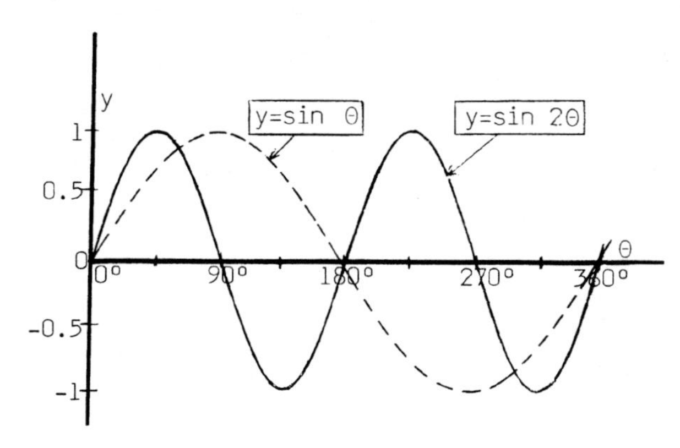

Use the graph to complete these:

a) The amplitude of $y = \sin 2\theta$ is _______.

b) From 0° to 360°, $y = \sin 2\theta$ has _______ basic cycles.

c) From 0° to 360°, $y = \sin 2\theta$ has positive peaks at _______° and _______°.

Answers to 38:

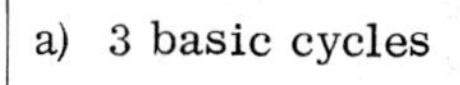

a) 3 basic cycles

b) 4 basic cycles

Answers to 39:

a) 1

b) 2 basic cycles

c) 45° and 225°

40. The graph of the harmonic $y = \sin 4\theta$ from 0° to 360° is given below with the graph of $y = \sin \theta$.

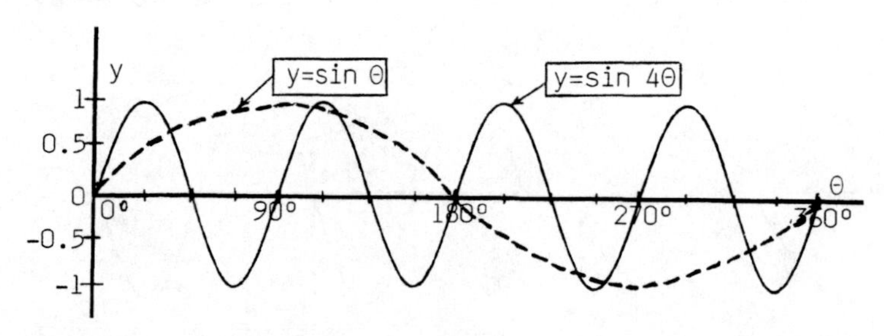

Use the graph to complete these:

a) The amplitude of $y = \sin 4\theta$ is ________.

b) From 0° to 360°, $y = \sin 4\theta$ has ________ basic cycles.

c) From 0° to 360°, $y = \sin 4\theta$ has positive peaks at ________°, ________°, ________°, and ________°.

---

41. The graph of the harmonic $y = 2 \sin 3\theta$ from 0° to 360° is given below with the graph of $y = 2 \sin \theta$.

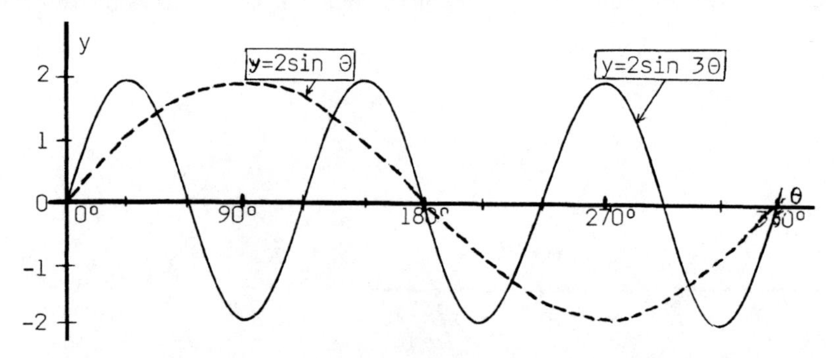

a) The amplitude of $y = 2 \sin 3\theta$ is ________.

b) From 0° to 360°, $y = 2 \sin 3\theta$ has ________ basic cycles.

c) From 0° to 360°, $y = 2 \sin 3\theta$ has positive peaks at ________°, ________°, and ________°.

a) 1

b) 4 basic cycles

c) $22\frac{1}{2}$°, $112\frac{1}{2}$°, $202\frac{1}{2}$°, and $292\frac{1}{2}$°

---

42. The general equation for harmonics of fundamental sine waves is:

$$y = A \sin n\theta$$

where: "A" is the amplitude.
"n" is the number of basic cycles from 0° to 360°.

a) 2

b) 3 basic cycles

c) 30°, 150°, 270°

Continued on following page.

42. Continued

a) If $y = \sin 5\theta$ were graphed, there would be ________ basic cycles from 0° to 360°.

b) If a sine wave has an amplitude of "1" and 10 basic cycles from 0° to 360°, its equation is ______________________.

c) If a sine wave has an amplitude of "4" and 6 basic cycles from 0° to 360°, its equation is ______________________.

Answers:
a) 5
b) $y = \sin 10\theta$
c) $y = 4 \sin 6\theta$

43. Any sine wave with two basic cycles between 0° and 360° is called a "second harmonic". Some examples are:

$$y = \sin 2\theta$$
$$y = 5 \sin 2\theta$$

Any sine wave with three basic cycles between 0° and 360° is called a "third harmonic". Some examples are:

$$y = \sin 3\theta$$
$$y = 9 \sin 3\theta$$

Similarly: a) $y = 10 \sin 4\theta$ is a __________ harmonic.

b) $y = \sin 6\theta$ is a __________ harmonic.

c) $y = 1.9 \sin 9\theta$ is a __________ harmonic.

Answers:
a) fourth
b) sixth
c) ninth

44. Though "n" is not explicitly shown in the equations of fundamental sine waves, it is really "1". For example:

$y = \sin \theta$ can be written: $y = \sin 1\theta$

$y = 5 \sin \theta$ can be written: $y = 5 \sin 1\theta$

The "1" makes sense because there is 1 basic cycle between 0° and 360°. However, though any fundamental sine wave is a first harmonic, the term "first harmonic" is not used.

Write the equation of each sine wave.

a) A fourth harmonic with an amplitude of 0.75 .

____

b) A tenth harmonic with an amplitude of 8.3 .

____

Answers:
a) $y = 0.75 \sin 4\theta$
b) $y = 8.3 \sin 10\theta$

45. Any fundamental sine wave has a 360° cycle or period. To find the number of degrees in a cycle or period of a harmonic of a fundamental sine wave, we divide 360° by n.

For $y = \sin 2\theta$, a cycle or period is $\frac{360°}{2} = 180°$

For $y = 5 \sin 3\theta$, a cycle or period is $\frac{360°}{3} = 120°$

How many degrees are there in a cycle or period for these?

a) $y = \sin 6\theta$ ________ b) $y = 0.54 \sin 10\theta$ ________

---

a) 60°, from: $\frac{360°}{6}$

b) 36°, from: $\frac{360°}{10}$

46. Harmonics can also be shifted horizontally. The general equation for harmonics of that type is:

$$y = A \sin(n\theta \pm \varphi)$$

where: 1. The amplitude is A.

2. The cycle or period is $\frac{360°}{n}$.

3. The phase shift is $\frac{\varphi}{n}$.

As an example, the graph of $y = \sin(2\theta + 90°)$ is given below with the graph of $y = \sin 2\theta$.

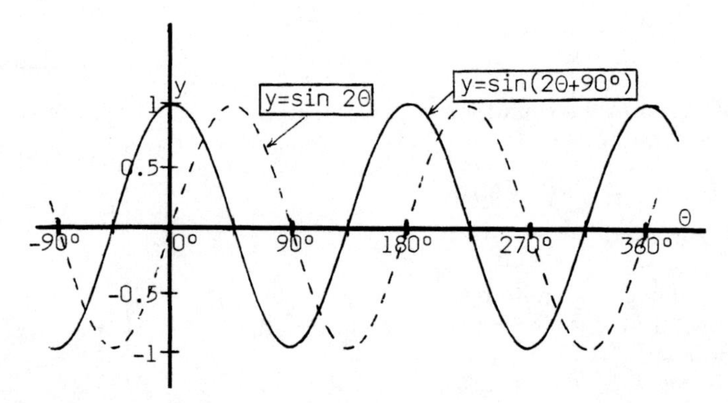

Note the following:

1. The amplitude A for each graph is "1".

2. The basic cycle or period is 180°, since $\frac{360°}{n} = \frac{360°}{2} = 180°$.

3. The phase shift is 45°, since $\frac{\varphi}{n} = \frac{90°}{2} = 45°$.

The graph of $y = \sin(2\theta + 90°)$ must be shifted how many degrees to the right to put it in-phase with the graph of $y = \sin 2\theta$? ________

45°

8-6 SKETCHING FUNDAMENTAL SINE WAVES

In this section, we will discuss a method for sketching fundamental sine waves.

---

47. The basic cycle of $\boxed{y = \sin\theta}$ from 0° to 360° is graphed below. The amplitude of the sine wave is "1".

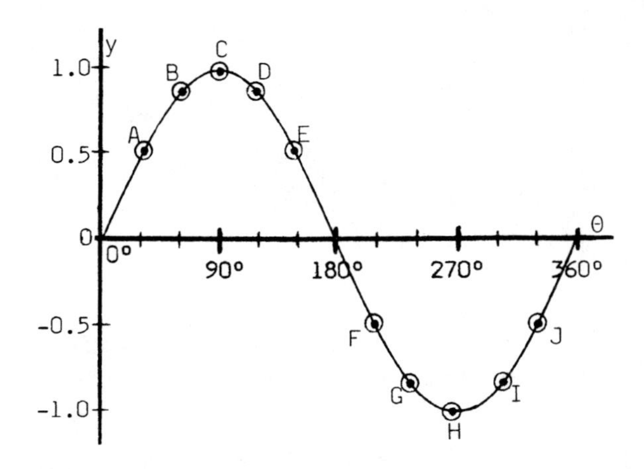

Notice the following about the basic cycle above:

1. Since each loop is 180°, the graph crosses the horizontal axis at 0°, 180°, and 360°.
2. There is a positive peak at 90° (point C) and a negative peak at 270° (point H).
3. The graph is halfway up to its positive peak at 30° (point A) and 150° (point E), since $\sin 30° = \sin 150° = 0.5$ or $\frac{1}{2}$.
4. The graph is approximately $\frac{9}{10}$ of the way up to its positive peak at 60° (point B) and 120° (point D), since $\sin 60° = \sin 120° = 0.866$ or approximately $\frac{9}{10}$.
5. The graph is halfway down to its negative peak at 210° (point F) and 330° (point J), since $\sin 210° = \sin 330° = -0.5$ or $-\frac{1}{2}$.
6. The graph is approximately $\frac{9}{10}$ of the way down to its negative peak at 240° (point G) and 300° (point I), since $\sin 240° = \sin 300° = -0.866$ or approximately $-\frac{9}{10}$.

48. The basic cycle of $y = 4 \sin \theta$ from 0° to 360° is graphed below. The amplitude of the sine wave is "4".

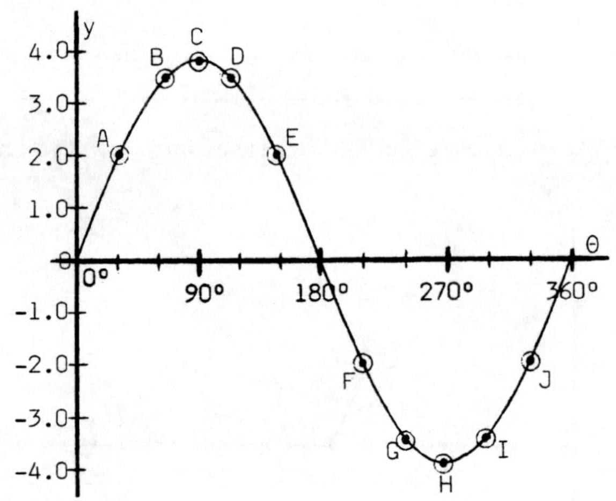

Notice the following about the basic cycle:

1. Since each loop is 180°, the graph crosses the horizontal axis at 0, 180°, and 360° with peaks at 90° (point C) and 270° (point H).

2. Halfway up or down is +2 or -2 since the amplitude is 4. It is halfway up at 30° (point A) and 150° (point E). It is halfway down at 210° (point F) and 330° (point J).

3. $\frac{9}{10}$ of the way up or down is $\frac{9}{10}(4) = 3.6$ since the amplitude is 4. It is approximately $\frac{9}{10}$ of the way up at 60° (point B) and 120° (point E). It is approximately $\frac{9}{10}$ of the way down at 240° (point G) and 300° (point I).

---

49. The diagram below shows the points that should be plotted to sketch one cycle of a fundamental sine wave.

In addition to the horizontal-axis crossings and the positive and negative peaks, the following points are plotted.

1. The points that are halfway up or halfway down.

   Note: These halfway points are $\frac{1}{3}$ of the horizontal distance from each axis-crossing to a peak.

2. The points that are $\frac{9}{10}$ of the way up or down.

   Note: These "$\frac{9}{10}$" points are $\frac{2}{3}$ of the horizontal distance from each axis-crossing to a peak.

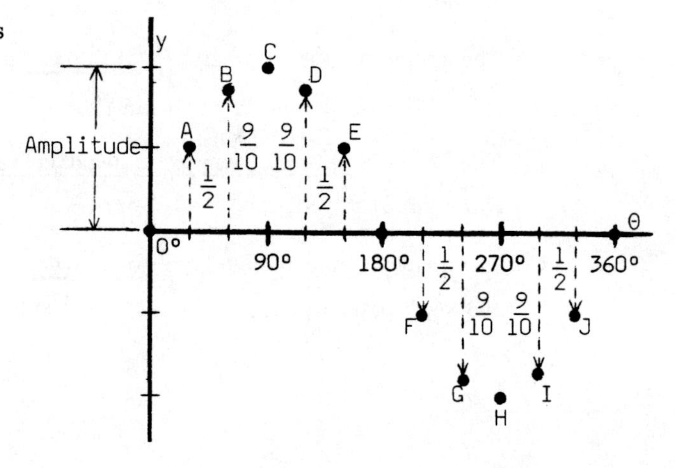

50. When plotting "peaks", "halfway" points, and "$\frac{9}{10}$ of the way" points, the amplitude of the sine wave must be taken into account.

For $y = 2 \sin \theta$, the amplitude is 2. Therefore:

The "peaks" are +2 and -2.

The "halfway" points are +1 and -1, since $\frac{1}{2}(2) = 1$.

The "$\frac{9}{10}$ of the way" points are +1.8 and -1.8, since $\frac{9}{10}(2) = 1.8$.

For $y = 10 \sin \theta$, the amplitude is 10. Therefore:

a) The "peaks" are ________ and ________.

b) The "halfway" points are ________ and ________.

c) The "$\frac{9}{10}$ of the way" points are ________ and ________.

---

a) +10 and -10

b) +5 and -5

c) +9 and -9

---

51. To sketch one basic cycle of the fundamental sine wave:

$y = 2 \sin \theta$

a) Plot the horizontal-axis crossings, peaks, "halfway" points, and "$\frac{9}{10}$" points at the right.

b) Then draw a smooth curve through the plotted points.

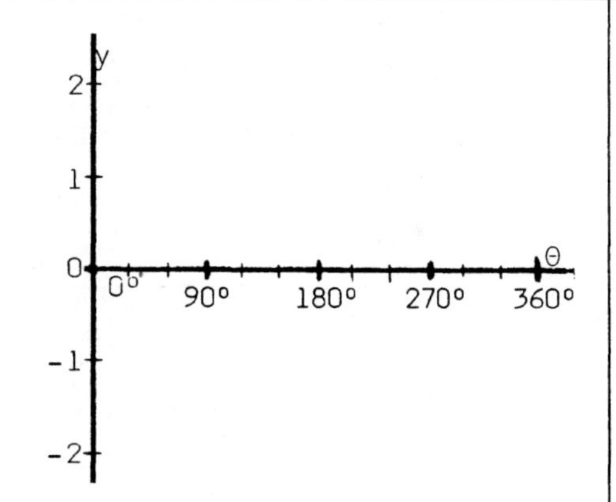

Answer for Frame 51:

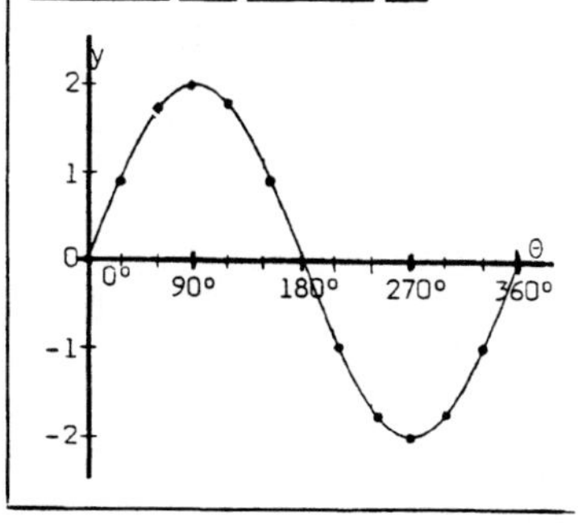

52. Let's sketch $\boxed{y = 10 \sin \theta}$ below. Begin by sketching the basic cycle from 0° to 360°. Then add similar cycles between -360° and 0° and between 360° and 720°.

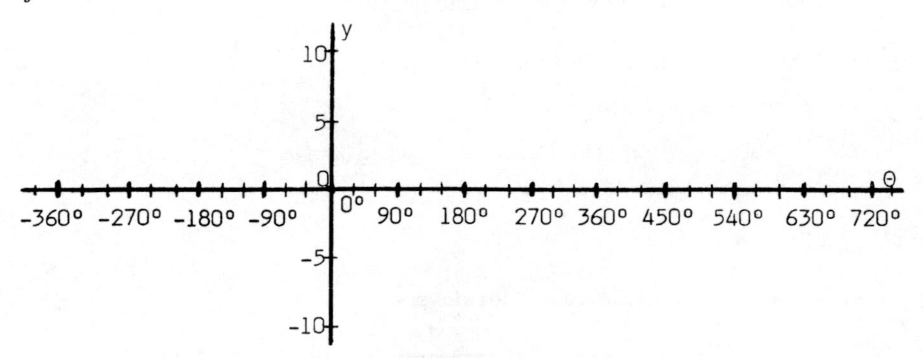

Answer for Frame 52:

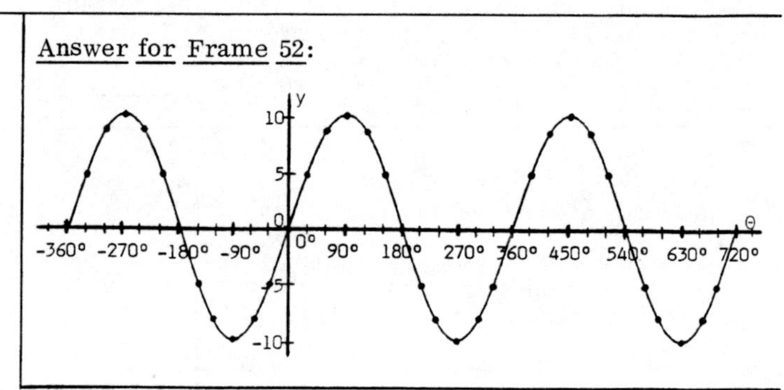

8-7 SKETCHING NON-FUNDAMENTAL SINE WAVES

In this section, we will discuss a method for sketching non-fundamental sine waves.

53. Basic cycles of two non-fundamental sine waves are graphed below.

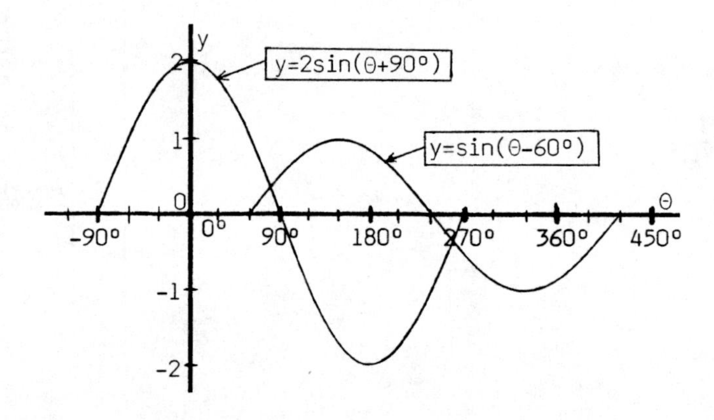

Neither graph has a basic cycle from 0° to 360°.

For $y = 2 \sin(\theta + 90°)$, a basic cycle begins at -90°.

For $y = \sin(\theta - 60°)$, a basic cycle begins at 60°.

Continued on following page.

53. Continued

A basic cycle begins when $y = 0$ at the beginning of a positive loop.

For $y = \sin\theta$, a basic cycle begins when $\theta = 0°$, since: $\sin 0° = 0$.

For $y = 2\sin(\theta + 90°)$, a basic cycle begins when $\theta = -90°$, since: $\sin(-90° + 90°) = \sin 0° = 0$.

For $y = \sin(\theta - 60°)$, a basic cycle begins when $\theta = +60°$, since: $\sin(60° - 60°) = \sin 0° = 0$.

---

54. To find the angle at which a basic cycle begins for a non-fundamental sine wave, we must find the value of θ that makes the expression in parentheses equal 0°. For example:

For $y = \sin(\theta + 30°)$, a basic cycle begins at $-30°$, since $(\theta + 30°) = 0°$ when $\theta = -30°$.

For $y = \sin(\theta - 45°)$, a basic cycle begins at $+45°$, since $(\theta - 45°) = 0°$ when $\theta = +45°$.

At what value of θ would a basic cycle begin for each of these?

a) $y = \sin(\theta - 90°)$ Basic cycle begins at ________°.

b) $y = \sin(\theta + 60°)$ Basic cycle begins at ________°.

c) $y = \sin(\theta - 120°)$ Basic cycle begins at ________°.

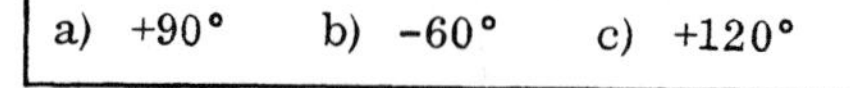
a) +90° b) -60° c) +120°

55. A basic cycle of $y = \sin(\theta + 60°)$ is sketched below.

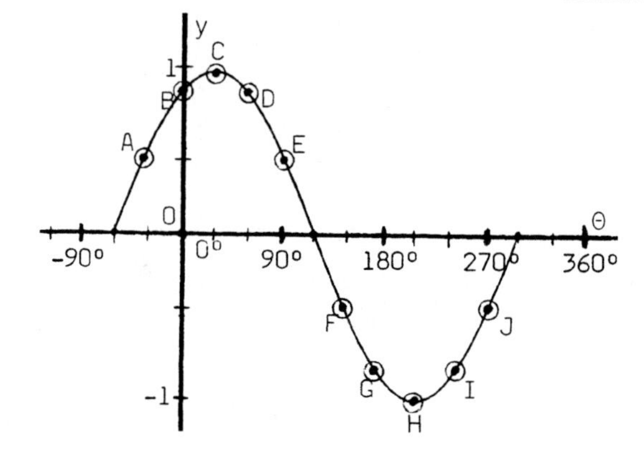

Notice the following about the basic cycle above:

1. The basic cycle contains 360°, from $-60°$ to 300°.
2. Since each loop is 180°, the graph crosses the horizontal axis at $-60°$, 120°, and 300°.
3. There is a positive peak at 30° (point C) and a negative peak at 210° (peak H).
4. The "halfway up" points (A and E) and "halfway down" points (F and J) are $\frac{1}{3}$ of the horizontal distance from each axis-crossing to a peak.
5. The "$\frac{9}{10}$ of the way up" points (B and D) and "$\frac{9}{10}$ of the way down" points (G and I) are $\frac{2}{3}$ of the horizontal distance from each axis-crossing to a peak.

56. To sketch one basic cycle of the non-fundamental sine wave:

$$y = \sin(\theta + 90°)$$

a) Plot the points needed to sketch the cycle from -90° to 270°.

b) Draw a smooth curve through the plotted points.

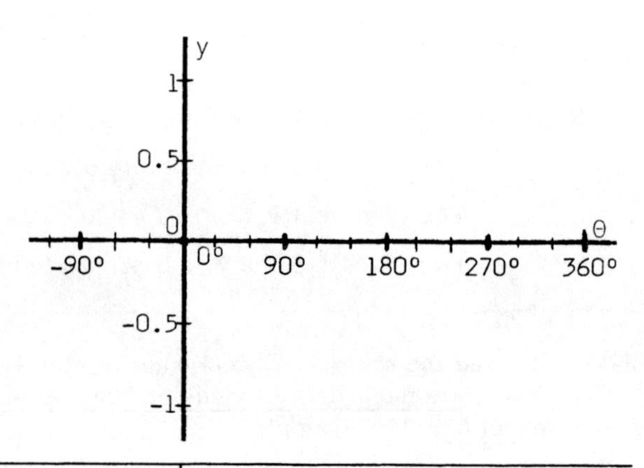

57. Let's sketch $y = 4 \sin(\theta - 90°)$ below. Begin by sketching the basic cycle from 90° to 450°. Then extend the curve in both directions to the ends of the horizontal axis.

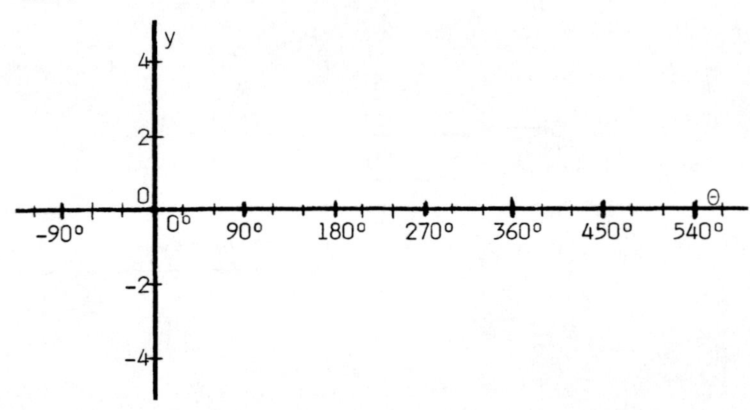

Answer for Frame 56:

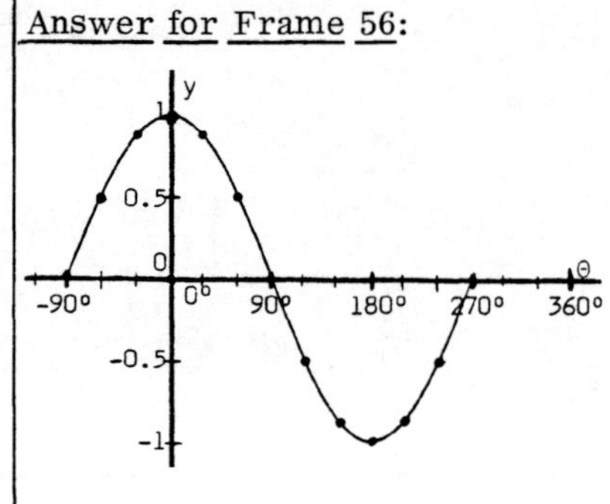

Answer for Frame 57:

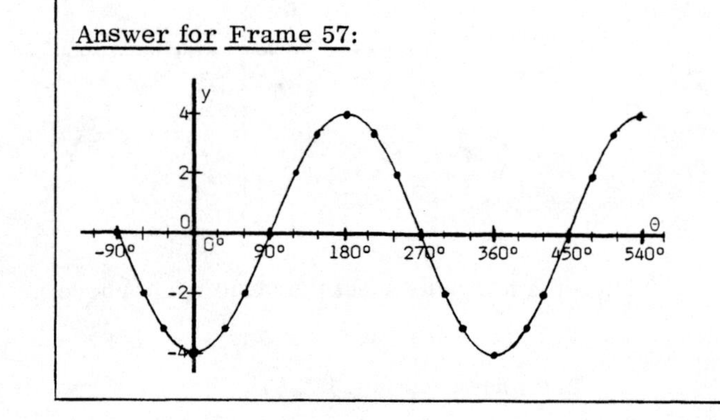

8-8 SKETCHING HARMONICS

In this section, we will discuss a method for sketching harmonics of fundamental sine waves.

58. As we saw earlier, the general equation for a harmonic of a fundamental sine wave is:

$$\boxed{y = A \sin n\theta}$$

where "A" is the amplitude and "n" is the number of basic cycles from 0° to 360°.

To find the number of degrees in each basic cycle, we divide 360° by "n". That is:

For $\boxed{y = 5 \sin 3\theta}$, the number of degrees in a basic cycle is $\frac{360°}{3} = 120°$

Find the number of degrees in each basic cycle for these harmonics.

a) $y = \sin 2\theta$ ________ b) $y = 10 \sin 4\theta$ ________

---

59. A sketch of $\boxed{y = 2 \sin 3\theta}$ from 0° to 360° is given below.

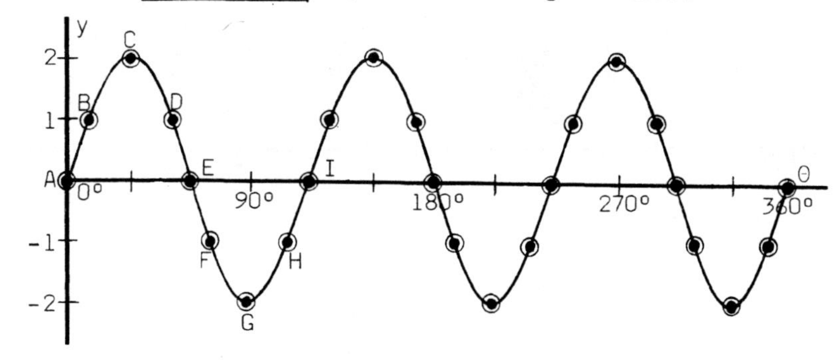

We began by sketching the first cycle from 0° to 120°. The steps are:

1. Plot the end points at 0° and 120° (A and I).

2. Plot the midpoint at 60° (E).

3. Plot the positive peak at 30° (C) and the negative peak at 90° (G).

4. Plot the "halfway up" points at 10° and 50° (B and D) and the "halfway down" points at 70° and 110° (F and H).

5. Draw a smooth curve through the plotted points.

Then we plotted similar points to sketch the cycles from 120° to 240° and from 240° to 360°.

---

a) 180°, from: $\frac{360°}{2}$

b) 90°, from: $\frac{360°}{4}$

60. To sketch the harmonic below from 0° to 360°:

$$y = \sin 2\theta$$

a) Plot the points needed to sketch the two cycles from 0° to 180° and from 180° to 360°.

b) Draw a smooth curve through the plotted points.

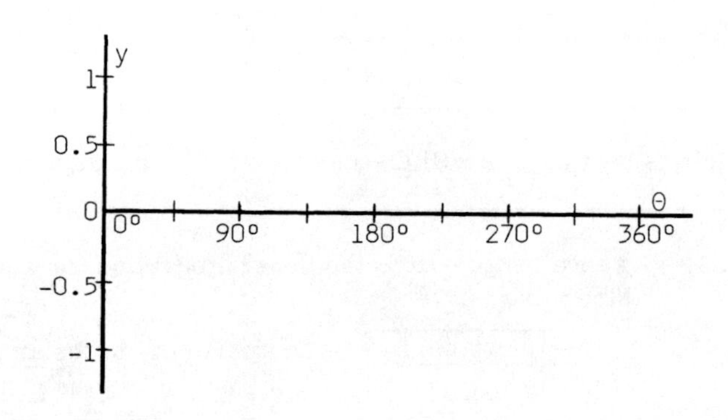

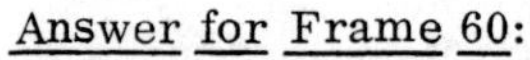

Answer for Frame 60:

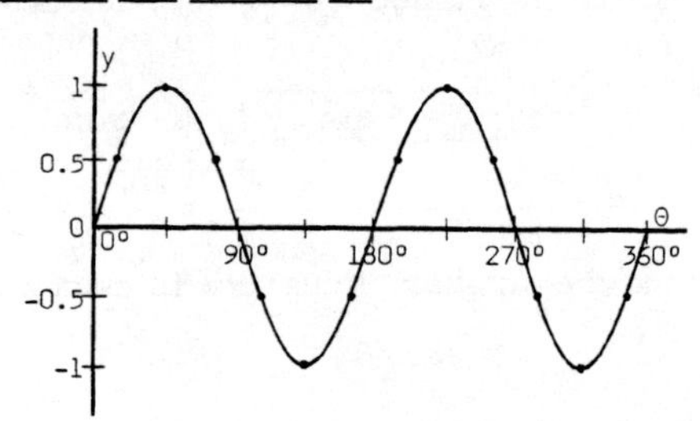

61. To sketch this harmonic from 0° to 360°:

$$y = 2 \sin 4\theta$$

a) Plot the points needed to sketch the four cycles.

b) Draw a smooth curve through the plotted points.

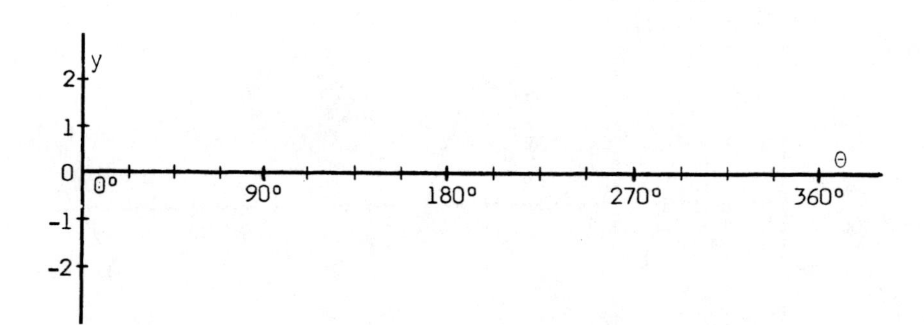

Answer for Frame 61:

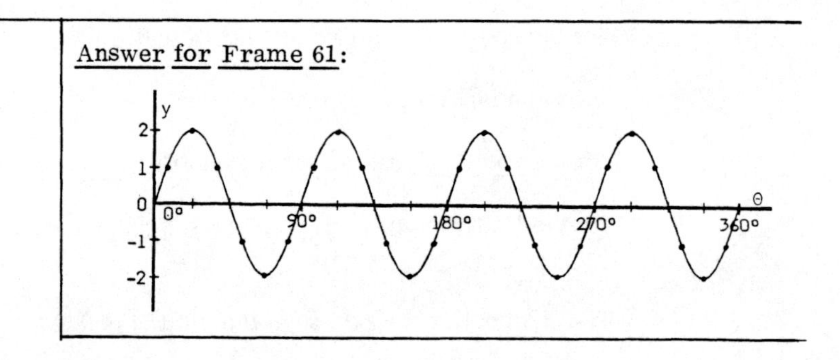

SELF-TEST 28 (pages 371-385)

If the harmonic $y = 8 \sin 5\theta$ is graphed:

1. How many basic cycles are these between 0° and 360°? ________
2. How many degrees are there in a cycle or period? ________
3. What is the maximum value of y? ________

4. Write the equation of a sixth harmonic whose amplitude is 1.82 . ________

5. Write the equation of the harmonic whose amplitude is 16 and whose period is 180°. ________

6. Sketch the graph of $y = 8 \sin \theta$ from -360° to 720°.

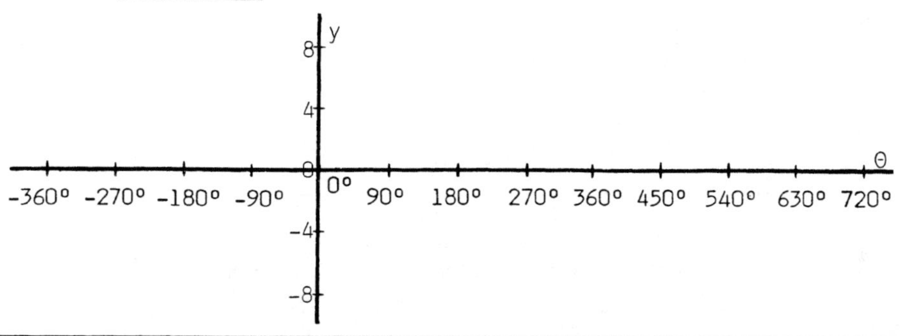

7. For $y = 2 \sin(\theta - 45°)$, at what angle θ does a basic cycle of its graph begin? ________

8. Sketch one basic cycle of $y = 10 \sin(\theta + 30°)$ from -30° to 330°.

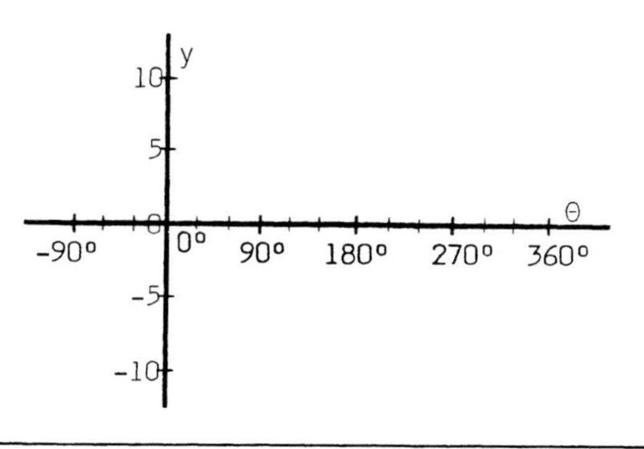

9. Sketch the harmonic $y = 4 \sin 3\theta$ from 0° to 360°.

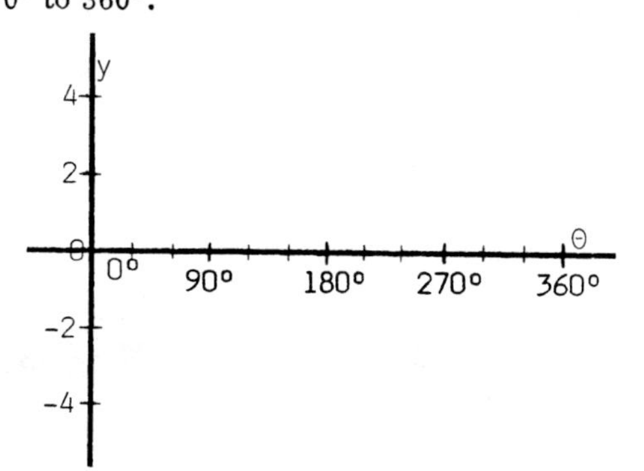

ANSWERS:

1. 5 cycles
2. 72°
3. 8
4. $y = 1.82 \sin 6\theta$
5. $y = 16 \sin 2\theta$
6. The graph of $y = 8 \sin \theta$ has basic cycles of amplitude 8 from -360° to 0°, 0° to 360°, and 360° to 720°. See Frame 52.
7. $\theta = +45°$
8. The graph of $y = 10 \sin(\theta + 30°)$ has a basic cycle of amplitude 10 beginning at -30° and ending at 330°. It crosses the horizontal axis at 150° and has a positive peak at 60° and a negative peak at 240°. See Frame 55 for a similar graph starting at -60° instead of -30°.
9. The graph of $y = 4 \sin 3\theta$ has three basic cycles from 0° to 360°. The cycles are from 0° to 120°, 120° to 240°, and 240° to 360°. Their amplitude is 4. See Frame 59.

8-9 SINE WAVES WITH NEGATIVE AMPLITUDES

In this section, we will show that any sine wave with a negative amplitude is the "opposite" of a sine wave with the opposite positive amplitude.

---

62. In each sine-wave equation below, the amplitude "A" is negative.

$y = -1 \sin \theta$ (or $y = -\sin \theta$)

$y = -2 \sin \theta$

$y = -3 \sin \theta$

To graph equations of that type, we must plot pairs of values for "y" and θ. For example:

If $\theta = 30°$: $y = -1 \sin \theta = -1(0.5) = -0.5$

$y = -2 \sin \theta = -2(0.5) = -1.0$

$y = -3 \sin \theta = -3(0.5) =$ ________

---

63. The tables below contain pairs of values for $y = \sin \theta$ and $y = -\sin \theta$. Their amplitudes (+1 and -1) are opposites. Therefore, the values of "y" for $y = -\sin \theta$ have the opposite sign of the corresponding values of "y" for $y = \sin \theta$.

Answer: -1.5

θ	y = sin θ	y = -sin θ
0°	0	0
30°	0.50	-0.50
60°	0.87	-0.87
90°	1	-1
120°	0.87	-0.87
150°	0.50	-0.50
180°	0	0

θ	y = sin θ	y = -sin θ
180°	0	0
210°	-0.50	0.50
240°	-0.87	0.87
270°	-1	1
300°	-0.87	0.87
330°	-0.50	0.50
360°	0	0

The two equations are graphed from 0° to 360° below. When the values of "y" are positive for $y = \sin \theta$, they are negative for $y = -\sin \theta$, and vice versa.

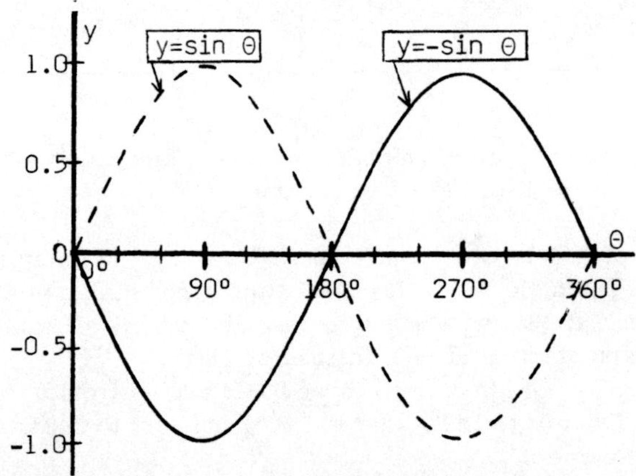

Continued on following page.

63. Continued

The two sine waves are "opposites" in the sense that their positive and negative peaks are interchanged. That is:

a) At 90°: $y = \sin\theta$ has a positive peak, and
$y = -\sin\theta$ has a __________ peak.

b) At 270°: $y = \sin\theta$ has a negative peak, and
$y = -\sin\theta$ has a __________ peak.

---

a) negative

b) positive

64. The graphs of $y = 4\sin(\theta - 90°)$ and $y = -4\sin(\theta - 90°)$ are shown below. The amplitudes of the two non-fundamental sine waves are opposites.

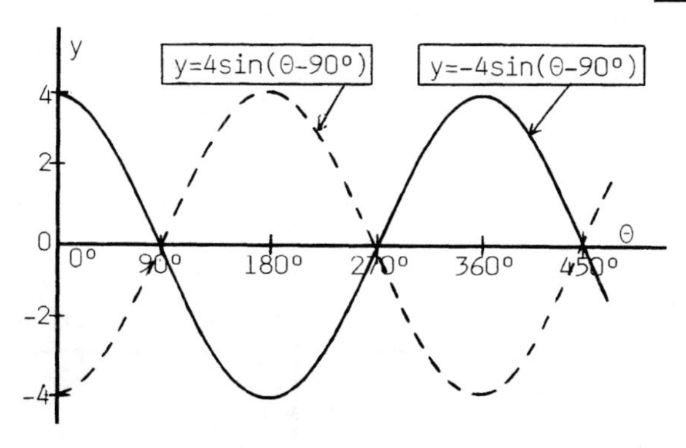

Notice that the two sine waves are "opposites" in the sense that their positive and negative peaks are interchanged. That is:

$y = 4\sin(\theta - 90°)$ has: a positive peak at 180°, and
a negative peak at 360°.

$y = -4\sin(\theta - 90°)$ has: a) a positive peak at ______°.

b) a negative peak at ______°.

---

a) 360°

b) 180°

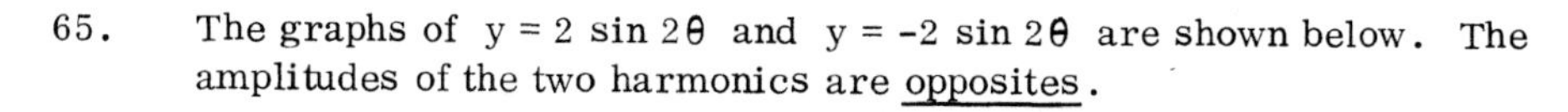

65. The graphs of $y = 2\sin 2\theta$ and $y = -2\sin 2\theta$ are shown below. The amplitudes of the two harmonics are opposites.

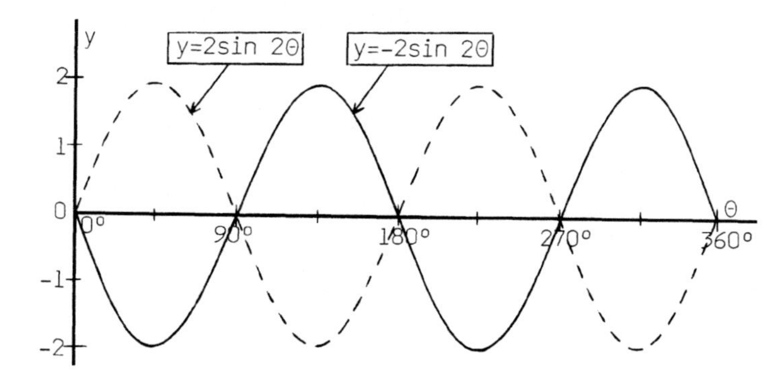

Continued on following page.

65. Continued

Notice again that the two sine waves are "opposites" because their positive and negative peaks are interchanged. That is:

$y = 2 \sin 2\theta$ has: positive peaks at 45° and 225°.
negative peaks at 135° and 315°.

$y = -2 \sin 2\theta$ has: a) positive peaks at ________ and ________.
b) negative peaks at ________ and ________.

---

a) 135° and 315°

b) 45° and 225°

66. The following two equations are graphed below.

Graph A: $y = \sin \theta$

Graph B: $y = -\sin \theta$

You can see that $y = -\sin \theta$ is 180° out-of-phase with $y = \sin \theta$. The phase angle or phase shift has the same number of degrees in both directions since:

$\varphi = +180°$ or $\varphi = -180°$

Since $y = -\sin \theta$ is 180° out-of-phase with a fundamental, we can write two equations with a positive amplitude that are equivalent to $y = -\sin \theta$.

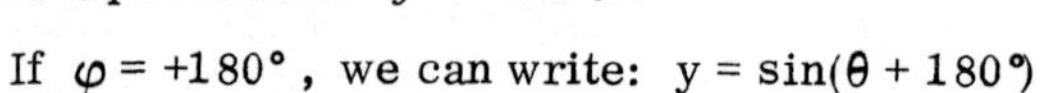

If $\varphi = +180°$, we can write: $y = \sin(\theta + 180°)$

If $\varphi = -180°$, we can write: ________________

$y = \sin(\theta - 180°)$

67. Sine waves with equations of the form $y = -A \sin \theta$ have <u>negative</u> amplitudes and are 180° out-of-phase with a fundamental. Therefore, we can write two equivalent non-fundamental equations with <u>positive</u> amplitudes for any equation of that form. That is:

For $y = -A \sin \theta$, the equivalent equations are: $y = A \sin(\theta + 180°)$, $y = A \sin(\theta - 180°)$

Write $y = -3 \sin \theta$ in two equivalent forms:

________________ ________________

---

a) $y = 3 \sin(\theta + 180°)$

b) $y = 3 \sin(\theta - 180°)$

68. Write each of these as an equivalent equation with a <u>negative</u> amplitude.

a) $y = 4 \sin(\theta + 180°)$ ________________

b) $y = 7.5 \sin(\theta - 180°)$ ________________

---

a) $y = -4 \sin \theta$

b) $y = -7.5 \sin \theta$

8-10 ADDING A CONSTANT TO A SINE WAVE

In this section, we will show how a sine wave graph moves up or down when we add a constant to its equation.

69. In each equation below, we have added a constant to the equation of a fundamental sine wave.

$$y = 3 + \sin\theta$$
$$y = 1 + 2\sin\theta$$
$$y = -2 + \sin\theta$$

To graph equations of that type, we must plot pairs of values for "y" and θ. For example:

If $\theta = 30°$: $y = 3 + \sin\theta = 3 + 0.5 = 3.5$

$y = 1 + 2\sin\theta = 1 + 2(0.5) = 2$

$y = -2 + \sin\theta = -2 + 0.5 =$ ________

-1.5

70. The graphs of the three equations from the last frame are shown below.

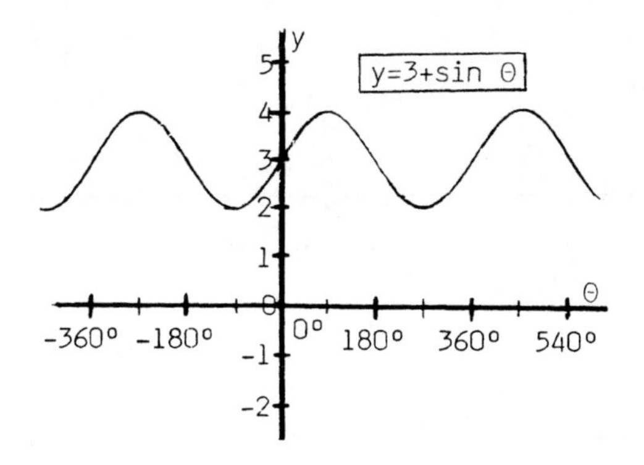

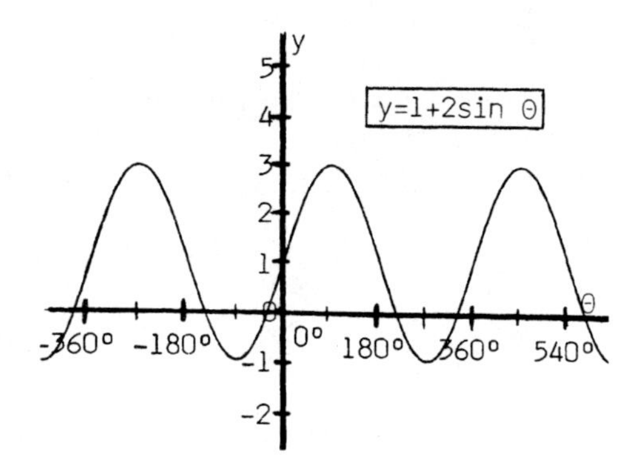

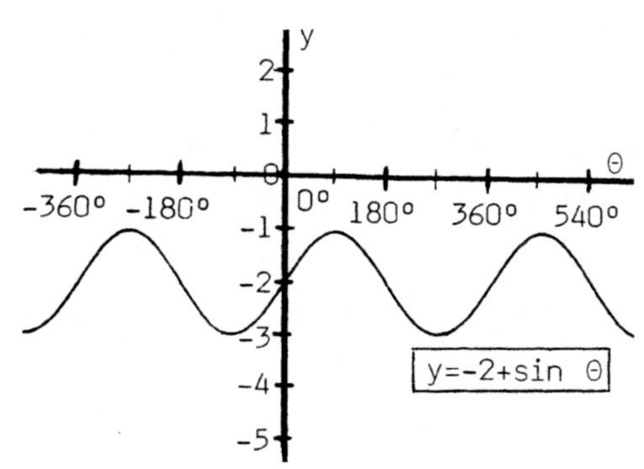

Continued on following page.

70. Continued

As you can see, the constant in each equation gives the value where the sine wave crosses the y-axis. That is:

$y = 3 + \sin\theta$ crosses the y-axis at $y = 3$.

$y = 1 + 2\sin\theta$ crosses the y-axis at $y = 1$.

$y = -2 + \sin\theta$ crosses the y-axis at $y =$ ______.

---

71. The general equation for sine waves like those in the last frame is:

Answer: -2

$\boxed{y = C + A\sin\theta}$ where: A is the amplitude. C is the y-intercept.

If each of the following were graphed, at what y-value would they cross the y-axis?

a) $y = 5 + 3\sin\theta$ Crosses at $y =$ ______.

b) $y = -4 + \sin\theta$ Crosses at $y =$ ______.

Answer: a) $y = 5$ b) $y = -4$

8-11 ADDING SINE WAVES

In this section, we will briefly discuss the addition of sine waves. We will show that the sum or resultant is either a sine wave or some other periodic function.

---

72. The equations of a fundamental and a non-fundamental sine wave are given below.

$$y = \sin\theta$$

$$y = \sin(\theta + 90°)$$

We can add the two sine waves and write their sum or "resultant" as a single equation. We get:

$$y = \sin\theta + \sin(\theta + 90°)$$

We found some pairs of values for "y" and θ for the resultant equation below. Notice that "y" for the resultant is the sum of the "y's" of the two original equations.

If $\theta = 0°$, $y = \sin 0° + \sin(0° + 90°) = \sin 0° + \sin 90° = 0 + 1 = 1$

If $\theta = 45°$, $y = \sin 45° + \sin(45° + 90°) = \sin 45° + \sin 135° = 0.71 + 0.71 = 1.42$

If $\theta = 135°$, $y = \sin 135° + \sin(135° + 90°) = \sin 135° + \sin 225° = 0.71 + (-0.71) = 0$

If $\theta = 270°$, $y = \sin 270° + \sin(270° + 90°) = \sin 270° + \sin 360° =$ ______ + ______ = ______

Answer: $-1 + 0 = -1$

73. The two original sine waves in the last frame were $y = \sin \theta$ and $y = \sin(\theta + 90°)$. They are graphed below.

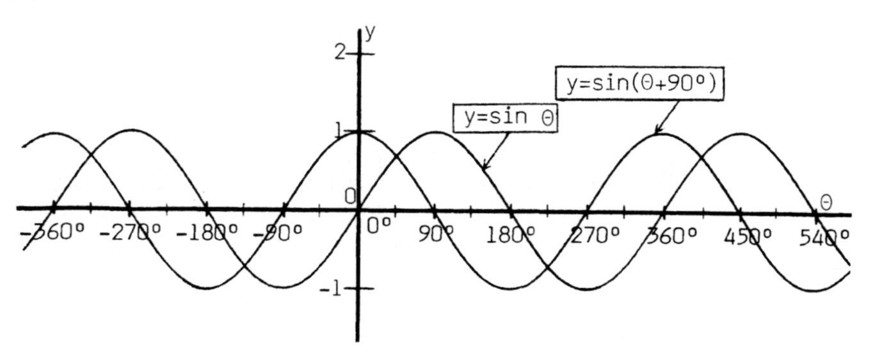

The resultant in the last frame was $y = \sin \theta + \sin(\theta + 90°)$. It is graphed below.

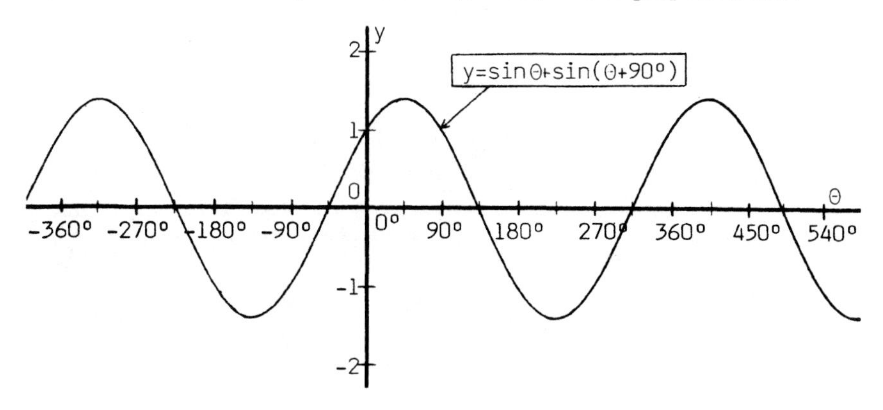

Notice these two points about the graph of the resultant.

1. It is a <u>sine wave</u>.
2. For each value of θ, the value of "y" is <u>the sum of the "y" values</u> of the two original sine waves at that value of θ.

---

74. The equations of another fundamental and another non-fundamental sine wave are given below.

$$y = 4 \sin \theta$$

$$y = 3 \sin(\theta + 90°)$$

If we add the two sine waves and write their sum or "<u>resultant</u>" as a single equation, we get:

$$y = 4 \sin \theta + 3 \sin(\theta + 90°)$$

We found some pairs of values for "y" and θ for the resultant below. Notice again that "y" for the resultant is the sum of the "y's" of the two original equations.

If $\theta = 0°$, $y = 4 \sin 0° + 3 \sin(0° + 90°) = 4 \sin 0° + 3 \sin 90° = 4(0) + 3(1) = 0 + 3 = 3$

If $\theta = 90°$, $y = 4 \sin 90° + 3 \sin(90° + 90°) = 4 \sin 90° + 3 \sin 180° = 4(1) + 3(0) = 4 + 0 = 4$

If $\theta = 270°$, $y = 4 \sin 270° + 3 \sin(270° + 90°) = 4 \sin 270° + 3 \sin 360° =$ ______

$4(-1) + 3(0) = -4 + 0 = -4$

75. The two original sine waves in the last frame were $y = 4 \sin \theta$ and $y = 3 \sin(\theta + 90°)$. They are graphed below.

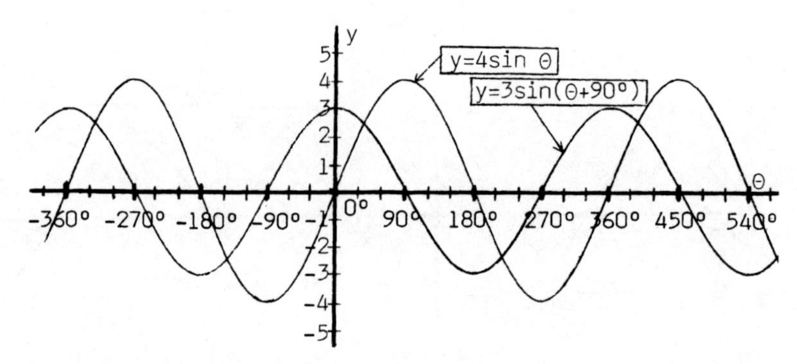

The resultant in the last frame was $y = 4 \sin \theta + 3 \sin(\theta + 90°)$. It is graphed below.

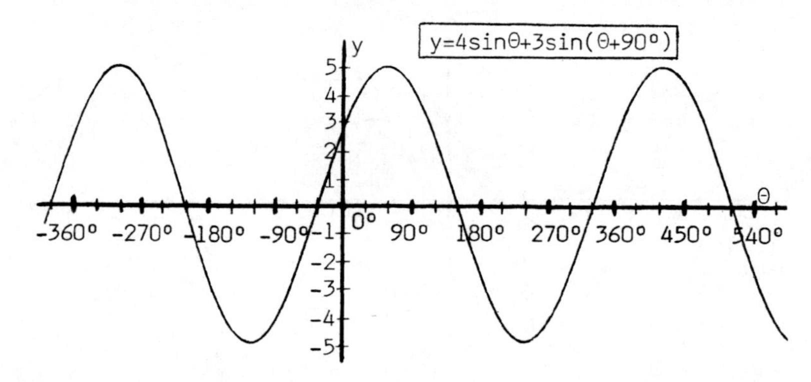

Notice these points about the graph of the resultant.

1. It is a sine wave.

2. For each value of θ, the value of "y" is the sum of the "y" values of the two original sine waves at that value of θ.

---

76. The equation of a fundamental sine wave and a harmonic of a fundamental are given below.

$$y = 2 \sin \theta$$

$$y = \sin 2\theta$$

If we add the two sine waves, the resultant is:

$$y = 2 \sin \theta + \sin 2\theta$$

We found some pairs of values for "y" and θ for the resultant below. Notice again that "y" for the resultant is the sum of the "y's" of the two original equations.

If $\theta = 0°$, $y = 2 \sin 0° + \sin(2)(0°) = 2 \sin 0° + \sin 0° = 2(0) + 0 = 0 + 0 = 0$

If $\theta = 90°$, $y = 2 \sin 90° + \sin(2)(90°) = 2 \sin 90° + \sin 180° = 2(1) + 0 = 2 + 0 = 2$

If $\theta = 270°$, $y = 2 \sin 270° + \sin(2)(270°) = 2 \sin 270° + \sin 540° =$ ______________

$2(-1) + 0 = -2 + 0 = -2$

77. The two original sine waves in the last frame were $y = 2 \sin \theta$ and $y = \sin 2\theta$. They are graphed below.

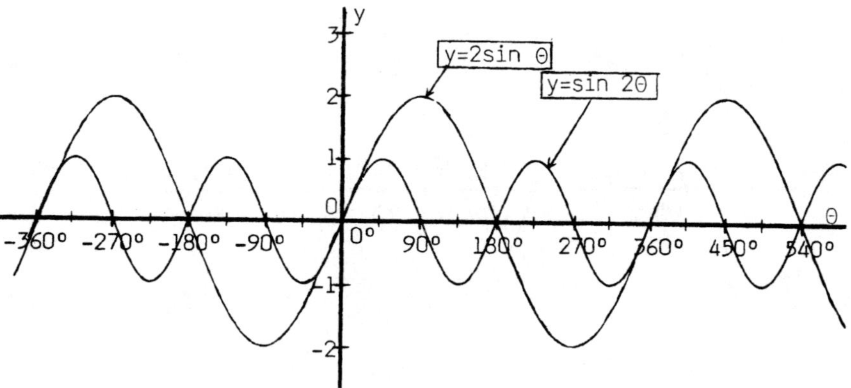

The resultant in the last frame was $y = 2 \sin \theta + \sin 2\theta$. It is graphed below.

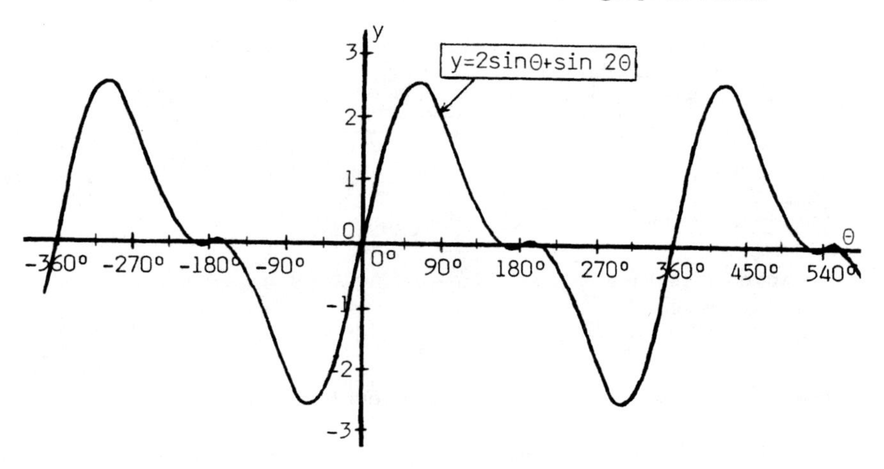

Notice these points about the graph of the resultant.

1. It is not a sine wave. However, it is a periodic function.
2. For each value of θ, the value of "y" is the sum of the "y" values of the two original sine waves at that value of θ.

8-12 COSINE WAVES

In this section, we will briefly discuss cosine waves and relate them to sine waves.

78. The graph of $y = \cos \theta$ is shown below. The graph of $y = \sin \theta$ is also shown.

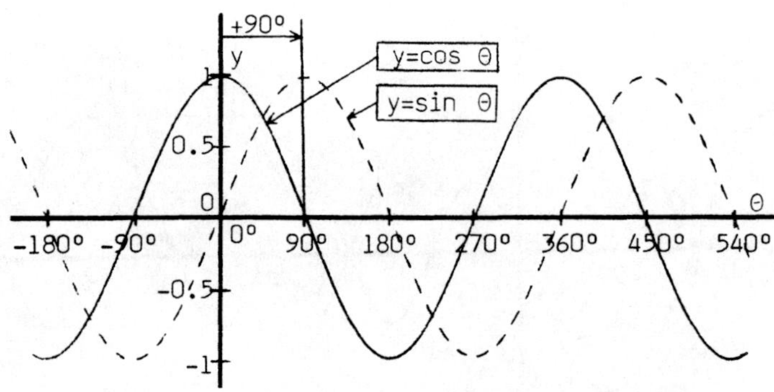

Note: 1. Both graphs have the same shape, period, and amplitude.

2. The graph of $y = \cos \theta$ is +90° out-of-phase with the graph of $y = \sin \theta$.

Since the cosine wave is equivalent to a sine wave that is +90° out-of-phase with the fundamental, the cosine wave is equivalent to what non-fundamental sine wave? ______________

$y = \sin(\theta + 90°)$

79. The general equation of a fundamental cosine wave is:

$\boxed{y = A \cos \theta}$ where "A" is the amplitude.

Since cosine waves of that type are +90° out-of-phase with a fundamental sine wave, the equivalent sine wave equation is:

$\boxed{y = A \sin(\theta + 90°)}$

a) Write a sine wave equation that is equivalent to:

$y = 5 \cos \theta$ ______________

b) Write a cosine wave equation that is equivalent to:

$y = 10 \sin(\theta + 90°)$ ______________

---

a) $y = 5 \sin(\theta + 90°)$

b) $y = 10 \cos \theta$

80. Sketch the graph of $\boxed{y = 4 \cos \theta}$ below. Since $y = 4 \cos \theta$ is equivalent to $y = 4 \sin(\theta + 90°)$, it has a basic cycle beginning at $\theta = -90°$.

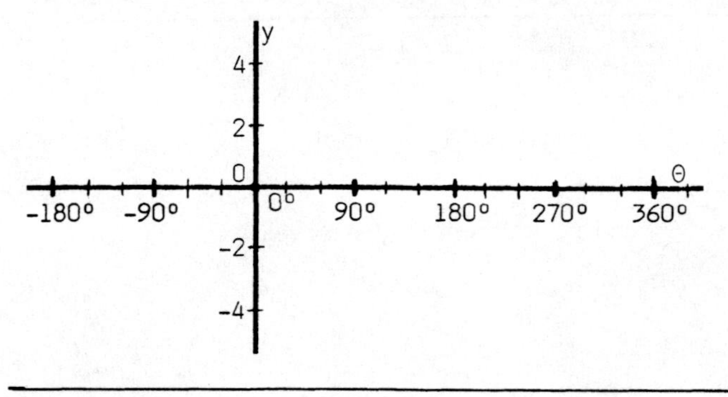

Answer for Frame 80:

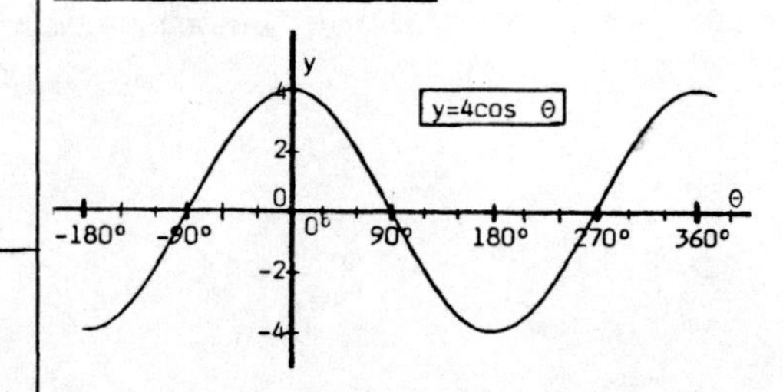

SELF-TEST 29 (pages 386-395)

1. $\boxed{y = -8 \sin \theta}$ If $\theta = -90°$, $y =$ ______.

2. $\boxed{y = -5 \sin(\theta + 90°)}$ If $\theta = 0°$, $y =$ ______.

3. Write the equation of the sine wave whose positive and negative peaks are equal to and opposite from those of $y = 12 \sin(\theta - 90°)$. ______________

Write two equations whose graphs are identical to the graph of $y = -20 \sin \theta$.

4. ______________

5. ______________

6. $\boxed{y = 5 + 3 \sin \theta}$ If $\theta = 30°$, $y =$ ______.

7. $\boxed{y = -2 + 4 \sin \theta}$ If $\theta = -90°$, $y =$ ______.

At what y-value does the graph of each equation below cross the y-axis?

8. $y = -8 + 4 \sin \theta$ ______

9. $y = 4 + 2 \sin \theta$ ______

10. $y = -7 \sin \theta$ ______

$\boxed{y = \sin \theta + 2 \sin(\theta + 90°)}$

11. If $\theta = 0°$, $y =$ ______.

12. If $\theta = -90°$, $y =$ ______.

$\boxed{y = 5 \sin \theta + 2 \sin 3\theta}$

13. If $\theta = 90°$, $y =$ ______.

14. If $\theta = -30°$, $y =$ ______.

15. Which <u>one</u> equation below has a graph that is <u>not</u> a sine wave? ______

a) $y = -5 \sin \theta$ b) $y = 3 \sin \theta + 2 \sin 2\theta$ c) $y = 2 \sin \theta + \sin(\theta + 90°)$

16. For $y = \cos \theta$, at what negative angle between $-360°$ and $0°$ does a basic cycle begin? ______

17. Using the "sine" function, write an equation that is equivalent to $y = 8 \cos \theta$. ______________

18. Using the "cosine" function, write an equation that is equivalent to $y = 75 \sin(\theta + 90°)$. ______________

ANSWERS:

1. 8
2. -5
3. $y = -12 \sin(\theta - 90°)$
4. $y = 20 \sin(\theta + 180°)$
5. $y = 20 \sin(\theta - 180°)$
6. 6.5
7. -6
8. -8
9. 4
10. 0
11. 2
12. -1
13. 3
14. -4.5
15. b
16. -90°
17. $y = 8 \sin(\theta + 90°)$
18. $y = 75 \cos \theta$

SUPPLEMENTARY PROBLEMS - CHAPTER 8

Assignment 27

Do these evaluations. If necessary, round to two decimal places.

$y = 8 \sin \theta$	$y = 5 \sin(\theta + 30°)$	$y = 12 \sin(\theta - 90°)$
1. If $\theta = 60°$, $y = ?$	3. If $\theta = 0°$, $y = ?$	5. If $\theta = 90°$, $y = ?$
2. If $\theta = 270°$, $y = ?$	4. If $\theta = 180°$, $y = ?$	6. If $\theta = 280°$, $y = ?$

Using y and θ, write the equation of the fundamental sine wave whose:

7. Amplitude is 28.
8. Maximum value of y is 8.25 .

9. A basic cycle of the graph of $y = 52 \sin \theta$ begins at $\theta = 360°$. At what value of θ does that basic cycle end?

10. State the amplitude of the graph of $y = 170 \sin(\theta + 60°)$.

11. State the maximum value of y for the graph of $y = \sin(\theta - 45°)$.

12. For $y = 30 \sin \theta$, for what angle θ between 0° and 360° does $y = 30$?

13. What is the period, in degrees, of the graph of $y = 5 \sin(\theta + 90°)$?

To put it in-phase with a fundamental sine wave:

14. The graph of $y = 2 \sin(\theta + 40°)$ must be shifted 40° to the __________ (right/left).

15. The graph of $y = 35 \sin(\theta - 90°)$ must be shifted 90° to the __________ (right/left).

With respect to a fundamental sine wave:

16. The signed phase angle of $y = 120 \sin(\theta - 30°)$ is ________.

17. The signed phase angle of $y = 0.75 \sin(\theta + 70°)$ is ________.

Using y and θ, write the equation of:

18. A sine wave of amplitude 12.6 which is in-phase with $y = \sin \theta$.

19. A sine wave of amplitude 275 which has a phase angle $\phi = +50°$ with a fundamental sine wave.

20. A sine wave whose maximum value is 20 which has a phase angle $\phi = -10°$ with a fundamental sine wave.

21. The graph of $y = 20 \sin(\theta - 30°)$ is in-phase with the graph of which equation below? ______

a) $y = 20 \sin \theta$ b) $y = \sin(\theta - 30°)$ c) $y = 10 \sin(\theta + 30°)$

22. To put the graph of $y = 4 \sin(\theta - 60°)$ in-phase with the graph of $y = 4 \sin(\theta + 30°)$, it would have to be shifted ______° to the ____________ (right/left).

Assignment 28

For the harmonic equation $\boxed{y = 6 \sin 2\theta}$:

1. How many basic cycles are there between 0° and 360°?
2. How many degrees are there in a cycle or period?
3. The first basic cycle begins at 0°. At what value of θ does the second cycle begin?
4. What is the maximum value of "y"?
5. For what angles between 0° and 360° is "y" a maximum value?

For the harmonic equation $\boxed{y = 2.8 \sin 6\theta}$:

6. How many basic cycles are there between 0° and 360°?
7. How many degrees are there in a cycle or period?
8. The first basic cycle begins at 0°. At what values of θ between 0° and 360° do the other basic cycles begin?
9. What is the maximum value of "y"?
10. For what angles between 0° and 360° is "y" a maximum?

Write the equation of each sine wave harmonic.

11. A tenth harmonic of amplitude 5.
12. Of amplitude 18 with five basic cycles between 0° and 360°.
13. Of amplitude 0.58 with one basic cycle from 0° to 30°.
14. Of amplitude 2 with period 120°.

State the number of degrees in a basic cycle for each harmonic below.

15. $y = 3 \sin 4\theta$ 16. $y = 24 \sin 5\theta$ 17. $y = 8 \sin 15\theta$

It is desired to sketch the graph of $y = 80 \sin \theta$ from -360° to 720°.

18. How many complete basic cycles will there be?
19. List the values of θ where each cycle will begin.
20. For what values of θ will "y" be its maximum value of 80?
21. For what values of θ will $y = -80$?
22. For what values of θ will $y = 40$?

At what value of θ between -90° and +90° would a basic cycle begin for:

23. $y = 3 \sin(\theta - 60°)$ 24. $y = 14 \sin(\theta + 30°)$ 25. $y = 60 \sin \theta$

The graph of $y = 10 \sin(\theta + 90°)$ has a basic cycle beginning at -90° and ending at 270°. For that cycle, find the value or values of angle θ when:

26. $y = 10$ 27. $y = -10$ 28. $y = 5$ 29. $y = -5$

Assignment 29

1. For $y = -20 \sin \theta$, find "y" when $\theta = 90°$.

2. For $y = -6 \sin(\theta - 90°)$, find "y" when $\theta = -90°$.

3. Write the equation of the sine wave whose amplitude is the opposite of the amplitude of $y = 16 \sin \theta$.

4. Write the equation of the sine wave whose positive and negative peaks are interchanged and opposite from those of $y = -12 \sin(\theta + 90°)$.

5. Write two equations that are equivalent to $y = -50 \sin \theta$.

6. For $y = -6 + 2 \sin \theta$, find "y" when $\theta = 270°$.

7. Does the graph of $y = 8 + 3 \sin \theta$ lie above or below the θ-axis ?

8. The graph of which one equation below crosses the θ-axis?

   a) $y = 2 + 4 \sin \theta$ b) $y = -5 + \sin \theta$ c) $y = 3 + 3 \sin \theta$

9. At what y-value does the graph of $y = -24 + 6 \sin \theta$ cross the y-axis?

For $y = 3 \sin \theta + 4 \sin(\theta - 90°)$, find "y" when:

10. $\theta = 0°$ 11. $\theta = 90°$ 12. $\theta = 180°$

For $y = 7 \sin \theta + 3 \sin 4\theta$, find "y" when:

13. $\theta = 0°$ 14. $\theta = 90°$ 15. $\theta = -90°$

16. Find the maximum value of "y" for $y = 8 \sin \theta + 6 \sin(\theta + 90°)$ if it occurs when $\theta = 37°$. Round to one decimal place.

State whether the graph of each equation below is or is not a sine wave.

17. $y = 3 \sin \theta + 4 \sin(\theta - 90°)$ 18. $y = \sin \theta + \sin 2\theta$ 19. $y = 5 \cos \theta$

20. What is the period, in degrees, of $y = \cos \theta$?

21. To put the graph of $y = \cos \theta$ in-phase with the graph of $y = \sin \theta$, it must be shifted ________° to the ____________ (right/left).

22. Using the "sine" function, write an equation that is equivalent to $y = 32 \cos \theta$.

23. Using the "cosine" function, write an equation that is equivalent to $y = 170 \sin(\theta + 90°)$.

ANSWERS FOR SUPPLEMENTARY PROBLEMS

CHAPTER 1 - RIGHT TRIANGLES

Assignment 1

1. $F = 78°$ 2. $P = 28°$ 3. $A = 65°$ 4. $t = 5,600$ m 5. $a = 32.8$ cm 6. $p = 5.52''$
7. $d = 26.2$ ft 8. $t = 630$ in 9. $w = 0.463$ cm 10. $G = 37°$, $H = 106°$ 11. $A = 30°$, $B = 30°$
12. $R = 45°$, $S = 45°$ 13. $A = 14,500\ m^2$ 14. $A = 254\ cm^2$ 15. $A = 20.1\ in^2$

Assignment 2

1. 0.2126 2. 0.9986 3. 0.3256 4. 1. 5. 0.5 6. 0.8660 7. 2.1445 8. 0.0175
9. $A = 14°$ 10. $P = 73°$ 11. $H = 89°$ 12. $Q = 72°$ 13. $T = 7°$ 14. $F = 82°$ 15. $A = 45°$
16. $B = 12°$ 17. 29.1 18. 11.9 19. 77.6 20. 41.2 21. 347 22. 275 23. 110
24. 516 25. $A = 23°$ 26. $G = 59°$ 27. $R = 71°$ 28. $E = 9°$ 29. $\sin R = \frac{r}{s}$ 30. $\cos R = \frac{t}{s}$
31. $\tan R = \frac{r}{t}$ 32. $\sin T = \frac{t}{s}$ 33. $\cos T = \frac{r}{s}$ 34. $\tan T = \frac{t}{r}$ 35. $\cos B = \frac{a}{c}$ 36. $\sin A = \frac{a}{c}$
37. $\tan B = \frac{b}{a}$ 38. $\tan A = \frac{a}{b}$ 39. $\cos A = \frac{b}{c}$ 40. $\sin B = \frac{b}{c}$

Assignment 3

1. $h = 3.63$ cm 2. $w = 3.31''$ 3. $p = 7.22$ m 4. $d = 438$ ft 5. $t = 105$ cm 6. $b = 268''$
7. $F = 37°$, $G = 53°$ 8. $A = 61°$, $B = 29°$ 9. $R = 41°$, $S = 49°$ 10. $h = 6.70''$, $v = 5.62''$
11. $A = 34°$ 12. $w = 11.2$ cm

CHAPTER 2 - OBLIQUE TRIANGLES

Assignment 4

1. a, b, f 2. 97° 3. 90° 4. 48° 5. 135° 6. b 7. c 8. a 9. b 10. p
11. d 12. G 13. K 14. $\frac{\boxed{\sin T}}{t} = \frac{\sin R}{\boxed{r}}$ 15. $\frac{s}{\boxed{\sin S}} = \frac{\boxed{t}}{\sin T}$ 16. $a = 401$ ft
17. $B = 57°$ 18. $b = 503$ ft 19. $H = 80°$ 20. $K = 43°$ 21. $k = 17.7$ cm

Assignment 5

1. Law of Sines 2. Law of Cosines 3. Law of Cosines 4. Law of Sines
5. $p^2 = r^2 + v^2 - 2rv \cos P$ 6. $v^2 = p^2 + r^2 - 2pr \cos V$ 7. $\cos A = \frac{b^2 + c^2 - a^2}{2bc}$
8. $\cos C = \frac{a^2 + b^2 - c^2}{2ab}$ 9. $f = 17.0$ cm 10. $D = 68°$ 11. $E = 78°$ 12. $G = 77°$
13. $P = 60°$ 14. $T = 43°$

Assignment 6

1. 0.4695 2. 0.9976 3. -0.2250 4. -0.9998 5. $P = 161°$ 6. $F = 132°$ 7. $A = 96°$
8. $s = 87.5$ cm 9. $N = 124°$ 10. $f = 962$ ft 11. $H = 136°$ 12. $d = 3,049$ m
13. $A=20°$, $h = 1.57$ cm

CHAPTER 3 - TRIGONOMETRIC FUNCTIONS

Assignment 7

1. 51°, 89° 2. 94°, 127° 3. 186°, 230° 4. 275°, 312° 5. 80° (Q4) 6. 85° (Q2) 7. 67° (Q3) 8. 72° (Q1) 9. 24° (Q4) 10. 18° (Q3) 11. 202° 12. 94° 13. 311° 14. 14° 15. 285° 16. 149° 17. 1 and 4 18. 1 and 2 19. 2 and 4 20. a, c, d 21. sin 27° (Q2) 22. cos 42° (Q4) 23. tan 75° (Q3) 24. -0.7431 25. 2.1445 26. -0.2924 27. -0.9563 28. -0.4226 29. -7.1154 30. 0.3443 31. 0.1736 32. 0.9877

Assignment 8

1. 0 2. +1 3. 0 4. -1 5. 180° 6. 90° 7. 90° and 270° 8. 278° 9. 195° 10. 30° 11. 206° 12. 23° (Q2) 13. 80° (Q4) 14. 45° (Q3) 15. 15° (Q1) 16. sin 234° = sin 54° (Q3) = -0.8090 17. cos 159° = cos 21° (Q2) = -0.9336 18. tan 70° = tan 70° (Q1) = 2.7475 19. cos 275° = cos 85° (Q4) = 0.0872 20. -0.7880 21. 0.1736 22. -0.7813 23. 0.4226 24. -1 25. increases 26. +1 27. decreases 28. increases 29. -180°, 0°, 180°

Assignment 9

1. 240° 2. 45° 3. $\frac{1}{360}$ 4. radius 5. 16 cm 6. 32 cm 7. 8 cm 8. $\theta = \frac{s}{r}$ 9. 2.47 radians 10. 0.61 radian 11. 3.77 radians 12. 60° 13. 270° 14. 315° 15. 720° 16. $\frac{\pi}{6}$ radian 17. $\frac{\pi}{2}$ radians 18. $\frac{2\pi}{3}$ radians 19. -2π radians 20. 57.296° 21. 21.3° 22. -105.4° 23. 150° 24. -405° 25. 1.45 radians 26. 0.61 radian 27. 4.50 radians 28. -5.39 radians 29. 0.9511 30. 0.8660 31. -0.3508 32. -0.9505 33. 180° 34. 90° 35. 0.5 radian 36. 2.5 radians

Assignment 10

1. $\cos\theta$ 2. $\tan\theta$ 3. $\sin\theta$ 4. -1 5. 0 6. undefined 7. (-0.7071, -0.7071) 8. (0.9455, 0.3256) 9. (0.5446, -0.8387) 10. (-0.1392, 0.9903) 11. 1 unit 12. $\frac{\pi}{3}$ units 13. π units 14. $\frac{7\pi}{6}$ units 15. $\frac{5\pi}{3}$ units 16. $\frac{\pi}{3}$ radians 17. π radians 18. $\frac{7\pi}{6}$ radians 19. $\frac{5\pi}{3}$ radians 20. $\frac{1}{\sqrt{2}}$ or $\frac{\sqrt{2}}{2}$ 21. $\frac{1}{\sqrt{2}}$ or $\frac{\sqrt{2}}{2}$ 22. $\frac{1}{1}$ or 1 23. $\frac{\sqrt{3}}{2}$ 24. $\frac{\sqrt{3}}{1}$ or $\sqrt{3}$ 25. $\frac{1}{2}$ 26. $\frac{\sqrt{3}}{2}$

CHAPTER 4 - VECTORS

Assignment 11

1. a, b, c, f 2. a) 23 miles b) 34° south of east 3. a) 444 lb b) 31°
4. a)

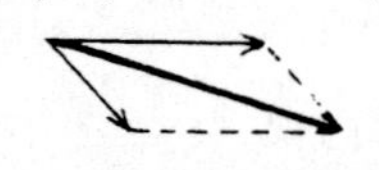

b)

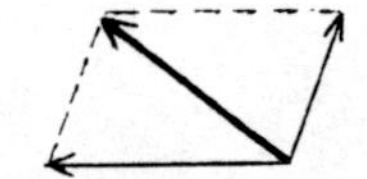

c)

5. a) 978 units b) 11° 6. a) 20.0 units b) 37° 7. a) 51 units b) 101° 8. a) 5.00 units b) -24°

Assignment 12

1. (-200, 350) 2. a) -483 units b) 619 units 3. a) 7.80 units b) -3.31 units
4. a) -49.1 units b) -16.9 units 5. 241° 6. 102° 7. 344° 8. -137° 9. -57°
10. a) 682 units b) 325° or -35° 11. a) 3.97 units b) 126° 12. a) 63.4 units b) 197° or -163°
13. a) -425 units b) 0 c) 425 units d) 180° or -180° 14. a) 0 b) -7.14 units c) 7.14 units

Assignment 13

1. a) (-19.0, 19.0) b) 26.9 units c) 135° 2. a) (470, -187) b) 506 units c) -22° or 338°
3. a) (-39.3, 26.7) b) 47.5 units c) 146° 4. a) (-595, -161) b) 616 units c) 195° or -165°
5. a) (2.96, -4.35) b) 5.26 units c) -56° or 304° 6.

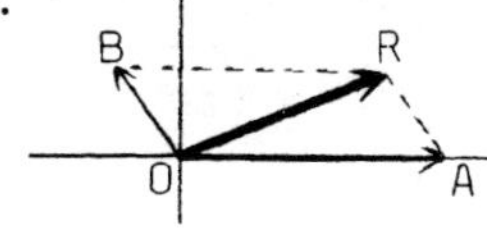

7.

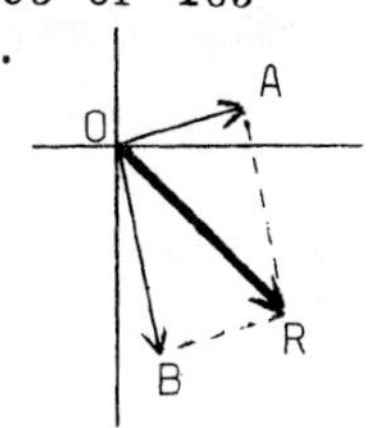

8. $\overrightarrow{OH}$: (25, 50) 9. $\overrightarrow{OF}$: (-620, -480)

Assignment 14

1. (4, 1) 2. (-425, 200) 3. a) (-110, -195) b) 224 units c) -119° or 241° 4. (-70, 50)
5. (0, 0) 6. (-30, 20) 7. (-280, -60) 8. a) (140, 280) b) 313 units c) 63°
9. 450 lb 10. a) 34 ma b) -58° or 302° 11. a) 473 mph b) 3° 12. 639 kg
13. a) 417 mv b) 30°

CHAPTER 5 - CIRCLE CONCEPTS

Assignment 15

1. a) 40° b) 12.5 cm 2. 7.96" 3. 23.4 cm 4. a) 3.56" b) 2.07" c) 102°
5. a) 90 cm b) 50 cm c) 214 cm 6. 26.6" 7. 83.1 cm

Assignment 16

1. 841 cm^2 2. a) 460 in^2 b) 218 in^2 c) 242 in^2 3. a) 96° b) 8.00 cm 4. 4.70"
5. a) 144° b) 446 mm c) 446 mm d) 469 mm 6. a) 18.7" b) 54° c) 126° d) 41.1"
7. 93.4 cm

Assignment 17

1. 900 rpm 2. 242 rps 3. a) 324 rpm b) 42 rps 4. 577 ft/min 5. 52.4 cm/sec
6. a) 26 cm/sec b) 107 ft/min 7. a) 40 rad/sec b) 6.37 rps 8. 92.2 in/sec
9. a) 50 rad/sec b) 580 cm/sec c) 477 rpm 10. a) 13.7 rps b) 157 rad/sec

Assignment 18

1. a) 52.034° b) 62.961° c) 75.482° 2. a) 127°33' b) 70°0' c) 84°28'17" d) 132°0'5"
3. a) 12°54' b) 72°31' c) 26°41'0" d) 62°49'20" 4. 71°44'40"
5. a) 51.622° b) 138.230° c) 2.906° 6. a) 15°8'14" b) 80°58'30" c) 127°40'53"
7. a) 2.95437 b) 0.68806 c) 0.96189 8. a) 44°44'3" b) 77°39'51" c) 9°34'11"

CHAPTER 6 - IDENTITIES, INVERSE NOTATION, AND EQUATIONS

Assignment 19

1. $\cot\theta$ 2. $\csc\theta$ 3. $\sec\theta$ 4. $\csc G = \frac{hyp}{opp}$ 5. $\sec G = \frac{hyp}{adj}$ 6. $\cot G = \frac{adj}{opp}$

7. $\sec P = \frac{r}{s}$ 8. $\cot P = \frac{s}{p}$ 9. $\csc P = \frac{r}{p}$ 10. $\csc S = \frac{r}{s}$ 11. $\cot S = \frac{p}{s}$ 12. $\sec S = \frac{r}{p}$

13. 1.4945 14. 2.7904 15. 0.2679 16. -1.8361 17. 2.0000 18. -2.6051

19. $\cot\theta$ 20. $\cos\theta$ 21. $\tan\theta$ 22. $\sin\theta$ 23. $\cos^2\theta$ 24. $\tan^2\theta$ 25. 1

26. $\cos\theta$ 27. $\cot\theta$ 28. $\sec\theta$ 29. $\sin\theta$ 30. $\sin\theta = \frac{\cos\theta}{\cot\theta}$

31. $\csc\theta = \frac{1}{\sin\theta}$ 32. $\cos\theta = \pm\sqrt{1 - \sin^2\theta}$ 33. $\sec\theta = \pm\sqrt{\tan^2\theta + 1}$

Assignment 20

1. $\cos\theta \csc\theta = \cot\theta$

$$\cos\theta\left(\frac{1}{\sin\theta}\right) =$$
$$\frac{\cos\theta}{\sin\theta} =$$
$$\cot\theta =$$

2. $\sin\theta \sec\theta \cot\theta = 1$

$$\sin\theta\left(\frac{1}{\cos\theta}\right)\left(\frac{\cos\theta}{\sin\theta}\right) =$$
$$1 =$$

3. $\tan\theta \csc\theta = \sec\theta$

$$\left(\frac{\sin\theta}{\cos\theta}\right)\left(\frac{1}{\sin\theta}\right) =$$
$$\frac{1}{\cos\theta} =$$
$$\sec\theta =$$

4. $\sin\theta = \csc\theta - \csc\theta\cos^2\theta$

$$= \csc\theta(1 - \cos^2\theta)$$
$$= \left(\frac{1}{\sin\theta}\right)(\sin^2\theta)$$
$$= \sin\theta$$

5. $\sec\theta = \cos\theta + \tan\theta\sin\theta$

$$= \cos\theta + \left(\frac{\sin\theta}{\cos\theta}\right)\sin\theta$$
$$= \frac{\cos^2\theta + \sin^2\theta}{\cos\theta}$$
$$= \frac{1}{\cos\theta}$$
$$= \sec\theta$$

6. $\cot\theta\cos\theta = \csc\theta - \sin\theta$

$$= \frac{1}{\sin\theta} - \sin\theta$$
$$= \frac{1 - \sin^2\theta}{\sin\theta}$$
$$= \frac{\cos^2\theta}{\sin\theta}$$
$$= \left(\frac{\cos\theta}{\sin\theta}\right)\cos\theta$$
$$= \cot\theta\cos\theta$$

7. $30° = \arcsin 0.5$ 8. $\theta = \arccos r$ 9. $A = \arctan 2.7384$ 10. $\cos 52° = 0.6157$

11. $\tan\theta = 1.2517$ 12. $\sin\alpha = w$ 13. $60° = \tan^{-1} 1.732$ 14. $12° = \sin^{-1} d$ 15. $\theta = \cos^{-1} N$

16. $\theta = 58°$ 17. $\alpha = 73°$ 18. $A = 83°$ 19. $T = 0.8098$ 20. $h = 0.9976$ 21. $x = 0.7071$

22. $\theta = 33°$ 23. $Z = 8,693$ 24. $X = 22,408$ 25. $Z = 570$

Assignment 21

1. $\theta = 32°$ and $148°$
2. $\theta = 80°$ and $260°$
3. $\theta = 84°$ and $276°$
4. $\theta = 139°$ and $319°$
5. $\theta = 195°$ and $345°$
6. $\theta = 135°$ and $225°$

7. $\theta = 71°$ and $289°$
8. $\theta = 247°$ and $293°$
9. $\theta = 112°$ and $292°$
10. $\theta = 120°$ and $240°$
11. $\theta = 26°$ and $154°$
12. $\theta = 44°$ and $316°$
13. $\theta = 105°$ and $285°$
14. $\theta = 90°$
15. $\theta = 63°$ and $243°$
16. $\theta = 270°$
17. $\theta = 104°$ and $256°$
18. $\theta = 108°$ and $288°$
19. $\theta = 217°$ and $323°$
20. $\theta = 116°$ and $244°$
21. $\theta = 135°$ and $315°$
22. $\theta = 180°$
23. $\theta = 30°$ and $150°$
24. $\theta = 82°$ and $278°$
25. $\theta = 104°$ and $284°$
26. $\theta = 132°$ and $228°$
27. $\theta = 7°$ and $173°$
28. $\theta = 37°$ and $323°$
29. $\theta = 79°$ and $259°$
30. $\theta = 56°$ and $236°$

CHAPTER 7 - COMPLEX NUMBERS

Assignment 22

1. b, d, e 2. 8j 3. -2j 4. 5.63j 5. 2 + 5j 6. -4 - 2j 7. 0 - 4j 8. -6 + 6j
9. -8 + 0j 10. 4 - 8j 11. a, d, f 12. b, c, e 13. -3 + j 14. 44 - 23j
15. 0 - 3.2j 16. 630 - 560j 17. 0 + 2j 18. -1 - j 19. -24.5 + 0j 20. 7 + 4j
21. 14 - 12j 22. -3.6 + 1.7j 23. 30 - 50j 24. 6 - 4j 25. 510 + 150j

Assignment 23

1. 2 - 9j 2. -30 + 28j 3. 10 + 70j 4. 3 + 5j 5. 32 - 16j 6. -300 - 200j
7. Vector A = 10 - 40j 8. Vector G = 460 + 0j 9. $709\angle 36^\circ$ 10. $40.3\angle 116^\circ$
11. $3.45\angle -59^\circ$ or $3.45\angle 301^\circ$ 12. $28,500\angle 208^\circ$ 13. 1.93 + 6.30j 14. 254 - 129j
15. -27.3 - 39.0j 16. -6,400 + 5,000j 17. $12\angle 90^\circ$ 18. $57\angle 180^\circ$
19. $210\angle -90^\circ$ or $210\angle 270^\circ$ 20. 6 + 0j 21. 0 + 150j 22. 0 - 35j 23. $583\angle 31^\circ$
24. $336\angle 48^\circ$ 25. $288\angle -56^\circ$ or $288\angle 304^\circ$ 26. $459\angle 131^\circ$

Assignment 24

1. -1 + 3j 2. -2 - 4j 3. 16 - 2j 4. 43 - 73j 5. -220 - 310j 6. -136 + 0j
7. 449 + 0j 8. 4.6 - 0.9j 9. 0 + 40j 10. 130 - 200j 11. -2500 + 0j 12. 66.2 + 1.1j
13. 2 + 0j 14. 34 + 0j 15. 113 +0j 16. 18,000 + 0j 17. $54\angle 103^\circ$
18. $300\angle 260^\circ$ or $300\angle -100^\circ$ 19. $8.33\angle 44^\circ$ 20. $40\angle 30^\circ$ 21. $300\angle -78^\circ$ or $300\angle 282^\circ$
22. $51.3\angle -113^\circ$ or $51.3\angle 247^\circ$ 23. $420\angle 252^\circ$ or $420\angle -108^\circ$ 24. $600\angle -30^\circ$ or $600\angle 330^\circ$
25. $35,900\angle 70^\circ$ 26. 0 + 6j 27. 0 - 160j 28. 5.4 + 0j 29. -1500 + 0j 30. 0 - 96j
31. 0 + 8.1j 32. $72\angle 20^\circ$ 33. $70\angle 300^\circ$ or $70\angle -60^\circ$ 34. $200\angle 135^\circ$ 35. 10 + 0j
36. 20 + 15j 37. -15 - 40j

Assignment 25

1. 3 + j 2. 2 - j 3. -1 + j 4. 2 - 3j 5. -3 - 2j 6. 1 - 3j 7. -2 + 3j
8. 0.3 - 0.7j 9. -.385 + 0.077j 10. -0.13 - 1.01j 11. -0.467 + 0.733j 12. -4 + 2j
13. 1.6 - 0.8j 14. 0 + 0.75j 15. 2.4 - 1.4j 16. -20 - 8j 17. -0.448 - 0.336j
18. $4\angle 90^\circ$ 19. $3.75\angle -90^\circ$ or $3.75\angle 270^\circ$ 20. $4.8\angle 180^\circ$ 21. $4\angle 0^\circ$ 22. $2.4\angle 90^\circ$
23. $3.4\angle -90^\circ$ or $3.4\angle 270^\circ$

Assignment 26

1. $1.6\angle 68^\circ$ 2. $5\angle 100^\circ$ 3. $0.348\angle -42^\circ$ or $0.348\angle 318^\circ$ 4. $0.75\angle 120^\circ$ 5. $32.4\angle 10^\circ$
6. $2.48\angle -54^\circ$ or $2.48\angle 306^\circ$ 7. $3.49\angle -45^\circ$ or $3.49\angle 315^\circ$ 8. $0.681\angle 132^\circ$
9. $4\angle -180^\circ$ (or $4\angle 180^\circ$) and -4 + 0j 10. $0.75\angle 90^\circ$ and 0 + 0.75j 11. $3\angle 0^\circ$ and 3 + 0j
12. $2.5\angle -90^\circ$ (or $2.5\angle 270^\circ$) and 0 - 2.5j 13. $I = 0.25\angle 22^\circ$ 14. $Z_t = 151\angle 41^\circ$
15. $E = 58\angle 55^\circ$ 16. $Z_2 = 2.15 - 7.23j$ 17. $E = 5 + 0j$

CHAPTER 8 - SINE WAVES

Assignment 27

1. $y = 6.93$ 2. $y = -8$ 3. $y = 2.5$ 4. $y = -2.5$ 5. $y = 0$ 6. $y = -2.08$
7. $y = 28 \sin \theta$ 8. $y = 8.25 \sin \theta$ 9. $\theta = 720^\circ$ 10. 170 11. 1 12. $\theta = 90^\circ$
13. 360° 14. right 15. left 16. -30° 17. +70° 18. $y = 12.6 \sin \theta$
19. $y = 275 \sin(\theta + 50^\circ)$ 20. $y = 20 \sin(\theta - 10^\circ)$ 21. b 22. 90° to the left

Assignment 28

1. 2 cycles 2. 180° 3. 180° 4. 6 5. 45° and 225° 6. 6 cycles 7. 60°
8. 60°, 120°, 180°, 240°, and 300° 9. 2.8 10. 15°, 75°, 135°, 195°, 255°, and 315°
11. $y = 5 \sin 10\theta$ 12. $y = 18 \sin 5\theta$ 13. $y = 0.58 \sin 12\theta$ 14. $y = 2 \sin 3\theta$ 15. 90°
16. 72° 17. 24° 18. 3 cycles 19. -360°, 0°, 360° 20. -270°, 90°, and 450°
21. -90°, 270°, and 630° 22. -330°, -210°, 30°, 150°, 390°, and 510° 23. 60° 24. -30°
25. 0° 26. 0° 27. 180° 28. -60° and 60° 29. 120° and 240°

Assignment 29

1. $y = -20$ 2. $y = 0$ 3. $y = -16 \sin \theta$ 4. $y = 12 \sin(\theta + 90°)$ 5. $y = 50 \sin(\theta + 180°)$ and $y = 50 \sin(\theta - 180°)$ 6. $y = -8$ 7. above 8. a 9. $y = -24$ 10. $y = -4$ 11. $y = 3$ 12. $y = 4$ 13. $y = 0$ 14. $y = 7$ 15. $y = -7$ 16. $y = 9.6$ 17. is 18. is not 19. is 20. 360° 21. 90° to the right 22. $y = 32 \sin(\theta + 90°)$ 23. $y = 170 \cos \theta$

INDEX